Service Systems Engineering and Management

The Operations Research Series

Series Editor: A. Ravi Ravindran
Professor Emeritus, Department of Industrial and Manufacturing Engineering
The Pennsylvania State University – University Park, PA

Published Titles:

Service Systems Engineering and Management
A. Ravi Ravindran, Paul M. Griffin, and Vittaldas V. Prabhu

Multiple Criteria Decision Analysis for Industrial Engineering: Methodology and Applications
Gerald William Evans

Multiple Criteria Decision Making in Supply Chain Management
A. Ravi Ravindran

Operations Planning: Mixed Integer Optimization Models
Joseph Geunes

Introduction to Linear Optimization and Extensions with MATLAB®
Roy H. Kwon

Supply Chain Engineering: Models and Applications
A. Ravi Ravindran & Donald Paul Warsing

Analysis of Queues: Methods and Applications
Natarajan Gautam

Integer Programming: Theory and Practice
John K. Karlof

Operations Research and Management Science Handbook
A. Ravi Ravindran

Operations Research Applications
A. Ravi Ravindran

Operations Research: A Practical Introduction
Michael W. Carter & Camille C. Price

Operations Research Calculations Handbook, Second Edition
Dennis Blumenfeld

Operations Research Methodologies
A. Ravi Ravindran

Probability Models in Operations Research
C. Richard Cassady & Joel A. Nachlas

Service Systems Engineering and Management

A. Ravi Ravindran
Paul M. Griffin
Vittaldas V. Prabhu

CRC Press is an imprint of the
Taylor & Francis Group, an **informa** business

CRC Press
Taylor & Francis Group
6000 Broken Sound Parkway NW, Suite 300
Boca Raton, FL 33487-2742

© 2018 by Taylor & Francis Group, LLC
CRC Press is an imprint of Taylor & Francis Group, an Informa business

No claim to original U.S. Government works

Printed on acid-free paper

International Standard Book Number-13: 978-1-4987-2306-0 (Hardback)

This book contains information obtained from authentic and highly regarded sources. Reasonable efforts have been made to publish reliable data and information, but the author and publisher cannot assume responsibility for the validity of all materials or the consequences of their use. The authors and publishers have attempted to trace the copyright holders of all material reproduced in this publication and apologize to copyright holders if permission to publish in this form has not been obtained. If any copyright material has not been acknowledged please write and let us know so we may rectify in any future reprint.

Except as permitted under U.S. Copyright Law, no part of this book may be reprinted, reproduced, transmitted, or utilized in any form by any electronic, mechanical, or other means, now known or hereafter invented, including photocopying, microfilming, and recording, or in any information storage or retrieval system, without written permission from the publishers.

For permission to photocopy or use material electronically from this work, please access www.copyright.com (http://www.copyright.com/) or contact the Copyright Clearance Center, Inc. (CCC), 222 Rosewood Drive, Danvers, MA 01923, 978-750-8400. CCC is a not-for-profit organization that provides licenses and registration for a variety of users. For organizations that have been granted a photocopy license by the CCC, a separate system of payment has been arranged.

Trademark Notice: Product or corporate names may be trademarks or registered trademarks, and are used only for identification and explanation without intent to infringe.

Visit the Taylor & Francis Web site at
http://www.taylorandfrancis.com

and the CRC Press Web site at
http://www.crcpress.com

To Our Wives

Bhuvana, Susan, and Radhika

Contents

List of Figures xix
List of Tables xxv
Preface xxxi
Acknowledgments xxxvii
Authors xxxix

1 **Service Systems: An Overview** 1
   1.1   Goods and Services 1
   1.2   Differences between Goods and Services 2
        1.2.1   Intangibility 2
        1.2.2   Perishability 2
        1.2.3   Proximity 3
        1.2.4   Simultaneity 3
   1.3   Challenges Posed by Services 3
        1.3.1   Impact of Technology 3
        1.3.2   Bundled Products 4
        1.3.3   Self-Service 4
        1.3.4   Service Quality 4
        1.3.5   Managing Capacity and Demand 4
   1.4   Classification of Services 5
        1.4.1   Service Industries 5
               1.4.1.1   Communications 5
               1.4.1.2   Education 6
               1.4.1.3   Entertainment 6
               1.4.1.4   Financial Services 6
               1.4.1.5   Government Services 7
               1.4.1.6   Healthcare 7
               1.4.1.7   Hospitality and Leisure 7
               1.4.1.8   Insurance 8
               1.4.1.9   Professional Services 8
               1.4.1.10 Retail and Wholesale 8
               1.4.1.11 Transportation 9
               1.4.1.12 Utilities 9
        1.4.2   Ancillary and Support Services 9
        1.4.3   Service Process Matrix 10
               1.4.3.1   Service Factory 11
               1.4.3.2   Service Shop 11
               1.4.3.3   Mass Service 11
               1.4.3.4   Professional Service 12
               1.4.3.5   Use of Service Process Matrix 12

vii

1.5	Role of Services in the Economy		12
	1.5.1	U.S. Employment by Sector	13
	1.5.2	U.S. GDP by Sector	13
1.6	Strength of American Manufacturing		14
	1.6.1	Decline in U.S. Manufacturing	14
	1.6.2	Revival of U.S. Manufacturing	15
1.7	What Is Service Systems Engineering?		16
	1.7.1	Engineering Issues in Service Systems	16
	1.7.2	Engineering Problems in Services	16
	1.7.3	Multiple-Criteria Decision Making in Services	17
1.8	Quality of Service		19
	1.8.1	QoS Measures	19
	1.8.2	Dimensions of Service Quality	20
1.9	Methods for Measuring Service Quality		21
	1.9.1	SERVQUAL	21
	1.9.2	Statistical Process Control	24
		1.9.2.1 Variable Control Charts	24
		1.9.2.2 Attribute Control Charts	28
		1.9.2.3 Selecting the Sample Size for Control Charts	30
	1.9.3	Other Service Quality Tools	31
		1.9.3.1 Run Chart	32
		1.9.3.2 Histogram	33
		1.9.3.3 Pareto Chart	34
		1.9.3.4 Scatterplot	34
		1.9.3.5 Flowchart	35
		1.9.3.6 Fishbone Diagram	36
1.10	Summary		38
Exercises			39
References			42

2 Forecasting Demand for Services ... 45

2.1	Role of Demand Forecasting in Services		45
2.2	Forecasting Process		46
2.3	Qualitative Forecasting Methods		47
	2.3.1	Executive Committee Consensus	47
	2.3.2	Delphi Method	48
	2.3.3	Survey of Sales Force	49
	2.3.4	Customer Surveys	49
2.4	Quantitative Forecasting Methods		49
	2.4.1	Time Series Forecasting	49
	2.4.2	Constant Level Forecasting Methods	51
	2.4.3	Last Value Method	52
	2.4.4	Averaging Method	52
	2.4.5	Simple Moving Average Method	53
	2.4.6	Weighted Moving Average Method	53

Contents

	2.4.7	Computing Optimal Weights by Linear Programming Model	54
	2.4.8	Exponential Smoothing Method	56
		2.4.8.1 Choice of Smoothing Constant (α)	57
2.5	Incorporating Seasonality in Forecasting	58
2.6	Incorporating Trend in Forecasting	62
	2.6.1	Simple Linear Trend Model	62
	2.6.2	Holt's Method	64
2.7	Incorporating Seasonality and Trend in Forecasting	66
	2.7.1	Method Using Static Seasonality Indices	66
	2.7.2	Winters Method	67
2.8	Forecasting for Multiple Periods	70
	2.8.1	Multi-Period Forecasting Problem	70
	2.8.2	Multi-Period Forecasting under Constant Level	71
	2.8.3	Multi-Period Forecasting with Seasonality	71
	2.8.4	Multi-Period Forecasting with Trend	72
	2.8.5	Multi-Period Forecasting with Seasonality and Trend	72
2.9	Forecasting Errors	73
	2.9.1	Uses of Forecast Errors	76
	2.9.2	Selecting the Best Forecasting Method	76
2.10	Monitoring Forecast Accuracy	77
	2.10.1	Tracking Signal	77
2.11	Forecasting Software	79
	2.11.1	Types of Forecasting Software	79
		2.11.1.1 Automatic Software	79
		2.11.1.2 Semi-Automatic Software	79
		2.11.1.3 Manual Software	80
	2.11.2	User Experience with Forecasting Software	81
2.12	Forecasting in Practice	81
	2.12.1	Real-World Applications	81
	2.12.2	Forecasting in Practice: Survey Results	82
2.13	Summary and Further Readings	83
	2.13.1	Demand Forecasting—Summary	83
	2.13.2	ARIMA Method	84
	2.13.3	Croston's Method	84
	2.13.4	Further Readings in Forecasting	85
Exercises	85
References	90

3 Design of Service Systems 93
   3.1 Modeling with Binary Variables 93
      3.1.1 Capital Budgeting Problem 93
      3.1.2 Fixed Charge Problem 96
      3.1.3 Constraint with Multiple Right-Hand-Side (RHS) Constants 97

		3.1.4	Quantity Discounts	98
			3.1.4.1 "All-Unit" Quantity Discounts	98
			3.1.4.2 Graduated Quantity Discount	100
		3.1.5	Handling Nonlinear Integer Programs	102
	3.2	Set Covering and Set Partitioning Models		104
		3.2.1	Set Covering Problem	105
		3.2.2	Set Partitioning Problem	106
		3.2.3	Application to Warehouse Location	106
		3.2.4	Application to Airline Scheduling	108
			3.2.4.1 Airline Scheduling	109
			3.2.4.2 Use of Set Covering and Partitioning Model	109
			3.2.4.3 Crew Scheduling Model	110
			3.2.4.4 Airline Applications	112
		3.2.5	Workforce Planning	112
			3.2.5.1 Workforce Planning Applications	114
		3.2.6	Real-World Applications	114
			3.2.6.1 Ford Motor Company	115
			3.2.6.2 United Parcel Service	115
			3.2.6.3 Wishard Memorial Hospital	117
			3.2.6.4 Mount Sinai Hospital	118
			3.2.6.5 United Airlines	118
	3.3	Queuing Models in Service Systems		119
		3.3.1	Introduction	119
		3.3.2	Poisson Processes	120
		3.3.3	The M/M/1 Model	121
		3.3.4	The M/M/c Model	125
		3.3.5	The M/M/c/K Model	127
		3.3.6	The G/G/c Model	129
		3.3.7	Queuing Networks	132
		3.3.8	Modeling a Call Center as a Queuing System	134
			3.3.8.1 Time-Varying Arrivals	137
			3.3.8.2 Service Times	137
			3.3.8.3 Customer Abandonment	139
			3.3.8.4 Staffing	139
			3.3.8.5 Discussion	142
	Exercises			142
	Steps of ISA			147
	References			147
4	**Evaluation of Service Systems**			**151**
	4.1	Multiple-Criteria Decision Making—An Overview		151
	4.2	Multiple-Criteria Selection Problems		152
		4.2.1	Concept of "Best Solution"	152
		4.2.2	Dominated Alternative	153

		4.2.3 Non-Dominated Alternatives	153
		4.2.4 Ideal Solution	153
	4.3	Multiple-Criteria Ranking Methods	153
		4.3.1 Rating Method	154
		4.3.2 Borda Method	154
		4.3.3 Pairwise Comparison of Criteria	156
		4.3.4 Scaling Criteria Values	158
		4.3.5 Analytic Hierarchy Process	158
	4.4	MCMP Problems	158
		4.4.1 MCMP Problem	159
		4.4.2 Efficient, Non-Dominated, or Pareto Optimal Solution	159
		4.4.3 Determining an Efficient Solution	161
		4.4.4 Test for Efficiency	162
	4.5	Classification of MCMP Methods	162
	4.6	Goal Programming	163
		4.6.1 Goal Programming Formulation	164
		4.6.2 Solution of Goal Programming Problems	167
		4.6.2.1 Linear Goal Programs	167
		4.6.2.2 Integer Goal Programs	167
		4.6.2.3 Nonlinear Goal Programs	167
	4.7	Method of Global Criterion and Compromise Programming	168
		4.7.1 Compromise Programming	168
	4.8	Interactive Methods	169
	4.9	MCDM Applications	170
	4.10	MCDM Software	171
	4.11	Data Envelopment Analysis	171
		4.11.1 Production Function	172
		4.11.2 Technology	175
		4.11.3 Nonparametric Models	176
	4.12	DEA Linear Programming Formulation	180
	4.13	Practical Considerations for DEA	183
	4.14	Allocative and Total Efficiency in DEA	184
	4.15	Stochastic DEA	188
	4.16	DEA Software	191
	Exercises		192
	References		199
5	**Supply Chain Engineering**		**201**
	5.1	Supply Chain Decisions and Design Metrics	201
		5.1.1 Flows in Supply Chains	203
		5.1.2 Meaning of SCE	203
		5.1.3 Supply Chain Decisions	204
		5.1.4 Enablers and Drivers of Supply Chain Performance	205
		5.1.4.1 Supply Chain Enablers	205
		5.1.4.2 Supply Chain Drivers	206

		5.1.5	Assessing and Managing Supply Chain Performance...208
			5.1.5.1 Supply Chain Efficiency 209
			5.1.5.2 Supply Chain Responsiveness............................ 210
			5.1.5.3 Supply Chain Risk.. 210
			5.1.5.4 Conflicting Criteria in Supply Chain Optimization... 211
	5.1.6	Relationship between Supply Chain Metrics and Financial Metrics... 211
		5.1.6.1 Supply Chain Metrics 212
		5.1.6.2 Business Financial Metrics................................. 213
	5.1.7	Importance of SCM.. 215
		5.1.7.1 Supply Chain Top 25... 216
5.2	Supply Chain Network Design and Distribution 217
	5.2.1	Supply Chain Distribution Planning................................... 217
	5.2.2	Location–Distribution Problem 220
	5.2.3	Location–Distribution with Dedicated Warehouses 223
	5.2.4	Supply Chain Network Design... 225
	5.2.5	Real-World Applications... 229
		5.2.5.1 Multinational Consumer Products Company ... 229
		5.2.5.2 Procter and Gamble 231
		5.2.5.3 Hewlett–Packard 232
		5.2.5.4 BMW.. 233
		5.2.5.5 AT&T 233
5.3	Outsourcing Decisions and Supplier Selection 234
	5.3.1	Importance of Supplier Selection 234
	5.3.2	Supplier Selection Process.. 235
	5.3.3	In-House or Outsource... 236
	5.3.4	Supplier Selection Methods.. 237
		5.3.4.1 Sourcing Strategy 237
		5.3.4.2 Criteria for Selection 238
		5.3.4.3 Prequalification of Suppliers 238
		5.3.4.4 Final Selection.. 241
	5.3.5	Single Sourcing Methods for Supplier Selection.............. 242
	5.3.6	Multi-Criteria Ranking Methods for Supplier Selection.... 245
		5.3.6.1 Case Study 1: Ranking of Suppliers................... 245
		5.3.6.2 Use of the L_p Metric for Ranking Suppliers...... 248
		5.3.6.3 Rating (Scoring) Method 249
		5.3.6.4 Borda Method 251
		5.3.6.5 Analytic Hierarchy Process 253
	5.3.7	Scaling Criteria Values 258
		5.3.7.1 Simple Scaling... 259
		5.3.7.2 Ideal Value Method 260
		5.3.7.3 Simple Linearization (Linear Normalization) ... 260
		5.3.7.4 Use of L_p Norm (Vector Scaling)....................... 260
		5.3.7.5 Illustrative Example of Scaling Criteria Values ... 261

		5.3.7.6	Simple Scaling Illustration 262
		5.3.7.7	Illustration of Scaling by Ideal Value 262
		5.3.7.8	Simple Linearization (Linear Normalization) Illustration 263
		5.3.7.9	Illustration of Scaling by L_p Norm 264
	5.3.8	Group Decision Making 265
	5.3.9	Comparison of Ranking Methods 265
	5.3.10	Multiple Sourcing Methods for Supplier Selection and Order Allocation 266
		5.3.10.1	Case Study 2 266
		5.3.10.2	Solution of Case Study 2 269
5.4	Supply Chain Logistics 273
	5.4.1	Mode Selection 273
	5.4.2	Vehicle Routing 274
	5.4.3	Inventory Modeling 277
5.5	Supply Chain Pooling and Contracting 280
	5.5.1	An Introduction to Pooling 280
		5.5.1.1	Location Pooling 281
		5.5.1.2	Postponement 281
		5.5.1.3	Product Pooling 281
		5.5.1.4	Correlation and Pooling 282
	5.5.2	Supply Chain Contracting 283

Exercises 287
References 298

6 Warehousing and Distribution 303
6.1 Warehouse Functions 303
6.2 Warehouse Equipment 307
- 6.2.1 Moving Product 308
- 6.2.2 Storing Product 309
- 6.2.3 Sorting Product 312
- 6.2.4 Tracking Product 312
6.3 Warehouse Design 314
- 6.3.1 Material Flow and Staffing 316
- 6.3.2 Storage Strategies 316
- 6.3.3 Warehouse Layout 318
- 6.3.4 Aisle Layout 321
6.4 Order Picking 322
6.5 Slotting 325
- 6.5.1 SKU Profiling 325
- 6.5.2 Prioritizing SKUs 328
- 6.5.3 Packing SKUs 330
6.6 Forward–Reserve 331
6.7 Line Balancing and Bucket Brigades 335
6.8 Cross Docking 338

	6.9	Warehouse Location and Inventory	343
	6.10	Further Readings	345
	Exercises		346
	References		347

7 Financial Engineering 349
	7.1	Basic Concepts in Investing	349
		7.1.1 Investing in Cash	350
		7.1.2 Investing in Bonds	350
		7.1.3 Investing in Stocks	350
		7.1.4 Asset Allocation	351
		7.1.5 Mutual Funds	352
	7.2	A Simple Model for Portfolio Selection	352
		7.2.1 Linear Programming Model	354
		7.2.2 Drawbacks of the LP Model	357
	7.3	Markowitz's Mean Variance Model for Portfolio Selection	358
		7.3.1 Quantifying Investment Risk	358
		7.3.2 Excel Functions to Compute Investment Risk	360
		7.3.3 Markowitz's Bi-Criteria Model	361
		7.3.4 Risk–Return Graph for Portfolio Selection	363
		7.3.5 Use of Portfolio Mean and Standard Deviation	364
		7.3.6 Illustration of Markowitz' Model	365
	7.4	Sharpe's Bi-Criteria Model for Portfolio Selection	371
		7.4.1 Market Risk of a Security	371
		7.4.2 Meaning of the S&P 500 Index	372
		7.4.3 Calculating Beta Risk	373
		7.4.4 Sharpe's Bi-Criteria Model	374
		7.4.5 Illustration of Sharpe's Model	375
	7.5	Investing in Bonds	379
		7.5.1 Understanding Bonds	379
		7.5.2 Price, Yield, and Coupons	381
		7.5.3 Types of Bonds	383
		7.5.4 Bond Ratings	384
		7.5.5 Bond Duration	384
		7.5.6 Bond Convexity	386
		7.5.7 Immunization	387
		7.5.8 Selecting a Bond Portfolio	387
	7.6	Principles of Investing	391
		7.6.1 Annualized Returns	391
		7.6.2 Dollar Cost Averaging Principle	393
		7.6.3 Investing in Mutual Funds	394
		7.6.3.1 Load versus No-Load Funds	394
		7.6.3.2 Stock Mutual Funds	395
		7.6.3.3 Bond Mutual Funds	395
		7.6.3.4 Balanced Mutual Funds	395

Contents   xv

                7.6.3.5   Index Funds 395
                7.6.3.6   Life Cycle or Target Date Funds 397
       7.6.4   Investment Strategies 397
   Exercises 398
   References 406

8   Revenue Management 407
   8.1   History of Revenue Management 407
   8.2   Difficulties in Managing Service Capacity 408
   8.3   Revenue Management Strategies in Services 409
       8.3.1   Proactive Strategies to Manage Customer Demand 409
                8.3.1.1   Use of Reservations/Appointment Systems 410
                8.3.1.2   Customer Segmentation 410
                8.3.1.3   Differential Pricing 411
                8.3.1.4   Sales Promotions 412
       8.3.2   Reactive Strategies for Capacity Planning 412
                8.3.2.1   Workforce Planning 412
                8.3.2.2   Use of Part-Time/Temporary Workers 413
                8.3.2.3   Cross-Training of Employees 413
                8.3.2.4   Promoting Self-Service 413
                8.3.2.5   Designing Adjustable Capacity 414
       8.3.3   Applicability of Revenue Management 414
   8.4   Optimization Models for Overbooking 415
       8.4.1   No-Show Problem 415
       8.4.2   Practice of Overbooking 415
       8.4.3   Optimization Problem 416
                8.4.3.1   Input Data 416
                8.4.3.2   Objective Function 417
                8.4.3.3   Overbooking Policy for the Continuous Case ... 418
                8.4.3.4   Overbooking Policy for the Discrete Case 419
                8.4.3.5   Illustrative Example 420
   8.5   Revenue Management in Airlines 424
       8.5.1   Airline Fare Classes 425
       8.5.2   Setting Limits on Two Fare Classes 426
                8.5.2.1   Optimization Problem 426
       8.5.3   Setting Limits on Multiple Fare Classes 428
       8.5.4   Bid Price Controls 432
       8.5.5   Booking Curves 434
       8.5.6   Remarks on Airline Revenue Management 436
   Exercises 437
   References 440

9   Retail Engineering 443
   9.1   Introduction 443
   9.2   Types of Retailers 446

9.3	Financial Strategy in Retailing		447
	9.3.1	Financial Reporting	448
	9.3.2	Assets, Liabilities, and Stockholder's Equity	448
	9.3.3	Balance Sheet	448
	9.3.4	Income Statement	450
	9.3.5	Cash Flow Statement	450
9.4	Financial Metrics in Retailing		451
9.5	Analysis of Balance Sheet		452
9.6	DuPont Model		455
9.7	Breakeven Analysis		464
9.8	Checkout Process Staffing		467
Exercises			469
References			471

10 Healthcare Delivery Systems ... 473

10.1	Introduction		473
	10.1.1	U.S. Healthcare Costs	474
	10.1.2	Healthcare Outcomes	476
10.2	Healthcare Financing		478
	10.2.1	Health Insurance	478
	10.2.2	Provider Compensation	480
	10.2.3	Provider Cost Allocation	481
10.3	Healthcare Components		482
	10.3.1	Hospital Units	484
	10.3.2	Patient Services	488
	10.3.3	Patient Flow	490
	10.3.4	Value Stream Mapping	492
10.4	Health Information Technology		494
	10.4.1	Medical Coding	494
	10.4.2	Electronic Health Record	495
	10.4.3	Health Information Exchange	496
	10.4.4	Claims and Billing	497
10.5	Resource Management		498
	10.5.1	Nurse Scheduling	498
	10.5.2	OR Scheduling	504
	10.5.3	Capacity Management for Imaging Equipment	507
	10.5.4	Bed Management	508
10.6	Health Analytics		510
	10.6.1	Randomized Controlled Trials	511
	10.6.2	Measuring Accuracy	513
	10.6.3	Prediction by Support Vector Machines	516
	10.6.4	Prediction Example: 30-Day Readmissions	522
10.7	Discussion		523
Exercises			523
References			526

11 Financial Services 529
- 11.1 Payment Systems 529
- 11.2 Banking 534
  - 11.2.1 Staffing and Customer Waiting Time 535
  - 11.2.2 Diversifying Banking Channels 537
- 11.3 Electronic Trading 540
- 11.4 Valuing Companies and Their Stock Prices 544
  - 11.4.1 Company Valuation Using Discounted Cash Flows 544
  - 11.4.2 Price–Earnings Ratio 549
  - 11.4.3 PE Growth Ratio 549
  - 11.4.4 Earnings Estimates 550
  - 11.4.5 Technical Analysis 550
  - 11.4.6 Cup with Handle 551
  - 11.4.7 Head and Shoulders Pattern 552
- 11.5 Derivatives 554
- 11.6 Options 555
  - 11.6.1 Call Option 555
  - 11.6.2 Put Option 556
  - 11.6.3 Option Valuation 557
    - 11.6.3.1 Binomial Option Pricing 557
    - 11.6.3.2 Black–Scholes Option Valuation 559
- Exercises 561
- References 562

Index 565

List of Figures

Figure 1.1 Service Process Matrix 11
Figure 1.2 $\bar{x}$ chart for customer service times (Example 1.1) 27
Figure 1.3 R chart for customer service times (Example 1.1) 28
Figure 1.4 p chart for Example 1.2 30
Figure 1.5 Run chart of no-shows (Example 1.3) 33
Figure 1.6 Histogram of billing errors (Example 1.3) 33
Figure 1.7 Pareto chart of number of problems in each area (Example 1.3) ... 34
Figure 1.8 Scatterplot between overtime days and medical errors (Example 1.3) 35
Figure 1.9 Flowchart of the medical billing process (Example 1.3) 36
Figure 1.10 Cause-and-effect diagram for patient no-shows 37
Figure 2.1 Constant level 50
Figure 2.2 Constant level with seasonality 50
Figure 2.3 Constant level with trend 50
Figure 2.4 Constant level with seasonality and trend 51
Figure 2.5 Plot of tracking signals (Example 2.12) 78
Figure 3.1 All-unit quantity discount model 99
Figure 3.2 Graduated quantity discount model 100
Figure 3.3 Supply chain network (Example 3.4) 107
Figure 3.4 UPS's next-day air service 116
Figure 3.5 A standard queuing model 119
Figure 3.6 Arrivals in a Poisson process 121
Figure 3.7 Birth–death process for an M/M/1 queuing system 122
Figure 3.8 Birth–death process for an M/M/c queuing system 126
Figure 3.9 Birth–death process for an M/M/c/K queuing system 128
Figure 3.10 Examples of an open network (a) and closed network (b) 133
Figure 3.11 Key components of a simple call center 135

xix

Figure 3.12 A simple call center modeled as a queuing system 136

Figure 3.13 Call volume over time 138

Figure 3.14 Staffing levels with respect to delay probability 142

Figure 4.1 Decision space (Example 4.3) 160

Figure 4.2 Objective space (Example 4.3) 161

Figure 4.3 Illustration of returns to scale 173

Figure 4.4 Cobb–Douglas input possibility set and isoquant ($\alpha = \beta = 0.5$, $A = 10$, $y = 20$) 174

Figure 4.5 Plot of (a) input possibility set, (b) assuming free disposability, and (c) assuming convexity and free disposability 177

Figure 4.6 H–R input possibility sets for $u = 19$ (a) and $u = 15$ (b) 178

Figure 4.7 Plot of scaled data (a), input possibility set for an output of 1 (b), and input possibility set for an output of 2 (c) 179

Figure 4.8 Boundary of input possibility set (curved line) and isocost line (straight line) for Example 4.9 185

Figure 4.9 Illustration of technical, allocative, and overall efficiency 186

Figure 5.1 Typical supply chain network 202

Figure 5.2 Responsiveness–efficiency trade-off frontier 208

Figure 5.3 Supply chain network for Example 5.2 221

Figure 5.4 XYZ's three-stage supply chain (Example 5.3) 225

Figure 5.5 Supplier selection factors 235

Figure 5.6 Supplier selection steps 235

Figure 5.7 Supplier selection problem 237

Figure 5.8 Two-phase supplier selection model 245

Figure 5.9 AHP hierarchy for supplier criteria (Case Study 1) 254

Figure 5.10 Illustration of route improvement by two-opt on arcs a and b 276

Figure 5.11 Two-echelon supply chain for a fashion good 284

Figure 5.12 Supplier selection criteria and subcriteria (Exercise 5.19) 293

Figure 6.1 Basic warehouse functions 304

Figure 6.2 Different types of handling units 306

Figure 6.3 Warehouse hierarchy 306

List of Figures

Figure 6.4	Warehouse with online fulfillment capabilities	307
Figure 6.5	Powered roller conveyor	308
Figure 6.6	Lift truck (also known as a forklift)	309
Figure 6.7	Block stacking pallets	310
Figure 6.8	Honeycombing in block stacking	311
Figure 6.9	Flow-through (or gravity) rack	311
Figure 6.10	Carousel system	312
Figure 6.11	Sortation system for a three-order example	313
Figure 6.12	Barcode label examples (1D and 2D)	313
Figure 6.13	Simple warehouse layout	315
Figure 6.14	Relationship chart example for five departments	319
Figure 6.15	Examples of activity relationship layouts	319
Figure 6.16	Examples of block layouts	320
Figure 6.17	Aisle layouts with differing number of cross lanes	321
Figure 6.18	Fishbone aisle layout	322
Figure 6.19	Example pick path for a picker in a warehouse	323
Figure 6.20	Example pick path using S-shape heuristic	324
Figure 6.21	Slotting layout for an aisle	329
Figure 6.22	General slotting arrangement for the pick face of flow racks	330
Figure 6.23	Six stack orientations for a carton	331
Figure 6.24	Forward and reserve areas	331
Figure 6.25	Consolidation benefit through cross docking	339
Figure 6.26	Cross dock layout	340
Figure 6.27	Staging in a cross dock (sort-at-shipping and sort-at-receiving)	341
Figure 6.28	The impact of interior and exterior corners for cross docks	341
Figure 6.29	Multiple stages in a warehouse network	345
Figure 6.30	Two possible ways to meet service level s in Figure 6.29	345
Figure 7.1	Risk–return graph for portfolio selection	363
Figure 7.2	Risk–return graph (Example 7.2)	370

Figure 7.3	Plot of security returns against market returns	373
Figure 7.4	Risk–return graph (Example 7.3)	379
Figure 7.5	Change in bond price versus time remaining to maturity	382
Figure 7.6	Change in bond price with varying yields for different coupon rates	382
Figure 7.7	Change in bond price with varying yield for different maturity periods	383
Figure 7.8	Weighted cash flows for Example 7.7	385
Figure 7.9	Convexity of bonds	386
Figure 8.1	Impact of Q on EC_1, EC_2, and TEC	418
Figure 8.2	Total expected cost for a discrete distribution	420
Figure 8.3	Point-to-point and hub-and-spoke networks	424
Figure 8.4	Multi-leg flight from Atlanta, Georgia to State College, Pennsylvania	425
Figure 8.5	Airline network with leg capacities	430
Figure 8.6	Examples of booking curve for non-business travelers (a) and business travelers (b).	434
Figure 8.7	S-shaped booking curve ($a = 40$, $b = 0.2$, $c = 1$, $d = 5$)	435
Figure 8.8	Cumulative demand exceeds the threshold	436
Figure 8.9	Capacity data for Exercise 8.13	439
Figure 9.1	Retail firms link consumers and their supply chains	444
Figure 9.2	Trend in cash and cash equivalents (Walmart)	454
Figure 9.3	Trend in debt–equity ratio (Walmart)	454
Figure 9.4	Trend in total current assets (Walmart)	455
Figure 9.5	Trend in cash and cash equivalents (Amazon)	457
Figure 9.6	Trend in debt–equity ratio (Amazon)	457
Figure 9.7	Trend in total current assets (Amazon)	458
Figure 9.8	Strategic profit model layout	459
Figure 9.9	Strategic profit model of Walmart for 2014	460
Figure 9.10	Strategic profit model of Walmart for 2017	461
Figure 9.11	Strategic profit model of Amazon.com for 2013	462

List of Figures xxiii

Figure 9.12 Strategic profit model of Amazon.com for 2016 463
Figure 9.13 Breakeven analysis (Example 9.1) 466
Figure 10.1 Accountable care organization 486
Figure 10.2 Hospital units and their relationships 487
Figure 10.3 Example process map of the Hershey Medical Center in Pennsylvania 493
Figure 10.4 Outcomes from using a cutpoint of a test to determine the presence or absence of a condition 514
Figure 10.5 ROC curve 515
Figure 10.6 Two-dimensional feature space example 517
Figure 10.7 Septic patients (squares) and non-septic patients (circles). On left-hand side, several separating hyperplanes will linearly separate. The classifier chosen is shown on the right 517
Figure 10.8 Support vectors and decision boundary 518
Figure 10.9 Example that is not linearly separable 519
Figure 10.10 Decision boundary linearly separates on the left. Decision boundary on the right does not completely separate, but has a large margin 520
Figure 10.11 Light and dark dots are not linearly separable. Transforming by squaring the value of the feature yields the figure on the right, which is linearly separable 521
Figure 11.1 Trends in payments in the United States during 2000–2012 530
Figure 11.2 Trends in payment modes around the world 531
Figure 11.3 The four-corner payment model 532
Figure 11.4 Payment process architecture 534
Figure 11.5 Customer arrival pattern at a bank (Example 11.1) 536
Figure 11.6 Overview of major entities in the securities industry 540
Figure 11.7 Order specifications for an electronic trade 542
Figure 11.8 The three-period model 548
Figure 11.9 Cup and handle chart for TTL stock 551
Figure 11.10 Head and shoulder chart for ETR 552
Figure 11.11 Head and shoulders mirror image 553
Figure 11.12 Binomial option pricing 557

List of Tables

Table 1.1	U.S. employment by sector	13
Table 1.2	Top 10 world economies	14
Table 1.3	U.S. GDP by sector	14
Table 1.4	SERVQUAL Survey (Part I: Expectations of Service)	22
Table 1.5	SERVQUAL Survey (Part II: Perceptions of Service)	23
Table 1.6	Data for Example 1.1—Customer service time in minutes	25
Table 1.7	Variable control chart constants	26
Table 1.8	Record of problems faced by EZ Access clinic (Example 1.3)	32
Table 1.9	Service time in minutes (Exercise 1.7)	40
Table 1.10	Customer wait times in minutes (Exercise 1.8)	41
Table 1.11	Data on daily mail delivery (Exercise 1.9)	41
Table 2.1	Demand data	52
Table 2.2	Laptop computer sales for 2014–2016 (Example 2.3)	59
Table 2.3	Forecasting with seasonality (Example 2.3)	61
Table 2.4	Data for Example 2.4	63
Table 2.5	Calculations for the Regression Model (Example 2.4)	63
Table 2.6	Holt's method (Example 2.5)	66
Table 2.7	Computations for Example 2.6	68
Table 2.8	Computations for Winters method (Example 2.7)	70
Table 2.9	Forecast errors (Example 2.11)	75
Table 2.10	Sales data for Exercise 2.7	86
Table 2.11	Sales data for Exercise 2.8	86
Table 2.12	Data on Freshman application (Exercise 2.10)	87
Table 2.13	Sales and forecast data for Exercise 2.14	89
Table 2.14	Quarterly sales data for 2012–2016 (Exercise 2.15)	89
Table 3.1	Data on Potential Projects (Example 3.1)	95

Table 3.2	Sample data for flight segments	110
Table 3.3	Sample crew schedules	110
Table 3.4	Nurse scheduling problem (Example 3.5)	113
Table 3.5	Supplier data for Exercise 3.3	143
Table 3.6	Data for Exercise 3.5	144
Table 3.7	Call volume data for Exercise 3.16	146
Table 4.1	Pay-off matrix of MCSP	152
Table 4.2	Data for Example 4.1 (Faculty recruiting)	155
Table 4.3	Pairwise comparison matrix (P) for Example 4.2	157
Table 4.4	Data for hospital network	172
Table 4.5	DMU data for Example 4.7	179
Table 4.6	DMU data for Example 4.8	182
Table 4.7	Distribution of input variable for Example 4.11	190
Table 4.8	Covariance data for Example 4.11	190
Table 4.9	Data for Exercise 4.11	195
Table 4.10	DMU data for Exercise 4.20	198
Table 4.11	DMU data for Exercise 4.22	198
Table 5.1	Top 10 supply chains (2017)	216
Table 5.2	Warehouse supplies (Example 5.1)	218
Table 5.3	Retailer demands (Example 5.1)	218
Table 5.4	Unit shipping cost in dollars (Example 5.1)	218
Table 5.5	Optimal distribution plan (Example 5.1)	220
Table 5.6	Warehouse data (Example 5.2)	221
Table 5.7	Warehouse capacities and investment cost (Example 5.3)	226
Table 5.8	Annual demand (Example 5.3)	226
Table 5.9	Unit shipping cost (Example 5.3)	226
Table 5.10	Optimal solution for Example 5.3	229
Table 5.11	Importance of supplier selection criteria	239
Table 5.12	Key criteria for supplier selection	241
Table 5.13	Supplier selection by the LWP method	243

List of Tables

Table 5.14	Supplier data for Example 5.5	244
Table 5.15	Supplier criteria values for Case Study 1	247
Table 5.16	Ideal values (H) for Case Study 1	249
Table 5.17	Supplier ranking using L_2 metric (Case Study 1)	249
Table 5.18	Criteria weights using rating method (Case Study 1)	250
Table 5.19	Supplier ranking using rating method (Case Study 1)	251
Table 5.20	Criteria weights using Borda count (Case Study 1)	252
Table 5.21	Supplier ranking using Borda count (Case Study 1)	252
Table 5.22	Degree of importance scale in AHP	255
Table 5.23	Pairwise comparison of criteria (Case Study 1)	256
Table 5.24	Final criteria weights using AHP (Case Study 1)	257
Table 5.25	AHP subcriteria weights for Case Study 1	258
Table 5.26	Supplier ranking using AHP (Case Study 1)	259
Table 5.27	Supplier criteria values for Example 5.6	261
Table 5.28	Scaled criteria values by the ideal value method (Example 5.6)	263
Table 5.29	Scaled criteria values by simple linearization (Example 5.6)	263
Table 5.30	Scaled criteria values using L_∞ norm (Example 5.6)	264
Table 5.31	Product demand	267
Table 5.32	Supplier capacities	267
Table 5.33	Lead time of products in days	267
Table 5.34	Quality of product (measured by percentage of rejects)	268
Table 5.35	Product prices (including shipping cost) of suppliers with quantity discounts and their corresponding price break points	268
Table 5.36	Average lead time requirements of the buyers	268
Table 5.37	Optimal order allocation (units)	273
Table 5.38	Data for Exercise 5.12	290
Table 5.39	Data for Mighty Manufacturing (Exercise 5.13)	290
Table 5.40	Data for Exercise 5.14	291
Table 5.41	Criteria rankings for Exercise 5.16	292
Table 5.42	Supplier data for Exercise 5.17	292

Table 5.43	Supplier data for Exercise 5.18	293
Table 5.44	Supplier criteria data (Exercise 5.19)	294
Table 5.45	Data for Exercise 5.28	296
Table 6.1	Mean and standard deviation of daily storage requirements for seven items	317
Table 6.2	Sample data for 14 orders (Example 6.1)	327
Table 6.3	Order-based metrics (Example 6.1)	327
Table 6.4	SKU-based metrics (Example 6.1)	328
Table 6.5	SKU prioritization result for Example 6.1	329
Table 6.6	Data for forward reserve (Example 6.2)	334
Table 6.7	Calculations for forward reserve (Example 6.2)	335
Table 6.8	Data for line balancing (Example 6.3)	337
Table 6.9	Solution to the line balancing problem (Example 6.3)	337
Table 7.1	Comparison of cash, bond, and stock investments	351
Table 7.2	Performance data of funds for Example 7.1	355
Table 7.3	Optimal solution to Example 7.1	357
Table 7.4	Twenty-year returns of investment securities (1997–2016)	358
Table 7.5	Twenty-year security returns and risk (1997–2016)	359
Table 7.6	Data on security returns and correlations between security returns (Example 7.2)	366
Table 7.7	Ideal solution (Example 7.2)	369
Table 7.8	Efficient portfolios (Example 7.2)	370
Table 7.9	Mutual fund data for Example 7.3	376
Table 7.10	LP optimal solutions (Example 7.3)	378
Table 7.11	Bond ratings from various rating agencies	384
Table 7.12	Calculating bond duration for Example 7.7	386
Table 7.13	Bond data for Example 7.8	389
Table 7.14	Maximum return portfolio (Example 7.8)	391
Table 7.15	Minimum risk portfolio (Example 7.8)	391

List of Tables

Table 7.16	Data on security returns (Exercise 7.6)	399
Table 7.17	Correlation between funds (Exercise 7.6)	400
Table 7.18	Retirement fund data (Exercise 7.7)	401
Table 7.19	Description of securities (stocks)	402
Table 7.20	Description of securities (bond and cash)	402
Table 7.21	Annual returns in percentages (2003–2013)	403
Table 7.22	Performance data on mutual funds (2002–2007)	405
Table 8.1	No-show data for Example 8.3	423
Table 8.2	Example coach fare classes	425
Table 8.3	Data for Example 8.5	430
Table 8.4	No-show data for Exercise 8.9	438
Table 8.5	Price and demand data for Exercise 8.13	439
Table 8.6	Ticket requests for Exercise 8.14	440
Table 8.7	Demand data for Exercise 8.16	440
Table 9.1	Major retail firms around the world (2015 Data)	444
Table 9.2	Store formats and their characteristics	445
Table 9.3	Typical items in a balance sheet	449
Table 9.4	Typical items in an income statement	450
Table 9.5	Typical items in a cash flow statement	451
Table 9.6	Balance sheet of Walmart stores, 2013–2017	453
Table 9.7	Balance sheet of Amazon.com for fiscal years 2012–2016	456
Table 9.8	Data for Exercise 9.4	470
Table 10.1	Healthcare spending per person (U.S. Dollars) in 1995 and 2015	475
Table 10.2	Population health outcomes and risk factors in OECD countries in 2013	477
Table 10.3	Delivery of healthcare services	483
Table 10.4	Stakeholder groups	485

Table 11.1 Payment modes in the United States, 2000–2012 530
Table 11.2 Selected banking infrastructure and access metrics 535
Table 11.3 Data for Example 11.2 538
Table 11.4 Variance–covariance matrix of returns on investments (Example 11.2) 539
Table 11.5 Binomial option pricing example 558

Preface

At its heart, this book is about the design and control of service systems, as evidenced by the words "engineering" and "management" in its title. Until World War II, *agriculture* and *manufacturing* sectors dominated the world economies. However, in the last 50 years, these two sectors have fallen, in terms of both labor force and gross domestic product (GDP). On the other hand, the *service* sector employment and GDP have increased steadily, all over the world, and we are now in a "service economy." For example, in the United States, service sector employment and GDP are close to 80%, dominating the other sectors.

This book grew out of a senior-level course in Service Systems Engineering, taught in the Department of Industrial Engineering at Pennsylvania State University since 2007. The course was developed in response to alumni input and the fact that more than 50% of the IE graduates are employed in service industries over the last 10 years. Service Systems Engineering emphasizes the use of engineering principles to the design and operation of service enterprises. It is a relatively new discipline in engineering. It relies on mathematical models and methods to solve problems in the service industries. It recognizes the special challenges in managing services—perishability, intangibility, proximity, and simultaneity.

This text emphasizes a quantitative approach to solving problems related to designing and operating service systems. Importantly, though, it is not so "micro" in its focus that the perspective on the larger business problems is lost, nor is it so "macro" in its treatment of that business context that it fails to develop students' appreciation for, and skills to solve, the tactical problems that must be addressed in effectively managing flows of goods and people in service industries. Economists often speak of the need to understand "first principles" before one can understand and solve larger problems. We share that view, and we have therefore structured this book to provide a grounding in the "first principles" relevant to the broad and challenging problem of managing services. We feel strongly that students of service systems engineering are best served by *first* developing a solid understanding of, and a quantitative toolkit for, tactical decision making in areas such as demand forecasting, service quality measurements, data envelopment analysis, multiple-criteria decision making, integer programming models, and queuing theory—*before* making any attempt to "optimize the service system," a task that is clearly much easier said than done.

Still, the idea of optimization is indeed prevalent throughout this book. This book is careful and deliberate in its approach to service system optimization. Indeed, the perspective taken is one that is well known to engineers of all types, namely, the perspective of *design*. Engineers design things. Some engineers design discrete physical items, and some design collections of items that operate together as systems. Engineers that design service systems take on the latter challenge.

Target Audience

This book is targeted to serve in the following contexts:

- A textbook for graduate-level and advanced undergraduate-level courses in industrial engineering.
- A textbook for, or reference book to support, advanced MBA elective courses in operations management, supply chain, or service systems management that emphasize quantitative analysis.
- A reference for technical professionals and researchers in industrial engineering, operations management, supply chain, and service systems engineering.

This book is primarily addressed to those who are interested in learning how to apply *operations research* (OR) models to managing service systems. This book assumes that the reader has a basic understanding of OR techniques, such as linear programming, nonlinear optimization, queuing models, and applied probability.

Organization of the Book

The flow of this book proceeds from a basic overview that defines service enterprises and establishes this book's emphasis on design and then presents several topics in service systems engineering—supply chain management, warehousing and distribution, financial engineering, revenue management, retail engineering, health systems engineering, and financial services. Each chapter concludes with a series of end-of-chapter exercises. Each set of exercises includes conceptual questions, numerical problems, and "mini-case studies." An Instructor's Manual, with solutions to the numerical problems and mini-case studies, is available for those adopting this book for classroom use.

Chapter Summaries

Chapter 1—Service Systems: An Overview

This chapter begins with the meaning of "goods" and "services." It explains the unique characteristics of services—intangibility, perishability, proximity, and simultaneity. Next, examples of service industries and the role of services in the economy are discussed. It also introduces the meaning of service quality and presents various methods for measuring service quality.

Chapter 2—Forecasting Demand for Services

The success of a service enterprise depends on its ability to accurately forecast customer demands and provide services to meet those demands. In this chapter, commonly used forecasting methods in industry (both qualitative and quantitative) are presented. It also includes a discussion of available forecasting software and real-world applications of forecasting methods.

Chapter 3—Design of Service Systems

Integer programming models with binary variables have been successfully used in practice to design service systems, solve resource allocation problems, schedule airline crews and aircrafts, and supply chain management. This chapter begins with a review of modeling with binary variables and applies it to location and distribution decisions in supply chains, airline crew scheduling, and workforce planning. The chapter ends with a discussion of basic queuing models in operations research and their applications to the service setting of staffing a call center.

Chapter 4—Evaluation of Service Systems

Evaluating the design and operation of a service system frequently involves cost and customer service as the main criteria. However, cost and customer service are conflicting criteria that require trade-offs between them so as to arrive at a satisfactory solution. In this chapter, two approaches are presented to evaluate the performance of services systems under multiple conflicting objectives—multiple criteria decision making and data envelopment analysis.

Chapter 5—Supply Chain Engineering

The focus of this chapter is on the design and operation of supply chains. It begins with a meaning of supply chain engineering and describes at a high level the type of decisions that are made in managing a supply chain. Metrics for measuring supply chain performance are introduced and their relationships to a company's financial measures are presented. Use of integer programming models for location and distribution decisions in supply chains is discussed. Multiple criteria ranking methods are used for selecting the best suppliers. The chapter ends with a discussion of supply chain logistics, risk pooling and contracting in supply chains.

Chapter 6—Warehousing and Distribution

This chapter first describes the basic functions of a warehouse, including warehouse type, followed by a discussion of the various equipment required to support these functions. Next, the basics of warehouse design, layout, and equipment required to support the operations are presented. Three aspects of item retrieval in a warehouse are then described and modeled, namely, order picking, slotting, and design of a picking forward area. This is followed by a discussion of line balancing and a self-organizing strategy called bucket brigades. The chapter concludes with a discussion on cross dock design and operations.

Chapter 7—Financial Engineering

This chapter begins with the basics of investing in general. It then presents in detail Markowitz's mean-variance quadratic programming model, known as the modern portfolio theory, and its applications in selecting a diversified portfolio. Next, extensions by Sharpe, who developed a bi-criteria linear programming model for portfolio optimization, are discussed. The chapter ends with a discussion of investing in bonds and prudent strategies for saving and investing in general.

Chapter 8—Revenue Management

This chapter discusses several strategies used by service industries under the umbrella of "revenue/yield management," to manage the variability in customer demand and supply. Mathematical models for "overbooking" and "differential pricing" strategies are presented. Since revenue management originated in the airline industry, the chapter ends with a discussion of different revenue management practices currently used by the airlines.

Chapter 9—Retail Engineering

This chapter starts with a discussion of different types of retailers and presents financial metrics used in managing retail industry. In this context, balance sheets, income statements, and cash-flow statements are reviewed along with the DuPont method for analyzing retail enterprises. Operational decision-making in retail industry is discussed in the context of staffing checkout processes in stores.

Chapter 10—Healthcare Delivery Systems

This chapter provides an overview of healthcare systems from an engineering perspective. There is first a discussion of the important structures of healthcare, including how it is financed, information technology, and basic operations. This is followed by techniques for more effective use of healthcare resources, including scheduling and capacity management. A brief discussion of health analytics is presented with applications to conclude the chapter.

Chapter 11—Financial Services

This chapter provides an overview of the major segments of financial service industries including payments, banking, electronic trading, and derivatives. Payment systems are discussed using the four-corner payment model. Examples of decision-making in banking are introduced by applying queuing models for teller staffing and in channel diversification. Basics of electronic trading are introduced along with valuation and technical trading techniques. The chapter concludes with an introduction to options and their valuations.

In summary, this book is a blend of two key perspectives—an introductory discussion that lays out the framework for the effective design of a service system and its supporting policies, and then studying the elemental problems in specific service sector applications, in the context of managing service enterprises. The result is what we believe to be a comprehensive treatment of the subject that we hope will serve many students and practitioners of the science of designing effective service systems for many years to come.

A. Ravi Ravindran
University Park, Pennsylvania

Paul M. Griffin
West Lafayette, Indiana

Vittaldas V. Prabhu
University Park, Pennsylvania

Acknowledgments

First and foremost, we express our sincere appreciation to Aswin Dhamodharan, an industrial engineering doctoral student at Penn State University, for serving as the editorial assistant for this book and for his outstanding help with the preparation of the final versions of the manuscript for the publishers.

We thank our former PhD students—Suchithra Rajendran and Sharan Srinivas—currently faculty members at the University of Missouri in Columbia, for their valuable contributions to the material presented in Chapters 1 and 7. Their reviews of the chapter drafts are also much appreciated. We also thank our students Achal Goel, Karthik Reddy, Mohan Tirupati, and Vikas Dachepalli for their contributions to Chapters 7, 9, and 11. In addition, several graduate students at Penn State, Georgia Tech, and Purdue University helped us by reviewing many chapters of this book and providing valuable comments that improved the presentation. In particular, we wish to acknowledge the reviews provided by Richard Titus, Lucas Servera, Bárbara Venegas, Chintan Patil, Mohan Chiriki, Shiyu Sun, and Chen (Chelsea) Feng. This book was supported in part with funding from Service Enterprise Engineering 360° (SEE 360) at Penn State and the Regenstrief Foundation at Purdue.

We thank Cindy Renee Carelli, executive editor at CRC Press, for her constant support and encouragement from inception to completion of this book. Finally, we thank our families for their support, love, understanding, and encouragement, when we were focused completely on writing this book.

Authors

A. Ravi Ravindran has been a professor and former department head of industrial and manufacturing engineering at the Pennsylvania State University (1997–2017). He is a professor emeritus now. Formerly, he was a faculty member in the School of Industrial Engineering at Purdue University (1969–1982) and at the University of Oklahoma (1982–1997). At Oklahoma, he served as the director of the School of Industrial Engineering for 8 years and as the associate provost for the university for 7 years. He received his BS in electrical engineering with honors from BITS, Pilani, India, and his MS and PhD in industrial engineering and operations research from the University of California, Berkeley. His research interests are in multiple-criteria decision making, financial engineering, healthcare delivery systems, and supply chain optimization.

Dr. Ravindran has published eight books (*Operations Research* [John Wiley and Sons, Inc., 1987], *Engineering Optimization* [John Wiley and Sons, Inc., 2006], *Operations Research and Management Science Handbook* [CRC Press, 2008], *Operations Research Methodologies* [CRC Press, 2009], *Operations Research Applications* [CRC Press, 2009], *Supply Chain Engineering* [CRC Press, 2013], *Multiple Criteria Decision Making in Supply Chain Management* [CRC Press, 2016], and *Big Data Analytics Using Multiple Criteria Decision-Making Models* [CRC Press, 2017]) and more than 150 journal articles in operations research. His recent book on *Supply Chain Engineering* received the Institute of Industrial Engineers (IIE) *Book-of-the-Year* Award in 2013. He is a Fellow of IIE and a Fulbright Fellow. In 2001, he was recognized by IIE with the *Albert G. Holzman Distinguished Educator Award* for significant contributions to the industrial engineering profession by an educator. In 2013, he received the *Outstanding Teaching Award* in the College of Engineering from the Penn State Engineering Alumni Society. He also received the *Excellence in Teaching Award* from the Logistics & Supply Chain Division of IIE in 2017. He has been a consultant to AT&T, CNH America, General Motors, IBM, Kimberly Clark, General Electric, U.S. Department of Transportation, the Cellular Telecommunications Industry Association, and the U.S. Air Force. He currently serves as the Operations Research Series editor for Taylor & Francis Group/CRC Press.

Paul Griffin is the St. Vincent Health chair and director of the Regenstrief Center for Healthcare Engineering at Purdue University and a professor in the Schools of Industrial Engineering and Biomedical Engineering (2017–present). Previously, he was the Virginia C. and Joseph C. Mello chair and professor in the School of Industrial & Systems Engineering (ISyE) at Georgia Tech (2015–2016). Before this, he served as the Peter and Angela Del Pezzo chair and department head of Industrial and Manufacturing Engineering at

Penn State (2009–2015). He began his academic career as an assistant professor in ISyE at Georgia Tech in 1988.

He received his BA in chemistry and BS in chemical engineering from the University of Texas in Austin, his MS in industrial engineering from the University of Texas at El Paso, and his PhD in industrial engineering from Texas A&M University. His research interests are in healthcare engineering, health analytics, cost and comparative effectiveness in public health, and supply chain coordination.

Vittaldas V. Prabhu is currently a professor in the Marcus Department of Industrial and Manufacturing Engineering at Penn State, where he started his academic career as an assistant professor in 1996. He also serves as the director of Service Enterprise Engineering 360. He received his PhD in mechanical engineering from the University of Wisconsin–Madison, where he also got his MS degrees in mechanical engineering and manufacturing systems engineering. He received his BE in instrumentation technology from Bangalore University. Vittal works in the area of distributed control systems with a focus on manufacturing and service enterprises consisting of discrete events, physical processes, and service processes. He teaches courses in manufacturing systems, service systems engineering, retail services engineering, and distributed controls. He is a Fellow of the Institute of Industrial Engineers. He is also an active member of the Society of Manufacturing Engineers and the International Federation for Information Processing—Working Group 5.7: Advances in Production Management Systems.

1

Service Systems: An Overview

At its heart, this book is about the design and control of service systems, as evidenced by the words "engineering" and "management" in its title. In this chapter, we begin with the meaning of "goods" and "services," their differences, examples of service systems, and the role of services in our economy. We also introduce the meaning of "service quality" and how to measure it.

1.1 Goods and Services

People all over the world use a combination of goods and services. Goods are tangible physical objects (big and small) that we use or consume daily, while services are processes, deeds, or performances (sometimes invisible) that we experience or enjoy. Goods are generally manufactured in distant locations and shipped all over the world, while most services are produced closer to home. The definition of goods and services is not black and white; there are a lot of gray areas. For example, when you buy a car, it is considered a "good"; if you lease a car, it becomes a "service"! Given below are some examples of goods that we use from morning till evening in our daily lives:

- Toothpastes, toothbrushes, and soaps
- Coffeepots and stoves
- Cereals, breads, and eggs for breakfast
- TVs to check the news and weather
- Smartphones and computers to read e-mail
- Cars to drive to work
- Furniture and equipment (including computers) to do our work in the office
- Vegetables and meats for the evening dinner

For the above activities, we also use the following services:

- Water supplied by the local utility
- Electricity supplied by a power company

- Natural gas for cooking
- Cable TV and Internet for entertainment and information
- Services by telecommunication companies
- Roads and highways maintained by government agencies (state and local)

In addition, retail services (drugstores and supermarkets) make the goods we buy for our daily needs available. Instead of a home-cooked evening meal, we may decide to dine at a local restaurant and go to a movie, availing ourselves the services provided by the hospitality industries. Instead of driving our car to work, we may use a taxi (yellow cab, Uber) or public transportation (buses or trains). When you use your credit card to pay for goods at the retailers, you are using the financial services to process those transactions.

1.2 Differences between Goods and Services (Daskin 2010; Davis and Heineke 2003; Fitzsimmons and Fitzsimmons 2006; Metters et al. 2006)

There are four key differences between goods and services:

1. Intangibility
2. Perishability
3. Proximity
4. Simultaneity

1.2.1 Intangibility

Goods, such as cars, TVs, computers, clothing, and food items, can be seen and touched and can be used or consumed. On the other hand, services are intangible. You cannot see the knowledge gained by a student during a professor's lecture. Most services provide some personal value to the customers. The value may be in the form of satisfaction after watching a movie, seeing a doctor for an illness, eating at a restaurant, or listening to a live concert.

1.2.2 Perishability

Goods can be produced ahead of time, stored, and then sold later to the customers. On the contrary, services are perishable. Empty seats in an airline have no value once the plane takes off, football tickets are worthless after the game is over, and today's movie tickets cannot be sold tomorrow!

Perishability of services is one of the biggest challenges in managing capacity and demand for services.

1.2.3 Proximity

Goods can be produced anywhere in the world and shipped over long distances to the customers. On the contrary, most services require closer proximity of the service provider and its customers. Examples include attending a ball game, going to a movie or concert, buying groceries from a local supermarket, and visiting a doctor for an illness. However, not all services require proximity. Online services and back office operations (e.g., payroll, reading medical x-rays, and insurance claims processing) do not require proximity. Many technical "help-desks" and reservation centers are located in India to serve customers in the United States.

1.2.4 Simultaneity

Simultaneity refers to the fact that most services are created and consumed at the same time. In other words, there is direct consumer interaction during service. On the other hand, consumer presence is not required when producing goods. Inventory of finished goods separates the producer and customer. For example, you do not have to be present when your car is manufactured at the plant. However, your car (as your surrogate) has to be present when it is serviced!

1.3 Challenges Posed by Services

The unique characteristics of services—intangibility, perishability, proximity, and simultaneity—have posed special challenges in designing and controlling service systems. Moreover, advances in digital technology and increased use of "self-service" have posed other challenges. They are discussed in this section.

1.3.1 Impact of Technology

Automation and advances in digital technology have blurred some of the differences between goods and services. In the past, customers have to go physically to a bank, during normal business hours, to complete their financial transactions. With the introduction of ATMs, the customers could do their transactions at select locations any time. Then came online banking, which the customers can use to complete their bank transactions 24/7 from home. Thus, "proximity" is no longer a difference. However, direct customer interaction (simultaneity) is still valid; the interaction is now between the customers and the bank's "surrogates"—ATMs and the bank's computers.

1.3.2 Bundled Products

All the products customers buy cannot be classified strictly as goods or services. Many are in fact a combination of goods and services. For example, many Americans buy their smartphones and the wireless "service" as a bundle from AT&T, Verizon, Sprint, and other companies. Even though retail stores are classified as services, customers buy mostly goods at the supermarkets and department stores. Major appliance manufacturers sell both goods and "after sales services" to their customers.

1.3.3 Self-Service

Introduction of self-service at gas stations, retail stores, airports, and other locations has resulted in customers interacting more actively with their services. In fact, customers are working as "unpaid" employees of the company! For example, in the past, store clerks checked out the grocery items, completed the cash/credit card transactions, and bagged the items. Now, customers do all the work at "self-check" lanes. Instead of four checkout clerks working at four checkout lanes, one clerk now supervises four "self-check" lanes. The same thing is true when customers make online airline reservations or check in at the airport kiosks. Unfortunately, in the calculation of the gross domestic product (GDP) of goods and services of a nation's economy, the value of "self-service" is not included!

1.3.4 Service Quality

Because services are intangible and require customer interaction, it is difficult to measure the quality of service (QoS), while designing service systems. It is easier to assess product quality, while designing a manufacturing process. However, customer satisfaction, which is a key component of service quality, is subjective and is difficult to measure. We will discuss service quality and its measurements more in detail in Section 1.8.

1.3.5 Managing Capacity and Demand

While designing the capacity of manufacturing systems, inventory is used as a buffer to account for the variability in customer demand and supply. However, perishability (no inventory) and simultaneity (direct customer interaction) make capacity management very difficult in service systems. Tools, such as "revenue management," are specifically designed to address this challenge. Revenue management will be discussed in detail in Chapter 8.

1.4 Classification of Services

Following Davis and Heineke (2003), services can be classified broadly into two sectors, with some overlap:

- Service industries
- Ancillary and support services

Service industries represent enterprises whose primary mission is to provide a particular type of service. *Ancillary and support services* are peripheral services performed in manufacturing or service organizations that are not their core functions.

1.4.1 Service Industries

Following Daskin (2010), service industries can be classified, based on their primary functions, as follows:

1. Communications
2. Education
3. Entertainment
4. Financial services
5. Government services
6. Healthcare
7. Hospitality and leisure
8. Insurance
9. Professional services
10. Retail and wholesale
11. Transportation
12. Utilities

Next, we will discuss each sector in detail.

1.4.1.1 Communications

Communication services refer to the transmission of voice, data, and video. They include the following:

- Post office/U.S. mail
- Landline phones
- Cell phones

- E-mail
- Text messages
- Internet service
- Videos

1.4.1.2 Education

This sector includes K-12 and college education, including online education as follows:

- Primary and secondary schools
- Private, parochial, and charter schools
- Two-year community colleges
- Four-year colleges
- National universities
- For-profit online schools

1.4.1.3 Entertainment

Providers of all forms of entertainment, both inside and outside the home, are included here:

- Movie theaters
- Concerts and plays
- Sports franchises
- Television stations
- Radio stations

1.4.1.4 Financial Services

Financial services include the following enterprises:

- Banks
- Brokerage services (e.g., Merrill Lynch)
- Investment companies, including mutual funds (Fidelity, Vanguard)
- Credit card companies (MasterCard, Visa)
- Mortgage services

Deregulation and advances in technology have blurred the industry lines in this sector. Many banks now offer investment products, credit card companies offer banking services, and airlines offer credit cards!

1.4.1.5 Government Services

Public services provided by local, state, and federal governments are included here. Given below are a few examples:

- Housing services
- Safety and security
- Emergency services
- Parks and recreation
- Roads and highways
- Revenue collection
- Judicial services
- Social services
- Research and development

1.4.1.6 Healthcare

Enterprises involved in people's health form the healthcare sector and they include the following:

- Hospitals
- Health maintenance organizations
- Walk-in clinic, minor/weekend emergency centers
- Specialty clinics
- Ambulance services
- Professional services of physicians, nurses, medical technicians, psychiatrists, and social workers
- Health research organizations, such as Centers for Disease Control and Prevention, the National Institutes of Health, the American Cancer Society, and the American Heart Association.

1.4.1.7 Hospitality and Leisure

Direct leisure industries and hospitality industries that support leisure travel are included. Examples include the following:

- Hotels
- Restaurants
- Amusement and theme parks (Six Flags, Disneyland)
- Car rental agencies

Hotels can be further subdivided into five-star luxury hotels to local motels. Restaurants can range from "fine-dining" to "fast-food" chains and cafeterias.

1.4.1.8 Insurance

The insurance sector covers insurance services by application areas as follows:

- Health
- Life
- Automobile
- Dental
- Home
- Personal property
- Personal liability

1.4.1.9 Professional Services

Professional services include services provided as independent contractors, each with their own client base, as follows:

- Architects
- Engineers
- Accountants
- Financial consultants
- Lawyers

1.4.1.10 Retail and Wholesale

Retail and wholesale sectors cover a wide spectrum of service industries as given below:

- Grocery stores that include large supermarkets to mom-and-pop stores
- Department stores—from large upscale stores (such as Saks and Nordstrom) to small boutiques
- Discount stores—Walmart, Sears, and Target
- Big-box stores—Sam's Club and Costco
- Office supplies—Staples and Office Depot
- Gas stations and auto repair shops
- Personal services such as hair salons, dry cleaners, and spas

Service Systems 9

1.4.1.11 Transportation

This sector includes industries involved in the transportation of people and goods as listed below:

- Airlines
- Buses
- Trains
- Taxis
- Trucks
- UPS, FedEx, and U.S. Postal Service
- Moving companies

1.4.1.12 Utilities

The utilities sector provides the following services:

- Electricity
- Water
- Natural gas
- Heating oil
- Telephones
- Cable TV
- Internet service

Note that telephones, cable TV, and Internet services overlap with communication and entertainment. In 2015, the U.S. Federal Communications Commission (FCC) ruled under its "Net Neutrality" Policy that all Internet providers should give equal access to websites, apps, and video services. Communications companies challenged that ruling in federal court. But the court ruled in FCC's favor in June 2016, agreeing that Internet service is in fact a utility. However, in late 2017, under the new FCC chairman, appointed by President Trump, the commission scrapped the "Net Neutrality" policy!

1.4.2 Ancillary and Support Services

Davis and Heineke (2003) define *ancillary and support services* as peripheral services performed in industries that are not their core functions. They can be in manufacturing or service enterprises. Examples include the following services:

- Residence halls and dining services in colleges and universities
- Security services in manufacturing plants and department stores

- Janitorial services
- Transportation services

There is a growing trend among companies to outsource some of the ancillary and support services so that they can concentrate on their core business. For example, many schools and colleges have outsourced cafeteria services to fast-food chains. Many automotive companies have outsourced their transportation services to third-party logistics providers to deliver parts and raw materials. Security services are outsourced to companies that specialize in providing security to plants, warehouses, and special sports and musical events. With globalization, procurement functions are also being outsourced. However, not all companies follow the outsourcing trend. For instance, Walmart runs its own fleet of trucks for distribution and has brought the global procurement function within the company.

Reasons to outsource ancillary and support services include focusing on core business, reducing cost, and getting better service. However, when services are outsourced, companies lose control and increase their risk. Poor customer service by a third party can affect the company's brand and its revenue.

1.4.3 Service Process Matrix (Fitzsimmons and Fitzsimmons 2006; Schmenner 1986)

In Sections 1.4.1 and 1.4.2, services were classified based on their functionality or application areas. Services can also be classified based on their operational characteristics. The *Service Process Matrix*, proposed by Schmenner (1986), classifies services based on their labor use and customer interaction as follows.

- Labor intensity
  - High
  - Low
- Customer interaction and customization
  - High
  - Low

This leads to a 2 × 2 matrix as shown in Figure 1.1. Labor intensity is defined as the ratio of labor cost to capital cost. Hence, "capital-intensive" services fall under low labor intensity (quadrants I and II), while "labor-intensive" services fall under quadrants III and IV in Figure 1.1. Customer interaction refers to the ability of the customer to personalize the services to his or her needs. In general, customization increases the cost of service. Next, we discuss each of the service quadrants in detail.

Service Systems

		Customer interaction	
		Low	High
Labor intensity	Low	Service factory (I)	Service shop (II)
	High	Mass service (III)	Professional service (IV)

FIGURE 1.1
Service Process Matrix. (Based on Schmenner, R.W. 1986. How can service business survive and prosper? *Sloan Management Review* 27(3): 21–32.)

1.4.3.1 Service Factory

The first quadrant (Figure 1.1) refers to services that are capital intensive, but low in customer interaction. They are more like factory assembly lines. Given below are examples of services that fall under "Service Factory," where most customers receive the same service with very little customization:

- Airlines, trains, and buses
- Amusement parks
- Hotels
- Car rental agencies
- Mass transit in major cities

1.4.3.2 Service Shop

The second quadrant (Figure 1.1) also represents services that are capital intensive, but they provide individualized customer service. Examples of such services are the following:

- Hospitals
- Auto repair shops
- X-ray and MRI services

1.4.3.3 Mass Service

The third quadrant (Figure 1.1) refers to labor-intensive services that provide uniform or undifferentiated services. Example of mass service include the following:

- Colleges, schools, and universities
- Retailers
- Wholesalers

- Professional sports teams
- Concerts and plays

1.4.3.4 Professional Service

Quadrant IV (Figure 1.1) represents customized services provided by highly trained specialists. Examples include the following:

- Lawyers
- Architects
- Accountants
- Mental health counselors
- Engineers

1.4.3.5 Use of Service Process Matrix

Classification of services by functional areas, given in Sections 1.4.1 and 1.4.2, helps service industries benchmark their services against their peers who are in the same industry classification, so that they can improve their service operations. The Service Process Matrix helps companies learn the best practices of other companies, whose services have similar operational characteristics. For example, American Airlines pioneered the revenue management techniques to reduce empty seats due to "No-shows" in order to increase its revenue. Other companies in the transportation sector (airlines and trains) quickly followed the revenue management strategies. The hotel industry, which belongs to the "recreation and leisure" sector, realized that it too has similar problems—empty rooms due to no-shows (quadrant I in the Service Process Matrix). Hence, they also began to apply the same revenue management strategies in hotel reservations to increase revenues.

1.5 Role of Services in the Economy

The economic activity of a nation can be broadly classified into three sectors:

1. Agriculture
2. Manufacturing including mining and construction
3. Service

Until World War II (WWII), agriculture and manufacturing sectors dominated the world economies. However, in the last 50 years, agriculture and manufacturing sectors have fallen, in terms of both labor force and GDP. On the other hand, the service sector employment and GDP have increased steadily all over the world and we are now in a "service economy." For example, during WWII, 40% of labor was employed in manufacturing; now, it is less than 15%.

1.5.1 U.S. Employment by Sector

In the year 2016, the top three countries comprising the world labor force were China (22%), India (14%), and the United States (4%). However, the distribution of employment by sector varied widely. Both in China and India, the agriculture sector was still dominating (50%–60%) and the service sector was between 30% and 40%. However, in the United States, the service sector employment was close to 80%, dominating the other two sectors. Table 1.1 illustrates the growth of the service sector employment in the United States.

1.5.2 U.S. GDP by Sector

The GDP is the total value of all goods and services produced in a country and is considered the broadest measure of a nation's economy. The top 10 economies in the world, based on their 2014 GDP, are listed in Table 1.2.

Similar to the employment data (Table 1.1), the share of GDP by the service sector has been steadily increasing in the last 40 years, particularly in the developed economies. Table 1.3 illustrates the U.S. GDP by sector since 1970. From 1970 to 2010, U.S. manufacturing, as a percentage of GDP, has decreased from 29% to 17%. During the same period, the service sector GDP has increased from 68% to 82%. Similar changes have happened at the emerging economies as well. For example, according to the World Bank data, service sector GDP in China increased from 19% in 1970 to 42% in 2015. During the same period, the service sector GDP in India increased from 44% to 53% of total GDP.

TABLE 1.1

U.S. Employment by Sector

Year	Agriculture	Manufacturing[a]	Service
1970	4.8%	30.9%	64.3%
1980	3.6%	25.8%	70.6%
1990	2.9%	21.0%	76.1%
2000	2.8%	18.2%	79.0%
2010	2.6%	13.3%	84.1%

Source: Adapted and modified from Haksever and Render. 2013. *Service Management: An Integrated Approach to Supply Chain Management and Operations*. Upper Saddle River, NJ: FT Press, p. 5.

[a] Includes mining and construction.

TABLE 1.2

Top 10 World Economies

Rank	Country	2014 GDP (in billions of U.S. dollars)
1	United States	$17,611
2	China	$10,097
3	Japan	$5339
4	Germany	$3854
5	France	$2844
6	United Kingdom	$2802
7	Brazil	$2430
8	Italy	$2102
9	India	$2028
10	Russia	$1931

Source: World Bank (2014 World Development Indicators).

TABLE 1.3

U.S. GDP by Sector

Year	Agriculture	Manufacturing[a]	Service
1970	2.6%	28.9%	68.5%
1980	2.2%	28.0%	69.8%
1990	1.6%	22.4%	76.0%
2000	1.0%	20.0%	79.0%
2010	1.1%	16.9%	82.0%

Source: Adapted and modified from Haksever and Render. 2013. *Service Management: An Integrated Approach to Supply Chain Management and Operations.* Upper Saddle River, NJ: FT Press, p. 7.

[a] Includes mining and construction.

1.6 Strength of American Manufacturing

In recent years, there have been reports in the media and speeches by politicians that the U.S. manufacturing is "either dead or dying." They point out the declining employment in the manufacturing sector (see Table 1.1). It is true that more than 6 million jobs were lost from 2000 to 2010 in the United States.

1.6.1 Decline in U.S. Manufacturing

The primary reasons for the decline are the following:

1. *Strong labor productivity:* Productivity measured as "output per employee hour" increased by 40% during the last 10 years.
2. *Automation:* Introduction of new manufacturing technologies—CNC machines and assembly robots—has reduced the labor needs.

3. *Off-shore outsourcing:* Globalization and cheap labor resulted in low-end manufacturing jobs to move to emerging economies, in particular to China and India. It is important to recognize that even though the manufacturing employment has declined in the last 40 years, the manufacturing share of *GDP in **dollars** has been increasing steadily!* Hence, it is not true that manufacturing is dead or dying. Basically, U.S. manufacturing has moved "upscale" by making high-value items. Inexpensive goods requiring manual labor are made overseas and the United States is still the leading manufacturer by *value of goods produced.* Boeing, General Electric, General Motors, HP, Intel, IBM, John Deere, and Lockheed Martin are the largest manufacturers by revenue. They make a wide array of high-end products including aircraft, missiles, fighter jets, computer chips, gas turbines, locomotives, farm and mining equipment, automobiles, and space-related equipment.

1.6.2 Revival of U.S. Manufacturing

There has been a steady uptick in the U.S. manufacturing employment since 2010. It is primarily due to a phenomenon called "insourcing" or "reshoring." Both the U.S. and foreign companies are establishing new plants or bringing off-shore plants back to America. Reshoring has resulted in an increase of U.S. manufacturing jobs from 11.5 million in 2010 to 12.5 million in 2016.

Reasons for "reshoring" are as follows:

- Productivity gains and slow wage growth have made the U.S. labor cost competitive with respect to other developed countries. For example, U.S. labor costs are 9% lower compared to the UK, 11% lower compared to Japan, 21% lower compared to Germany, and 24% lower compared to France.
- Wages are rising at 15%–20% per year in the emerging markets (China and India) and have steadily eroded the cost differential. For example, China's labor cost is only 15% lower than that of the U.S. labor cost. When shipping and inventory cost are added, China's labor advantage reduces to single digits!
- U.S. companies are realizing that overextended global supply chains have increased the risk of disruptions, particularly after the 2011 Japanese tsunami.
- Low-cost energy and better transportation infrastructure are making U.S. an attractive manufacturing location.

A Boston consulting group found, in a 2014 survey of companies, that 21% of U.S. manufacturers with $1 billion or more in sales were actively "reshoring" and 54% said that they were seriously considering it. In 2014,

Walmart announced that it would sell $50 billion more of "Made-in-America" products over the next 10 years.

1.7 What Is Service Systems Engineering?

Service Systems Engineering emphasizes the use of engineering principles to the design and operation of service enterprises. It is a relatively new discipline. It relies on numbers and uses operations research models and methods to solve problems in the service industries. It recognizes the special challenges in services—intangibility, perishability, proximity, and simultaneity.

1.7.1 Engineering Issues in Service Systems

In order to provide greater customer satisfaction, service systems have to be engineered effectively. They have to be designed and operated efficiently. They should have real-time problem-solving capabilities, namely, the ability to provide quick responses to fix problems, such as overcrowding, service disruptions, and so on.

Engineering issues in service systems can be broadly classified as *design issues* that deal with the design of the service systems and *control issues* that deal with the day-to-day operation of service systems. Examples of *design issues* in service include facility location, capacity planning, pricing decisions, and revenue optimization. Examples of *control issues* in services include resource allocation and management, security, logistics, and admission control. The methodologies used in the design and control of service systems include statistics, quality control, optimization, queuing, and simulation.

1.7.2 Engineering Problems in Services

Examples of design and control problems in service systems engineering are given below:

- Design Problems
  - Distribution network of a supermarket chain—the number, location, and capacity of the warehouses and the location of retail outlets
  - Design of "hub and spoke" airport network by airlines
  - Location of transshipment terminals by trucking companies
  - Selection of suppliers
  - Admission policies at universities

- Resource Allocation and Scheduling Problems
    - Scheduling physicians, nurses, and patients in hospitals
    - Aircraft and crew scheduling by commercial airlines
    - Workforce planning at fast-food restaurants
    - Allocation of investment capital to different financial products
    - Allocation of classroom spaces in colleges and universities
- Routing and Logistics Problems
    - Routing policies by package delivery companies
    - Aircraft routing strategies
    - Routing protocols in communication networks
    - Pickup and delivery schedules by moving companies
- Inventory Problems
    - Inventory policies at warehouses and retailers
    - Stocking of spare parts for repairs by airlines
    - Ordering of vegetables and meat at fast-food restaurants
- Revenue Management Problems
    - Reservation policies in airlines, hotels, and car rental companies
    - Pricing in healthcare, telephone, and restaurants
    - Admission control in Internet, theaters, and parks
- Queuing Problems
    - Waiting lines at airports, banks, doctor's offices, and amusement parks
    - Congestion in transportation and communication networks
    - Traffic congestion in highways

The focus of this book is to present how industrial engineering and operations research techniques can be used to design, control, and operate service systems efficiently. Forecasting the demand for services is critical for service systems engineering and management. Chapter 2 discusses the various forecasting methods. The other chapters in this book will discuss the service systems engineering problems and solution methods for specific service sector applications.

1.7.3 Multiple-Criteria Decision Making in Services

As Daskin (2010) points out, services are provided by both private and public sectors. The public sector primarily refers to government agencies at the local, state, or federal levels. The private sector includes both profit and nonprofit companies. The objectives of the private and public sector companies may be quite different. Profit maximization is the main objective of private sector companies, with customer service as a secondary objective. For nonprofits and government agencies, cost containment and customer service are the key objectives. They may also have other objectives, such as service equity among different socioeconomic groups and number of people served. In all such cases, the objectives are conflicting in nature, which lead to Multiple-Criteria Decision Making (MCDM) in services. We discuss the challenges posed by MCDM problems in general and their solution approaches in Chapter 4. Given below are few examples of multiple conflicting objectives in the service sector.

- Healthcare Sector

    Cost and accessibility to medical care are the two conflicting criteria. Cost criterion may be broken into several subcriteria that include daily cost of hospital care, cost of outpatient surgery, and cost of doctor's visit and emergency care. The subcriteria for accessibility may include distance to a hospital, physicians per 1000 population, waiting time for an elective surgery, and others.

- Financial Industries

    Investment decisions by banks, mutual funds, and insurance companies have two competing objectives—*return on investment* and *risk of investment*. Investment return can be measured by 1-year, 3-year, 5-year, and 10-year annualized returns. Investment risk can be measured by market risk, industry-specific risk, and loss of capital.

- Airlines

    Maximizing revenue and customer service are the two conflicting objectives. Revenue is a function of ticket price and cost of fuel, equipment, and personnel. Customer service can be measured in terms of number of flights per day, maximum time between scheduled flights, and number of flights during peak periods.

- Supply Chain Management

    In the design of supply chain networks, profit maximization, customer responsiveness, and supply chain risk are the conflicting criteria. Profit includes revenue from sales, cost of facilities, personnel, transportation, and operation. Customer responsiveness may include demand fulfillment (percentage of customer demand satisfied), delivery time, and quality. Supply chain risk may include

disruptions to the supply chain network due to natural and man-made events.
- Supplier Selection

    Procurement cost, delivery time, and quality are the three commonly considered criteria in supplier selection. Other criteria for supplier selection may include supplier's financial position, technical capability, geographical location, and disruption risk.

1.8 Quality of Service

QoS is more difficult to specify and measure than the quality of goods/products. Product quality can be specified in terms of dimensions and performance metrics. As long as the product meets those within certain tolerance limits, they meet the quality standards. However, service quality depends more on the expectation of the customer and the perception of how well the service is delivered. Hence, there is a lot of customer subjectivity involved in the service quality, which makes the assessment more difficult. The textbooks by Davis and Heineke (2003), Fitzsimmons and Fitzsimmons (2006), and Metters et al. (2006) have excellent discussions of service quality.

1.8.1 QoS Measures

A set of measures that users "want" from the service system is called QoS measures. These measures are the *"Expectations"* of the customer. Customer expectations may come from advertisements or prior experience. After the service is performed, the assessment of QoS measures by the customer is called *"Perception."* The difference between "Perception" and "Expectation" is "Customer Experience" as defined below:

$$\text{Customer Experience} = \text{Perception} - \text{Expectation} \quad (1.1)$$

The customer is satisfied as long as the experience is positive. Thus, there are two potential outcomes for Equation 1.1:

1. Perception exceeds the expectation, resulting in a positive customer experience. In this case, the customer may not only pay return visits but also influence others to visit the service establishment.

2. Perception is lower than the expectation, resulting in a negative customer experience. Here, the customer may not only return back but also inform others about the poor service. Negative experience spreads quickly in the social media now!

Customer expectation in Equation 1.1 may come from "word-of-mouth," social media, advertisements, or prior experience. Another complicating factor with the customer expectation is that it is not static, even for the same service establishment. A customer who has a positive experience may increase the expectation for the next visit.

The difference between customer perception and customer expectation in Equation 1.1 is also called the *Service Quality Gap*. It occurs due to the following reasons:

- Management's lack of understanding of customer expectations
- Management's failure to articulate the customer expectations to a set of service standards
- Improper job design to translate service standards to service delivery
- Management's failure in the employee recruitment and training
- Lack of a reward structure to motivate better employee performance
- Lack of uniformity in providing the services by different employees
- Insufficient supervision of employees
- Inadequate communication between "marketing" and "operations," resulting in inflated marketing promises that cannot be met by service personnel

QoS measures can be both qualitative and quantitative. For a restaurant, qualitative measures include ease of parking, cleanliness, staff professionalism, ambience, and quality of food. Quantitative measures may include wait time for a table, time to take the order, and waiting time for the food. Different customers may perceive the same qualitative measure differently. Even for the quantitative measures, what is set as acceptable by the service provider may not be so for the customer. For example, the restaurant may decide that the acceptable wait time for a table should be less than 5 minutes. However, every customer may not feel that way. A 5-minute wait time may be acceptable for a leisurely dinner, but may not be acceptable if you are trying to grab a quick lunch before a meeting.

1.8.2 Dimensions of Service Quality

QoS measures can be classified into five categories, called the *five dimensions of service quality*. They are *reliability, responsiveness, assurance, empathy*, and *tangibles*. A brief description of the five dimensions is given below:

1. *Reliability*: This refers to performing the service operations correctly and dependably. Delivery dates are met and the order fulfillments are accurate. Reliability is critical to online retailers.
2. *Responsiveness*: This represents the willingness of employees to help customers when there are problems and provide quick resolutions without any "red tape."
3. *Assurance*: This refers to the ability of service providers to convey trust and confidence to their customers about the quality of their services. Service competency is a major factor in assurance.
4. *Empathy*: This covers service customization, individual attention, and willingness to listen. For example, in the healthcare sector, the "bedside manners" of the physician are key elements of empathy.
5. *Tangibles*: The physical facilities of the service providers are called tangibles. Since customers are generally at the physical facilities (e.g., hospitals, banks, supermarkets, and restaurants) during service operations, tangibles are important elements of service establishments. Tangibles may also include location, ease of parking, cleanliness, and waiting room furniture.

1.9 Methods for Measuring Service Quality

As previously discussed, service quality measures can be both qualitative and quantitative. For the qualitative measures, Parasuraman et al. (1988) have developed a survey instrument, called SERVQUAL, which measures the customer expectations and perceptions of service. For the quantitative measures, statistical process control charts, which are used in the manufacturing industries, can be used for the service industries. In this section, we will discuss both qualitative and quantitative tools that are available to measure service quality.

1.9.1 SERVQUAL

SERVQUAL is a two-part survey instrument based on the five dimensions of service quality: reliability, responsiveness, assurance, empathy, and tangibles. Part I of the survey, which contains 22 items, measures the customer expectations, while Part II measures the customer perceptions for the same 22 items.

Table 1.4 illustrates the items included in part I of SERVQUAL. The customer rates each expectation item on a 1 to 7 scale, where "1" represents "strong disagreement" and "7" represents "strong agreement." Items E1 to E4 represent the service dimension *tangibles*, E5 to E9 represent *reliability*, E10 to E13 represent *responsiveness*, E14 to E17 represent *assurance*, and E18 to E22 represent *empathy*. In order to get a fair and honest response, some items use a positive scale and some use a negative scale. The negative items are "reverse scored" during the analysis. The average rating of each item gives the relative importance of that item. By aggregating the items under each dimension of service quality, a service provider can determine the relative importance of that dimension from the customer's perspective.

TABLE 1.4

SERVQUAL Survey (Part I: Expectations of Service)

DIRECTIONS: This survey deals with your opinions of _____ services. Please show the extent to which you think firms offering _____ services should possess the features described by each statement. Do this by picking one of the seven numbers next to each statement. If you strongly agree that these firms should possess a feature, circle the number 7. If you strongly disagree that these firms should possess a feature, circle 1. If your feelings are not strong, circle one of the numbers in the middle. There are no right or wrong answers—all we are interested in is a number that best shows your expectations about firms offering _____ services.

E1. They should have up-to-date equipment.
E2. Their physical facilities should be visually appealing.
E3. Their employees should be well dressed and appear neat.
E4. The appearance of the physical facilities of these firms should be in keeping with the type of services provided.
E5. When these firms promise to do something by a certain time, they should do so.
E6. When customers have problems, these firms should be sympathetic and reassuring.
E7. These firms should be dependable.
E8. They should provide their services at the time they promise to do so.
E9. They should keep their records accurately.
E10. They shouldn't be expected to tell customers exactly when services will be performed. (−)
E11. It is not realistic for customers to expect prompt service from employees of these firms. (−)
E12. Their employees don't always have to be willing to help customers. (−)
E13. It is OK if they are too busy to respond to customer requests promptly. (−)
E14. Customers should be able to trust employees of these firms.
E15. Customers should be able to feel safe in their transactions with these firms' employees.
E16. Their employees should be polite.
E17. Their employees should get adequate support from these firms to do their jobs well.
E18. These firms should not be expected to give customers individual attention. (−)
E19. Employees of these firms cannot be expected to give customers personal attention. (−)
E20. It is unrealistic to expect employees to know what the needs of their customers are. (−)
E21. It is unrealistic to expect these firms to have their customers' best interests at heart. (−)
E22. They shouldn't be expected to have operating hours convenient to all their customers. (−)

Source: Parasuraman, A., V A. Zeithaml, and L. L. Berry. 1988. SERVQUAL: A multiple-item scale for measuring consumer perceptions of service quality. *Journal of Retailing* 64(1): 38–40.

Service Systems   23

Table 1.5 illustrates Part II of the SERVQUAL instrument that measures the perceptions of the customer. It uses the same 22 items that are in Part I and the rating scale is similar. The difference between the expectation and perception of each item measures the Service Quality Gap (SQG) as defined below:

$$SQG_i = E_i - P_i \quad \text{for } i = 1, \ldots, 22 \tag{1.2}$$

where E_i is the average rating for expectation and P_i is the average rating of perception of that service quality measure. Larger positive SQG_i values denote serious problems with the customer service and need prompt management attention in order to improve service quality. Both SQG_i and E_i values should be used together to make sure that scarce resources are used first

TABLE 1.5

SERVQUAL Survey (Part II: Perceptions of Service)

DIRECTIONS: The following set of statements relate to your feelings about XYZ. For each statement, please show the extent to which you believe XYZ has the feature described by the statement. Once again, circling a 7 means that you strongly agree that XYZ has that feature, and circling a 1 means that you strongly disagree. You may circle any of the numbers in the middle that show how strong your feelings are. There are no right or wrong answers—all we are interested in is a number that best shows your perceptions about XYZ.

P1. XYZ has up-to-date equipment.
P2. XYZ's physical facilities are visually appealing.
P3. XYZ's employees are well dressed and appear neat.
P4. The appearance of the physical facilities of XYZ is in keeping with the type of services provided.
P5. When XYZ promises to do something by a certain time, it does so.
P6. When you have problems, XYZ is sympathetic and reassuring.
P7. XYZ is dependable.
P8. XYZ provides its services at the time it promises to do so.
P9. XYZ keeps its records accurately.
P10. XYZ does not tell customers exactly when services will be performed. (-)
P11. You do not receive prompt service from XYZ's employees. (-)
P12. Employees of XYZ are not always willing to help customers. (-)
P13. Employees of XYZ are too busy to respond to customer requests promptly. (-)
P14. You can trust employees of XYZ.
P15. You feel safe in your transactions with XYZ's employees.
P16. Employees of XYZ are polite.
P17. Employees get adequate support from XYZ to do their jobs well.
P18. XYZ does not give you individual attention. (-)
P19. Employees of XYZ do not give you personal attention. (-)
P20. Employees of XYZ do not know what your needs are. (-)
P21. XYZ does not have your best interests at heart. (-)
P22. XYZ does not have operating hours convenient to all their customers. (-)

Source: Parasuraman, A., V A. Zeithaml, and L. L. Berry. 1988. SERVQUAL: A multiple-item scale for measuring consumer perceptions of service quality. *Journal of Retailing* 64(1): 38–40.

to address those items that are very important to the customers and have large positive gaps as per Equation 1.2.

SERVQUAL has seen many applications in service industries. It has proven to be reliable and has good validity over the years. Companies have used SERVQUAL to track overall service quality trends, as well as along each quality dimension. Multi-unit retail chains have used SERVQUAL successfully to compare the service quality at different locations. Companies have also used SERVQUAL to determine the relative importance of the five dimensions of the service quality that are relevant to their industries. Averaging across all the service industries, it has been found that *reliability* is the most important dimension overall and *tangibles* are the least important. However, for a particular service industry, the relative importance may be different.

1.9.2 Statistical Process Control

Control charts are used in quality control to determine whether a process (manufacturing or service) is *"in-control"* or *"out-of-control."* If a process is in-control, no corrective action is necessary. An out-of-control process requires immediate corrective action. Use of control charts could result in four different outcomes, as listed below:

1. Process is in-control and no corrective action is taken
2. Process is in-control, but corrective action is taken, resulting in *Type I Error*
3. Process is out-of-control and corrective action is taken to fix it
4. Process is out-of-control, but no corrective action is taken, resulting in *Type II Error*

Type I Error (outcome 2) is sometimes referred to as the producer's risk and Type II Error (outcome 4) is referred to as the consumer's risk. When control charts are used to monitor the service process, it is important to minimize both Type I and Type II errors, which can occur due to measurement errors or small sample sizes.

Two types of control charts are commonly used in practice:

- Variable control chart
- Attribute control chart

1.9.2.1 Variable Control Charts

Variable control charts are used to monitor *measurable data*; examples include waiting time for service, customer service time, number of ambulance calls,

Service Systems

length of the queue, and others. We shall illustrate the construction of a variable control chart with a numerical example.

Example 1.1

A grocery store is interested in monitoring customer service times and the efficiency of checkout operations in order to improve the overall customer experience. Five customer service times in minutes were randomly recorded for each day, for a period of 7 days. Table 1.6 illustrates the data collected for a week.

In Table 1.6, the daily mean service time ($\bar{X}$) and its range (R) are also shown. In controlling the service process, both the mean service time and its variability are important. Hence, two different variable control charts are used in practice:

- $\bar{X}$ chart to monitor the mean
- R chart to monitor the range or variability

CONSTRUCTION OF $\bar{X}$ CHART

To construct the $\bar{X}$ chart, the following formulas are used to calculate the *Upper Control Limit (UCL)* and the *Lower Control Limit (LCL)*:

$$\text{UCL} = \bar{\bar{x}} + A_2 \bar{R} \tag{1.3}$$

$$\text{LCL} = \bar{\bar{x}} - A_2 \bar{R} \tag{1.4}$$

where
$\bar{\bar{x}}$ = estimate of the population mean
$\bar{R}$ = estimate of the population range
A_2 = control chart constant for sample size n

TABLE 1.6

Data for Example 1.1—Customer Service Time in Minutes

Sample	Day 1	Day 2	Day 3	Day 4	Day 5	Day 6	Day 7
1	2	1	5	2	5	3	1
2	5	1	4	1	1	4	3
3	1	2	4	2	2	5	2
4	3	2	2	1	2	3	2
5	1	1	1	4	4	3	3
Mean ($\bar{X}$)	2.4	1.4	3.2	2	2.8	3.6	2.2
Range (R)	4	1	4	3	4	2	2

The values of control chart constants for different sample sizes are given in Table 1.7. For the service time data given in Table 1.6 (Example 1.1), we calculate the UCL and LCL values as shown below:

$$\bar{\bar{x}} = (2.4 + 1.4 + 3.2 + 2 + 2.8 + 3.6 + 2.2)/7$$
$$= 17.6/7 = 2.51 \text{ minutes}$$

$$\bar{R} = (4 + 1 + 4 + 3 + 4 + 2 + 2)/7 = 20/7 = 2.86$$

From Table 1.7, the control chart constant for sample size 5 is $A_2 = 0.577$. Thus,

$$\text{UCL} = 2.51 + (0.577)(2.86) = 4.16$$

$$\text{LCL} = 2.51 - (0.577)(2.86) = 0.86$$

Figure 1.2 illustrates the $\bar{X}$ chart for Example 1.1. The three horizontal lines in Figure 1.2 represent the LCL, population mean, and the UCL values. After the $\bar{X}$ chart is constructed, the service process is monitored

TABLE 1.7

Variable Control Chart Constants

Sample Size	$\bar{X}$ Chart	R Chart	
n	A_2	D_3	D_4
2	1.880	0	3.267
3	1.023	0	2.574
4	0.729	0	2.282
5	0.577	0	2.114
6	0.483	0	2.004
7	0.419	0.076	1.924
8	0.373	0.136	1.864
9	0.337	0.184	1.816
10	0.308	0.223	1.777
12	0.266	0.283	1.717
14	0.235	0.328	1.672
16	0.212	0.363	1.637
18	0.194	0.391	1.608
20	0.180	0.415	1.585
22	0.167	0.434	1.566
24	0.157	0.451	1.548

Source: Adapted from Table 27 of *ASTM Manual on Presentation of Data and Control Chart Analysis,* copyright 1976, Philadelphia; American Society for Testing and Materials.

Service Systems

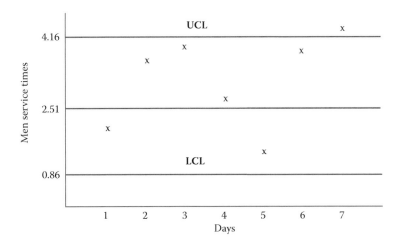

FIGURE 1.2
$\bar{X}$ chart for customer service times (Example 1.1).

daily by taking five sample service times and computing their means. As long as the mean service times are within the LCL and UCL, the process is in-control and no corrective action is needed. As an illustration, compare the mean service times for the next 7 days, after the control chart has been established. The mean values are within the UCL and LCL for days 1 through 6. However, for day 7, the mean value is above the UCL. It does not automatically mean that the service operation is out-of-control and needs corrective action. Instead, an investigation should be carried out to see whether there are any unusual environmental conditions that resulted in a large mean value. For instance, a severe snowstorm might have been forecast for day 8, which resulted in several customers stocking up on essential groceries on day 7, just before the storm. In that case, there is no need for any corrective action and the process should continue to be monitored for the following week.

CONSTRUCTION OF R CHART

In addition to monitoring the mean values of a variable through the $\bar{X}$ chart, it is also important to monitor their range or variability. For this, an R chart is constructed using the following formulas:

$$\text{UCL} = D_4 \bar{R} \tag{1.5}$$

$$\text{LCL} = D_3 \bar{R} \tag{1.6}$$

where
  $\bar{R}$ = estimate of the population range
  D_3, D_4 = control chart constants for sample size n (Table 1.7)

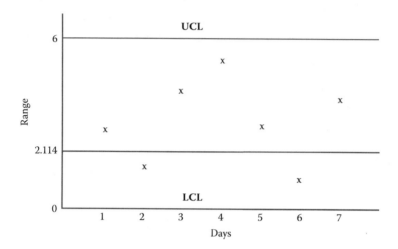

FIGURE 1.3
R chart for customer service times (Example 1.1).

For the service time data given in Table 1.6 (Example 1.1), the R chart is computed as follows:

$$\text{UCL} = D_4 \bar{R} = (2.114)(2.86) = 6.05$$

$$\text{LCL} = D_3 \bar{R} = (0)(2.86) = 0$$

Figure 1.3 illustrates the R chart for Example 1.1. The R chart is used in the same manner as the $\bar{X}$ chart. When five daily sample service times are taken, its range is also computed. As long as the daily ranges are within the LCL and UCL, no corrective action is necessary, as illustrated in Figure 1.3.

1.9.2.2 Attribute Control Charts

$\bar{X}$ and R charts are used to monitor variables that are measurable. Attribute control charts, also known as p charts, are used to monitor specific attributes of a service process. Examples include error rate in fulfilling customer orders, fraction of customers who were put on "hold" before talking to service technician, fraction of airline flights that departed late, and percentage of lost luggage by an airline. The p chart uses the following formulas to determine the UCL and LCL for that attribute:

$$\text{UCL} = \bar{p} + \sqrt{\frac{\bar{p}(1-\bar{p})}{n}} \tag{1.7}$$

Service Systems

$$\text{LCL} = \bar{p} - \sqrt{\frac{\bar{p}(1-\bar{p})}{n}} \tag{1.8}$$

where
$\bar{p}$ = average probability of the attribute
n = sample size

It is to be noted that sometimes the LCL value, computed by Equation 1.8, could be negative. In such cases, the LCL values are set equal to zero. We shall illustrate the construction of a p chart with another example.

Example 1.2

Even though the Internal Revenue Service (IRS) encourages online filings of federal tax returns, many households still file paper returns. IRS employs several data entry operators to convert the paper returns to digital form for checking the accuracy of the returns and selecting returns for auditing. Converting the paper returns to digital form introduces potential input errors. In fact, to reduce the input errors, IRS generally uses two operators to enter the same data independently. The two digital returns are then compared for accuracy. IRS employs several new operators during a tax filing season and train them as data entry operators.

Consider a regional office of the IRS, which wants to reduce the error rates and monitor the efficiency of its data entry operators. It took several samples of 100 returns each from its experienced operators and determined the number of returns with errors. The overall average error rate of the samples was 3%; that is, on average, 3 out of 100 returns contained errors in transferring the data.

Using Equations 1.7 and 1.8, the UCL and LCL values are computed as follows:

$$\text{UCL} = 0.03 + \sqrt{\frac{(0.03)(0.97)}{100}} = 0.03 + 0.017 = 0.047$$

$$\text{LCL} = 0.03 - \sqrt{\frac{(0.03)(0.97)}{100}} = 0.03 - 0.017 = 0.013$$

Figure 1.4 illustrates the p chart for Example 1.2. After the p chart is established, new operators can be monitored for their accuracies. The x's in Figure 1.4 illustrates the error rates for 100 returns for one of the new operators. Initially, the operator has more errors, but beginning week 3, with experience, the operator's error rates are within the acceptable control limits.

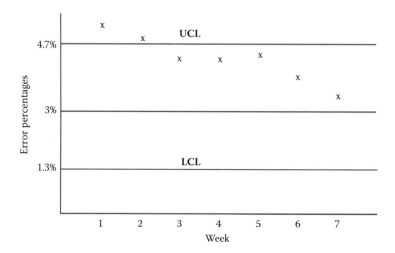

FIGURE 1.4
p chart for Example 1.2.

1.9.2.3 Selecting the Sample Size for Control Charts

Selecting the right sample size (n) is important for the use of $\bar{X}$, R, and p charts. The value of n is used in calculating the UCL and LCL values, which, in turn, indicate whether or not the service process is in-control or needs corrective action.

1.9.2.3.1 Attribute Control Charts

Selecting the correct sample size is very critical to attribute control charts. Since the attributes (usually errors and delays) occur less frequently, it is extremely important that the sample size is sufficiently large that the attribute will be observed in the limited sample. In Example 1.2, if we had chosen a smaller sample size, say 20 returns to examine, then the average number of erroneous returns in a sample will be (0.03) (20) = 0.6. In other words, the chances of finding a return with errors are very low, if only 20 returns are examined and the process may always appear to be in-control, even though the IRS operators are making mistakes in entering the data. This causes Type II errors in the input process, resulting in more returns being selected for auditing due to IRS errors, even though the paper returns have been filed correctly!

The general rule in using p charts is to have a sample size sufficiently large that at least there are two occurrences of the error over the long run in the sample. Mathematically, the rule is to select the sample size n such that

$$n\bar{p} \geq 2 \tag{1.9}$$

In Example 1.2, Equation 1.9 dictates that $n \geq \dfrac{2}{0.03} = 67$. Hence, our use of 100 returns for each sample meets the criterion for the minimum sample size.

1.9.2.3.2 Variable Control Charts

Sample sizes are not critical for $\bar{X}$ and R charts. Since the variables are measurable, small values of n can be used. Sample sizes between 5 and 10 are more common in practice. Larger sample sizes require more measurements, and depending on the variable being measured, it may be costly, taking more time and resources.

With the measurable data, a service process can go out-of-kilter due to shifts in the mean or range or both. Hence, it is recommended that both $\bar{X}$ and R charts should be constructed for a variable and both charts be used simultaneously to monitor the process. As long as both the mean and range values are within the control limits, no corrective action is needed. However, when either one or both are out of control, an examination of the causes is necessary.

1.9.3 Other Service Quality Tools

In addition to the control charts ($\bar{X}$, R, and p charts) that we discussed in the previous section, there are some statistical tools that can be used to monitor the service quality. They can also be used to identify the service problems and develop strategies for their solution. Fitzsimmons and Fitzsimmons (2006) present an excellent discussion of the service quality tools. We shall briefly discuss some of the service quality tools in this section with illustrative examples.

Given below are the various service quality tools that can be used to analyze quality problems in service industries:

- Variable and attribute control charts
- Run chart
- Histogram
- Pareto chart
- Scatterplot
- Flowchart
- Fishbone diagram

We have already discussed the control charts in Section 1.9.2. We shall discuss the other quality tools here with the help of Example 1.3.

Example 1.3

EZ Access is a for-profit family medicine clinic that serves more than 100,000 patients annually. Recently, there has been a decrease in clinic revenue and the clinic administrators have decided to address this

TABLE 1.8

Record of Problems Faced by EZ Access Clinic (Example 1.3)

Month	No-Shows	Overtime Days	Medical Errors	Billing Errors	Other
January	3	16	3	3	1
February	3	4	1	2	1
March	4	6	2	0	0
April	1	2	0	4	0
May	2	10	2	2	1
June	5	14	2	1	0
July	3	18	4	2	3
August	7	12	3	2	3
September	3	10	3	4	1
October	11	8	1	0	2
November	12	12	3	2	0
December	16	19	3	5	3
Total	70	131	27	27	15

problem as their top priority. To begin the analysis and problem identification, the clinic administrators have identified key problem areas by interacting with the staff, physicians, and patients. Table 1.8 presents the record of problems faced by the clinic during the previous year. The problems are grouped into four major categories—no-shows, overtime days, medical errors, and billing errors. Any problem that does not fall into one of those four categories are listed under "Other." No-shows represent patients, who make appointments but do not show up at the clinic. Overtime days represent situations when nurses and doctors have to stay longer to take care of all the patients in the waiting room. Medical errors result in actual revenue loss to the clinic and may potentially affect future business. Billing errors result in the rejection of claim payments by the insurance companies. This would result in more follow-up work and additional cost to the clinic.

1.9.3.1 Run Chart

A run chart is a plot of the values of a service quality variable over time. It is a quick way to observe any trends or patterns in the data; for example, whether the service quality is getting worse or is cyclical. Another use of run charts is to plot them before and after a corrective action has been implemented to determine the effectiveness of the solution strategy.

Figure 1.5 illustrates the run chart of the number of patient no-shows over time for Example 1.3. It can be seen that from January to September, only normal process variations are observed on the patient no-shows. However, there is an increasing trend in the number of no-shows from September to December.

Service Systems

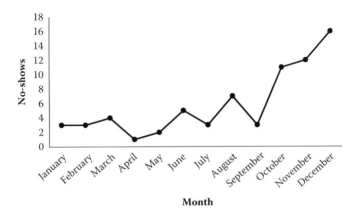

FIGURE 1.5
Run chart of no-shows (Example 1.3).

1.9.3.2 Histogram

A histogram is essentially a frequency distribution of a service quality variable. It is commonly known as the *bar chart* and can be obtained using the "Chart" command in Microsoft Excel. A histogram can be used to study unusual features, such as symmetry or skewness in the observed data. Histograms are very useful for running simulation models of service systems.

Figure 1.6 illustrates the histogram of the billing errors for Example 1.3. It can be observed from the histogram that monthly billing errors varied from 0 to a maximum of 5. The most prevalent occurrence was 2 billing errors per month and the frequency distribution is nearly symmetric on either side. It has only one mode that happens at 2 errors per month.

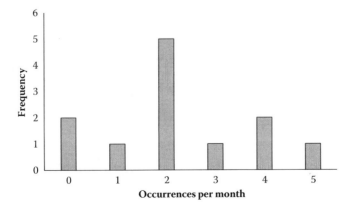

FIGURE 1.6
Histogram of billing errors (Example 1.3).

1.9.3.3 Pareto Chart

A Pareto chart rank orders the service quality problems by their relative frequencies of occurrence. It helps to identify and prioritize the most important problems a service organization should address first. Strategies to solve service quality problems generally require the use of additional resources. When the company resources are scarce, it is critical to use them strategically to solve the most important problems first so as to get the maximum payoff.

Figure 1.7 gives the relative frequencies of the problems in rank order for Example 1.3. It can be observed that 49% of the problems (131 out of 270) are due to overtime days. The next highest problem area is patient no-shows, causing 26% of the problems. Thus, overtime days and patient no-shows account for 75% of all the problems and addressing them first could substantially improve the clinic's revenue.

1.9.3.4 Scatterplot

A scatterplot, also known as an "X–Y chart," visually shows whether there is a correlation between two variables. Usually, the Y-axis represents a service quality problem and the X-axis represents a factor that may cause the quality problem. If an increase in the X values always results in an increase in Y values, then there exists a positive correlation between the two variables. Hence, controlling the value of X may likely reduce the potential service quality problem.

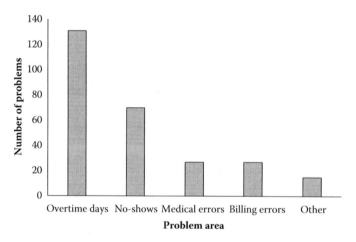

FIGURE 1.7
Pareto chart of number of problems in each area (Example 1.3).

Service Systems

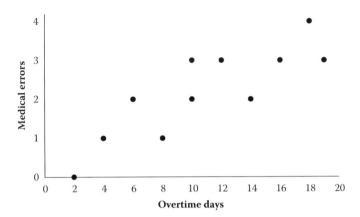

FIGURE 1.8
Scatterplot between overtime days and medical errors (Example 1.3).

Scatterplots can also be used to study whether there exists a correlation between two factors that affect service quality. If there is a positive relationship between two factors, then controlling one effectively controls the other factor also, with the net result of a greater decrease in the service quality problems. For example, Figure 1.8 gives a scatterplot of the relationship between overtime days and medical errors. Based on the scatterplot, it appears that an increase in overtime days leads to more medical errors (i.e., there appears to be a positive correlation between the two variables). Even though the relationship is not perfect, the clinic should consider reducing the number of overtime days in order to minimize the medical errors in the future. Since both factors affect the clinic's revenue, controlling them could produce significant payoff.

1.9.3.5 Flowchart

A flowchart is a pictorial representation of all the inputs and outputs, at different stages, of a service process. It helps to identify the "bottleneck" areas in the service process, where most quality problems may arise. With the help of a flowchart, different strategies to prevent the bottlenecks can be developed and analyzed for their effectiveness.

Figure 1.9 shows the flowchart of the medical billing process for Example 1.3 and can be used by the management to check for potential bottlenecks in the medical billing process.

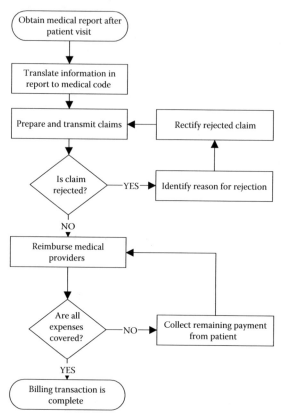

FIGURE 1.9
Flowchart of the medical billing process (Example 1.3).

1.9.3.6 Fishbone Diagram

Fishbone diagram, also known as "root cause analysis" or cause-and-effect diagram, is a visualization method to identify the root causes of a service quality problem. The objective is to identify the top 3 to 5 causes so that appropriate solutions can be developed and studied for their effectiveness. Frequently, it may require "brainstorming" with the personnel in those areas for their inputs and additional data collection on those causes.

In Example 1.3, the Pareto chart (Figure 1.7) identified patient no-shows as the second most significant problem causing revenue loss to the clinic. Figure 1.10 is the fishbone diagram for no-shows and it can be used to identify the most important causes that lead to patients missing their appointments.

Service Systems

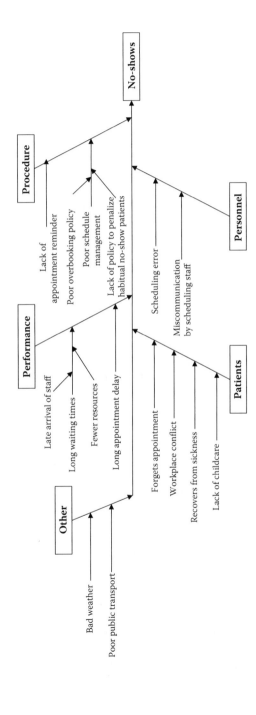

FIGURE 1.10
Cause-and-effect diagram for patient no-shows.

1.10 Summary

In this chapter, we began with the definition of a service system and its two key elements, namely, service provider(s) and users. Users, also called customers if money is involved, receive the service from the service provider. Users can be single individuals or groups (e.g., students at different locations taking online courses). Similarly, for a given service system, there can be one or many providers (e.g., multiple checkout clerks at a department store). Most services are generally owned by a single entity (e.g., barbershops and car mechanics). Sometimes, multiple entities may own the service system, such as the Internet.

Next, we discussed the differences between manufacturing and service systems. The distinctive characteristics of a service system included customer participation in the service, simultaneity, perishability, intangibility, and heterogeneity. We then provided a classification of service systems. Service systems can be single individual units (e.g., theaters, hotels, repair shops, supermarkets, and barbershops) or complex interconnected networks, such as computer-communication networks, hospitals, transportation systems, and theme parks. We discussed the importance of the service sector to the nation's economy and its contribution to the employment and the GDPs of various countries. In most developed countries, the service sector employs more than half of the workforce and accounts for more than two-thirds of the economic output. The key engineering problems in managing the services were also discussed. Textbooks by Daskin (2010) and Haksever and Render (2013) are good sources for further readings on these topics.

In the second half of this chapter, we introduced the concept of QoS and how difficult it is to measure compared to manufacturing quality. QoS is the difference between the perception of service performance and the expectation by the user. Five dimensions of service quality were introduced to measure QoS. They included reliability, responsiveness, assurance, empathy, and tangibles. We then introduced a survey instrument, called SERVQUAL, to measure the five dimensions of service quality. Other service quality tools, such as variable and attribute control charts, run chart, histogram, Pareto chart, scatterplot, flowchart, and fishbone diagram, were also discussed with illustrative examples. Textbooks by Davis and Heineke (2003), Fitzsimmons and Fitzsimmons (2006), and Metters et al. (2006) are good sources for additional readings on service quality and its measurement.

We conclude this chapter with the observation that the science behind the service sector is an emerging area of study. Industrial engineers can play a vital role in the design of service systems and improve the service processes. Human factors, operations research, and other industrial engineering tools are valuable to the design and operation of service systems.

Exercises

1.1 Illustrate the "Distinctive" characteristics of services (customer participation, simultaneity, perishability, intangibility, heterogeneity) for the *United States Postal Service* Operation.

1.2 Discuss the pros and cons of the following on the performance of service systems:
   a. Increasing self-service
   b. Allowing employees to exercise judgment in meeting customer needs
   c. Introducing an advanced reservation system for service

1.3 Illustrate the five dimensions of service quality for a service of your choice.

1.4 As discussed in Section 1.9.1, SERVQUAL is a survey tool that was designed in the late 1980s that measures the customer's perception of service quality in terms of reliability, responsiveness, assurance, empathy, and tangibles. It is very useful with services that involve direct customer interaction. However, it might not be very effective to measure service quality for computer-mediated services like video hosting websites. Thus, as part of this exercise, identify methods to measure the service quality of video hosting websites and compare the service quality of www.youtube.com and any other video hosting website of your choice. A detailed list of different video hosting websites can be found in the following link: http://en.wikipedia.org/wiki/List_of_video_hosting_services, accessed Feb. 23, 2018.

1.5 Explain in one or two sentences the differences between the following terms:
   a. Type I and Type II errors
   b. Run chart and histogram
   c. Flowchart and fishbone diagram
   d. X–Y plot and Pareto chart

1.6 State whether each of the following statements is true or false, or whether the answer depends on other factors (which you should identify):
   i. Airlines fall under the category of "high labor/low customization."
   ii. Value of self-service is included in the GDP.
   iii. Financing is an example of value-added service.
   iv. Public universities are examples of services with "high degree of customer interaction and high degree of labor intensity."

v. An empty hotel room illustrates the characteristic of "Perishability" of a service.
vi. Legal services fall under "high labor/low customization."
vii. Fast-food restaurants can reduce cost by increasing more self-service operations.
viii. Grocery stores represent services with one of the highest proportion of goods.
ix. Under the five dimensions of service quality, the willingness to help customers and provide prompt service is called "empathy."
x. According to SERVQUAL, "reliability" is the most important service quality.

1.7 A family-owned convenience store is interested in monitoring the quality of customer service at the checkout counter. Five service times in minutes have been randomly recorded daily for a week and the data are given in Table 1.9. Develop the variable control charts for the mean and range for monitoring the QoS.

1.8 A call center, providing software support for customers, is concerned about the amount of time a customer has to wait before talking to a service technician. They have randomly selected 10 calls each day over a period of 5 days and noted the customer wait times. They are given in Table 1.10.

a. Develop a variable control chart ($\bar{X}$ chart and p chart) for customer wait times.
b. After the control chart has been developed, the company has taken 10 samples and the recorded wait times are 6, 7, 15, 15, 9, 4, 1, 8, 11, and 5 minutes. Is the process in control or needs attention? Why or why not?

1.9 Company XYZ, a Six Sigma listed company, receives and delivers mails across the United States. Once each day, a sample of 125 mails

TABLE 1.9

Service Time in Minutes (Exercise 1.7)

	Day						
	1	2	3	4	5	6	7
Sample 1	2	2	3	5	3	3	5
Sample 2	4	2	5	4	1	3	5
Sample 3	5	1	2	1	2	5	5
Sample 4	2	2	1	4	2	1	3
Sample 5	5	2	3	5	1	4	5

Service Systems   41

TABLE 1.10

Customer Wait Times in Minutes (Exercise 1.8)

Sample 1	Day 1	Day 2	Day 3	Day 4	Day 5
1	15	6	3	12	5
2	9	11	11	1	7
3	2	14	5	10	4
4	4	10	9	4	15
5	3	3	5	2	12
6	7	1	10	7	6
7	5	5	5	5	15
8	15	1	7	2	12
9	10	3	13	14	5
10	7	9	7	9	6

delivered is drawn from the "Delivery" log that the company maintains and is examined by the Logistics Manager of that company. The Delivery log consists of all mails that are delivered each day by the company. Those mails that have been delivered to the wrong address or not delivered on-time are classified as defective and the rest are classified as satisfactory. This is one of the ways the manager assesses the quality of their operation. Table 1.11 gives the data on samples of 125 mails collected on 9 different days.

The company wishes to monitor the proportion of defective mail deliveries on a daily basis.

a. Prepare a p chart, with $n = 125$, for daily monitoring of the defective mail delivery. Use the established control limits to visually represent the chart.

b. After determining the control limits, another sample of 125 mails is taken and is observed to have 95 satisfactory mails. Is corrective action necessary?

c. Is the sample size sufficient for monitoring the defective mail delivery on a daily basis?

1.10 Late departure statistics of all airlines are available online to the travelling public. A major carrier is concerned about its late departures at one of its hub airports. The airline has 100 flights daily from

TABLE 1.11

Data on Daily Mail Delivery (Exercise 1.9)

Sample no.	1	2	3	4	5	6	7	8	9
# Satisfactory	110	112	109	114	112	111	105	100	104

the airport. They recorded delayed departures for 10 days and the data are given below:

5, 7, 4, 15, 12, 6, 15, 12, 5, 6

a. Prepare a p chart with a sample size consisting of 1 week's delayed departure percentage.

b. Is the sample size sufficient to monitor delays on a daily basis?

1.11 Consider the Record of Problems faced by EZ Access Clinic given in Table 1.8 (Example 1.3, Section 1.9.3). Draw the run charts for the following problem areas:

a. Overtime days
b. Medical errors
c. Billing errors

What can you learn about the problems based on the above run charts?

1.12 Consider the data given in Table 1.8 (Example 1.3) again. Draw the histograms for the following problem areas:

a. No-shows
b. Overtime days
c. Medical errors

What can you conclude based on the above histograms?

1.13 For the data given in Table 1.8, draw the scatterplots of the following variables and discuss your conclusions:

a. No-shows and overtime
b. Medical errors and billing errors
c. Overtime and billing errors

1.14 Draw the flowchart of no-shows for Example 1.3.

1.15 Draw the fishbone diagram of billing errors for Example 1.3.

References

American Society for Testing and Materials. 1976. *ASTM Manual on Presentation of Data and Control Chart Analysis*. Philadelphia: American Society for Testing and Materials.

Daskin, M. S. 2010. *Service Science*. Hoboken, NJ: John Wiley.

Davis, M. M. and J. Heineke. 2003. *Managing Services: Using Technology to Create Value*. New York, NY: McGraw-Hill.

Fitzsimmons, J. A. and M. J. Fitzsimmons. 2006. *Service Management: Operations, Strategy, Information Technology*, 5th Edition. New York, NY: McGraw-Hill.

Haksever, C. and B. Render. 2013. *Service Management: An Integrated Approach to Supply Chain Management and Operations*. Upper Saddle River, NJ: FT Press.

Metters, R., K. King-Metters, M. Pullman, and S. Walton. 2006. *Successful Service Operations Management*, 2nd Edition. Mason, OH: Thompson South-Western.

Parasuraman, A., V. A. Zeithaml, and L. L. Berry. 1988. SERVQUAL: A multiple-item scale for measuring consumer perceptions of service quality. *Journal of Retailing*. 64(1): 38–40.

Schmenner, R. W. 1986. How can service business survive and prosper? *Sloan Management Review*. 27(3): 21–32.

2

Forecasting Demand for Services*

The success of a service enterprise depends on its ability to accurately forecast customer demands and provide services to meet those demands. Thus, **forecasting** is the starting point for most service management decisions. Based on the demand forecasts, distribution and allocation of resources (equipment, labor, etc.) are done to service the customers efficiently. In this chapter, we will discuss commonly practiced forecasting methods in industry.

2.1 Role of Demand Forecasting in Services

Forecasts form the basis of planning in service enterprises. Strategic plans (e.g., where to locate stores, warehouses, etc.) are based on long-term forecasts covering several years. Allocation of labor and capital resources for near-term operations are usually made a few months in advance based on medium-range forecasts. Thus, forecasting is a key activity that influences many aspects of service system design, planning, and operations.

Simply stated, a forecast is the best estimate of a random variable (demand, price, etc.) based on the available information. Clearly, forecasting is as much an art as a science. Forecasting is different from estimating the probability distribution of demand. Since the demand distributions may change over the *forecast horizon*, forecasting handles *non-stationary* data by estimating the mean of the probability distribution at a given time. Some common features of forecasts are as follows (Nahmias 1993, Foote and Murty 2008, 2009):

- *Forecasts are generally wrong*—Forecasts are estimates of the mean of the random variable (demand, price, etc.) and actual outcomes may be very different.
- *Aggregate forecasts are more accurate*—Demand forecasts for the entire product family are more accurate compared to those made for each member of the product family.

---

* Adapted with permission from Ravindran, A. R. and D. P. Warsing, Jr. 2013. *Supply Chain Engineering: Models and Applications*. Chapter 2. Boca Raton, FL: CRC Press.

- *Forecasting errors increase with the forecast horizon*—Demand forecasts for the next month will be more accurate than those for next year.
- *Garbage-In Garbage-Out Principle*—Source and accuracy of data used for forecasting are very important.

Demand forecasts are used by all business functions. Given below are a few examples:

- *Accounting:* Cost/profit estimates
- *Finance:* Cash flow and investment decisions
- *Human Resources:* Hiring/recruiting/training
- *Marketing:* Sales force allocation, pricing, and promotion strategy
- *Operations:* Production and workforce planning, inventory management

2.2 Forecasting Process

The elements of a good forecast are as follows:

- Timeliness
- Reliability
- Accuracy
- Regular reviews
- Equal chance of being over and under
- Good documentation
- Easy to use

The major steps in the forecasting process are the following:

Step 1: Determine the purpose of the forecast.

Step 2: Establish a *forecast* horizon, namely, how far into the future we would like to make prediction.

Step 3: Select the appropriate forecasting technique(s).

Step 4: Get past data and analyze. Validate the chosen forecasting technique.

Step 5: Prepare the final forecast.

Step 6: Monitor the forecasts regularly and make adjustments when needed.

Step 3 is an important and difficult step in the forecasting process. There are a number of forecasting methods that are available in the literature. They can be broadly classified into two categories.

1. Qualitative or judgmental methods
2. Quantitative or statistical methods

Qualitative forecasting methods are usually based on subjective judgments and past historical data are not needed. They are most useful for new products or services. Section 2.3 discusses in detail the various qualitative forecasting methods.

Quantitative forecasting methods require historical data that are accurate and consistent. They assume that past represents the future, namely, history will tend to repeat itself. Most quantitative methods fall in the category of *time series analysis* and are discussed in detail in Section 2.4.

2.3 Qualitative Forecasting Methods

These methods are based on subjective opinions obtained from company executives and customers. These methods are ideally suited for forecasts where there are no past data or past data are not reliable due to changes in environment (e.g., peacetime data on spare part needs for military aircrafts are not useful for forecasts during wartime). Qualitative methods are commonly used for strategic decisions where long-term forecasts are necessary. The qualitative approaches vary in sophistication from scientifically conducted consumer surveys to intuitive judgments from top executives. Quite frequently, qualitative methods are also used as supplements to quantitative forecasts. In a 1994 survey of forecasting-in-practice, 78% of top 500 companies responded that they always or frequently used qualitative methods for forecasting (Sanders and Manrodt 1994).

A brief review of the most commonly used qualitative methods is given below. For a detailed discussion of qualitative methods, the reader is referred to the textbook by Lilien et al. (2007).

2.3.1 Executive Committee Consensus

In this approach, a group of senior executives of the company from various departments meet, discuss, and arrive at consensus forecasts for products and services. This is primarily a *"top-down"* approach. Once the forecasts are agreed upon by the top management, they are communicated to the various departments for use in planning their activities. Because of the face-to-face meetings used in this approach, the forecasts may be skewed due to the dominance of authoritative figures, "bandwagon" effect, and persuasiveness of some individuals. The next approach, *Delphi Method*, avoids these pitfalls.

2.3.2 Delphi Method

The Delphi Method was developed in 1950 at the RAND Corporation as a long-term forecasting tool (Delbecq et al. 1975). It uses a scientifically conducted anonymous survey of key experts who are knowledgeable about the item being forecasted.

The Delphi Method requires the following:

1. *A Panel of Experts*

    The panel members can be geographically displaced in multiple countries. They remain anonymous throughout the process. The panel members are generally top management people with different backgrounds and expertise. The selection of expert panel is key to the success of the Delphi Method.

2. *A Facilitator*

    Only the facilitator interacts directly with the panel members, obtains the individual responses, and presents a statistical summary of the responses to the group.

The Delphi process is conducted as follows:

Step 1: The facilitator sends a survey to the panel members asking for their expert opinions on the forecast.

Step 2: The panel's responses are analyzed by the facilitator, who then prepares a statistical summary of the forecast values, for example, the median value and the two quartiles (75th and 25th percentile values).

Step 3: The statistical summary is shared with the panel members. They will then be asked to revise their forecasts based on the group's response. The experts are also asked to provide comments on why they did or did not change their original forecasts.

Step 4: A statistical summary of the panel's response and their comments are anonymously shared with the experts again.

The process is continued until a consensus among the experts is reached. The process can take several rounds to reach consensus. In the past, the Delphi process used to be very time consuming and took several weeks. However, using e-mail and online surveys, the Delphi process can be done in a matter of days. Because of the statistical feedback, there is no need to restrict the size of the panel.

Forecasting Demand for Services

2.3.3 Survey of Sales Force

Unlike the previous two methods, this method is a *"bottom-up"* approach. Here, the regional sales people, who directly interact with the customers, are asked to estimate the sales for their regions. They are reviewed at the managerial level and an aggregate forecast is then developed by the company.

2.3.4 Customer Surveys

This approach uses scientifically designed customer surveys to determine their needs for products and services. The survey results are tabulated at the corporate level and forecasts are prepared. This method is referred to as the "grassroots" approach since it directly involves the end users. This approach is frequently employed for estimating demands for brand new product lines or services.

2.4 Quantitative Forecasting Methods

Quantitative forecasting methods require historical data on past demand. They assume that the "conditions" that generated the past demand will generate the future demand; that is, history will tend to repeat itself. Most of the quantitative forecasting methods fall under the category of *time series analysis*.

2.4.1 Time Series Forecasting

A *time series* is a set of values for a sequence of random variables over time. Let $X_1, X_2, X_3, \ldots, X_n$ be random variables denoting demands for periods $1, 2, \ldots, n$. The forecasting problem is to estimate the demand for period $(n + 1)$ given the observed values of demands for the last n periods, $D_1, D_2, \ldots, D_n$. If F_{n+1} is the forecast of demand for period $(n + 1)$, then F_{n+1} is the predicted mean of the random variable X_{n+1}. In other words,

$$F_{n+1} \approx E[X_{n+1}].$$

In quantitative forecasting, we assume that the time series data exhibit a *systematic component*, superimposed by a *random component* (noise). The systematic component may include the following:

- Constant level
- Constant level with seasonal fluctuations
- Constant level with trend (growth or decline)
- Constant level with seasonality and trend

Figures 2.1 through 2.4 illustrate the four different time series patterns with random fluctuations or noise.

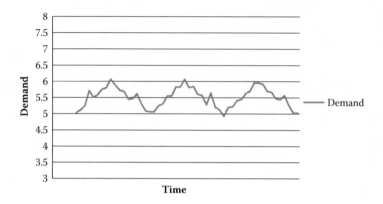

FIGURE 2.1
Constant level.

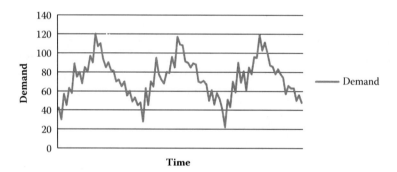

FIGURE 2.2
Constant level with seasonality.

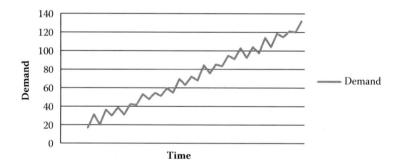

FIGURE 2.3
Constant level with trend.

Forecasting Demand for Services

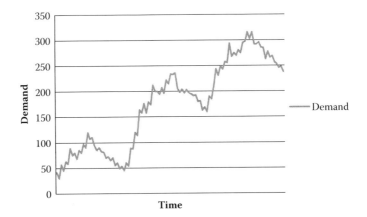

FIGURE 2.4
Constant level with seasonality and trend.

2.4.2 Constant Level Forecasting Methods

Under constant level forecasting methods, it is assumed that the random variable X_t for period t is given by

$$X_t = L_t + e_t \quad \text{for} \quad t = 1,2,...n,$$

where L_t is the constant level and e_t is the random fluctuation or noise. We assume that e_t is normally distributed with mean zero and negligible variance. Given the values of $X_t = D_t$ for $t = 1, 2, ..., n$, the forecast for period $(n + 1)$ is given by

$$F_{n+1} = E(X_{n+1}) = L_{n+1}.$$

There are five commonly used forecasting methods for estimating the value of F_{n+1}:

- Last Value Method
- Simple Averaging Method
- Moving Average Method
- Weighted Moving Average Method
- Exponential Smoothing Method

We will discuss them in detail in the following sections. We will then incorporate seasonality and trend into the constant level forecasting methods.

2.4.3 Last Value Method

Under this method, also known as the *Naive Method*, the next forecast is the last value observed for that variable. In other words,

$$F_{n+1} = D_n.$$

This is not a bad method for new and innovative products whose demands are changing rapidly during the growth phase. Historical values are not relevant for fast-moving products.

Consider the demand data for a service during the past 6 months given in Table 2.1.

Under the naive method, the forecast for month 7 = F_7 = Demand for month 6 = 340.

2.4.4 Averaging Method

Here, the entire historical data are used by computing the average of all the past demand for the forecast.

$$F_{n+1} = \sum_{t=1}^{n} \frac{D_t}{n}$$

For the demand data given in Table 2.1,

$$F_7 = \frac{270 + 241 + 331 + 299 + 360 + 340}{6} \approx 307.$$

The main drawback of this method is that it assumes that all the past data are relevant for the current forecast. This may not be true if the past data cover several years.

TABLE 2.1

Demand Data

Month	1	2	3	4	5	6
Demand	270	241	331	299	360	340

2.4.5 Simple Moving Average Method

In this method, only the most recent data are relevant for estimating future demand. For the *m-month moving average method* ($m < n$), *only* the average of the past m months' data is used:

$$F_{n+1} = \frac{(D_n + D_{n-1} + D_{n-2} + \cdots + D_{n-m+1})}{m}$$

Using the 3-month moving average method for the example data given in Table 2.1, we get

$$F_7 = (D_6 + D_5 + D_4)/3 = (340 + 360 + 299)/3 = 333.$$

Note that all the demands used in the moving average method are given *equal weights*.

2.4.6 Weighted Moving Average Method

Under this method, increasing weights are given to more recent demand data. Let $W_1, W_2, \ldots, W_m$ be the weights in the m-period weighted moving average method, with W_1 assigned to the oldest demand and W_m assigned to the most recent demand. Then, the forecast for period ($n + 1$) is computed as follows:

$$F_{n+1} = W_m D_n + W_{m-1} D_{n-1} + \cdots + W_1 D_{n-m+1} \qquad (2.1)$$

$$W_m \geq \cdots \geq W_1 \geq 0$$

$$W_1 + W_2 + \cdots + W_m = 1.$$

For the example data given in Table 2.1, let the weights $(W_1, W_2, W_3) = (0.2, 0.3, 0.5)$. Then, the 3-month weighted moving average forecast for month 7 will be

$$F_7 = (0.5)(340) + (0.3)(360) + (0.2)(299) \approx 338.$$

2.4.7 Computing Optimal Weights by Linear Programming Model

To determine the best weights for the weighted moving average method, a Linear Programming (LP) model can be used. The objective of the LP model is to find the optimal weights that minimize the forecast error. Let e_t denote the forecast error in period t. Then, e_t is given by

$$e_t = F_t - D_t,$$

where F_t is given by Equation 2.1 and D_t is the actual demand in period t. Note that e_t can be positive or negative. Define the total forecast error as

$$Z = \sum |e_t|. \qquad (2.2)$$

In fact, Z/n is called the *Mean Absolute Deviation* (*MAD*) in forecasting and is used as a key measure of forecast error in validating the forecasting method (refer to Section 2.9). For the LP model, the unrestricted variable e_t will be replaced by the difference of two non-negative variables as follows:

$$e_t = e_t^+ - e_t^-.$$

Since at most one of the non-negative variables, e_t^+ or e_t^-, can be positive in the LP optimal solution, we can write $|e_t|$ as

$$|e_t| = e_t^+ + e_t^-, \text{ for } t = 1,...n.$$

Thus, the complete LP model for finding the weights for the weighted m-period moving average method will be as follows:

$$\text{Minimize } Z = \sum_{t=m+1}^{n} (e_t^+ + e_t^-)$$

Subject to

$$e_t^+ - e_t^- = \left(W_m D_{t-1} + W_{m-1} D_{t-2} + ... + W_1 D_{t-m} \right) - D_t \text{ for } t = m+1, m+2,....,n$$

$$W_m \geq W_{m-1} \geq W_{m-2} ... \geq W_1 \geq 0$$

Forecasting Demand for Services

$$\sum_{t=1}^{m} W_t = 1$$

$$e_t^+, e_t^- \geq 0 \text{ for } t = m+1, m+2, \ldots, n$$

NOTE: In the objective function Z, the index t begins at $t = m + 1$, because m-demand values are necessary for the m-period moving average.

Example 2.1

For the sample data given in Table 2.1, formulate the LP model for determining the optimal weights (W_1, W_2, W_3) for the 3-month moving average. Solve the LP problem and determine the forecast for month 7.

SOLUTION (LP MODEL)

$$\text{Minimize } Z = \sum_{t=4}^{6} (e_t^+ + e_t^-)$$

Subject to

$$e_4^+ - e_4^- = 331W_3 + 241W_2 + 270W_1 - 299$$

$$e_5^+ - e_5^- = 299W_3 + 331W_2 + 241W_1 - 360$$

$$e_6^+ - e_6^- = 360W_3 + 299W_2 + 331W_1 - 340$$

$$W_3 \geq W_2 \geq W_1 \geq 0$$

$$W_1 \geq W_2 \geq W_3 = 1$$

$$e_t^+, e_t^- \geq 0 \quad \text{for} \quad t = 4, 5, 6.$$

The optimal solution is given by

$$W_1 = 0, W_2 = 0.356, W_3 = 0.644$$

$$e_4^+ = e_4^- = e_5^+ = e_6^+ = 0$$

$$e_5^- = 49.62, \quad e_6^- = 1.69$$

Minimum $Z = 51.31$

Then, the forecast for month 7 is given by

$$F_7 = 340\,(0.644) + 360\,(0.356) + 299\,(0) = 347.$$

2.4.8 Exponential Smoothing Method

Perhaps the most popular forecasting method in practice is the *exponential smoothing method*, introduced and popularized by Brown (1959). It is basically a weighted averaging method with weights decreasing exponentially on older demands. Unlike the weighted moving average method, it uses all the data points. Under this method, given demands $D_1, D_2, \ldots, D_n$, the forecast for period $(n+1)$ is given by

$$F_{n+1} = \alpha D_n + \alpha(1-\alpha)D_{n-1} + \alpha(1-\alpha)^2 D_{n-2} + \ldots, \qquad (2.3)$$

where α is between 0 and 1 and is called the *Smoothing Constant*. Note that $\alpha > \alpha(1-\alpha) > \alpha(1-\alpha)^2 > \ldots$

Thus, the most recent demand is given the highest weight α and the weights are decreased by a factor $(1-\alpha)$ as the data get older.

Equation 2.3 can be rewritten as

$$F_{n+1} = \alpha D_n + (1-\alpha)[\alpha D_{n-1} + \alpha(1-\alpha)D_{n-2} + \alpha(1-\alpha)^2 D_{n-3}\ldots]$$
$$F_{n+1} = \alpha D_n + (1-\alpha)F_n \qquad (2.4)$$

Thus, the forecast for period $(n+1)$ uses the forecast for period n and the actual demand for period n. The value of α is generally chosen between 0.1 and 0.4. In other words, the weights assigned to the actual demand is less than that of the forecasted demand, the reason being, the actual demands fluctuate a lot, while the forecast has smoothed the fluctuations.

Forecasting Demand for Services

Example 2.2

Using the sample data given in Table 2.1, determine the forecast for month 7 using the exponential smoothing method.

SOLUTION

In order to use this method, the value of α and the initial forecast for month 1 (F_1) are necessary. The value of F_1 can be chosen by averaging all the demands or using the most recent demand value. Assuming $\alpha = 0.2$, $F_1 = 307$ (averaging method) and using Equation 2.4, we get

$$F_2 = 0.2\,(270) + 0.8\,(307) = 299.6$$

$$F_3 = 0.2\,(241) + 0.8\,(299.6) = 287.9$$

$$F_4 = 0.2\,(331) + 0.8\,(287.9) = 296.5$$

$$F_5 = 0.2\,(299) + 0.8\,(296.5) = 297$$

$$F_6 = 0.2\,(360) + 0.8\,(297) = 309.6$$

Thus, the forecast for month 7 is given by

$$F_7 = (0.2)\,(340) + (.8)\,(309.6) = 315.7 \approx 316.$$

NOTE: The initial value chosen for F_1 will have negligible effect on the forecast as the number of data points increases. However, the choice of α is very important and will affect the forecast accuracy.

2.4.8.1 Choice of Smoothing Constant (α)

Typically, values of α between 0.1 and 0.4 are used in practice. Smaller values of α (e.g., $\alpha = 0.1$) yield forecasts that are relatively smooth (low variance). However, it takes longer to react to changes in the demand process. Higher values of α (e.g., $\alpha = 0.4$) can react to changes in data quicker, but the forecasts have significantly higher variations, resulting in forecast errors with high variance. One disadvantage of both the method of moving averages and the exponential smoothing method is that when there is a definite trend in the demand process (either growing or falling), the forecasts obtained by them lag behind the trend. Holt (1957) developed a modification to the exponential smoothing method by incorporating trend. This will be discussed in Section 2.6.

2.5 Incorporating Seasonality in Forecasting

The constant level forecasting methods discussed so far assume that the values of demand in the various periods form a stationary time series. In some applications, this series may be seasonal; that is, it has a pattern that repeats after every few periods. For example, retail sales during Christmas season are usually higher. For most retailers, sales during Christmas season may account for as much as 40% of their annual sales. For some companies, sales are arranged by sales agents, who operate on quarterly sales targets. In those cases, the demands for products tend to be higher during the third month of each quarter as the agents work much harder to meet their quarterly targets. It is relatively easy to incorporate seasonality in the constant level forecasting methods by computing the *seasonality index* for each period.

The basic steps to incorporate seasonality in forecasting are as follows:

Step 1: Compute the seasonality index for any period given by

$$\text{Seasonality Index} = \frac{\text{Average demand during that period}}{\text{Overall average of demand for all periods}}$$

Step 2: Seasonally adjust the actual demands in the time series by dividing by the seasonality index, to get *deseasonalized demand* data.

Step 3: Select an appropriate time series forecasting method.

Step 4: Apply the forecasting method on the *deseasonalized demand forecast*.

Step 5: Compute the *actual forecast* by multiplying the deseasonalized forecast by the seasonality index for that period.

Let us illustrate the basic steps with an example.

Example 2.3

The quarterly sales of laptop computers for 3 years (2014–2016) are given in Table 2.2.

The problem is to determine the forecast for quarter 1 of year 2017 (period 13).

Let us use the five basic steps given earlier for forecasting under seasonality.

Step 1: Compute the *seasonality* index for each quarter.
— Compute the *overall average* of quarterly demand using the demand values for 12 quarters

Overall average = $(540 + 522 + 515 + \ldots + 550 + 629 + 785)/12 = 7236/12 = 603$.

Forecasting Demand for Services

TABLE 2.2

Laptop Computer Sales for 2014–2016 (Example 2.3)

Year	Quarter	Period (t)	Actual Demand (D_t)
2014	1	1	540
2014	2	2	522
2014	3	3	515
2014	4	4	674
2015	1	5	574
2015	2	6	569
2015	3	7	616
2015	4	8	712
2016	1	9	550
2016	2	10	550
2016	3	11	629
2016	4	12	785

— Compute the quarterly average using the three demand values for each quarter as follows:

$$\text{Quarter 1 average} = (540 + 574 + 550)/3 = 555$$

$$\text{Quarter 2 average} = (522 + 569 + 550)/3 = 547$$

$$\text{Quarter 3 average} = (515 + 616 + 629)/3 = 587$$

$$\text{Quarter 4 average} = (674 + 712 + 785)/3 = 724$$

Note that the third quarter average is slightly higher primarily due to "Back-to-School" sales to students. The fourth quarter average is much higher due to "Christmas Sales."

The *Seasonality Indices* (*SI*) are then computed as follows:

$$SI \text{ for quarter 1} = 555/603 = 0.92$$

$$SI \text{ for quarter 2} = 547/603 = 0.907$$

$$SI \text{ for quarter 3} = 587/603 = 0.973$$

$$SI \text{ for quarter 4} = 724/603 = 1.2$$

Step 2: Compute the deseasonalized sales data $\left(\overline{D_t}\right)$ by dividing the actual sales data $\left(D_t\right)$ by its appropriate seasonality index. For example, deseasonalized sales for 2014 quarter 1 = 540/.92 = 587.

Table 2.3 gives the deseasonalized data for all quarters.

Step 3: Select any time series forecasting method. For illustration, we will use the *exponential smoothing* forecasting method with $\alpha = 0.2$. For the initial forecast for quarter 1 of year 2014, we will use 600.

Step 4: Apply the exponential smoothing method on the *deseasonalized demand* to get *deseasonalized* forecast as shown in Table 2.3. For example, the deseasonalized forecast for quarter 1 (2017) is given by

$$\overline{F}_{13} = (0.2)(654.2) + (0.8)(610.2) = 619.$$

Step 5: Compute the actual forecast for year 2017, quarter 1 as follows:

$$F_1(2017) = \overline{F}_1(2017) \times SI_1 = (619)(0.92) = 569.5.$$

Note that we could have used any of the time series forecasting method discussed in Section 2.4 for Step 3.

For the sake of illustration, the forecasts for the first quarter of 2017, using the naive method and the four-quarter moving average method are also given below:

Under naive method

$$\overline{F}_{13} = \text{Deseasonalized demand for quarter 4 of 2010} = \overline{D}_{12} = (654.2)$$

$$F_{13} = (654.2)(0.92) \approx 602$$

Under four-quarter moving average

$$\overline{F}_{13} = \frac{(597.8 + 606.4 + 646.5 + 654.2)}{4} = 626.2$$

$$F_{13} = (626.2)(0.92) \approx 576$$

NOTE: Seasonality indices are not static. They can be updated as more data become available.

TABLE 2.3
Forecasting with Seasonality (Example 2.3)

Year	Quarter	Period (t)	Actual Demand (D_t)	Seasonality Index (SI)	Deseasonalized Demand ($\bar{D}_t$)	Deseasonalized Forecast ($\bar{F}_t$)	Actual Forecast (F_t)
2014	1	1	540	0.92	587.0	600.0	552.0
2014	2	2	522	0.907	575.5	597.4	541.8
2014	3	3	515	0.973	529.3	593.0	577.0
2014	4	4	674	1.2	561.7	580.3	696.3
2015	1	5	574	0.92	623.9	576.6	530.4
2015	2	6	569	0.907	627.3	586.0	531.5
2015	3	7	616	0.973	633.1	594.3	578.2
2015	4	8	712	1.2	593.3	602.0	722.5
2016	1	9	550	0.92	597.8	600.3	552.3
2016	2	10	550	0.907	606.4	599.8	544.0
2016	3	11	629	0.973	646.5	601.1	584.9
2016	4	12	785	1.2	654.2	610.2	732.2
2017	1	13		0.92		619.0	569.5

2.6 Incorporating Trend in Forecasting

One disadvantage of both the moving average and exponential smoothing methods is that when there is a definite trend in the demand process (either growing or falling), the forecasts obtained by them lag behind the trend. There are ways to accommodate trend in forecasting from a very *simple linear trend model* to Holt's model (Holt 1957). Holt's method uses variations of the exponential smoothing method to track linear trends over time.

2.6.1 Simple Linear Trend Model (Srinivasan 2010)

Here, we assume that the level and trend remain constant over the forecast horizon. Thus, $X_t = a + bt + \varepsilon$ for $t = 1, 2, \ldots, n$, where X_t is the random variable denoting demand at period t, a and b are constant level and trend, and ε is the random error. Given the observed values of $X_t = D_t$ for $t = 1, \ldots, n$, the forecast for period t is given by

$$F_t = E(X_t) = a + bt.$$

Then, the forecast error is

$$e_t = F_t - D_t, \quad \text{for } t = 1, 2, \ldots, n.$$

From the given data, we then have n linear equations relating the actual and forecasted demands. Following the approach given in Srinivasan (2010), we can use the least square regression method to estimate the level (a) and trend (b) parameters. The unconstrained optimization problem is to determine a and b such that

$$\sum_{t=1}^{n} e_t^2 = \sum_{t=1}^{n} \left[(a + bt) - D_t \right]^2 \qquad (2.5)$$

is minimized.

Note that the minimization function given by Equation 2.5 is a convex function. Hence, a *stationary point* will give the absolute minimum. The stationary point can be obtained by setting the partial derivatives with respect to a and b to zero. Thus, we get the following two equations to solve for a and b:

$$\sum_{t=1}^{n} D_t = na + b \sum_{t=1}^{n} t \qquad (2.6)$$

Forecasting Demand for Services

$$\sum_{t=1}^{n} tD_t = a\sum_{t=1}^{n} t + b\sum_{t=1}^{n} t^2 \qquad (2.7)$$

Solving the two linear equations, we determine the values of a and b. Then, the forecast for period $t = n + 1$ is given by

$$F_{n+1} = a + b(n+1).$$

Example 2.4

Consider the monthly demand for a product for the past 6 months given in Table 2.4. The demand data clearly show an increasing trend. The problem is to determine the forecast for month 7 using the *simple linear trend model*.

SOLUTION

Table 2.5 gives the necessary computations for the linear regression model. Using the last row (sum) in Table 2.5 and substituting them in Equations 2.6 and 2.7, we get the following:

$$1700 = 6a + 21b$$

$$6355 = 21a + 91b$$

TABLE 2.4

Data for Example 2.4

t	1	2	3	4	5	6
D_t	220	250	280	295	315	340

TABLE 2.5

Calculations for the Regression Model (Example 2.4)

	t	t^2	D_t	tD_t
	1	1	220	220
	2	4	250	500
	3	9	280	840
	4	16	295	1180
	5	25	315	1575
	6	36	340	2040
Sum	21	91	1700	6355

Solving the two equations for a and b, we get

$$a = 202.333$$

$$b = 23.143$$

Then, the forecast for month 7 is given by

$$F_7 = 202.33 + (23.14)\,7 \approx 364.$$

The main drawback of the *simple linear trend* model is that it assumes that the level and trend remain constant throughout the forecast horizon.

2.6.2 Holt's Method (Chopra and Meindl 2013; Hillier and Lieberman 2001; Holt 1957)

Regular exponential smoothing model estimates the constant level (L) to forecast future demands. Holt's model improves it by estimating both the *level* (L) and *trend factor* (T). It adjusts both the level and trend factor using exponential smoothing. Holt's method is also known as *double exponential smoothing* or *trend adjusted exponential smoothing* method.

Given the actual values of the random variable $X_t = D_t$ for $t = 1, \ldots, n$, we are interested in computing the forecast for period $n + 1$, that is, $F_{n+1} \approx E(X_{n+1})$. Given the actual demand D_t, forecast F_t and the estimates of level (L_t) and trend (T_t) for period t, the forecast for period ($t + 1$) is given by

$$F_{t+1} = L_{t+1} + T_{t+1}. \tag{2.8}$$

Under the exponential smoothing method, the estimate of the level for ($t + 1$) is given by

$$L_{t+1} = \alpha D_t + (1-\alpha)F_t. \tag{2.9}$$

The same approach is used to estimate the trend factor for ($t + 1$) using another smoothing constant β as follows:

$$T_{t+1} = \beta[L_{t+1} - L_t] + (1-\beta)T_t, \tag{2.10}$$

where ($L_{t+1} - L_t$) is the latest trend based on two recent level estimates and T_t is the most recent estimate of the trend.

Forecasting Demand for Services

The term $(L_{t+1} - L_t)$ in Equation 2.10 can be rewritten using Equation 2.9 as follows:

$$L_{t+1} - L_t = \alpha(D_t - D_{t-1}) + (1-\alpha)(F_t - F_{t-1}). \tag{2.11}$$

The term $(D_t - D_{t-1})$ in Equation 2.11 represents the latest trend based on the actual values observed for the last two periods, while $(F_t - F_{t-1})$ is the latest trend based on the last two forecasts. Once L_{t+1} and T_{t+1} are determined, using Equations 2.9 and 2.10, respectively, the forecast for period $(t + 1)$ is given by their sum, as in Equation 2.8.

Example 2.5 (Holt's Method)

Consider again the 6-month time series data on demand given in Table 2.4. We will apply Holt's method to determine the forecast for month 7. In order to get started with Holt's method, we need the values of the smoothing constants, α and β, and the initial estimates for the level and trend for month 1, namely L_1 and T_1. Let us assume that $\alpha = \beta = 0.3$.

SOLUTION

For illustration, L_1 is assumed to be equal to D_1 and T_1 is the linear slope using D_1 and D_6. In other words,

$$T_1 = \frac{D_6 - D_1}{5} = \frac{340 - 220}{5} = 24.$$

Hence, $L_1 = 220$ and $T_1 = 24$.
Then, the forecast for month 1 is given by

$$F_1 = L_1 + T_1 = 220 + 24 = 244.$$

For month 2, L_2 and T_2 are computed using Equations 2.9 and 2.10 as follows:

$$L_2 = \alpha D_1 + (1-\alpha)F_1 = (0.3)\,220 + (0.7)\,244 = 236.8$$

$$T_2 = \beta(L_2 - L_1) + (1-\beta)T_1 = (0.3)(236.8 - 220) + (0.7)\,24 = 5.04 + 16.8 = 21.8$$

The remaining calculations are done on a spreadsheet and are shown in Table 2.6. Note that the cell values in the table have been displayed with one decimal accuracy.

TABLE 2.6

Holt's Method (Example 2.5)

Month (t)	Demand (D_t)	Estimate of Level (L_t)	Estimate of Trend (T_t)	Forecast (F_t)
1	220	220.0	24.0	244.0
2	250	236.8	21.8	258.6
3	280	256.0	21.1	277.1
4	295	278.0	21.3	299.3
5	315	298.0	20.9	318.9
6	340	317.8	20.6	338.3
7		338.8	20.7	359.6

Note: $\alpha = \beta = 0.3$; initial values $L_1 = 220$ and $T_1 = 24$.

Thus, the forecast for month 7 is given by

$$F_7 = L_7 + T_7 = 338.8 + 20.7 \approx 360$$

2.7 Incorporating Seasonality and Trend in Forecasting

2.7.1 Method Using Static Seasonality Indices (Hillier and Lieberman 2001)

A straightforward approach to incorporate seasonality and trend is to combine the approaches for seasonality discussed in Section 2.5 and one of the methods for trend discussed in Section 2.6. The basic steps of such an approach are as follows:

Step 1: Compute the *seasonality indices* as discussed in Section 2.5.

Step 2: Using the seasonality indices, compute the *deseasonalized demands*.

Step 3: Apply one of the methods discussed in Section 2.6 to incorporate trend, on the deseasonalized demand data.

Step 4: Determine the deseasonalized forecasts with trend.

Step 5: Compute the actual forecasts using the seasonality indices again.

We will illustrate the above method using Example 2.6.

Example 2.6 (Holt's Method with Seasonality)

Consider the demand data for laptop computer sales for 3 years (2014–2016) given in Table 2.2. Using Holt's method, determine the forecast for the first quarter of 2017 incorporating both seasonality and trend.

Forecasting Demand for Services 67

SOLUTION

First, we compute the seasonality indices and the deseasonalized demands using the approach given in Section 2.5.

Step 1: Compute seasonality indices (SI) as described in Example 2.3. The SI values are 0.92, 0.907, 0.973, and 1.2 for quarters 1, 2, 3, and 4, respectively.

Step 2: Compute the deseasonalized demands by dividing the actual quarterly demands by their respective SI values as shown in Table 2.7.

In Example 2.3, we then used the exponential smoothing method with $\alpha = 0.2$ to determine the deseasonalized forecasts. Instead, we will apply Holt's method to also incorporate trend.

Step 3: Apply Holt's method on the deseasonalized demand data with $\alpha = 0.2$, $\beta = 0.2$, $L_1 = 600$, $T_1 = 6$.

Step 4: Compute the estimates of level and trend and the deseasonalized forecast. These calculations are done on a spreadsheet and are shown in Table 2.7. Note that the displayed values are shown with one decimal accuracy.

Step 5: Compute the actual forecast by multiplying the deseasonalized forecast by the SI values.

Thus, the deseasonalized forecast for quarter 1 of year 2011 is 634.4 and the actual forecast is $(634.4)(0.92) \approx 584$.

2.7.2 Winters Method (Chopra and Meindl 2013; Winters 1960)

A drawback of the previous approach is that the seasonality indices remain static and are not updated during the forecast horizon. Winters (1960) has extended Holt's method by updating seasonality indices also using exponential smoothing.

Given the actual demand D_t, the estimates of level L_t, trend T_t, and seasonality indices $SI_1, SI_2, \ldots, SI_{t+p-1}$, the forecast for period $(t + 1)$ is given by

$$F_{t+1} = (L_{t+1} + T_{t+1})SI_{t+1}, \quad (2.12)$$

where

$$L_{t+1} = \alpha\left(D_t / SI_t\right) + (1-\alpha)(L_t + T_t) \quad (2.13)$$

$$T_{t+1} = \beta(L_{t+1} - L_t) + (1-\beta)T_t \quad (2.14)$$

$$SI_{t+p} = \gamma(D_t/L_t) + (1-\gamma)SI_t, \quad (2.15)$$

TABLE 2.7
Computations for Example 2.6

Year	Period	Actual Demand	SI	Deseasonalized Demand	Estimate of Level (L)	Estimate of Trend (T)	Deseasonalized Forecast	Actual Forecast
2014	1	540	0.92	587	600.0	6.0	606.0	557.5
2014	2	522	0.907	576	602.2	5.2	607.4	550.9
2014	3	515	0.973	529	601.0	4.0	605.0	588.7
2014	4	674	1.2	562	589.9	0.9	590.8	709.0
2015	5	574	0.92	624	585.0	−0.2	584.7	538.0
2015	6	569	0.907	627	592.6	1.3	593.9	538.7
2015	7	616	0.973	633	600.6	2.7	603.3	587.0
2015	8	712	1.2	593	609.2	3.9	613.1	735.7
2016	9	550	0.92	598	609.1	3.1	612.2	563.2
2016	10	550	0.907	606	609.3	2.5	611.8	554.9
2016	11	629	0.973	646	610.8	2.3	613.0	596.5
2016	12	785	1.2	654	619.7	3.6	623.3	748.0
2017	13		0.92		629.5	4.9	634.4	583.6

Forecasting Demand for Services

where α, β and γ are smoothing constants between 0 and 1 for level, trend, and seasonality, respectively. Note that Equations 2.13 and 2.14 are very similar to the Holt's model (Equations 2.9 and 2.10). In Equation 2.15, the index p refers to the *periodicity*, periods after which the seasonal cycle repeats itself. Equation 2.15 updates the seasonality index by weighting the observed value and the current estimate of seasonality. We shall illustrate Winters method using Example 2.7.

Example 2.7 (Winters Method)

Consider the quarterly laptop computer sales data for 3 years (2014–2016) given in Table 2.2. The problem is to determine the forecast for quarter 1 of year 2017 using Winters method with α = β = 0.2 and γ = 0.3.

SOLUTION

To begin Winters method, we need the initial estimates for level (L_1) and trend (T_1) for quarter 1 of 2014 and the quarterly seasonality indices (SI) for year 1 (SI_1, SI_2, SI_3, SI_4). Let us assume that $L_1 = 600$, $T_1 = 6$, and the initial SI values are $SI_1 = 0.92$, $SI_2 = 0.907$, $SI_3 = 0.973$, and $SI_4 = 1.2$, as estimated in Example 2.3 (Section 2.5). Note that the *periodicity* "p" is equal to 4 in this example, since the demands are quarterly and the seasonal cycle repeats after every four quarters.

Given the actual demand $D_1 = 540$, the SI value for period 5 (SI_5) will be updated. Using $t = 1$ and $p = 4$ in Equation 2.15 and γ = 0.3, we get the SI value for quarter 1 (2015) as follows:

$$SI_5 = \gamma(D_1/L_1) + (1-\gamma)(SI_1) = (0.3)(540/600) + (0.7)(0.92) = 0.914.$$

The forecast for period 2 is given by Equation 2.12 as

$$F_2 = (L_2 + T_2)SI_2,$$

where L_2 and T_2 are given by Equations 2.13 and 2.14.
Thus,

$$L_2 = \alpha(D_1/SI_1) + (1-\alpha)(L_1 + T_1) = (0.2)(540/0.92) + (0.8)(600+6) = 602.2$$

$$T_2 = \beta(L_2 - L_1) + (1-\beta)T_1 = 0.2(602.2 - 600) + (0.8)6 = 5.24$$

$$F_2 = (602.2 + 5.24)(0.907) = 550.9.$$

At this time, the seasonality index SI_6 (quarter 2 of 2015) will be updated using Equation 2.15 as

$$SI_6 = \gamma(D_2/L_2) + (1-\gamma)(SI_2) = (0.3)(522/602.2) + (0.7)(0.907) = 0.895.$$

TABLE 2.8

Computations for Winters Method (Example 2.7)

Year	Period	Actual Demand (D_t)	SI	Level	Trend	Forecast
2014	1	540	0.920	600.0	6.0	557.4
2014	2	522	0.907	602.2	5.2	551.0
2014	3	515	0.973	601.1	4.0	588.6
2014	4	674	1.200	589.9	0.9	709.0
2015	5	574	0.914	585.0	−0.2	534.4
2015	6	569	0.895	593.4	1.5	532.5
2015	7	616	0.938	603.1	3.1	568.7
2015	8	712	1.183	616.3	5.2	735.1
2016	9	550	0.934	617.5	4.4	580.9
2016	10	550	0.914	615.3	3.0	565.3
2016	11	629	0.963	615.0	2.4	594.6
2016	12	785	1.175	624.5	3.8	738.0
2017	13		0.921	636.3	5.4	591.1

The remaining calculations are shown in Table 2.8 using a spreadsheet. Comparing the forecast of Example 2.6 (Holt's method) and Example 2.7 (Winters method), we note that the forecasts are the same for the first four quarters (2014). However, they are different from 2015 onward due to the updating of the seasonality indices. The forecast for the first quarter of 2017 is given by

$$F_{13} = (L_{12} + T_{12})SI_{12} = (636.3 + 5.4)(0.921) \approx 591.$$

2.8 Forecasting for Multiple Periods

So far in our discussion, we were concerned with forecasting for one period only, namely, F_{n+1}, given the actual demands $D_1, D_2, ..., D_n$. In practice, one has to forecast for several periods in the future. Methods for forecasting for multiple periods depend on the type of forecasting method selected—constant level, constant level with seasonality, constant level with trend, or constant level with seasonality and trend.

2.8.1 Multi-Period Forecasting Problem

Given the actual demands for the last n periods as $D_1, D_2, ..., D_n$, determine the forecasts for the next m periods, $F_{n+1}, F_{n+2}, ..., F_{n+m}$.

Forecasting Demand for Services 71

2.8.2 Multi-Period Forecasting under Constant Level

Under the constant level forecasting approach, the level forecasts are updated based only on the observed demands. Since there is no new demand information beyond period n, the forecast for the next m periods will remain the same.

$$F_{n+i} = F_{n+1} \text{ for } i = 1, 2, \ldots, m. \tag{2.16}$$

In other words, the best forecasts for periods $n + 2$, $n + 3$ … is just F_{n+1}!

To illustrate, consider Example 2.2 where we forecasted the demand for month 7 as 316 based on the last 6 months of demand, using exponential smoothing. Forecasts beyond month 7 are given by

$$F_{6+i} = F_7 = 316 \text{ for } i = 2, 3, \ldots.$$

2.8.3 Multi-Period Forecasting with Seasonality

Under seasonality, forecasts for multiple periods are given as follows:

$$F_{n+i} = (\overline{F}_{n+1}) SI_{n+i} \text{ for } i = 1, 2, \ldots \tag{2.17}$$

where $\overline{F}_{n+1}$ is the deseasonalized forecast for period $(n + 1)$ and SI_{n+i} is the seasonality index for period $(n + i)$. Since the deseasonalized level forecast does not change, the forecasts for future period are only affected by their respective seasonality indices.

Example 2.8

Consider Example 2.3, where laptop sales by quarter for years 2014, 2015, and 2016 were given and we forecasted the demand for the first quarter of 2017 using exponential smoothing method with seasonality. Suppose the problem now is to forecast the sales for all the four quarters of 2017.

SOLUTION

In Example 2.3 (Table 2.3), the deseasonalized forecast for the first quarter of 2017, namely, $\overline{F}_{13}$, was computed as 619. Thus, the actual forecasts for year 2017 are as follows:

$$F_{13} = (619)(0.92) \approx 570 \text{ (quarter 1)}$$

$$F_{14} = (619)(0.907) \approx 561 \text{ (quarter 2)}$$

$$F_{15} = (619)(0.973) \approx 602 \text{ (quarter 3)}$$

$$F_{16} = (619)(1.2) \approx 743 \text{ (quarter 4)}$$

2.8.4 Multi-Period Forecasting with Trend

Using Holt's method for trend (Section 2.6.2), forecasts for multiple periods are given by

$$F_{n+i} = L_{n+1} + (i)T_{n+1} \text{ for } i = 1, 2, \ldots, m, \quad (2.18)$$

where L_{n+1} and T_{n+1} are the latest estimates of level and trend, respectively. Note that we assume a linear trend for future forecasts.

Example 2.9

Consider Example 2.5, where Holt's method was used to forecast demand for month 7, based on the actual demands for the first 6 months. Determine the forecast of demands for months 7, 8, and 9.

SOLUTION

In the solution to Example 2.5 (Table 2.6), the latest estimates of level and trend for month 7 were computed as

$$L_7 = 338.8; \; T_7 = 20.7.$$

Hence, the forecasts for the future months are as follows:

$$F_7 = L_7 + T_7 = 338.8 + 20.7 \approx 360$$

$$F_8 = L_7 + 2T_7 = (338.8) + 2(20.7) \approx 380$$

$$F_9 = L_7 + 3T_7 = (338.8) + 3(20.7) \approx 401$$

2.8.5 Multi-Period Forecasting with Seasonality and Trend

Using Winters method to incorporate seasonality and trend (Section 2.7.2), forecasts for multiple periods are given by

$$F_{n+i} = [L_{n+1} + (i) T_{n+1}] SI_{n+i} \text{ for } i = 1, 2, \ldots \quad (2.19)$$

Note that the latest estimates of level, trend, and seasonality indices are used in Equation 2.19.

Example 2.10

Consider Example 2.7, where Winters method was used to forecast the demand for the first quarter of 2017. Determine the demand forecasts for all four quarters of 2017.

Forecasting Demand for Services

SOLUTION

In Example 2.7 (Table 2.8), the latest estimates for level (L_{13}), trend (T_{13}), and seasonality index (SI_{13}) were computed as

$$L_{13} = 636.3, \ T_{13} = 5.4, \ \text{and} \ SI_{13} = 0.921,$$

and the forecast for the first quarter of 2017 was

$$F_{13} = (L_{13} + T_{13}) \, SI_{13} = (636.3 + 5.4)(0.921) \approx 591.$$

Since no new data are available on demand, the estimates for level and trend will remain the same as 636.3 and 5.4, respectively, for periods 14, 15, and 16 (quarters 2, 3, and 4 of 2017). However, their seasonality indices can be updated using the data for periods 10, 11, and 12 in Table 2.8. Using Equation 2.15, we get

$$SI_{14} = \gamma(D_{10}/L_{10}) + (1-\gamma)(SI_{10}) = (0.3)(550/615.3) + (0.7)(0.914) = 0.908$$

$$SI_{15} = (0.3)(629/615) + (0.7)(0.963) = 0.981$$

$$SI_{16} = (0.3)(785/624.5) + (0.7)(1.175) = 1.2$$

Thus, the forecasts for the remaining quarters of 2017 are computed as follows:

Quarter 2: $F_{14} = [636.3 + 2(5.4)](0.908) \approx 588$

Quarter 3: $F_{15} = [636.3 + 3(5.4)](0.981) \approx 640$

Quarter 4: $F_{16} = [636.3 + 4(5.4)](1.2) \approx 790$

2.9 Forecasting Errors

The accuracy of the forecast depends on *forecast errors*, which measure the differences between the forecasted demands and their actual (observed) values. As discussed in Section 2.4.7, the *forecast error*, for period t, denoted by e_t, is given by

$$e_t = F_t - D_t \ \text{for} \ t = 1, 2, \ldots, n,$$

where F_t is the forecast and D_t is the actual demand in period t. Note that e_t can be positive or negative depending on whether the method is over-forecasting or underforecasting in period t. There are several measures of forecast errors used in practice for determining the accuracy of the chosen forecasting method as given below:

- Mean absolute deviation (MAD)
- Mean squared error (MSE)
- Standard deviation of forecast errors (STD)
- Bias
- Mean absolute percentage error (MAPE)

Let us discuss each of the measures in detail, assuming we have n values of forecast errors, $e_1, e_2, \ldots e_n$.

1. Mean absolute deviation (MAD)

    MAD measures the dispersion of the forecast errors.

    $$\text{MAD} = \frac{1}{n} \sum_{t=1}^{n} |e_t|$$

2. Mean squared error (MSE)

    MSE also measures the dispersion of the forecast errors, but larger errors get penalized more due to squaring.

    $$\text{MSE} = \frac{1}{n} \sum_{t=1}^{n} e_t^2$$

    MSE estimates the variance of forecast errors.

3. Standard deviation of forecast error (STD)

    $$\text{STD} = \sqrt{\text{MSE}} = \sigma$$

    The value of σ can be used to set confidence limits on the mean forecast. For example, if normality is assumed, then we can say that there is 68% probability that the actual forecast will be within one σ of the mean forecast.

4. Bias

    Bias measures whether the forecast is overestimating or underestimating the actual demand over the forecast horizon.

    $$\text{Bias} = \sum_{t=1}^{n} e_t$$

Forecasting Demand for Services

Note that the values of e_t can be positive, negative, or zero. If Bias > 0, the method is overestimating, and if Bias < 0, the method is underestimating. Ideally, a good method should have a Bias close to zero. Bias is frequently preferred by managers for measuring forecast accuracy.

5. Mean absolute percentage error (MAPE)

MAPE measures the relative dispersion of forecast errors and is given by

$$\text{MAPE} = \frac{1}{n}\sum_{t=1}^{n}\left|\frac{e_t}{D_t}\right|100$$

MAPE is better than MAD since it takes into account the relative magnitude of the actual demand and is also frequently used in practice.

Let us illustrate all the five measures of forecast errors with an example.

Example 2.11

Consider the actual demand, its forecast, and the errors for five periods given in Table 2.9. The forecasts are obtained using exponential smoothing method with $\alpha = 0.1$. Compute MAD, MSE, STD, Bias, and MAPE for measuring the accuracy of the forecasting method.

SOLUTION

$$\text{MAD} = \frac{20 + 42 + 18 + 76 + 48}{5} = 40.8$$

$$\text{MSE} = \frac{20^2 + 42^2 + 18^2 + 76^2 + 48^2}{5} = 2113.6$$

TABLE 2.9

Forecast Errors (Example 2.11)

Period (*t*)	Forecast (*F_t*)	Actual (*D_t*)	Error ($e_t = F_t - D_t$)
1	180	160	+20
2	178	220	−42
3	182	200	−18
4	184	260	−76
5	192	240	−48

$$\text{STD} = \sqrt{\text{MSE}} = \sqrt{2113.6} \approx 46$$

$$\text{Bias} = 20 - 42 - 18 - 76 - 48 = -164$$

$$\text{MAPE} = \frac{1}{5}\left(\left(\frac{20}{160}\right) + \left(\frac{42}{220}\right) + \left(\frac{18}{200}\right) + \left(\frac{76}{260}\right) + \left(\frac{48}{240}\right)\right) 100 \approx 18\%$$

We can make the following observations from the forecast error calculations:

- Large negative value for the Bias indicates that the method is consistently underestimating the actual demands.
- Large values of MAD, STD, MAPE, and MSE indicate that the method is not forecasting well.

Looking at the time series data, it is clear that the demand is generally increasing. Use of a low value of $\alpha = 0.1$ makes the method less reactive to the actual demands. Hence, an increase of α value is definitely warranted.

2.9.1 Uses of Forecast Errors

There are several uses of the different forecast errors as described below:

1. To select a forecasting method by retrospective testing on past demands.
2. To determine the best values of the parameters for a given forecasting method, for example, selecting the values of α and β in Holt's method.
3. To monitor how well the selected forecasting method is performing based on new data as they become available.

2.9.2 Selecting the Best Forecasting Method

It is recommended that multiple measures of forecast errors be used in selecting the best forecasting method. If a particular method consistently does well in all the measures, then it is clearly the best method to use. However, it is possible that some methods may do well in some measures and poorly in other measures. In such situations, it is quite common to choose the two or three best methods and use the average of their forecasts as the forecast for the future.

Forecasting Demand for Services 77

2.10 Monitoring Forecast Accuracy

Forecasts are dynamic and they are updated as more information becomes available. Hence, after selecting an appropriate forecasting method and the forecasts based on that method, it is important to continuously monitor the forecast accuracy. For this, one can use one or more of the forecast errors discussed in Section 2.9. In practice, another measure, called *Tracking Signal*, is also commonly used for monitoring forecast accuracy.

2.10.1 Tracking Signal

Tracking signal at period k, denoted by TS_k, is the ratio of Bias and MAD up to period k. In other words,

$$TS_k = \frac{\text{Bias}_k}{\text{MAD}_k} = \frac{\sum_{t=1}^{k} e_t}{\frac{1}{k}\sum_{t=1}^{k} |e_t|} \quad \text{for } k = 1, 2, \ldots n \quad (2.20)$$

NOTES:

i. The numerator of Equation 2.20 can be positive, negative, or zero, while the denominator is always positive. Hence, TS_k can be positive, negative, or zero for any k. Note that TS_1 is always equal to +1 or −1.

ii. Unlike the forecast error defined in Section 2.9, tracking signal is not a single number. Instead, it is a series of numbers, which can be used to detect changes in the pattern of the forecast.

iii. The generally acceptable values of TS_k are ±6. When the tracking signals go outside these limits, the forecaster should be notified to determine the cause of these limit violations.

iv. Tracking signals outside the limits do not automatically imply that the forecasting method is not working. Environmental conditions, such as local economy, sales promotions, new competition, and so on, can cause sudden fluctuations in tracking signals.

Example 2.12

Consider the data given in Table 2.9 (Example 2.11). Compute the tracking signals and plot.

SOLUTION

$$TS_1 = \frac{20}{20} = +1$$

$$TS_2 = \frac{20-42}{(20+42)/2} = -0.7$$

$$TS_3 = \frac{20-42-18}{(20+42+18)/3} = -1.5$$

$$TS_4 = \frac{20-42-18-76}{(20+42+18+76)/4} = -3$$

$$TS_5 = \frac{-164}{40.8} = -4$$

Since we have data for five time periods, we have five tracking signals. A plot of the tracking signals is shown in Figure 2.5. A downward trend of tracking signals points to forecasting problems ahead.

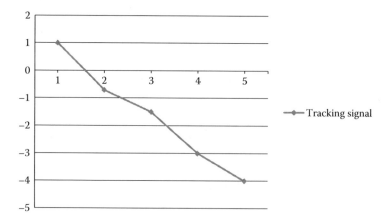

FIGURE 2.5
Plot of tracking signals (Example 2.12)

2.11 Forecasting Software

There are a number of free and commercial forecasting software available in the market. The website www.forecastingprinciples.com has extensive information for both academic and industry users about forecasting. The magazine *OR/MS Today* publishes biannual surveys of forecasting software. For a most recent survey, refer to Fry and Mehrotra (2016). Some of the forecasting software are stand-alone dedicated programs, while others are add-ins to Microsoft Excel.

2.11.1 Types of Forecasting Software

In general, forecasting software can be grouped into three categories, depending on the degree to which the software can select the forecasting method and its optimal parameters without user input:

- Automatic (more expensive)
- Semi-automatic (moderate cost)
- Manual (inexpensive/free)

2.11.1.1 Automatic Software

These are designated stand-alone programs and may cost several hundred dollars. The user does not have to be proficient in forecasting. The software asks the user to enter the time series data. It selects the appropriate forecasting method based on the analysis of the data and recommends a particular method to use. It also computes the optimal values of the parameters (e.g., smoothing constants α, β, and γ for Winters method) using forecast errors. It then gives the forecasts and provides measures of accuracy, such as confidence interval, MAD, MAPE, Bias, and so on. The user can override the recommended forecasting method and investigate other methods. These software programs are quite user friendly. However, tests have shown that they are not very reliable. Note that most automatic software can also operate under semi-automatic and manual mode allowing user intervention.

2.11.1.2 Semi-Automatic Software

These are moderately priced, but they require the user to have some basic knowledge of forecasting principles and techniques. Here, the user has to select an appropriate forecasting technique based on the analysis of time series data. The software will then compute the optimal parameters for the chosen method using some measure of forecast error. It also gives the forecasts and all the statistics, such as MAD, MAPE, MSE, Bias, and so on. The software makes no recommendation as to which forecasting technique is appropriate for the given data.

2.11.1.3 Manual Software

Some of these are free or inexpensive. The user is expected to have very good knowledge of forecasting principles and methods. The user has to select an appropriate forecasting method and its parameters. The program will then do the tedious calculations and provide the forecasts and its accuracy in terms of forecast errors. The user has to experiment with different values of the parameters for a particular forecasting method, as well as try different forecasting methods. The user has to decide which method is the best for the given data. The process can be tedious and time consuming.

For a complete listing of forecasting software, refer to the article by Fry and Mehrotra (2016). They also provide the requirements, available methods, and capabilities of each software. Given below is a partial list of the forecasting software discussed by Fry and Mehrotra (2016):

- *Forecast Pro XE* (Automatic/Semi-automatic/Manual)

    A stand-alone leading forecasting software for large problems (cost: $1300)

    URL: www.forecastpro.com

- *Crystal Ball Predictor* (Automatic/Semi-automatic/Manual)

    by Oracle Corporation, Windows based (cost: $995).

    URL: www.oracle.com/crystalball

- *NCSS* (Automatic/Semi-automatic/Manual)

    Windows-based *EXCEL* Add-in (cost: $479)

    URL: www.ncss.com

- *PEER Forecaster* (Automatic/Semi-automatic)

    Excel Add-in for Windows from Delphus (Free)

    URL: www.delphus.com

- *Minitab Statistical Software* (Semi-automatic/Manual)

    Includes forecasting software (cost: $1495)

    URL: www.minitab.com

- *DPL* (Manual)

    Windows-based software by Syncopation Inc. (cost: $495)

    URL: www.syncopation.com

- *XL Miner Analysis Tool Pak* (Manual)

    by Frontline systems, developer of Excel Solver (Free)

    URL: www.solver.com

NOTE: Most vendors have special educational discounts for classroom and student use.

2.11.2 User Experience with Forecasting Software

To evaluate the use and satisfaction with the various forecasting software, Sanders and Manrodt (2003) did a survey of 2400 U.S. corporations. The survey asked what software they used for forecasting, how satisfied they were with the software, and how well the software did. About 10% (240 companies) responded to the survey. Highlights of the findings by Sanders and Manrodt (2003) are given below:

- Only 11% of the respondents used commercial software.
- 48% used Excel spreadsheets.
- 25% used forecasting software developed in-house.
- 6% used outside vendors to develop custom made software for their companies.
- 10% reported not using any software for forecasting; they primarily used qualitative methods.
- 80% of the firms manually adjusted the forecasts provided by the software.
- 60% of the respondents expressed dissatisfaction with the forecasting software they were using.

2.12 Forecasting in Practice

2.12.1 Real-World Applications

There are several *published* results of real-world forecasting. We discuss briefly a few of the forecasting applications in practice. For more details and additional applications, the reader is referred to the cited references and the textbook by Taylor (2010).

1. **Taco Bell** (Hueter and Swart 1998)

   Lunchtime sales were forecasted using the moving average method and were used for estimating labor needs at Taco Bell restaurants.

2. **Dell** (Kapuscinski et al. 2004)

   Monthly forecasts of parts requirements for Dell's suppliers were done using moving average with seasonality. They were used by the suppliers for inventory management.

3. **FedEx** (Xu 2000)

   Exponential smoothing with trend and seasonality was used to forecast FedEx's call center loads.

4. **National Car Rental** (Geraghty and Johnson 1997)

   Time series forecasting models with seasonality was used by National Car Rental to forecast customer demands for cars and estimate revenues. They were used in the company's revenue management system.

5. **Nabisco** (Amrute 1998; Barash and Mitchell 1998)

   Sales forecasts of Nabisco products were done using time series forecasting. Moving average method was used for existing products and exponential smoothing with trend was used for new products. The forecasts were then used in production planning decisions.

6. **NBC** (Bollapragada et al. 2008)

   National Broadcasting Company used a qualitative approach to predict advertising demand for pricing and revenue projections. It used a combination of Delphi method and "grassroots" approach that used customer surveys by sales personnel.

2.12.2 Forecasting in Practice: Survey Results

Sanders and Manrodt (1994) surveyed 500 U.S. companies about their forecasting practices. The objectives of the survey were to ascertain the following:

1. Familiarity with different forecasting methods
2. Choice of forecasting method for short-term, medium-term, and long-term forecasting
3. Reasons for choosing certain forecasting methods
4. Satisfaction with the forecasting methods used

The survey included both manufacturing and service industries with annual sales between $6 million and $15 billion with the median at $1.5 billion. Hence, the results are somewhat skewed toward larger firms. About 20% of the companies (~100) responded to the survey and their responses were used in the analysis. Highlights of the survey results, reported by Sanders and Manrodt (1994), include the following:

- There was no significant difference in the use of the forecasting methods between the small and large companies.
- Not much difference was noted in the responses by manufacturing and service firms. However, manufacturing firms used more quantitative methods (15.5%) compared to service firms (9.6%).
- Overall, *moving average method* was the most preferred quantitative method for all firms.

- Most respondents were familiar with the different forecasting methods.
- For long-term (>1 year) and medium-term (6 months to a year) forecasting, most firms preferred qualitative methods, particularly *executive committee consensus*.
- Among the quantitative methods, *moving average* was the most frequently cited approach for short- and medium-term forecasting. However, for long-term forecasting, regression was the most popular choice.
- There was a high degree of satisfaction with *qualitative methods* in general and *executive committee consensus in* particular.
- Among the quantitative methods, there was a higher level of satisfaction with *moving average, exponential smoothing, regression*, and *simulation*.
- Difficulty in obtaining data, ease of use, and cost were cited as main reasons for not using sophisticated quantitative methods.

Most managers prefer to *underforecast* (70%) compared to *overforecast* (15%) due to the company's reward structure. A company's incentive system generally rewards those who exceed "expectations." Those who overforecast are generally in the manufacturing departments since shortages are expensive and they are held accountable.

2.13 Summary and Further Readings

Forecasting customer demands is essential for managing service systems. Based on the forecast, worker schedules are generated to meet the customer demands. In this chapter, we discussed commonly used forecasting methods in practice. In this section, we shall summarize the forecasting methods discussed in this chapter and suggest further readings about other forecasting methods and related topics available in the literature.

2.13.1 Demand Forecasting—Summary

We discussed two broad categories of forecasting methods—*qualitative or judgmental methods* and *quantitative or statistical methods*. Qualitative methods are usually based on human judgments and do not require past data. They are most useful for new products or services. Under qualitative methods, we discussed *executive committee census, Delphi method, survey of sales force,* and *customer surveys*. These methods are quite frequently used by companies, particularly for making strategic decisions, as well as to supplement quantitative forecasts. Quite often, the companies choose the final forecast as the average of the qualitative and quantitative forecasts. Empirical evidence (Frances 2011) suggests that the averaging strategy generally produces a much better forecast.

Under quantitative methods, we primarily discussed *time series forecasting* methods. They included *naïve method, averaging method, moving average method, weighted moving average method,* and *exponential smoothing method.* We also discussed modifications to the time series methods to incorporate seasonality and trend (*Holt's method* and *Winters method*).

2.13.2 ARIMA Method

One of the time series methods that we have not discussed, but very popular in the academic circles, is called *ARIMA (Auto Regressive Integrated Moving Average)* method developed by Box and Jenkins (1976). In this method, a mathematical model is fitted that is optimal with respect to the historical time series data. In fact, exponential smoothing method becomes a special case of the ARIMA model. ARIMA requires an enormous amount of past data. It has been found to be good for short-term (<3 months) and medium-term (3 months to 2 years) forecasting. Because of its mathematical sophistication and huge data requirements, it has not been used widely in practice. In the "Forecasting in practice" survey (Sanders and Manrodt 1994), less than 5% of the survey respondents have reported using the ARIMA method, even though nearly half are familiar with the method. Those interested in learning more about ARIMA can refer to the textbook by Box and Jenkins (1976) and the practical guide to ARIMA by Hoff (1983).

2.13.3 Croston's Method (Croston 1972)

Another quantitative forecasting method that we did not discuss was Croston's method. This is applicable for intermittent erratic or slow-moving demand. For example, a machine may not fail for several periods and may not need any spare parts. Croston's method considers separately the two random events—the time between (non-zero) demands and the size of the demand when it happens. It then combines the two events, using exponential smoothing, to generate the forecast. Under Croston's method, forecasts change only after a demand happens and is otherwise constant between demands. Forecast increases when there is a large demand or shorter time between demands. Forecast decreases when there is small demand or a long time between demands.

Interested readers can refer to the original paper by Croston (1972). Willemain et al. (1994) did a comparative evaluation of Croston's method in forecasting intermittent demand in manufacturing using industrial data. They concluded that Croston's method was superior to the exponential smoothing method and was robust even under situations when Croston's model assumptions were violated.

2.13.4 Further Readings in Forecasting

The forecasting website www.forecastingprinciples.com has extensive information to those interested in learning more about forecasting. There are separate sections for researchers and practitioners. The researchers' section discusses the current state-of-the-art forecasting knowledge and the research needs in forecasting. It also contains a list of current research papers on new forecasting principles. The practitioners' section has information on how to select a forecasting method and essential reading materials for the practitioners. It also has a list of consultants available for forecasting, a discussion group to exchange ideas and a list of forecasting courses offered by universities and companies.

The forecasting website also contains a list of textbooks, trade books, and journals in forecasting. The textbook by Armstrong (2001) is a good handbook on forecasting principles, written for both practitioners and researchers. It is a collection of articles by several authors. A structured approach to demand-driven forecasting is discussed in Chase (2009). In addition to the various time series forecasting methods, the book also discusses *causal forecasting* including regression and Box–Jenkin's ARIMA methods. The textbook by Shim (2000) addresses strategic forecasting issues. Essentials of forecasting methods are presented with examples and real-life cases. In addition to demand forecasting, applications to forecasting cost, earnings, cash flows, interest rates, and foreign exchanges are discussed. The textbook on business forecasting by Evans (2010) is available as paperback and e-book editions. It is appropriate for a wide range of readers with diverse backgrounds. Case studies are used to discuss both short-term and long-term forecasting approaches.

Exercises

2.1 Explain the differences between *qualitative* and *quantitative* forecasting methods. Which method is applicable under what conditions?

2.2 Compare and contrast the two qualitative forecasting methods—executive committee consensus and Delphi method.

2.3 Under what conditions may the "Last value" method be better than the "Averaging" method?

2.4 What are the advantages and disadvantages of the *exponential smoothing* method?

2.5 Name at least three practical uses of forecast errors.

2.6 Why do most companies "underforecast" than "overforecast"? Cite a situation when "overforecasting" may be better.

2.7 Monthly sales data for this year for QuikSell Realty are given in Table 2.10. The company wants a forecast of sales for the first quarter of next year (months 13, 14, and 15). Compute the forecasts using the following methods:

   a. Last-value method
   b. Averaging method
   c. Three-month moving average method
   d. Exponential smoothing with $\alpha = 0.25$
   e. Holt's method with $\alpha = 0.4$ and $\beta = 0.5$
   f. Given the sales pattern for this year, do any of these methods seem inappropriate for obtaining next year's forecasts? Is there another method you would recommend to get a better forecast for months 13, 14, and 15?

2.8 Joe Kool has decided to use the *3-month weighted moving average* method for one of his company's products. Last year's monthly sales data are given in Table 2.11.

TABLE 2.10

Sales Data for Exercise 2.7

Month	Sales	Month	Sales
1	34	7	34
2	33	8	33
3	42	9	43
4	34	10	31
5	36	11	35
6	43	12	41

TABLE 2.11

Sales Data for Exercise 2.8

Month	Sales	Month	Sales
1	5325	7	5375
2	5405	8	5585
3	5200	9	5460
4	5510	10	4905
5	5765	11	5755
6	5210	12	6320

Joe is trying to decide the appropriate weights to select for his forecasting method that will minimize the sum of the absolute errors between the forecasts and the actual values. Show that the determination of the optimal weights can be formulated as an LP problem. Solve the linear program using any optimization software and determine the forecast for month 13.

2.9 Consider the sales forecasts determined for months 13, 14, and 15 by the various methods in Exercise 2.7. Compute their forecast errors using MAD and Bias. Based on the forecast errors, which method would you recommend and why?

2.10 A community college is planning to use the exponential smoothing method to forecast freshman applications for the fall semester of each year. Past 5-year data on the actual number of applications received during 2011–2015 are given in Table 2.12.

  a. Beginning with an initial forecast of 5000 for 2011 and $\alpha = 0.2$, determine the forecasts for the years 2012, 2013, 2014, and 2015.
  b. Calculate MAD and Bias for the 5-year period 2011–2015.
  c. Calculate the tracking signals for the 5 years.
  d. Determine the forecasts for the years 2016, 2017, and 2018.

2.11 Consider the freshman applications data in Exercise 2.10 (Table 2.12). Suppose the community college wants to try the following forecasting methods as well:

- 3-month moving average
- Averaging method
- Simple Linear Trend

Determine the forecasts for the years 2016, 2017, and 2018. Based on MAD, compare the forecasts of all the methods, including the exponential smoothing method done in Exercise 2.10. Which method would you recommend and why?

2.12 An electronics company is interested in forecasting the quarterly sales for next year for one of its products. Based on the past sales

TABLE 2.12

Data on Freshman Application (Exercise 2.10)

Year	Number of Applications
2011	4650
2012	5200
2013	6000
2014	5700
2015	6200

data, they found that the quarterly sales have seasonality and two forecasting methods look promising:
1. Four-quarter moving average with seasonality
2. Last value method with seasonality

The seasonality indices have been computed as 0.93, 0.9, 0.99, and 1.18 for quarters 1, 2, 3, and 4, respectively. The actual sales for this year for quarters 1, 2, 3, and 4 are 6992, 6822, 7949, and 9690, respectively.

   a. Determine the deseasonalized forecasts for quarters 1, 2, 3, and 4 for next year by both forecasting methods.
   b. Determine the actual forecasts for quarters 1, 2, 3, and 4 for next year by both forecasting methods.

   NOTE: You can round off the solutions to the nearest integers.

2.13 ABC company is interested in forecasting quarterly sales for year 2017 for one of its products. Based on past 3 years of quarterly sales data (2014, 2015, and 2016), two forecasting methods look promising:
1. Exponential smoothing with seasonality with $\alpha = 0.5$.
2. Holt's method with seasonality with $\alpha = \beta = 0.3$.

The seasonality indices have been computed as 0.93, 0.9, 0.99, and 1.18 for quarters 1, 2, 3, and 4, respectively. The *actual demand* for the *last quarter* (quarter 4 of 2016) was 9650. You are also given the following data for the last quarter for the two methods:

- For exponential smoothing, the deseasonalized forecast was 7795.
- For Holt's method, the forecasts of level and trend were 7778 and 80, respectively.

Determine the *actual forecasts* for quarters 1, 2, 3, and 4 for the year 2017 by both methods.

NOTE: You can round off the solutions to the nearest integers.

2.14 Ephemeral Apparel is experimenting with two forecasting methods for its new jeans. Table 2.13 gives the actual sales and the forecasts obtained by the two methods during the past 6 months.
   a. Calculate MAD, MSE, and Bias for the two methods.
   b. Calculate and plot the tracking signals for the two methods.
   c. Are the methods underforecasting or overforecasting? Which method would you recommend for forecasting future sales and why?

2.15 Jill Smith has joined recently as the forecasting manager for a consumer electronic retailer BigBuy. She is interested in developing quarterly sales forecasts for one of the company's key products.

Forecasting Demand for Services

TABLE 2.13

Sales and Forecast Data for Exercise 2.14

Month	Actual Sales	Forecast-1	Forecast-2
1	558	532	521
2	490	541	538
3	576	520	546
4	632	550	542
5	515	575	555
6	610	590	575

TABLE 2.14

Quarterly Sales Data for 2012–2016 (Exercise 2.15)

Quarter	2012	2013	2014	2015	2016
1	800	1700	2100	2400	3600
2	750	1100	2200	3060	3900
3	600	680	1300	1800	1500
4	1500	2000	3100	4000	3320

She has collected data on quarterly sales for this product for the past 5 years and they are given in Table 2.14.

Jill decides to use Holt's method with seasonality and trend (discussed in Example 2.6, Section 2.7) with the initial estimates of level and trend as 600 and 50, respectively. She also decides to use the first 4 years of data (2012–2015) for determining the smoothing constants (α and β) and use the 2016 data for validating the forecasting method and the smoothing constants.

a. Using the data for the first 4 years, 2012–2015, prepare the initial estimates of the seasonal factors for each quarter.

b. Using a spreadsheet, develop and select a set of good smoothing constants for both α and β for Holt's method. Use the forecast error measures Bias and STD to test your parameters. Run your tests using the data for years 2012–2015. Use the following combinations of the smoothing constants for the tests:

$(\alpha, \beta) = (0.1, 0.1); (0.1, 0.2); (0.1, 0.3); (0.2, 0.2); (0.2, 03); (0.3, 0.3)$

c. Select the *two best* pairs of values for the smoothing constants (α and β) obtained from part (b) to prepare the quarterly forecasts for 2016. Using the 2016 actual sales, validate these forecasts. Use the forecast errors Bias, STD, and tracking signals for the validation.

d. Suppose Jill wants to try out a simple *moving average* forecasting method, by averaging the past four quarter's deseasonalized demands. For example, 2016 quarter 1 deseasonalized forecast will be the average of the 2015's deseasonalized quarterly demands. Using the moving average method, prepare the quarterly forecasts for 2016. Compare these forecasts with those obtained in part (c) using the three forecast errors. Are they better than the forecasts obtained using Holt's method? Why or why not?

e. What quarterly sales forecasts, including estimates of their accuracy, should Jill present to her company for the year 2017?

References

Amrute, S. 1998. Forecasting new products with limited history: Nabisco's experience. *Journal of Business Forecasting.* 17(3): 7–11.

Armstrong, J. S. 2001. *Principles of Forecasting: A Handbook for Researchers and Practitioners.* Springer Series in Operations Research and Management Science. Vol. 30. NY: Springer.

Barash, M. and D. Mitchell. 1998. Account based forecasting at Nabisco biscuit company. *Journal of Business Forecasting.* 17(2): 3–6.

Bollapragada, S., S. Gupta, B. Hurwitz, P. Miles, and R. Tyagi. 2008. NBC–Universal uses a novel qualitative forecasting technique to predict advertising demand. *Interfaces.* 38(2): 103–111.

Box, G. E. P. and G. M. Jenkins. 1976. *Time Series Analysis, Forecasting and Control.* San Francisco, CA: Holden Day.

Brown, R. G. 1959. *Statistical Forecasting for Inventory Control.* New York City McGraw-Hill.

Chase, C. 2009. *Demand-Driven Forecasting: A Structured Approach to Forecasting.* Hoboken, NJ: John Wiley & Sons.

Chopra, S. and P. Meindl. 2013. *Supply Chain Management: Strategy, Planning and Operation.* 5th Edition. Eaglewood Cliff, NJ: Prentice Hall.

Croston, J. D. 1972. Forecasting and stock control for intermittent demand. *Operational Research Quarterly.* 23(3): 1970–1977.

Delbecq, A. L., A. H. VandeVen, and D. Gustafson. 1975. *Group Techniques for Program Planning.* Glenview, IL: Scott Foresman.

Evans, M. K. 2010. *Practical Business Forecasting.* UK: Blackwell Publishers.

Foote, B. L. and K. G. Murty. 2008. Production systems. In *Operations Research and Management Science Handbook.* ed. A. Ravi Ravindran, 18-1–18-30. Boca Raton, FL: CRC Press.

Foote, B. L. and K. G. Murty. 2009. Production systems. In *Operations Research Applications.* ed. A. Ravi Ravindran, Chapter 4. Boca Raton, FL: CRC Press.

Frances, P. H. 2011. Averaging model forecasts and expert forecasts: Why does it work? *Interfaces.* 41(2): 177–181.

Fry, C. and V. Mehorotra. 2016. Forecasting Software Survey. *ORMS Today*. 43(3): 44–53.

Geraghty, M. K. and E. Johnson. 1997. Revenue management saves National Car Rental. *Interfaces*. 27(1): 107–127.

Hillier, F. S. and G. J. Lieberman. 2001. *Introduction to Operations Research*. 7th Edition. New York City: McGraw-Hill.

Hoff, J. C. 1983. *A Practical Guide to Box–Jenkins Forecasting*. Belmont, CA: Lifetime Learning Publications.

Holt, C. C. 1957. Forecasting seasonality and trends by exponentially weighted moving averages. *Office of Naval Research*. Report no. 52.

Hueter, J. and W. Swart. 1998. An integrated labor-management system for Taco Bell. *Interfaces*. 28(1): 75–91.

Kapuscinski, R., R. Q. Zhang, P. Carbonneau, R. Moore, and B. Reeves. 2004. Inventory decisions in Dell's supply chain. *Interfaces*. 34(3): 191–205.

Lilien, G. L., A. Rangaswamy, and A. D. Bruyn. 2007. *Principles of Marketing Engineering*. Victoria, BC, Canada: Trafford Publishing.

Nahmias, S. 1993. *Production and Operations Analysis*. 2nd Edition. Boston, MA: Irwin.

Ravindran, A. R. and D. P. Warsing Jr. 2013. *Supply Chain Engineering: Models and Applications*. Chapter 2. Boca Raton. FL: CRC Press.

Sanders, N. R. and K. B. Manrodt. 1994. Forecasting practices in US corporations: Survey Results. *Interfaces*. 24(2): 92–100.

Sanders, N. R. and K. B. Manrodt. 2003. Forecasting software in practice: Use, satisfaction and performance. *Interfaces*. 33(5): 90–93.

Shim, J. K. 2000. Strategic *Business Forecasting: The Complete Guide to Forecasting Real World Company Performance, Revised Edition*. Boca Raton, FL: CRC Press.

Srinivasan, G. 2010. *Quantitative Models in Operations and Supply Chain Management*. India: Prentice Hall.

Taylor, B. 2010. *Introduction to Management Science*. 10th Edition. Eaglewood Cliff, NJ: Prentice Hall.

Willemain, T. R., C. N. Smart, J. H. Shockor, and P. A. DeSautels. 1994. Forecasting intermittent demand in manufacturing: A comparative evaluation of Croston's method. *International Journal of Forecasting*. 10(4): 529–538.

Winters, P. R. 1960. Forecasting sales by exponentially weighted moving average. *Management Science*. 6: 324–342.

Xu, W. 2000. Long range planning for call centers at FedEx. *The Journal of Business Forecasting Methods & Systems*. 18(4): 7–11.

3

Design of Service Systems

Integer programming (IP) models with binary variables have been successfully used in practice to design service systems and solve resource allocation problems, scheduling of airline crews, and aircrafts and supply chain management. We begin this chapter with a review of modeling with binary variables. We will then apply the IP models for location and distribution decisions in supply chains, airline scheduling, and workforce planning. We end this chapter with a discussion of queuing models in service systems. We review the basic queuing models in Operations Research and then apply them to the service setting of staffing at a call center.

3.1 Modeling with Binary Variables*

This section will be devoted to the use of binary (0–1) variables in modeling real-world problems. Linear programming (LP) models with binary decision variables are called *integer programming* (IP) models. Of course, a general IP model may include regular integer variables (non 0–1), as well as continuous variables. Such IP models are called *mixed integer programming* (MIP) models. LP models with just binary and/or integer variables are called *pure integer programs*.

3.1.1 Capital Budgeting Problem

A company is planning its capital spending for the next T periods. There are N projects that compete for the limited capital B_i, available for investment in period i. Each project requires a certain investment in each period once it is selected. Let a_{ij} be the required investment in project j for period i. The value of the project is measured in terms of the associated cash flows in each period discounted for inflation. This is called the net present value (NPV). Let v_j denote the NPV for project j. The problem is to select the proper projects for investment that will maximize the total value (NPV) of all the projects selected.

---

* Portions of this section have been adapted with permission from Ravindran, A. R. and D. P. Warsing, Jr. 2013. *Supply Chain Engineering: Models and Applications.* Boca Raton, FL: CRC Press.

Formulation

To formulate this as an integer program, we introduce a binary variable for each project to denote whether it is selected or not.
Let,

$$x_j = 1, \quad \text{if project } j \text{ is selected}$$
$$x_j = 0, \quad \text{if project } j \text{ is not selected}$$

It is then clear that the following pure integer program will represent the capital budgeting problem:

Maximize

$$Z = \sum_{j=1}^{N} v_j \cdot x_j$$

Subject to

$$\sum_{j=1}^{N} a_{ij} \cdot x_j \leq B_i, \quad \text{for all } i = 1, 2, \ldots, T$$

$$0 \leq x_j \leq 1, \; x_j \text{ is a binary variable} \quad \text{for all } j = 1, 2, \ldots, N$$

Example 3.1*

Centre County has received $200,000 from the Commonwealth of Pennsylvania for community development projects. The county commissioners are reviewing four potential projects to fund from the planning committee. The cost of each project, required land, and potential usage by county residents are given in Table 3.1.

The basketball court will be built inside the park. Hence, no additional land space is needed. However, the basketball court cannot be built unless the park project is selected for funding. The county has 15 acres of land. The commissioners want to select those projects that will benefit the county residents the most (maximize daily usage) subject to the budget and land constraints.

---

* Adapted from Taylor, B. 2010. *Introduction to Management Science*, 10th Edition. Upper Saddle River, NJ: Prentice Hall.

Design of Service Systems

TABLE 3.1

Data on Potential Projects (Example 3.1)

Project	Daily Usage	Cost	Land Space (Acres)
Park with children's playground and walking trails	600	$50,000	8
Basketball court	100	$20,000	0
Recreation center	300	$150,000	4
Swimming pool	500	$70,000	5

SOLUTION

Binary variables

$$x_j = 1, \text{ if project } j \text{ is selected and 0 otherwise for } j = 1, 2, 3, 4$$

where
- $j = 1$: Build park
- $j = 2$: Build basketball court
- $j = 3$: Build recreation center
- $j = 4$: Build swimming pool

Constraints

1. (Budget) No more than $200,000 are available

$$50x_1 + 20x_2 + 150x_3 + 70x_4 \leq 200$$

2. (Space) No more than 15 acres of land are available

$$8x_1 + 4x_3 + 5x_4 \leq 15$$

3. (Conditional Constraint) Basketball court requires the selection of park project

$$x_2 \leq x_1$$

Note that the basketball court can be built ($x_2 = 1$) only when the park is developed ($x_1 = 1$). However, the park can be developed ($x_1 = 1$), with or without the basketball court ($x_2 = 1$ or 0). Building the basketball court without the park, namely, the solution, $x_1 = 0$, $x_2 = 1$, violates the conditional constraint and is infeasible.

Objective function: The total daily usage is to be maximized.

$$\text{Max } Z = 600x_1 + 100x_2 + 300x_3 + 500x_4$$

Optimal solution: The optimal solution to this integer program was obtained using Microsoft's Excel Solver software and is given below:

$$x_1 = 1, \; x_2 = 1, \; x_3 = 0, \; x_4 = 1$$

Optimal value of $Z = 1200$

The optimal decision for the county is to build the swimming pool and the park, with the basketball court. The facilities would be used by 1200 county residents daily.

3.1.2 Fixed Charge Problem

Consider a production planning problem with N products such that the jth product requires a fixed production or setup cost K_j, independent of the amount produced, and a variable cost C_j per unit, proportional to the quantity produced. Assume that every unit of product j requires a_{ij} units of resource i and there are M resources. Given that the product j, whose sales potential is d_j, sells for $\$p_j$ per unit and no more than b_i units of resource i are available ($i = 1, 2, \ldots, M$), the problem is to determine the optimal product mix that maximizes the net profit.

Formulation

The total cost of production (fixed plus variable) is a nonlinear function of the quantity produced. But with the help of binary (0–1) integer variables, the problem can be formulated as an integer linear program.

Let the binary integer variable δ_j denote the decision to produce or not to produce product j.

In other words,

$$\delta_j = \begin{cases} 1, & \text{if product } j \text{ is produced} \\ 0, & \text{otherwise} \end{cases}$$

Let x_j (≥ 0) denote the quantity of product j produced. Then, the cost of producing x_j units of product j is $K_j.\delta_j + C_j.x_j$, where $\delta_j = 1$ if $x_j > 0$ and $\delta_j = 0$ if $x_j = 0$. Hence, the objective function is

$$\text{Maximize } Z = \sum_{j=1}^{N} p_j.x_j - \sum_{j=1}^{N}(K_j.\delta_j + C_j.x_j)$$

Design of Service Systems

The constraints of the model are described below.
The supply constraint for the ith resource is given by

$$\sum_{j=1}^{N} a_{ij} \cdot x_j \leq b_i, \quad \text{for } i = 1, 2, \ldots M$$

The demand constraint for the jth product is given by

$$x_j \leq d_j \cdot \delta_j \quad \text{for } j = 1, 2, \ldots N$$

$$x_j \geq 0 \text{ and } \delta_j = 0 \text{ or } 1 \quad \text{for all } j$$

Note that x_j can be positive only when $\delta_j = 1$, in which case its production is limited by d_j and the fixed production cost Kj is included in the objective function.

3.1.3 Constraint with Multiple Right-Hand-Side (RHS) Constants

Consider a problem, where the constraint $a_1 x_1 + a_2 x_2 + \ldots + a_n x_n$ must be less than or equal to one of the RHS values b_1, b_2, b_3. In other words, the constraint becomes

$$a_1 x_1 + a_2 x_2 + \ldots + a_n x_n \leq b_1, b_2 \text{ or } b_3. \tag{3.1}$$

Constraints similar to Equation 3.1 arise in supply chain network design, where a company has the option to build warehouses of different capacities. In that case, x_i is the quantity of product i stored at that location, a_i is the square footage occupied by one unit of product i and b_1, b_2, b_3 are the three potential warehouse capacities. Using binary variables, one for each b_i value, we can represent Equation 3.1 as a linear constraint. Define $\delta_1, \delta_2, \delta_3$ as the binary variables such that when $\delta_i = 1$, the RHS value is b_i. Then, Equation 3.1 can be written as

$$a_1 x_1 + a_2 x_2 + \ldots + a_n x_n \leq \delta_1 b_1 + \delta_2 b_2 + \delta_3 b_3 \tag{3.2}$$

$$\delta_1 + \delta_2 + \delta_3 = 1 \tag{3.3}$$

$$\delta_i \in (0, 1), \quad \text{for } i = 1, 2, 3 \tag{3.4}$$

Equations 3.3 and 3.4 guarantee that only one of the δ_i's will be one and the rest will be zero. Hence, the RHS value of Equation 3.2 can only be b_1, b_2, or b_3. If Constraint 3.1 represents the decision to build a warehouse of capacity b_1, b_2, or b_3 only, then the fixed cost of building the warehouse, say K_1, K_2, and K_3, can be easily included in the objective function using the binary variables δ_1, δ_2, and δ_3, respectively. If we want to include the option of *not building the warehouse* at all at this location, we can write Equation 3.3 as

$$\delta_1 + \delta_2 + \delta_3 \leq 1 \tag{3.5}$$

Equation 3.5 allows the possibility for all the δ_i's to be zero, in which case no warehouse will be built at this location. This would make the RHS value of Equation 3.2 zero, which will force all the storage variables at this location ($x_1, x_2, \ldots, x_n$) to be equal to zero. An application of this model for designing a supply chain network is discussed in Chapter 5 (Example 5.3).

3.1.4 Quantity Discounts

Quantity discounts refer to the practice of offering lower prices for large-volume purchases. There are two types of quantity discounts offered by vendors:

1. "All-unit" discount
2. "Graduated" discount

Both types of quantity discount models result in nonlinear cost functions. However, they can be modeled as linear functions with the help of binary variables. We shall discuss the integer programming formulations of the "quantity discount" models next.

3.1.4.1 "All-Unit" Quantity Discounts

Under this scenario, the entire purchase will be charged at a lower price based on the order quantity. Figure 3.1 illustrates an example of "all-unit" quantity discount price structure:

The mathematical representation of the cost function $f(X)$, where X is the quantity purchased, is as follows:

$$\text{Cost } f(X) = \begin{cases} 7X & 0 \leq X \leq 200 \\ 5X & 201 \leq X \leq 500 \\ 3X & X \geq 500 \end{cases}$$

Design of Service Systems

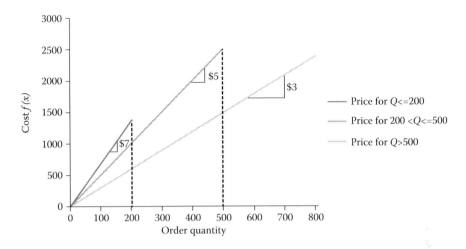

FIGURE 3.1
All-unit quantity discount model.

The cost function $f(X)$ is nonlinear in X since the slope changes depending on the value of X. However, we can linearize it using three binary variables, δ_1, δ_2, and δ_3, one for each price range as follows:

$$\delta_i = \begin{cases} 1, & \text{if price range } i \text{ is used} \\ 0, & \text{otherwise} \end{cases}$$

Let X_i = order quantity under price range "i," $i = 1, 2, 3$.
The mixed integer LP formulation will be as follows:

$$\text{Cost } f(X) = 7.X_1 + 5.X_2 + 3.X_3 \tag{3.6}$$

Subject to

$$X = X_1 + X_2 + X_3 \tag{3.7}$$

$$0 \leq X_1 \leq 200.\delta_1 \tag{3.8}$$

$$201.\delta_2 \leq X_2 \leq 500.\delta_2 \tag{3.9}$$

$$501.\delta_3 \leq X_3 \leq M.\delta_3 \tag{3.10}$$

$$\delta_1 + \delta_2 + \delta_3 = 1 \quad (3.11)$$

$$\delta_i \in (0,1) \quad (3.12)$$

Equations 3.11 and 3.12 will force exactly one of the δ_i's to be one and the others zero in any solution. Hence, only one of the X_i's will be positive in Equation 3.6. For example, if $\delta_2 = 1$, ($\delta_1 = \delta_3 = 0$), then X_2 will be positive between 201 and 500 by Equation 3.9. Since δ_1 and δ_3 will be zero, X_1 and X_3 will be forced to zero by Equations 3.8 and 3.10. Note that M is a large positive number in Equation 3.10. In case there is a capacity limit on the maximum order quantity, then M can be replaced by that capacity limit. Note that all the constraints and the objective function are now linear functions.

All unit quantity discount models are commonly used by trucking companies for the freight rate structure in road transportation.

3.1.4.2 Graduated Quantity Discount

Under this, the entire order quantity does not get the lower price; only the additional amount over a price range gets the lower cost. This is illustrated in Figure 3.2.

Under this model, the *first* 200 units are *always* charged at $7/unit, the *next* 300 units are charged at a lower price of $5/unit, and only the amount over 500 is charged at $3/unit. For example, if the order quantity is 300 units, then the cost is given by

$$f(X) = 200(\$7) + 100(\$5) = \$1900.$$

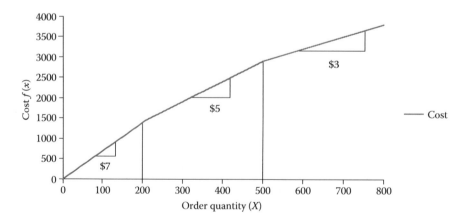

FIGURE 3.2
Graduated quantity discount model.

Design of Service Systems

Graduated quantity discount models are frequently used by utility companies for energy usage. Also, the U.S. Postal Service uses graduated rate for its first-class mail. The first ounce is always charged at a higher rate and successive ounces carry a lower cost per ounce.

Figure 3.2 also represents a nonlinear cost function. It can be linearized using binary variables, similar to the all unit quantity discount model. Let δ_1, δ_2, and δ_3 be the binary variables for each price range and X_1, X_2, and X_3 are the quantity purchased under price ranges 1, 2, and 3. Then, the linear integer programming formulation becomes

$$\text{Cost } f(X) = 7.X_1 + 5.X_2 + 3.X_3 \tag{3.13}$$

Subject to

$$X = X_1 + X_2 + X_3 \tag{3.14}$$

$$0 \leq X_1 \leq 200.\delta_1 \tag{3.15}$$

$$X_1 \geq 200.\delta_2 \tag{3.16}$$

$$0 \leq X_2 \leq (500 - 200).\delta_2 \tag{3.17}$$

$$X_2 \geq (500 - 200).\delta_3 \tag{3.18}$$

$$0 \leq X_3 \leq M.\delta_3 \tag{3.19}$$

$$\delta_1, \delta_2, \delta_3 \in (0, 1) \tag{3.20}$$

Unlike the all-unit discount model, one or more of the δ_i's can be one in a solution. Hence, one or more of the X_i's can be positive and their sum equals to the total order quantity X, as given by Equation 3.14. Note that X_3 can be positive only if $\delta_3 = 1$ (Equation 3.19). When $\delta_3 = 1$, $X_2 \geq 300$ by Equation 3.18. Since X_2 cannot be positive unless $\delta_2 = 1$, it will force $\delta_2 = 1$ by Equation 3.17. Now, both Equations 3.17 and 3.18 reduce to

$$X_2 \leq 300$$

$$X_2 \geq 300$$

Hence, the only solution is $X_2 = 300$, its maximum value under the second price range. Similarly, when $\delta_2 = 1$, it will force $\delta_1 = 1$ and $X_1 = 200$. Thus, for X_3 to be positive (to buy at the lowest price), we have to reach the maximum under the previous two price ranges, namely, $X_1 = 200$ and $X_2 = 300$.

Application: A case study illustrating the integer programming formulations of quantity discount models in supplier selection and order allocation is given in Chapter 5.

3.1.5 Handling Nonlinear Integer Programs

Nonlinear integer programming problems can be converted into linear integer programs using binary variables. We shall illustrate this with a numerical example.

Example 3.2 (Ravindran et al. 1987)

Consider a nonlinear (binary) integer programming problem:
Maximize

$$Z = x_1^2 + x_2 x_3 - x_3^3$$

Subject to

$$-2x_1 + 3x_2 + x_3 \leq 3$$

$$x_1, x_2, x_3 \in (0, 1)$$

The above nonlinear integer problem can be converted into a linear integer programming problem for solution. Observe the fact that for any positive k and a binary variable x_j, $x_j^k = x_j$. Hence, the objective function immediately reduces to $Z = x_1 + x_2 x_3 - x_3$. Now, consider the product term $x_2 x_3$. For binary values of x_2 and x_3, the product $x_2 x_3$ is always 0 or 1. Now, introduce a binary variable y_1 such that $y_1 = x_2 x_3$. When $x_2 = x_3 = 1$, we want the value of y_1 to be 1, while all other combinations of y_1 should be zero. This can be achieved by introducing the following two constraints:

$$x_2 + x_3 - y_1 \leq 1$$

$$-x_2 - x_3 + 2y_1 \leq 0$$

Note that when $x_2 = x_3 = 1$, the above constraints reduces to $y_1 \geq 1$ and $y_1 \leq 1$, implying $y_1 = 1$. When $x_2 = 0$ or $x_3 = 0$ or both are zero, the second constraint, $y_1 \leq \left(\dfrac{x_2 + x_3}{2}\right)$, forces y_1 to be zero.

Thus, the equivalent linear (binary) integer program becomes

Maximize
$$Z = x_1 + y_1 - x_3$$

Subject to
$$-2x_1 + 3x_2 + x_3 \leq 3$$

$$x_2 + x_3 - y_1 \leq 1$$

$$-x_2 - x_3 + 2y_1 \leq 0$$

$$x_1, x_2, x_3, y_1 \in (0,1)$$

A drawback of the above procedure for handling product terms is that an integer variable is introduced for each product term. It has been observed in practice that the solution time for integer programming problems increases with the number of integer variables. An alternate procedure has been suggested by Glover and Woolsey (1974) that introduces a continuous variable rather than an integer variable. This procedure replaces $x_2 x_3$ by a continuous variable x_{23} and introduces three new constraints as follows:

$$x_2 + x_3 - x_{23} \leq 1$$

$$x_{23} \leq x_2$$

$$x_{23} \leq x_3$$

$$x_{23} \leq 0$$

where x_{23} replaces the product term $x_2 x_3$.

Whenever x_2, x_3, or both are zero, the last two constraints force x_{23} to be zero. When $x_2 = x_3 = 1$, all the three constraints together force x_{23} to be 1. The primary disadvantage of this procedure is that it adds more constraints than the previous method.

REMARKS

The procedure for handling the product of two binary variables can be easily extended to the product of any number of variables. For example, consider the product terms $x_1 x_2 \ldots x_k$. Under the first procedure, a binary variable y_1 will replace $x_1 x_2 \ldots x_k$ and the following two constraints will be added:

$$\sum_{j=1}^{k} x_j - y_1 \leq k-1$$

$$-\sum_{j=1}^{k} x_j + k y_1 \leq 0$$

$$y_1 \in (0,1)$$

$$x_j \in (0,1)$$

Under the second procedure, the product terms $x_1 x_2 \ldots x_k$ will be replaced by a nonnegative variable x_0 and the following $(k + 1)$ constraints will be added:

$$\sum_{j=1}^{k} x_j - x_0 \leq k-1$$

$$x_0 \leq x_j \text{ for all } j = 1,2,\ldots, k$$

3.2 Set Covering and Set Partitioning Models*

Set covering and set partitioning models are linear integer programs with binary variables. They have been successfully applied in warehouse location decisions, airline crew scheduling, aircraft scheduling, supply chain

---

* Portions of this section have been adapted with permission from Ravindran, A. R. and D. P. Warsing, Jr. 2013. *Supply Chain Engineering: Models and Applications.* Boca Raton, FL: CRC Press.

Design of Service Systems

network design, and package delivery problems. We shall first look at the formal mathematical statement of the set covering problem.

3.2.1 Set Covering Problem

Consider an $(m \times n)$ matrix A, called the *set covering matrix*, whose elements a_{ij}'s are either 0 or 1. If $a_{ij} = 1$, we say that column j "covers" row i. If not, $a_{ij} = 0$. The set covering problem is to select the minimum number of columns such that *every row* is covered by *at least one column*.

To formulate the set covering problem as an IP model, we define a binary variable for each column such that

$X_j = 1$ if column j is selected and 0 otherwise, for $j = 1, 2, 3, \ldots n$.

The IP model is given by

$$\text{Minimize } Z = \sum_{j=1}^{n} X_j \qquad (3.21)$$

Subject to

$$\sum_{j=1}^{n} a_{ij} X_j \geq 1 \quad \text{for } i=1, 2,\ldots, m \qquad (3.22)$$

$$X_j \in (0,1)$$

Constraints denoted by Equation 3.32 guarantee that every row is "covered" by at least one column. In other words, for row i, when at least one $a_{ij} = 1$, the corresponding X_j must be one. The objective function, given by Equation 3.21, guarantees that the minimum number of columns are selected to cover all the rows.

Example 3.3

Consider the set covering matrix given by

$$A = \begin{bmatrix} 1 & 1 & 0 & 1 & 1 & 1 & 0 \\ 0 & 1 & 0 & 0 & 1 & 1 & 1 \\ 1 & 0 & 1 & 1 & 0 & 1 & 0 \\ 0 & 0 & 1 & 0 & 1 & 0 & 1 \end{bmatrix}$$

The set covering problem becomes

$$\text{Minimize } Z = \sum_{j=1}^{7} X_j$$

Subject to

$$X_1 + X_2 + X_4 + X_5 + X_6 \geq 1 \quad (3.23)$$

$$X_2 + X_5 + X_6 + X_7 \geq 1 \quad (3.24)$$

$$X_1 + X_3 + X_4 + X_6 \geq 1 \quad (3.25)$$

$$X_3 + X_5 + X_7 \geq 1 \quad (3.26)$$

$$X_j \in (0,1) \quad \text{for } i = 1,2,\ldots,7$$

Equation 3.23 guarantees that row 1 will be covered by at least one column. Similarly, Equations 3.24, 3.25, and 3.26 guarantee that rows 2, 3, and 4 will be covered respectively by at least one column.

3.2.2 Set Partitioning Problem

In the set covering problem, every row has to be covered by *at least* one column. In the set partitioning problem, every row has to be covered by *exactly* one column. Otherwise, the two problems are the same. Thus, the only change in the IP model is that constraints given by Equation 3.22 will now become equalities:

$$\sum_{j=1}^{n} a_{ij} X_j = 1 \quad \text{for } i = 1,2,\ldots,m$$

The objective function given by Equation 3.21 remains the same.

3.2.3 Application to Warehouse Location

In applying the set covering model to the warehouse location problem in supply chain, we treat the potential warehouse locations as "columns" and the customer regions as "rows" of the set covering matrix A. We construct the matrix A, by setting its elements a_{ij} as follows:

Design of Service Systems

$$a_{ij} = \begin{cases} 1, & \text{if customer region } i \text{ can be supplied by warehouse location } j, \\ & \text{considering service level and distance criteria} \\ 0, & \text{otherwise} \end{cases}$$

By including the cost of building a warehouse at location j as K_j, we will minimize the total cost of building warehouses such that every customer region can be supplied by at least one warehouse. We will illustrate this with Example 3.4.

Example 3.4 (Ravindran et al. 1987)

A firm has four possible sites for locating its warehouses. The cost of locating a warehouse at site i is $\$K_i$. There are nine retail outlets, each of which must be supplied by at least one warehouse. It is not possible for any one site to supply all the retail outlets as shown in Figure 3.3. The problem is to determine the location of the warehouse such that the total cost is minimized.

SOLUTION

The warehouse location problem is basically a set covering problem. The first step is to define the *set covering matrix* (A) based on the network configuration shown in Figure 3.3. The rows of the matrix will be the nine retail outlets and the columns will be the four potential warehouse locations. The elements of matrix A, a_{ij}, will be set to 1 if retailer i (R_i) can be supplied by warehouse location j (W_j); that is, there is a direct link between R_i and W_j. Otherwise, we set $a_{ij} = 0$. For example, W_1 can supply $R_1, R_2, R_3, R_4,$ and R_5. Hence, we set $a_{11} = a_{21} = a_{31} = a_{41} = a_{51} = 1$ and $a_{61} = a_{71} = a_{81} = a_{91} = 0$. The complete set covering matrix A is given below:

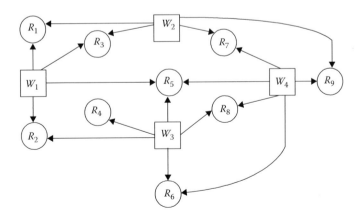

FIGURE 3.3
Supply chain network (Example 3.4).

Set Covering Matrix

$$A_{(9\times 4)} = \begin{array}{c} R_1 \\ R_2 \\ R_3 \\ R_4 \\ R_5 \\ R_6 \\ R_7 \\ R_8 \\ R_9 \end{array} \begin{bmatrix} W_1 & W_2 & W_3 & W_4 \\ 1 & 1 & 0 & 0 \\ 1 & 0 & 1 & 0 \\ 1 & 1 & 0 & 0 \\ 1 & 0 & 1 & 0 \\ 1 & 0 & 1 & 1 \\ 0 & 0 & 1 & 1 \\ 0 & 1 & 0 & 1 \\ 0 & 0 & 1 & 1 \\ 0 & 1 & 0 & 1 \end{bmatrix}$$

Model Formulation

Define binary variables X_j for $j = 1, 2, 3, 4$ such that

$$X_j = \begin{cases} 1, & \text{if site } j \text{ is selected for a warehouse location} \\ 0, & \text{otherwise} \end{cases}$$

The integer programming formulation of the warehouse selection problem becomes

$$\text{Minimize } Z = K_1 X_1 + K_2 X_2 + K_3 X_3 + K_4 X_4$$

Subject to

$X_1 + X_2 \geq 1$	(R_1)	
$X_1 + X_3 \geq 1$	(R_2)	
$X_1 + X_2 \geq 1$	(R_3)	(Redundant)
$X_1 + X_3 \geq 1$	(R_4)	(Redundant)
$X_1 + X_3 + X_4 \geq 1$	(R_5)	
$X_3 + X_4 \geq 1$	(R_6)	
$X_2 + X_4 \geq 1$	(R_7)	
$X_3 + X_4 \geq 1$	(R_8)	(Redundant)
$X_2 + X_4 \geq 1$	(R_9)	(Redundant)

Note that several of the set covering constraints are redundant and can be omitted before solving the integer program.

3.2.4 Application to Airline Scheduling

Before 1978, U.S. airlines were highly regulated with regard to routes and ticket prices. The Airline Deregulation Act of 1978 stimulated competition in the industry to benefit the customers. Because of deregulations, low-cost carriers entered the market, resulting in discount fares to passengers. To

Design of Service Systems

be competitive, airlines began using optimization models and methods to increase efficiency and reduce cost. Since an airline seat is one of the most perishable commodities in the world, American Airlines developed the revenue management system with dynamic pricing to predict and react to customer behavior to fill airline seats and maximize revenue. Revenue management methods will be discussed in detail in Chapter 8 of this textbook.

According to Snowdon and Paleologo (2008), scheduling of aircrafts and crews is one of the most complex problems faced by the airline industry. Two important airline scheduling applications that use set covering and partitioning models are as follows:

1. *Fleet Assignment*: assigning aircrafts to specific flights
2. *Crew Scheduling*: assignment of crews to specific flights

For both fleet assignment and crew scheduling problems, airlines use "flight segments" instead of flight numbers to denote specific flights. Each segment corresponds to a pair of takeoff and destination airports for a plane. For example, United Airlines flight 752 from Frankfurt to Chicago with a stopover in New York will be treated as two flight segments—the first segment is from Frankfurt to New York and the second segment is from New York to Chicago. The advantage of using flight segments, instead of flight numbers, is that both the plane and the crew can be changed when the plane is on the ground. In this example, a larger aircraft and crew can be used for the trans-Atlantic flight and a smaller aircraft and crew for the domestic route.

3.2.4.1 Airline Scheduling

Given a set of flight segments and aircrafts of different types, the fleet assignment problem is to determine which type of aircraft should be assigned to a specific flight segment in order to minimize the total cost of operations. The crew scheduling problem is to determine the optimal allocation of crews to flight segments to minimize cost, taking into account several operational constraints, including FAA regulations and union rules.

3.2.4.2 Use of Set Covering and Partitioning Model

For both fleet assignment and crew scheduling, a set covering matrix A has to be developed. The rows of the matrix will be flight segments and the columns are different aircraft or crew schedules depending on the problem. To illustrate the development of the set covering matrix for the crew scheduling problem, consider the flight data given in Table 3.2. Using the data in Table 3.2, a partial list of crew schedules for the set covering matrix is given in Table 3.3, where the rows are flight segments and columns are feasible crew schedules.

TABLE 3.2
Sample Data for Flight Segments

Flight #	Flight Segment	Origin–Destination
345	1	JFK–DFW
345	2	DFW–ORD
346	3	ORD–ATL
347	4	ORD–JFK
348	5	DFW–JFK
349	6	ATL–JFK

Note: JFK—New York; DFW—Dallas; ORD—Chicago; ATL—Atlanta.

TABLE 3.3
Sample Crew Schedules

	S1	S2	S3
1	1	1	1
2	1	0	1
3	0	0	1
4	1	0	0
5	0	1	0
6	0	0	1
Cost	C_1	C_2	C_3

For example, the first schedule (column 1 in Table 3.3) refers to a New York–based crew traveling from JFK to DFW and ORD by flight 345 and returning to New York the same day by flight 347. The cost of this feasible schedule is C_1. Because of FAA regulations on how long a crew can work, schedule 3 (column 3 in Table 3.3) may involve a long layover for the crew either in Chicago or Atlanta. This schedule will be more expensive since it requires overnight stay in a hotel for the crew. The complete set covering matrix A is then developed by generating all possible feasible schedules for the crews. Once all feasible crew schedules are developed (matrix A), the crew scheduling problem can be formulated as a set covering/partitioning problem.

3.2.4.3 Crew Scheduling Model

Given a set of m flight segments and n crew schedules, the set covering matrix is given by an $(m \times n)$ matrix, denoted by $A = [a_{ij}]$, where

$a_{ij} = 1$, if flight segment i is covered by crew schedule j and 0 otherwise.

Let C_j denote the cost of crew schedule j.

Design of Service Systems

Decision Variables

$$X_j = 1, \text{ if crew schedule } j \text{ is chosen}$$
$$= 0, \text{ otherwise, for } i = 1,\ldots m \text{ and } j = 1,\ldots n.$$

Constraints

$$\sum_{j=1}^{n} a_{ij} X_j \geq 1 \text{ for } i = 1, 2, \ldots m \qquad (3.27)$$

Objective Function

$$\text{Minimize } Z \sum_{j=1}^{n} C_j X_j$$

Equation 3.27 guarantees that every flight segment has an assigned crew. It allows the possibility of more than one crew to be assigned to a flight segment. In that case, one of the crews is working in the flight and the other crew is traveling as passengers. If that is not allowed, then Equation 3.27 can be written as strict equalities and the problem becomes a set partitioning model. For the fleet assignment problem, exactly one aircraft has to be assigned to each flight segment, and hence, it would be modeled as a set partitioning problem.

While developing the feasible crew schedules (columns of the A matrix), other operational constraints can be imposed as follows:

- Minimum connecting time between flights segments
- Sufficient layover time for crews
- Hours of flying time
- Total length of a crew schedule
- Crews returning to their home bases within 1–5 days
- Other FAA and union rules

In addition, if schedules 1, 2, and 3 in Table 3.3 refer to the same crew out of New York, then the following constraint can be added:

$$X_1 + X_2 + X_3 = 1$$

so that only one of the three schedules is selected. Similarly, if there are only k crews available ($k < n$), then we can add the following constraint:

$$\sum_{j=1}^{n} X_j \leq k.$$

3.2.4.4 Airline Applications

Successful applications of set covering and set partitioning models have been reported by several airlines. For example, Delta Airlines, before it merged with Northwest, had 2500 domestic flight segments daily and 450 planes of different types. The integer programming model had 40,000 constraints and 20,000 binary variables. Use of the optimal fleet assignment model saved Delta $300 million over 3 years (Subramanian et al. 1994).

Abara (1989) reported an application of integer programming model for fleet assignment at American Airlines, which handled 2300 daily flights, covering 150 cities, utilizing 500 aircrafts. Use of IP models resulted in increasing the utilization of bigger aircrafts for high demand flight segments from 76% to 90% and reducing the operating cost by 0.5%. This was equivalent to a savings of $75 million annually. Anbil et al. (1991) also used IP models for crew assignment at American Airlines. Called the Trip Evaluation and Improvement Program (TRIP), the IP model developed monthly crew schedules taking into account FAA and union work rules. At that time, the airline employed more than 8000 pilots and 16,000 flight attendants at a cost of $1.3 billion annually. The use of TRIP saved the airlines over $20 million annually.

United Parcel Service (UPS) has used the fleet assignment model for scheduling its aircrafts in the United States for its next-day air service (Armacost et al. 2004). This will be discussed in Section 3.2.6.2. General integer programming models have also been applied successfully for manpower planning by airlines. They will be discussed in Sections 3.2.5 and 3.2.6.

3.2.5 Workforce Planning

Workforce planning is to determine the optimal number of workers to have at different periods to meet customer demand. This is a complex problem, particularly in service industries, where simultaneous presence of customer and worker is generally required and service is perishable. For example, in fast-food restaurants, it is essential to have sufficient number of workers at different times of the day, where the customer demand may vary by time of the day and day of the week. Insufficient number of workers would result in long lines, customer dissatisfaction, and loss of business. More workers lead to idle time and cost. An essential requirement for workforce planning is a good demand forecast. Using the forecast, integer programming models have been successfully used

Design of Service Systems

to determine the optimal number of workers to employ at different time periods. We shall illustrate this with an example of scheduling nurses in a hospital.

Example 3.5 (Nurse Scheduling Problem)

A hospital administrator has minimal daily requirement for nursing personnel at different time periods as shown in Table 3.4. Nurses can report to the hospital wards at the beginning of each period and work for 8 consecutive hours. The hospital wants to determine the minimal number of nurses to employ so that there will be a sufficient number of nursing personnel available for each period.

SOLUTION

Decision Variables

x_j—Number of new nurses reporting to work at the beginning of period j; $j = 1, 2, \ldots, 6$

Constraints

For any period, the total number of nurses working (new plus carried over from the previous period) should be at least equal to the minimum required. Thus, we get the following six constraints, one for each period.

(Period 1)	$x_6 + x_1 \geq 60$
(Period 2)	$x_1 + x_2 \geq 70$
(Period 3)	$x_2 + x_3 \geq 60$
(Period 4)	$x_3 + x_4 \geq 50$
(Period 5)	$x_4 + x_5 \geq 20$
(Period 6)	$x_5 + x_6 \geq 30$

$x_1, x_2, x_3, x_4, x_5, x_6$ are non-negative integers.

Objective Function: The total number of nurses employed by the hospital is to be minimized.

$$\text{Minimize } Z = x_1 + x_2 + x_3 + x_4 + x_5 + x_6$$

TABLE 3.4

Nurse Scheduling Problem (Example 3.5)

Period	Clock Time (24-Hour Day)	Minimal Number of Nurses Required
1	6 a.m. to 10 a.m.	60
2	10 a.m. to 2 p.m.	70
3	2 p.m. to 6 p.m.	60
4	6 p.m. to 10 p.m.	50
5	10 p.m. to 2 a.m.	20
6	2 a.m. to 6 a.m.	30

Optimal Solution

The optimal solution to the above integer program is given below:

$$x_1 = 60,\ x_2 = 10,\ x_3 = 50,\ x_4 = 0,\ x_5 = 30,\ x_6 = 0$$

$$\text{Minimum } Z = 150$$

NOTE: The above model can be extended to multiple days (weekly scheduling) and the use of part-time nurses, who may work for less than 8 hours, as discussed in Chapter 10.

The nurse scheduling model can also be applied to workforce planning problems in other industries. United Airlines successfully applied this model to improve the utilization of personnel at its reservation offices and airports by matching work schedules to customer needs more closely (Holloran and Byrn 1986). Taco Bell developed an integrated labor-management system to determine the number of workers needed in its restaurants at different times (Hueter and Swart 1998). Demand varied widely during the day, with 52% of sales occurring at the lunch period (11 a.m. to 2 p.m.). First, a forecast of lunch sales over 15-minute intervals, by time of the day and day of the week, was developed using the Moving Average method. The forecast was then used as an input to an integer programming model for estimating the labor needs. Taco Bell achieved labor savings of over $40 million over 3 years by using the integrated labor-management system.

3.2.5.1 Workforce Planning Applications

The workforce scheduling model has been applied in several industries. A partial list is given below:

- Airlines
- Fast-food restaurants
- Call centers
- Retailers
- Hospitals

3.2.6 Real-World Applications

There are several published results of real-world applications using integer programming models. We briefly discuss a few of the applications in practice. For interested readers, the cited references will provide more details on the case studies.

3.2.6.1 Ford Motor Company

Before introducing new model cars into the market, automobile companies go through a time-consuming and expensive process of developing prototypes of the model and subjecting them to multiple tests to check for any design flaws. Since the prototypes are generally one of a kind, they are very expensive to make and may cost more than $250,000 per prototype. Complex vehicle design programs may require between 100 and 200 prototypes for product development. Naturally, it is imperative to keep this cost down but, at the same time, have enough prototype vehicles to conduct all the required tests. Ford, with the help of the Engineering Management students at Wayne State University in Detroit, developed a Prototype Optimization Model (POM) to solve this problem.

POM was an integer programming model based on the set covering model formulation we discussed in Section 3.2.1. The rows of the set covering matrix are the various tests that need to be conducted, while the columns represent the different buildable vehicle configurations (prototypes). The elements of the set covering matrix, called *Buildable Combination Matrix* (BCM), indicated whether a particular test can be done on a certain vehicle configuration. The objective function is to determine the minimum number of prototypes needed to complete all the required tests.

The set covering model was applied to the European Transit Vehicle program. Even though the BCM matrix initially contained 38,800 columns, the POM's optimal solution identified just 27 prototypes needed to cover all the required vehicle tests in the design-verification plan, resulting in a cost savings of $12 million. Based on that success, Ford used POM to manage other complex vehicle design programs for Taurus, Windstar, Explorer, and Ranger. The total cost savings achieved for prototype vehicles were $250 million from 1995 to 2000.

Interested readers should refer to the paper by Chelst et al. (2001) for more details on this case study.

3.2.6.2 United Parcel Service

Began as the American Messenger Service in Seattle in 1907, UPS was established in 1919 to deliver packages, first in the United States and then worldwide. In 2016, UPS delivered more than 19 million packages and documents daily to more than 220 countries and territories. UPS competes with FedEx and others to provide next-day air express delivery service to its customers. The daily U.S. air volume is over 2.7 million packages and documents, which contributes to the total revenue of $51 billion for UPS in 2016. Figure 3.4 illustrates UPS's next-day air service.

UPS has more than 300 aircrafts of nine different types. It serves 374 domestic airports (including 7 hubs) and 313 international airports. With the tight pickup and delivery window for the next-day air service, the

116　　*Service Systems Engineering and Management*

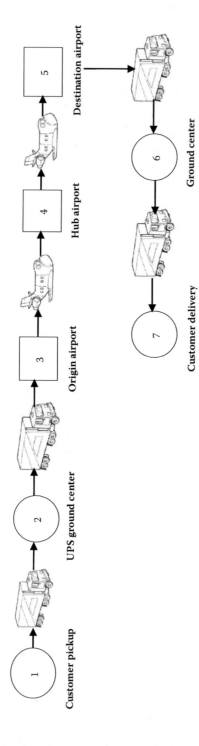

FIGURE 3.4
UPS's next-day air service.

Design of Service Systems

planning of the aircraft routes is a very complex problem. For example, each type of aircraft has different cargo capacity, maximum flying range and speed, and can only fly to certain airports. The maximum number of airports a plane can serve is two (excluding the hub). Until 1993, UPS used manual planners, who used to take 9 months to produce a single plan! In 1994, UPS's Operations Research team, working jointly with the MIT faculty, developed an optimization model, called *VOLCANO—Volume, Location and Aircraft Network Optimizer* (Armacost et al. 2004).

VOLCANO was basically an integer program, based on the set partitioning model discussed in Section 3.2.2. The "rows" of the set covering matrix were the "flight legs." Each flight leg was a pair of takeoff and destination airports for a plane. UPS flies an average of 2167 flight legs (1077 domestic and 1090 international). The "columns" of the set covering matrix were the feasible routes for the different aircrafts. There could be more than one feasible route for each aircraft based on its cargo capacity, speed, flying range, permissible airports, and so on. The elements of the set covering matrix, a_{ij}'s, were set to one, if flight leg i could be served by route j and zero otherwise. The cost of each feasible route was also determined. The objective was to cover all the flight legs such that the total cost of the aircraft routes was minimized. The set partitioning problem was solved using the commercial optimization software, ILOG-CPLEX solver.

Use of VOLCANO saved UPS more than $87 million between 2000 and 2002. UPS estimated additional savings of more than $189 million for the next 10 years.

3.2.6.3 Wishard Memorial Hospital

Arthur and Ravindran (1981) present an application of integer programming for scheduling nurses for the Wishard Memorial Hospital in Indianapolis, Indiana. The nurse scheduling problem was to determine the days and shifts when each member of the nursing staff was to report to work in a 2-week scheduling horizon. This is basically an extension of the workforce planning model illustrated in Example 3.5. The hospital had three nurse classes—Registered Nurse (RN), Licensed Practical Nurse (LPN), and Nurse's Aide (AIDE). Each nurse worked full time, five 8-hour shifts a week, and got every other weekend off. Nurses' preferences for the day and shift were also incorporated in the model.

The scheduling problem was solved in two phases because of computational complexity. In the first phase, the nurses were assigned their day-on/day-off pattern for the 2-week scheduling horizon. For this, a multiple objective integer programming model with binary variables was developed. The constraints included minimum staffing needs, desired staffing needs, nurses' preferences, and special requests. The second phase took the optimal day-on/day-off pattern generated by the IP model and made specific shift assignments by a heuristic method.

3.2.6.4 Mount Sinai Hospital

Blake and Donald (2002) report an actual application of integer programming for scheduling operating rooms at the Mount Sinai Hospital in Toronto. The hospital had 14 operating rooms that served five surgical departments—Surgery, Gynecology, Ophthalmology, Otolaryngology, and Oral Surgery. The Department of Surgery included five subspecialties: Orthopedics, General Surgery, Plastic Surgery, Vascular Surgery, and Urology. A daily operating schedule included the surgeries to be performed with their start and end times and names of the surgeons and anesthetists. An integer programming model was used to develop the weekly master schedule for each day of the week. The hospital had nearly 400 hours of operating time per week. The Department of Surgery was the largest user consuming nearly half of the total operating time. The hospital had targets for hours assigned to each of the five major surgical department. The objective was to minimize the deviations between the target assignments and the actual assignment times.

Before the use of the integer programming model, the master schedule used to be done manually taking several days by a committee of surgeons, anesthetists, and the nurse manager. It took just 1–2 hours to get the optimal schedule by the IP model. It resulted in labor savings of over $20,000 per year. Moreover, the schedules obtained by the IP model were perceived to be fair, objective, and more flexible by the surgical departments.

3.2.6.5 United Airlines

United Airlines successfully applied integer programming models, called the Station Manpower Planning System (SMPS), to improve the utilization of personnel at its reservation offices and airports by matching work schedules to customer needs more closely (Holloran and Byrn 1986). SMPS developed the work schedules for United's personnel as listed below:

- 4000 phone reservation operators and support personnel working 24/7 at 11 offices
- 1000 customer service agent at 10 largest airports
- Full-time workers with 8- and 10-hour shifts
- Part-time workers with 2- to 8-hour shifts

Based on the passenger traffic, there were considerable variations in demand during the 24-hour day. After forecasting the demand for 30-minute intervals, United used the workforce planning model to determine the number of employees of each shift length (full time and part time) to report to work at different time periods, with the objective of minimizing the total number of employees needed and the labor cost. The integer programming model had more than 20,000 variables and millions of LP matrix coefficients. The IP model was solved using IBM's MPSX and MIP/370 Optimization

Design of Service Systems

software. The model was run monthly to determine the new work schedules, saving United more than $6 million annually in direct salary and benefits.

3.3 Queuing Models in Service Systems

In this section, we present models that can be used to characterize randomness in a process. Specifically, we present models that characterize waiting lines, called queues. We will first present an introduction to some basic queuing models and then apply them to the service setting of staffing at a call center. Note that we only present a few basic queuing results here. The reader interested in more complete coverage of queuing may consult a text such as Gross et al. (2009).

3.3.1 Introduction

The basic structure of a standard queuing process is shown in Figure 3.5. Customers (represented as squares) arrive to the system for service according to a known arrival distribution. The service process is made up of n servers (represented as black circles) that have independent and identically distributed (i.i.d.) service times. If all of the servers in the service process are serving customers, the customers wait in a queue. Customers waiting in a queue are served according to a queue discipline; the most common being first come first served (FCFS). In Figure 3.5, the bottom server has just completed service on a customer, and so the customer at the front of the queue will be served next by that server.

There are many variations on this standard model. For example, the queue may have limited capacity. In this case, a customer arriving when the queue

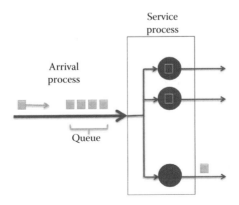

FIGURE 3.5
A standard queuing model.

is at capacity will exit the system without being served. Further, servers may work at a faster rate as the queue length increases (called state-dependent service). The system may also support customers of different types, each type requiring different levels of service.

We will use the following notation to define a queuing system: A1/A2/A3/A4/A5. In this case, A1 represents the distribution of the interarrival times of customers. The most commonly used distributions are exponential (M),* deterministic (D), and general (G). A2 represents the service time distribution, with the same distribution symbols as for interarrivals. A3 represents the number of parallel servers (1, 2, ..., ∞). A4 represents the restriction on capacity (1, 2, ..., c, ..., ∞). The default for A4 is ∞ (i.e., no restriction). Finally, A5 represents the queue discipline. The most common are FCFS, last come first served (LCFS), priority discipline (PD), and general discipline (GD). The default for A5 is FCFS. For example, M/G/3 means that customer interarrival times are exponentially distributed, service times are according to a general distribution, there are three parallel servers, there is no capacity restriction, and the priority discipline is FCFS.

3.3.2 Poisson Processes

Fundamental to a queuing system is characterizing arrivals and services. A Poisson process is one that counts the number of events that occur over time. We will let $N(t)$ represent the number of events that occurs in time interval $[0,t]$. There are several important assumptions in a Poisson process:

I. $N(0) = 0$.
II. $N(t)$ is integer valued and non-decreasing in t and is right continuous.
III. The number of occurrences in disjoint intervals are independent of each other (i.e., if $(t_i, t_j] \cap (t_k, t_l] = \emptyset$ then $N(t_j) - N(t_i)$ is independent of $N(t_k) - N(t_l)$).
IV. The distribution of $N(t + h) - N(t)$, where $h > 0$ is independent of t.
V. The probability of simultaneous occurrences is zero (i.e., $P[N(t + h) - N(t) \geq 2] = o(h^2)$ as $h \to 0$).

The arrival process is illustrated in Figure 3.6. In this figure, we define the time gap between arrival $j - 1$ and j as τ_j. It may be shown (proof omitted) under the stated assumptions that the following theorem holds.

THEOREM 3.1

Under the given assumptions for the Poisson process, the following properties hold:

---

* The use of M is traditional, and based on the Markovian property of the exponential distribution, which is discussed in Section 3.3.2.

Design of Service Systems

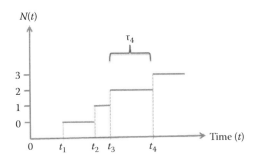

FIGURE 3.6
Arrivals in a Poisson process.

a. $N(t)$ is a step function that increases in steps of size 1.
b. There exists a number $\lambda > 0$, where $N(t + s) - N(s)$ has a Poisson distribution with parameter λt. That is,

$$P[N(t+s) - N(t) = k] = \frac{e^{-\lambda s}(\lambda s)^k}{k!}, \quad k = 0, 1, \ldots \quad (3.28)$$

where k is the number of arrivals in time interval $(t, t + s]$.

c. The time gaps (τ_j) between successive arrivals are i.i.d. random variables with exponential distribution. That is,

$$P\{\tau_j > t\} = e^{-\lambda t}, \quad t \geq 0. \quad (3.29)$$

■

Theorem 3.1 shows an important relationship between the Poisson and exponential distributions, namely, that interarrival times of a Poisson process are exponentially distributed. The exponential distribution has the very important *memoryless property*,

$$P(X - y > x | X > y) = P(X > x), \quad x \geq 0, y \geq 0. \quad (3.30)$$

For example, suppose the time between arrivals of a bus stop outside a student's apartment building is exponentially distributed. The memoryless property implies that the probability that a bus arrives in the next 5 minutes is independent of how long the student has been waiting for the bus.

3.3.3 The M/M/1 Model

Consider the case where service and interarrival times to a single server queuing system are exponentially distributed with rates μ and λ, respectively.

We can describe this system through the use of a birth–death process as shown in Figure 3.7. In the figure, the system states are represented as circles. For example, the circle with a 3 represents 3 customers in the system. This implies that 1 customer is being served while 2 are waiting in the queue. We move to state 4 if an arrival occurs before the service is completed and to state 2 if the service is completed before an arrival.

We can solve for the steady-state probability of being in a state through the use of balance equations. Namely, the probability of being in state i (P_i) is defined by the flows into and out of that state. The balance equations for the first three states are as follows:

$$\lambda P_0 = \mu P_1 \tag{3.31}$$

$$(\lambda + \mu) P_1 = \lambda P_0 + \mu P_2 \tag{3.32}$$

$$(\lambda + \mu) P_2 = \lambda P_1 + \mu P_3 \tag{3.33}$$

We can rewrite Equation 3.31 as

$$P_1 = \frac{\lambda}{\mu} P_0. \tag{3.34}$$

Substituting into Equation 3.32 gives

$$(\lambda + \mu) \frac{\lambda}{\mu} P_0 = \lambda P_0 + \mu P_2 \rightarrow P_2 = \left(\frac{\lambda}{\mu}\right)^2 P_0. \tag{3.35}$$

In general, we get

$$P_i = \left(\frac{\lambda}{\mu}\right)^i P_0. \tag{3.36}$$

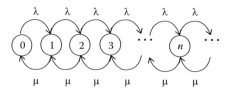

FIGURE 3.7
Birth–death process for an M/M/1 queuing system.

Design of Service Systems

Further, we know that the sum of the probabilities must equal 1. This gives

$$\sum_{i=0}^{\infty} \left(\frac{\lambda}{\mu}\right)^i P_0 = 1. \tag{3.37}$$

Solving Equation 3.37 for P_0 gives

$$P_0 = 1 - \frac{\lambda}{\mu}. \tag{3.38}$$

Note that we have made an assumption in this analysis; namely, we have assumed that the arrival rate is strictly less than the service rate. If this assumption does not hold, then over time, the system will move to the state of ∞; that is, the system will become unstable. For the M/M/1 system, we will define ρ as the ratio of arrival rate to service rate, which is known as the system utilization.

Since we have defined the probability of being in any state completely in terms of parameters λ and μ, we can develop service measures for the system. For example, the average number of customers in the system (L) may be determined through the following expectation:

$$\begin{aligned} L &= \sum_{i=0}^{\infty} iP_i = \sum_{i=0}^{\infty} iP^i(1-\rho) = (1-\rho)\rho\frac{d}{d\rho}\sum_{i=0}^{\infty}\rho^i \\ &= (1-\rho)\rho\frac{d}{d\rho}\left(\frac{1}{1-\rho}\right) = \frac{\rho}{1-\rho} \end{aligned} \tag{3.39}$$

A very useful relationship for queuing systems is *Little's law*. Little's law states that under steady-state conditions, the average number of customers in a queuing system equals the average rate at which customers arrive multiplied by the average time a customer spends in the system (W). That is,

$$L = \lambda W. \tag{3.40}$$

Little's law is quite general, and we will make use of it several times throughout this chapter. We can use it directly to determine the average waiting time in the system by applying it to Equation 3.39 as shown below:

$$W = \frac{L}{\lambda} = \frac{1}{\mu - \lambda}. \tag{3.41}$$

The average waiting time in the queue (W_q) equals the average waiting time in the system minus the average service time. This gives

$$W_q = W - \frac{1}{\mu} = \frac{\lambda}{\mu(\mu - \lambda)}. \tag{3.42}$$

Applying Little's law to Equation 3.42 gives the average number of customers in the queue (L_q):

$$L_q = \lambda W_q = \frac{\lambda^2}{\mu(\mu - \lambda)}. \tag{3.43}$$

It is worth noting that all of the results presented to this point in this section do not depend on the queue discipline. If we do restrict the queue discipline to be FCFS, we can develop the waiting time distribution for the M/M/1 system. Let T_i be the service time for customer i. We know from Theorem 3.1 that these are i.i.d. exponential. Let S be the conditional waiting time given that there are already n customers in the system:

$$S = T_1 + T_2 + \cdots + T_{n+1}, \tag{3.44}$$

Further, define $w(S|n+1)$ as the conditional density function of S given n customers already in the system. The sum of n i.i.d. exponential random variables follows a gamma distribution with parameters μ and n. We can therefore define the waiting time distribution by

$$\begin{aligned} w(S) &= \sum_{i=0}^{\infty} w(S|i+1) P_i \\ &= \sum_{i=0}^{\infty} \frac{\mu(\mu S)^i e^{-\mu S}}{i!} \left(1 - \frac{\lambda}{\mu}\right)\left(\frac{\lambda}{\mu}\right)^i. \end{aligned} \tag{3.45}$$

After some manipulation, we obtain

$$w(S) = (\mu - \lambda) e^{-(\mu - \lambda) S}. \tag{3.46}$$

That is, $w(S)$ is exponentially distributed with mean $1/(\mu - \lambda)$. The waiting time distribution is useful in developing service measures on the system. In particular, we can determine the proportion of customers whose waiting time exceeds a given value Z. That is,

Design of Service Systems

$$P(S > Z) = 1 - \int_0^Z w(S)dS = e^{-(\mu-\lambda)Z}. \quad (3.47)$$

Note that the cumulative waiting time distribution in the queues $W_q(S)$ may be derived similarly. In this case:

$$W_q(S) = \begin{cases} 1-\rho, & S = 0 \\ 1-\rho e^{-\mu(1-\rho)S}, & S \ne 0 \end{cases}. \quad (3.48)$$

Example 3.6

Consider a retail store with a single cashier. Customers arrive according to a Poisson process at a rate of $\lambda = 10$ per hour. The service time of the cashier is exponentially distributed with a rate of $\mu = 11$ per hour. We are interested in finding (i) the average queue length, (ii) the average time in the system for a customer, and (iii) the proportion of customers that need to wait at least 20 minutes.

From Equation 3.43, we get that

$$L_q = \frac{10^2}{11(11-10)} = 9.09 \text{ customers.}$$

Further, Equation 3.41 gives

$$W = \frac{1}{11-10} = 1 \text{ hour.}$$

Finally, Equation 3.47 yields

$$P\left(S > \frac{20}{60}\right) = e^{-(11-10)\frac{20}{60}} = 0.717.$$

3.3.4 The M/M/c Model

We extend the model in Section 3.3.4 to include c servers. It is assumed that the servers work in parallel and that they are independent of one another. In this case, the birth–death process is shown in Figure 3.8. Notice that the service transitions increase up to state c, beyond which the rate stays at $c\mu$.

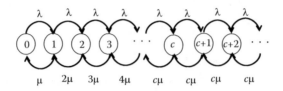

FIGURE 3.8
Birth–death process for an M/M/c queuing system.

The utilization (ρ) of this system is defined as $\lambda/c\mu$, and we require that $\rho < 1$ for stability. The steady-state probabilities can be determined by the balance equations as done in the previous sections. We leave the details to a practice problem at the end of this chapter and simply present the results here.

$$P_0 = \left[\sum_{i=0}^{c-1} \frac{\rho^i}{i!} + \frac{\rho^c}{c!}\left(\frac{c\mu}{c\mu - \lambda}\right) \right]^{-1} \tag{3.49}$$

$$P_n = \begin{cases} \dfrac{\rho^n}{n!} P_0, & 1 \leq n < c \\[6pt] \dfrac{\rho^n}{c^{n-c} c!} P_0, & n \geq c \end{cases} \tag{3.50}$$

$$L_q = \left[\frac{\rho^c \lambda \mu}{(c-1)!(c\mu - \lambda)^2} \right] P_0 \tag{3.51}$$

$$W_q = \frac{L_q}{\lambda} \tag{3.52}$$

$$L = \rho + L_q \tag{3.53}$$

$$W = \frac{L}{\lambda} \tag{3.54}$$

$$w(S) = \frac{\mu e^{-\mu S}\left(\lambda - c\mu + \mu W_q(0)\right) - \left(1 - W_q(0)\right)(\lambda - c\mu)\mu e^{-(c\mu - \lambda)S}}{\lambda - (c-1)\mu} \tag{3.55}$$

Design of Service Systems

$$W_q(S) = \left[\frac{\rho^c(1-e^{-(\mu c-\lambda)S})}{(c-1)!(c-\rho)}\right]P_0 + \left[1 - \frac{c\rho^c}{c!(1-\rho)}P_0\right] \quad (3.56)$$

$$W_q(0) = 1 - \frac{c\left(\dfrac{\lambda}{\mu}\right)^c}{c!\left(c - \dfrac{\lambda}{\mu}\right)}P_0 \quad (3.57)$$

Example 3.7

Suppose that a second cashier is added to the previous example, and all other factors remain the same. Determine the percent decrease in average queue length as well as the probability that a customer waits in the queue no more than 20 minutes. In this case, $\rho = 10/[2(11)] = 0.455$.

From Equation 3.49, we get that

$$P_0 = \left[1 + \frac{.455}{1!} + \frac{.455^2}{2!}\left(\frac{2(11)}{2(11)-10}\right)\right]^{-1} = 0.608.$$

Further, Equation 3.51 gives

$$L_q = \frac{.455^2(10)(11)}{1!(2(11)-10)^2} = 0.158 \text{ customers}$$

This is a significant reduction compared to Example 3.6. Finally, Equation 3.47 yields

$$W_q\left(\frac{20}{60}\right)P\left(S \le \frac{20}{60}\right) = \frac{0.455^2\left(1-e^{-(2(11)-10)(0.33)}\right)}{1!(2-0.455)}(0.608)$$

$$+ \left[1 - \frac{2(0.455^2)(0.608)}{2!(1-0.455)}\right] = 0.849.$$

3.3.5 The M/M/c/K Model

In many situations, there is a limit (K) to the total number of customers allowed in the system. An example that we will consider later in this chapter is an intensive care unit, which has a fixed number of beds. We assume that when the system is full, an arriving customer is turned away from the system. This is shown in Figure 3.9. Note that for this system, we do not put limits on the arrival rate for stability, since the system naturally stabilizes itself by turning away customers.

For this case, the performance measures for the system are given as follows. As in the previous section, we present the results without derivation. It is important to note that for this system, arrivals are not truly a Poisson process due to truncation. Bayes theorem was therefore used in the derivation of the waiting time in the queue.

$$P_0 = \begin{cases} \left[\dfrac{\left(\dfrac{\lambda}{\mu}\right)^c \left[1 - \left(\dfrac{\lambda}{c\mu}\right)^{K-c+1}\right]}{c!\left(1 - \dfrac{\lambda}{c\mu}\right)} + \sum_{i=0}^{c-1} \dfrac{1}{i!}\left(\dfrac{\lambda}{\mu}\right)^i \right]^{-1}, & \dfrac{\lambda}{c\mu} \neq 1 \\[2em] \left[\dfrac{\left(\dfrac{\lambda}{\mu}\right)^c [K - c + 1]}{c!} + \sum_{i=0}^{c-1} \dfrac{1}{i!}\left(\dfrac{\lambda}{\mu}\right)^i \right]^{-1}, & \dfrac{\lambda}{c\mu} = 1 \end{cases} \qquad (3.58)$$

$$P_n = \begin{cases} \dfrac{1}{n!}\left(\dfrac{\lambda}{\mu}\right)^c P_0, & 1 \leq n < c \\[1em] \dfrac{1}{c^{n-c}c!}\left(\dfrac{\lambda}{\mu}\right)^n P_0, & c \leq n \leq K \end{cases} \qquad (3.59)$$

$$L_q = \dfrac{P_0 \left(\dfrac{\lambda}{\mu}\right)^c \left(\dfrac{\lambda}{c\mu}\right)}{c!\left(1 - \dfrac{\lambda}{c\mu}\right)^2} \left[1 - \left(\dfrac{\lambda}{c\mu}\right)^{K-c+1} - \left(1 - \dfrac{\lambda}{c\mu}\right)(K - c + 1)\left(\dfrac{\lambda}{c\mu}\right)^{K-c} \right] \qquad (3.60)$$

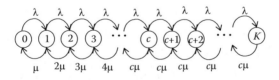

FIGURE 3.9
Birth–death process for an M/M/c/K queuing system.

Design of Service Systems

$$L = L_q + c - \sum_{i=0}^{c-1}(c-i)P_i \qquad (3.61)$$

$$W = \frac{L}{\lambda(1-P_K)} \qquad (3.62)$$

$$W_q = W - \frac{1}{\mu} \qquad (3.63)$$

$$W_q(S) = 1 - \sum_{i=c}^{K-1} \frac{P_i}{1-P_K} \sum_{j=0}^{i-c} \frac{(c\mu S)^j e^{-c\mu S}}{j!} \qquad (3.64)$$

3.3.6 The G/G/c Model

To this point, we have assumed that interarrival times and service times are exponentially distributed. In practice, this assumption may be overly restrictive. Unfortunately, such systems are quite difficult to characterize in closed form.

One exception is the M/G/1 system, where the well-known *Pollaczek–Khintchine* (P–K) formulas apply. In this case, we assume that the service process can be characterized by an average time μ and standard deviation σ_S. The P–K formulas are given below:

$$L = \frac{\lambda}{\mu} + \frac{\left(\frac{\lambda}{\mu}\right)^2 + \lambda^2 \sigma_S^2}{2\left(1 - \frac{\lambda}{\mu}\right)} \qquad (3.65)$$

$$L_q = L - \frac{\lambda}{\mu} \qquad (3.66)$$

$$W = \frac{L}{\lambda} \qquad (3.67)$$

$$W_q = W - \frac{1}{\mu} \qquad (3.68)$$

In order to model the more general G/G/c queuing system, we follow the approximation presented in Hopp and Spearman (2011). We assume that the arrival and service rates can be characterized by their mean and standard deviation. Further, the coefficient of variation of arrivals $\left(C_a^2\right)$ and service $\left(C_s^2\right)$ is a standardized measure of variation, which is given as follows:*

$$C_a^2 = \left(\frac{\sigma_a}{1/\lambda}\right)^2 \tag{3.69}$$

$$C_s^2 = \left(\frac{\sigma_s}{1/\mu}\right)^2 \tag{3.70}$$

The average waiting times are then given by the following approximations:

$$W_q = \left(\frac{C_a^2 + C_s^2}{2}\right)\left(\frac{\left(\frac{\lambda}{c\mu}\right)^{\sqrt{2(c+1)}-1}}{c\left(1-\frac{\lambda}{c\mu}\right)}\right)\frac{1}{\mu} \tag{3.71}$$

$$W = W_q + \frac{1}{\mu} \tag{3.72}$$

The corresponding number of customers in the system and in the queue may be found by applying Little's law. We illustrate by the following example.

Example 3.8

Interarrival times of customers using a bank with 4 tellers is on average 1 minute with a standard deviation of 1.3 minutes. Service times are 3 minutes on average with a standard deviation of 6 minutes. We can determine the average waiting time in the system as follows. The utilization of the system is given by $u = (60/((4)(20)) = 0.75$. From Equations 3.69 and 3.70, we have

$$C_a^2 = \left(\frac{1.3}{1}\right)^2 = 1.69$$

$$C_s^2 = \left(\frac{6}{4}\right)^2 = 2.25$$

---

* Note that for the exponential distribution, the mean and standard deviation are equal, and hence the squared coefficient of variation is 1.0.

Design of Service Systems

Equation 3.72 gives

$$W_q = \left(\frac{1.69 + 2.25}{2}\right)\left(\frac{(0.75)^{\sqrt{2(4+1)}-1}}{4(1-0.75)}\right)(3) + 3 = 6.17 \text{ minutes.}$$

Whitt (1993) showed that the above G/G/c approximation works well when $\lambda/c\mu \to 1$. When this does not hold, he developed the following improved approximation:

$$W_q = \phi\left(\frac{C_a^2 + C_s^2}{2}\right)\left(\frac{u^{\sqrt{2(c+1)}-1}}{c(1-u)}\right)\frac{1}{\mu} \tag{3.73}$$

where $u = \lambda/c\mu$ and

$$\phi = \begin{cases} \left(\dfrac{4(C_a^2 + C_s^2)}{4C_a^2 - 3C_s^2}\right)\phi_1 + \left(\dfrac{C_s^2}{4C_a^2 - 3C_s^2}\right)\psi, & C_a^2 > C_s^2 \\[2mm] \left(\dfrac{(C_a^2 + C_s^2)}{2C_a^2 + 2C_s^2}\right)\phi_3 + \left(\dfrac{C_s^2 + 3C_a^2}{2C_a^2 + 2C_s^2}\right)\psi, & C_a^2 \leq C_s^2 \end{cases} \tag{3.74}$$

$$\phi_1 = 1 + \gamma \tag{3.75}$$

$$\gamma = \min\left\{0.24, \frac{(1-u)(c-1)[(4+5c)^{0.5} - 2]}{16cu}\right\} \tag{3.76}$$

$$\psi = \begin{cases} 1, & \left(\dfrac{C_a^2 + C_s^2}{2}\right) \geq 1 \\[2mm] \phi_4\left[2\left(1 - \dfrac{C_a^2 + C_s^2}{2}\right)\right], & 0 \leq \left(\dfrac{C_a^2 + C_s^2}{2}\right) < 1 \end{cases} \tag{3.77}$$

$$\phi_4 = \min\left\{1, \frac{\phi_1 + \phi_3}{2}\right\} \tag{3.78}$$

$$\phi_3 = \phi_2 e^{-2(1-u)/3u} \tag{3.79}$$

$$\phi_2 = 1 - 4\gamma \tag{3.80}$$

3.3.7 Queuing Networks

In several applications, queues are connected in a network. For example, a customer may call a bank to inquire about making a loan. The caller is queued (i.e., put on hold) until answered by an operator. Based on the answer to some questions, the caller may be routed to another operator with a particular expertise and then queued until that operator becomes available. A patient arriving to a hospital by ambulance may be routed through the emergency department, imaging department, intensive care unit, step down unit, and discharge. Each of these resources can be considered as a queue that is connected in a network.

Modeling a network of queues is typically done through the use of simulation. However, some rather elegant analytical results exist for networks of M/M/1* queues with the following properties:

i. External arrivals into the system to node i occur according to a Poisson process with rate γ_i.

ii. Service times at node i are independent and exponentially distributed with rate μ_i.

iii. The probability that a customer completing service at node i will go to node j (r_{ij}) is independent of the state of the system. (Note that r_{i0} represents the probability that a customer will depart from the system from node i).

iv. There is no limit on queue capacity at any node.

If it is the case that $\gamma_i = 0$ and $r_{i0} = 0$ for all i, then this is referred to as a closed network. Figure 3.10 shows the difference between an open network (a) and a closed network (b). In this section, we only consider open networks.

As in previous sections, we are interested in the steady-state probabilities of the number of customers at each node, that is, $P\{N_1 = n_1, N_2 = n_2, ..., N_k = n_k\}$, that we represent as $P_{n_1,n_2,...,n_k}$. Note that in a network, arrivals to a node can come from multiple sources, both external and internal. The effective arrival rate to a node, then, is defined as follows:

$$\lambda_i = \gamma_i + \sum_{j=1}^{k} r_{ji}\lambda_j \tag{3.81}$$

Jackson (1957) showed that solving the stochastic balance equations for the network yields the following solution (where $\rho_i = \lambda_i / \mu_i$).

$$P_{n_1,n_2,...n_k} = \prod_{i=1}^{k}(1-\rho_i)\rho_i^{n_i} \tag{3.82}$$

---

* We will generalize this to a network of M/M/c queues later in this section.

Design of Service Systems

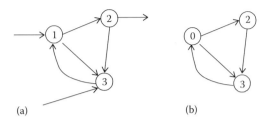

FIGURE 3.10
Examples of an open network (a) and closed network (b).

In other words, the joint probability is simply a product of the marginal probabilities. This should not be interpreted, however, to imply that a flow to each node is a Poisson process. The product form of the solution, however, does mean that the expected performance measures for each node in the network are equivalent to the M/M/1 measures found earlier. That is,

$$L_i = \frac{\rho_i}{1-\rho_i} \tag{3.83}$$

$$W_i = \frac{L_i}{\lambda_i} \tag{3.84}$$

A very nice aspect of the results is that they also hold for a network of M/M/c queues, where node i has c_i parallel servicers, each with independent and exponentially distributed service times with rate μ_i. In this case, we get the following steady-state probabilities:

$$P_{n_1,n_2,\ldots,n_k} = \prod_{i=1}^{k} \frac{\rho_i^{n_i}}{a_i} P_{0i}, \tag{3.85}$$

where P_{0i} is given by Equation 3.49, $\rho_i = \lambda_i / \mu_i$, and

$$a_i = \begin{cases} n_i!, & n_i \le c_i \\ c_i^{n_i-c_i} c_i!, & n_i > c_i \end{cases} \tag{3.86}$$

The expected performance measures for the M/M/c still apply, due to the product form in Equation 3.85. Finally, the total expected number of customers in the system may be found by summing over all nodes.

Example 3.9

Consider the network shown in Figure 3.10 with the following transition probabilities: $r_{12} = 0.7$, $r_{13} = 0.3$, $r_{21} = 1.0$, $r_{20} = 0.5$, $r_{23} = 0.5$. Further, nodes 1 and 3 are M/M/1 queues with expected service times of 1.5 and 1.8 minutes, respectively, and node 2 is an M/M/2 queue with expected service time of 10 minutes. Finally, the external arrival rate to node 1 is 4 customers per hour and that to node 3 is 2 customers per hour. We want to find the expected time a customer spends in the system. The effective arrival rate equations give

$$\lambda_1 = 4 + \lambda_3$$

$$\lambda_2 = 0.3\lambda_1$$

$$\lambda_3 = 2 + 0.7\lambda_1 + 0.5\lambda_2$$

Solutions of these simultaneous equations yield $\lambda_1 = 36$, $\lambda_2 = 10.8$, and $\lambda_3 = 32$. We get the average number of customers at each station from

$$L_1 = \frac{36/40}{1 - 36/40} = 9$$

$$L_2 = \frac{10.8}{6} + \left[\frac{(10.8/6)^3/2}{2!(1-(10.8/6))^2}\right]\left[1 + \frac{1}{1!}\left(\frac{10.8}{6}\right) + \frac{1}{2!}\left(\frac{10.8}{6}\right)^2\left(\frac{2(6)}{2(6)-10.8}\right)\right]^{-1} = 9.47$$

$$L_3 = \frac{32/33.3}{1 - 32/33.3} = 32$$

We get a total average number in the system by summing the three averages, which gives 50.47 customers. The average time in the system can be determined by Little's law, where $\lambda = 4 + 2 = 6$. Therefore, the average time in the system is $W = 50.47/6 = 8.41$ hours.

3.3.8 Modeling a Call Center as a Queuing System

Many organizations use call centers as a way to interact with their customers. Traditionally, phones are used as the mode of interaction. However, this

Design of Service Systems

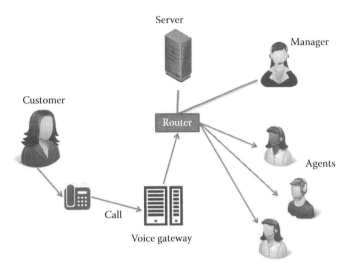

FIGURE 3.11
Key components of a simple call center.

model also supports many other channels including e-mail and Internet chat.* Interactions may include sales (ordering a product or service through the center), help desk (i.e., technical support), or scheduling appointments.

Traditionally, call centers have been physical locations made up of a few to several thousand agents. Recent technological advances, however, have allowed for agents to be geographically distributed (e.g., work from home). A simplified view of a call center is shown in Figures 3.11. In this case, a customer can call the business, though an equivalent structure may be found if the customer desires to initiate an Internet-based chat. The call goes into the voice gateway where typically the caller is given a menu of options to choose from, including the self-service function. An example of self-service may be a caller to a bank that is interested in their bank balance. An automated self-service system allows for that type of request to be handled without the need for an agent, which can save significant expense. If an agent is required, then the call is sent to the router, where it is queued until an appropriate agent becomes available. Note that a phone call may require transfers between agents, and so the network can become much more complicated in practice than shown here. An excellent overview of call center operations may be found in Gans et al. (2003).

A queuing representation of a call center is shown in Figure 3.12. In this system, if the queue is filled, then the customer gets a busy signal, and they may retry their call later. If the customer is on hold (i.e., queued) for a significant length of time, they may become inpatient and hang up without being served. This is called abandonment. Abandoned customers may retry at a

---

* Centers that include interactions such as web and social media are often called "contact" centers.

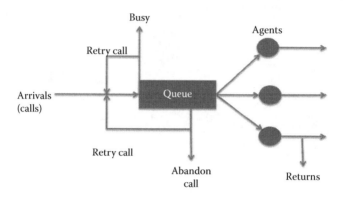

FIGURE 3.12
A simple call center modeled as a queuing system. (Based on Koole, G. and A. Mandlebaum. 2002. Queueing models of call centers: An introduction, *Annals of Operations Research*. 113(1–4): 41–59.)

later time. Finally, if the agent did not fully satisfy the customer request, they may also return at a later time.

Several of the queuing performance measures we have previously discussed are also appropriate here, including the total length of time that a customer needs to spend in the system. Call centers use several other measures, however, to evaluate their performance. Some of the more important metrics include the following:

- *Service level*—the percentage of calls that are answered within a given time period (e.g., 2 minutes)
- *Call abandonment*—the percentage of customers that hang up before talking with an agent
- *First call resolution*—the percentage of customer requests that are resolved during the first call to the center
- *Adherence to schedule*—the percentage of time during an agent's shift that they are handling contacts (or available to handle contacts)
- *Self-service accessibility*—the percentage of calls that are handled by the self-service system
- *Customer satisfaction*—an assessment of performance based on the customer perspective
- *Average handle time*—the average of the time a customer spends with an agent during a call

Many of these cannot be estimated during the design of the call center but are pure performance measures of the actual system.

When designing a call center, the key trade-off is the cost of the resources required to support the center (e.g., number of agents) versus the effectiveness of the service provided to the customer. Although there are several

Design of Service Systems 137

design factors including information technology, computer equipment, and space, we will focus on agent staffing in this section.

We could consider modeling this system as an M/M/c queue with calls arriving at rate λ. In this case, we would capture if a call was busy, but not abandoned or retry calls or returns. In addition, in practice, the arrivals may vary over time. Further, although modeling arrivals as a Poisson process may be reasonable, the service times typically are not exponentially distributed (Brown et al. 2005). We briefly discuss how we can modify the M/M/c model to be more appropriate. Note that throughout this section, it is assumed that customers are handled in an FCFS fashion and that there is no limit to the queue size (i.e., no customer receives a busy signal when they call).

3.3.8.1 Time-Varying Arrivals

Figure 3.13 shows an example of call volume to a call center. Note that there are peak periods from 8:00 a.m. to 9:00 a.m. and from 6:00 p.m. to 11:00 p.m. One way to deal with this is to define time intervals with average arrival rates. In this case, we could define six intervals: 2:00 a.m. to 8:00 a.m. ($\lambda_1 = 407$ calls per hour), 8:00 a.m. to 10:00 a.m. ($\lambda_2 = 845$), 10:00 a.m. to 4:00 p.m. ($\lambda_3 = 454$), 4:00 p.m. to 6:00 p.m. ($\lambda_4 = 845$), 6:00 p.m. to 12:00 a.m. ($\lambda_5 = 1517$), and 12:00 a.m. to 2:00 a.m. ($\lambda_6 = 665$).

We can determine staffing requirements for each interval and then use staggered shifts to handle the different capacity requirements. For this case, if we assume that a shift must be for 8 hours, then we could use four staggered shifts: 2:00 a.m. to 10:00 a.m., 8:00 a.m. to 4:00 p.m., 4:00 p.m. to 12:00 a.m., and 6:00 p.m. to 1:00 a.m. To illustrate how this would work, suppose that we staffed simply to satisfy the average, and that an agent handled on average 10 calls per hour. The minimum number of agents to handle the call volume (on average) can be solved with a simple integer programming model (given as an end of chapter homework problem) and yields 41 agents for shift 1, 46 agents for shift 2, 85 agents for shift 3, and 67 agents for shift 4.

It is important to point out that the approach to break arrivals into intervals and averaging over the intervals is reasonable when the variability of arrivals over an interval is not too large, the average service time is relatively short (e.g., less than 10 minutes), and the service requirements on the system are high (Green et al. 2007). If these assumptions do not hold, however, then this simplification does not hold. Although there has been significant work in modeling time-varying arrivals under a range of conditions (e.g., Avramidis et al. 2004), it is beyond the scope of this chapter.

3.3.8.2 Service Times

As mentioned previously, in practice, service times often are not exponentially distributed. Therefore, the squared coefficient of variation for service times is not equal to 1. If we assume that service times are i.i.d., then we can

138 ◆ *Service Systems Engineering and Management*

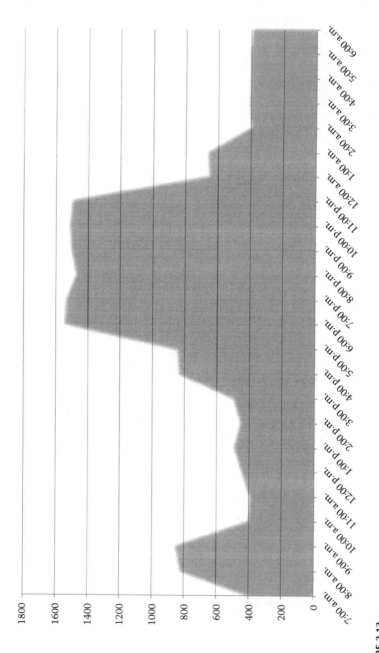

FIGURE 3.13
Call volume over time.

Design of Service Systems

use the $M_t/G/c$ model. We use M_t since the arrival rate depends on the time interval (i.e., is a non-homogeneous Poison process). We can use Equations 3.44 and 3.45 for this system, where $C_a^2 = 1$. We will use the notation $M_t/GI/c_t$ to emphasize the independence of the servers and that the number of parallel servers can change over time.

One common measure based on service times is what is known as the *offered load* (OL). If we let $E[S]$ be the expected value of the service time, then OL is defined as OL= $\lambda E[S]$.

3.3.8.3 Customer Abandonment

If customers have to wait for an agent for too long, they will hang up. That is, they will abandon the system. Behaviorally, this is different from balking (i.e., where customers become discouraged due to the length of the queue and hence the arrival rate decreases) since they do not see how many customers are ahead of them.

One way to model customer abandonment is through the use of the hazard function. If a customer has been waiting for time t, then the instantaneous rate to abandon is given by the hazard function ($h(t)$):

$$h(t) = \frac{f(t)}{1-F(t)}, \quad t \geq 0, \tag{3.87}$$

where $f(t)$ is the time to abandon density and $F(t)$ is the cumulative density. Note that if the time to abandon density is exponential, the hazard function is constant. In practice, estimating the abandon density is difficult without significant assumptions since it is a function of the system and the data are therefore censored (i.e., we observe customers that abandon, but don't know at what point other customers would have abandoned if their wait had not ended).

It is a common notation to write the model as $M_t/GI/c_t$ + GI. The +GI represents that customers have i.i.d. times to abandon with a general probability distribution. This model, unfortunately, proves quite challenging to analyze.

3.3.8.4 Staffing

We look at a special case to determine staffing levels. In particular, we will assume that the call center has a large number of agents (in the hundreds and thousands) and hence there are not a significant number of customers that wait. This is known as the *quality and efficiency driven* (QED) operational regime. We will also assume that the offered load is relatively high. It turns out that, under these conditions, the number of customers in the system is approximately normally distributed. Because of this, the well-known square root staffing formula (Halfin and Whitt 1981) can be a good approximation. It is given by

$$c = R + \beta\sqrt{R} \qquad (3.88)$$

Where R is the offered load. The solution to c must be rounded to an integer. In this expression, the first term is just the offered load and the second term is the excess capacity needed to achieve a service level in the presence of the variability faced by the system. The value β is a constant based on the desired service level. The higher the value of β is, the higher the quality of service will be. If we approximate the queuing system as an M/GI/∞ system, then

$$P(\text{delay}) \cong P(Q \geq s) = P\left(\frac{Q-R}{\sqrt{R}} \geq \frac{c-R}{\sqrt{R}}\right) \cong 1 - \Phi(\beta) \qquad (3.89)$$

where $\Phi(\cdot)$ is the cumulative density of the standard normal distribution and Q is the number of busy servers. Kolesar and Green (1998) provide a simple rule of thumb, namely, that if the aim is to avoid congestion without maintaining excessive capacity, then set $\beta = 2$. This will produce a probability of delay of approximately 0.02.

There are several other delay functions that have been developed that can give increased precision under certain conditions. We mention two here. The first was developed by Halfin and Whitt (1981) for an M/M/c model without customer abandonment:

$$P(\text{delay}) = \frac{1}{1 + \left(\dfrac{\beta\Phi(\beta)}{\phi(\beta)}\right)}, \qquad (3.90)$$

where $\phi(\cdot)$ is the probability density function and $\Phi(\beta)$ is the cumulative density function for the standard normal distribution. Garnett et al. (2002) developed a delay function for an M/M/c + M model, with abandonment rate θ. Setting $\theta_{rat} = \theta/\mu$ gives

$$P(\text{delay}) = \left[1 + \sqrt{\theta_{rat}} \cdot \frac{\phi\left(\dfrac{\beta}{\sqrt{\theta_{rat}}}\right) \Big/ \left(1 - \Phi\left(\dfrac{\beta}{\sqrt{\theta_{rat}}}\right)\right)}{\phi(-\beta) \big/ (1 - \Phi(-\beta))}\right]^{-1} \qquad (3.91)$$

Design of Service Systems

Another simple way to determine an appropriate service value β was proposed by Borst et al. (2004), which considers the cost of customer delay and the staffing cost. If it is assumed that they are both linear in time, then we can define r as the ratio of customer delay cost per hour to the cost of staffing an agent per hour. The expression is given by

$$\beta(r) = \begin{cases} \sqrt{\dfrac{r}{\left(1+r\left(\sqrt{\pi/2}-1\right)\right)}} & 0 \le r < 10 \\ \sqrt{2\ln\left(r/\sqrt{\pi}\right) - \ln\left(2\ln\left(r/\sqrt{\pi}\right)\right)} & 10 \le r \end{cases} \qquad (3.92)$$

Example 3.10

Consider a call center that operates 24 hours a day. In the range of 1 p.m. to 4 p.m., interarrival times are approximately exponentially distributed with a mean of 0.6 seconds. The service times are log-normally distributed with $\mu = 10$ and $\sigma = 0.01$. This gives a mean and variance of the service time of

$$E[S] = e^{0.5(\mu+\sigma^2)} = 148.4 \text{ seconds}$$

$$\text{Var}[S] = \left(e^{\sigma^2} - 1\right)e^{2\mu+\sigma^2} = 49006.5$$

This gives a squared coefficient of variation of 2.22. For this interval, the offered load is

$$R = \frac{1}{0.6}(148.4) = 247.33$$

Using the rule of thumb of $\beta = 2$ gives a staffing level of

$$c = 247.33 + 2\sqrt{247.33} = 278.8$$

which we would round to 279 service agents. Using Equation 3.89 to determine β, we can plot the number of service agents required for a given probability of delay. This is shown in Figure 3.14.

Suppose, in Example 3.10, the cost to staff an agent was $10/hour and delay cost per customer was estimated to be $130/hour, then from Equation 3.92, we get that β(13) = 1.61. This would result in a staffing level of 273 agents from Equation 3.88 and a corresponding probability of delay of 0.054 from Equation 3.89 and 0.067 from Equation 3.90. The challenge in using Equation 3.92 is in developing a reasonable estimate for the customer hourly delay cost.

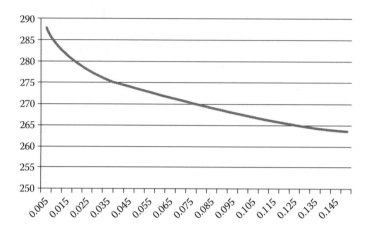

FIGURE 3.14
Staffing levels with respect to delay probability.

3.3.8.5 Discussion

As mentioned previously, analyzing the $M_t/GI/c_t + GI$ queuing system is challenging. There have been significant advances in this area using diffusion and fluid approximations, but the technical requirements are well beyond the level of this chapter. A good survey of this work may be found in Dai and He (2012).

Because of the limitations of queuing analysis, simulation offers another approach for analyzing call centers. Two examples are given in Feldman et al. (2008) and Whitt (2005). Feldman et al. (2008), for example, use a simulation-based iterative staffing algorithm. The algorithm iterates between the staffing function and number of customers in the system until convergence for an $M_t/GI/c_t + GI$ queuing system.

It is important to remember that even with accurate analysis of the $M_t/GI/c_t + GI$ system, this is still an approximation of a service call center. For example, there are typically several customer types that may have different priorities. In addition, there may be specific agent types required. For example, an international call center will need agents that have specific language capabilities. For these reasons, detailed simulation models may be the only reasonable approach to model such a system.

Exercises

3.1 Explain the differences between *set covering* and *set partitioning* models. Give two applications of each.

3.2 Discuss the differences between the two types of quantity discounts. Is one better than the other? If so, under what conditions?

3.3 Consider a supplier order allocation problem under multiple sourcing, where it is required to buy 2000 units of a certain product from three different suppliers. The fixed setup cost (independent of the order quantity), variable cost (unit price), and the maximum capacity of each supplier are given in Table 3.5 (two suppliers offer quantity discounts).

The objective is to minimize the total cost of purchasing (fixed plus variable cost). Formulate this as a linear integer programming problem. You must define all your variables clearly, write out the constraints to be satisfied with a brief explanation of each and develop the objective function.

3.4 Reformulate the problem in Exercise 3.3 under the assumption that both suppliers 1 and 3 offer "all units" discount, as described below:

- Supplier 1 changes $10/unit for orders up to 300 units, and for orders more than 300 units, the *entire order* will be priced at $7 per unit.
- Supplier 3 changes $6/unit for orders up to 500 units, and for orders more than 500 units, the *entire order* is priced at $4 per unit.

3.5 (From Ravindran et al. 1987) A company manufactures three products A, B, and C. Each unit of product A requires 1 hour of engineering service, 10 hours of direct labor, and 3 pounds of material. Producing one unit of product B requires 2 hours of engineering, 4 hours of direct labor, and 2 pounds of material. Each unit of product C requires 1 hour of engineering, 5 hours of direct labor, and 1 pound of material. There are 100 hours of engineering, 700 hours of direct labor, and 400 pounds of materials available. The cost of production is a nonlinear function of the quantity produced as shown in Table 3.6. Given the unit selling prices of products A, B, and C as $12, $9 and $7, respectively, formulate a linear mixed integer program to determine the optimal production schedule that will maximize the total profit.

TABLE 3.5

Supplier Data for Exercise 3.3

Supplier	Fixed Cost	Capacity	Unit Price
1	$100	600 units	$10/unit for the first 300 units
			$7/unit for the remaining 300 units
2	$500	800 units	$2/unit for all 800 units
3	$300	1200 units	$6/unit for the first 500 units
			$6/unit for the remaining 700 units

TABLE 3.6
Data for Exercise 3.5

Product A		Product B		Product C	
Production (Units)	Unit Cost ($)	Production (Units)	Unit Cost ($)	Production (Units)	Unit Cost ($)
0–40	10	0–50	6	0–100	5
41–100	9	51–100	4	Over 100	4
101–150	8	Over 100	3		
Over 150	7				

Note: If 60 units of A are made, the first 40 units cost $10 per unit and the remaining 20 units cost $9 per unit.

3.6 Explain how the following conditions can be represented as linear constraints using binary variables.
  a. Either $x_1 + x_2 \leq 3$ or $3x_1 + 4x_2 \geq 10$
  b. Variable x_2 can assume values 0, 4, 7, 10 and 12 only
  c. If $x_2 \leq 3$, then $x_3 \leq 6$; otherwise, $x_3 \leq 4$ (assume x_2 and x_3 are integers)
  d. At least two out of the following five constraints must be satisfied:
  $$x_1 + x_2 \leq 7$$
  $$x_1 - x_2 \geq 3$$
  $$2x_1 + 3x_2 \leq 20$$
  $$4x_1 - 3x_2 \geq 10$$
  $$x_2 \leq 6$$
  $$x_1, x_2 \geq 0$$

3.7 (From Ravindran et al. 1987) Convert the following nonlinear integer program to a linear integer program.

Minimize	$Z = x_1^2 - x_1 x_2 + x_2$
Subject to	$x_1^2 + x_1 x_2 \leq 8$
	$x_1 \leq 2$
	$x_2 \leq 7$
	$x_1, x_2 \geq 0$ and integer

Hint: Replace x_2 by $2^0\delta_1 + 2^1\delta_2$ and x_1 by $2^0\delta_3 + 2^1\delta_4 + 2^2\delta_5$, where $\delta_i \in (0,1)$ for $i = 1, 2, \ldots 5$

Design of Service Systems   145

3.8 An M/M/1 queuing system has an arrival rate of 10 customers per hour. The average service time is 9 minutes.
   a. What is the system utilization?
   b. What is the average queue length?
   c. What is the probability that there are more than 3 customers in the queue?
   d. What is the probability that a customer will wait for more than 3 minutes for service?

3.9 Using the birth–death model for the M/M/c queue, derive Equations 3.50 and 3.51.

3.10 Customers arrive to a service center according to a Poisson process with a rate of 30 per hour. Service times are exponentially distributed with a mean of 12 minutes.
   a. What is the minimum number of servers required for system stability?
   b. Using your answer in (a), what is the probability that the system is empty?
   c. What is the average waiting time in the queue?
   d. What is the probability that a customer will wait in the queue for more than 10 minutes?

3.11 Make a plot of the probability that a customer has to wait for more than 2 minutes as a function of number of servers, starting from the c you found in 3a and going up to $c + 5$.

3.12 Repeat Exercise 3.10 if there are only 10 customers allowed in the system.

3.13 Suppose the service time is not exponentially distributed for Exercise 3.8, but is instead log-normally distributed with $\mu = 8$ and $\sigma = 0.015$. Use the P–K formulas to estimate the average waiting time in the queue.

3.14 Consider a G/G/3 queuing system. Interarrival times are on average 5 minutes with a standard deviation of 6 minutes. Service times are on average 12 minutes with a standard deviation of 15 minutes. Use Equation 3.70 and Little's law to determine average waiting time in the queue and average number of customers in the system.

3.15 Repeat Exercise 3.14 using the Whitt refinement given in Equations 3.73 through 3.80.

3.16 The call volume data for Figure 3.13 is given in Table 3.7. Develop and solve an integer programing formulation to find the minimum number of service agents to staff the offered load. An agent handles on average 10 calls per hour.

TABLE 3.7

Call Volume Data for Exercise 3.16

Time	Calls/hour
7:00 a.m.	405
8:00 a.m.	830
9:00 a.m.	860
10:00 a.m.	410
11:00 a.m.	400
12:00 p.m.	450
1:00 p.m.	500
2:00 p.m.	460
3:00 p.m.	503
4:00 p.m.	840
5:00 p.m.	850
6:00 p.m.	1550
7:00 p.m.	1540
8:00 p.m.	1480
9:00 p.m.	1510
10:00 p.m.	1520
11:00 p.m.	1503
12:00 a.m.	660
1:00 a.m.	670
2:00 a.m.	405
3:00 a.m.	415
4:00 a.m.	410
5:00 a.m.	411
6:00 a.m.	398

3.17 How would your answer to Exercise 3.16 differ if you were allowed to staff agents for less than 8 hours if they are paid 10% more per hour compared to permanent agents?

3.18 Using the data in Table 3.7, determine the staffing levels required for each interval using the rule of thumb of $\beta = 2$.

3.19 For Exercise 3.18, what is the probability of delay using Equations 3.89 and 3.90?

3.20 A call center has 300 calls (arrivals) per hour. Average service time for a call is 3 minutes. Assuming an M/M/c + M approximation, determine the service value β from Equation 3.91, if we want the probability of delay to be 0.05. Assume an abandonment rate of 2%. Use the square root staffing formula to determine the appropriate staffing level.

3.21 A call center has 600 calls (arrivals) per hour. Average service time for a call is 6 minutes. Management has determined that the cost to staff an agent is $9.50 per hour. In addition, the cost of delaying a customer 1 hour is roughly $158. Use Equation 3.92 to determine the service value β. Use the square root staffing formula to determine the appropriate staffing level.

3.22 For Exercise 3.21, management estimates that the customer delay cost ranges from $120 to $160. How much difference will the estimate effect the staffing levels?

Steps of ISA

0. Intialize: Pick a large value for each $s_i(t)$, $\{s_i(t): 0 \leq t \leq T\}$.

1. Given the ith staffing function $\{s_i(t): 0 \leq t \leq T\}$, evaluate the distribution of $L_i(t)$, for all t, using simulation.

2. For each t, $0 \leq t \leq T$, let $s_i + 1(t)$ be the least number of servers so that the delay-probability constraint is met at time t; that is, let $s_i + 1(t) = \arg\min \{c \in N: P\{L_i(t) \geq c\} < \alpha\}$.

3. If there is negligible change in the staffing from iteration i to iteration $i + 1$, then stop; that is, if $||s_i + 1(\cdot) - s_i(\cdot)||_\infty \equiv \max\{|s_i + 1(t) - s_i(t)|: 0 \leq t \leq T\} \leq \tau$, then stop and let $s_i + 1(\cdot)$ be the proposed staffing function. Otherwise, advance to the next iteration, that is, replace i by $i + 1$ and go back to step 1. (We let $\tau = 1$.)

References

Abara, J. 1989. Applying integer linear programming to the fleet assignment problem. *Interfaces*. 19(4): 20–28.

Anbil, R., E. Gelman, B. Patty, and R. Tanga. 1991. Recent advances in crew-pairing optimization at American Airlines. *Interfaces*. 21(1): 62–74.

Armacost, A. P., C. Barnhart, K. A. Ware, and A. M. Wilson. 2004. UPS optimizes its air network. *Interfaces*. 34(1): 15–25.

Arthur, J. L. and A. Ravindran. 1981. A multiple objective nurse scheduling model. *AIIE Transactions*. 13(1): 55–60.

Avramidis, A. N., Alexandre D., and P. L'Ecuyer. 2004. Modeling daily arrivals to a telephone call center. *Management Science* 50(7): 896–908.

Blake, J. T. and J. Donald. 2002. Mount Sinai hospital uses integer programming to allocate operating room times. *Interfaces*. 32(2): 63–73.

Borst, S., A. Mandelbaum, and M. I. Reiman. 2004. Dimensioning large call centers. *Operations Research.* 52(1): 17–34.

Brown, L., N. Gans, A. Mandelbaum, A. Sakov, S. Zeltyn, L. Zhao, and S. Haipeng. 2005. Statistical analysis of a telephone call center: A queueing-science perspective. *Journal of the American Statistical Association.* 100(1): 36–50.

Chelst, K., J. Sidelko, A. Przebienda, J. Lockedge, and D. Mihailidis. 2001. Rightsizing and management of prototype vehicle testing at Ford motor company. *Interfaces.* 31: 91–107.

Dai, J. G. and S. He. 2012. Many-server queues with customer abandonment: A survey of diffusion and fluid approximations. *Journal of Systems Science and Systems Engineering.* 21(1): 1–36.

Feldman, Z., A. Mandelbaum, W. A. Massey, and W. Whitt. 2008. Staffing of time-varying queues to achieve time-stable performance. *Management Science.* 54(2): 324–338.

Gans, N., Koole, G., and A. Mandelbaum. 2003. Telephone call centers: Tutorial, review, and research prospects. *Manufacturing & Service Operations Management.* 5(2): 79–141.

Garnett, O., A. Mandelbaum and M. I. Reiman. 2002. Designing a call center with impatient customers. *Manufacturing & Service Operations Management.* 4(3): 208–227.

Glover, F. and E. Woolsey. 1974. Converting 0-1 polynominal programming problem to a 0-1 linear program. *Operations Research.* 22: 180–182.

Green, L., P. J. Kolesar, and W. Whitt. 2007. Coping with time-varying demand when setting staffing requirements for a service system. *Production & Operations Management.* 16(1): 13–39.

Gross, D., J. F. Shortle, J. M. Thompson, and C.M. Harris. 2009. *Fundamentals of Queueing Theory*, 4th Edition. Hoboken, NJ: John Wiley & Sons.

Halfin, S. and W. Whitt. 1981. Heavy-traffic limits for queues with many exponential servers. *Operations Research.* 29(3): 567–587.

Holloran, T. J. and J. E. Byrn. 1986. United Airlines station manpower planning system. *Interfaces.* 16(1): 39–50.

Hopp, W. J. and M. L. Spearman. 2011. *Factor Physics*, 3rd Edition. Long Grove, IL: Waveland Press Inc.

Hueter, J. and W. Swart. 1998. An integrated labor-management system for Taco Bell. *Interfaces.* 28(1): 75–91.

Jackson, J. R. 1957. Networks of waiting lines. *Operations Research.* 5(4): 518–521.

Kolesar, P. J. and L. V. Green. 1998. Insights on service system design from a normal approximation to Erlang's delay formula. *Production and Operations Management.* 7(3): 282–293.

Koole, G. and A. Mandlebaum. 2002. Queueing models of call centers: An introduction. *Annals of Operations Research.* 113(1–4): 41–59.

Ravindran, A., D. T. Philips, and J. Solberg. 1987. *Operations Research: Principles and Practice*, 2nd Edition. NY: John Wiley & Sons, Inc.

Ravindran A. R. and D. P. Warsing Jr. 2013. *Supply Chain Engineering: Models and Applications.* Boca Raton, FL: CRC Press.

Snowdon, J. L. and G. Paleologo. 2008. Airline optimization. In *Operations Research and Management Science Handbook.* ed. A. R. Ravindran, Chapter 20. Boca Raton, FL: CRC Press.

Subramanian, R., R. P. Scheff, J. D. Quillinan, D. S. Wiper, and R. E. Marsten. 1994. Coldstart: Fleet assignment at Delta airlines. *Interfaces*. 24(1): 104–120.

Taylor, B. 2010. *Introduction to Management Science*, 10th Edition. Upper Saddle River, NJ: Prentice Hall.

Whitt, W. 1993. Approximations for the GI/G/m queue. *Production and Operations Management*. 2(2): 114–161.

Whitt, W. 2005. Engineering solution of a basic call-center model. *Management Science*. 51(2): 221–235.

4

Evaluation of Service Systems

Evaluating the design and operation of a service system frequently involves cost and customer service as the main criteria. However, cost and customer service are conflicting objectives that require trade-offs between them to arrive at a satisfactory solution. In this chapter, we present two approaches to evaluate the performance of service systems under multiple objectives:

1. Multiple-Criteria Decision Making (MCDM)
2. Data Envelopment Analysis (DEA)

MCDM models and methods have been applied in financial engineering to select optimal investment portfolios by banks, mutual funds, insurance companies, and others (Chapter 7). They are also used in supply chain management to design optimal distribution networks (Chapter 5). DEA models and methods have been used to compare and evaluate the performance of several units/branches of a company in different locations under multiple inputs and outputs. Examples include fast-food restaurants, banks, hospitals, retail stores, and others. In the following sections, we present an overview of MCDM and DEA models and their similarities and differences in evaluating the performance of service systems.

4.1 Multiple-Criteria Decision Making—An Overview*

Managers are frequently called upon to make decisions under multiple criteria that conflict with one another. For example, supply chain engineers have to consider conflicting criteria—supply chain costs, customer responsiveness, and supply chain risk—in making decisions. The general framework of a multiple-criteria optimization problem is to simultaneously optimize several criteria, usually conflicting, subject to a system of constraints that define the feasible alternatives. MCDM problems are categorized on the basis of whether (i) the constraints are implicit, that is, the feasible alternatives

---

* Portions of this section have been adapted with permission from Ravindran, A. R. and Warsing, D. P. 2013. *Supply Chain Engineering: Models and Applications.* Chapter 6 and Appendix A. Boca Raton, FL: CRC Press.

are *finite and known*, or (ii) the constraints are explicit given by a set of linear and nonlinear inequalities or equations; in this case, the feasible alternatives are *infinite and unknown*. MCDM problems with finite (known) alternatives are called *Multiple-Criteria Selection Problems* (MCSP). MCDM problems with infinite (unknown) alternatives are called *Multiple-Criteria Mathematical Programming* (MCMP) problems. In this section, we will give an overview of some of the methods that are available for solving both MCSP and MCMP problems. For a more detailed discussion of the multi-criteria optimization methods, the reader is referred to Masud and Ravindran (2008, 2009).

4.2 Multiple-Criteria Selection Problems

For MCSP, the alternatives are finite and their criteria values are known *a priori* in the form of a *pay-off matrix*. Table 4.1 illustrates the pay-off matrix for an MCSP with n alternatives (1, 2, ..., n) and p criteria ($f_1, f_2, ..., f_p$). The matrix element f_{ij} denotes the value of criterion j for alternative i.

4.2.1 Concept of "Best Solution"

In a single-objective optimization problem, the "best solution" is defined in terms of an "optimal solution" that maximizes (or minimizes) the objective, compared to all other feasible solutions (alternatives). In MCSP, because of the conflicting nature of the objectives, the optimal values of the various criteria do not usually occur at the same alternative. Hence, the notion of an optimal or best alternative does not exist in MCSP. Instead, decision-making in MCSP is equivalent to choosing the *most preferred alternative* or the *best compromise solution* based on the preferences of the *decision maker* (DM). Thus, the objective of the MCSP method is to rank order the alternatives from the *best* to the *worst*, based on the DM's preference structure.

TABLE 4.1

Pay-Off Matrix of MCSP

	Criteria/Objectives (Max)					
	f_1	f_2	f_3	....	f_p
Alt. 1	f_{11}	f_{12}	f_{13}		f_{1p}
2	f_{21}	f_{22}	f_{23}		f_{2p}
–					
–					
–					
n	f_{n1}	f_{n2}	f_{n3}		f_{np}
Max f_i	f_1^*	f_2^*	f_3^*		f_p^*

Evaluation of Service Systems 153

We begin the discussion with some key definitions and concepts in solving MCSP. We will assume that *all the criteria in Table 4.1 are to be maximized.*

4.2.2 Dominated Alternative

Alternative i is dominated by alternative k if and only if $f_{kj} \geq f_{ij}$, for all $j = 1, 2, \ldots, p$ and for at least one j, $f_{kj} > f_{ij}$. In other words, the criteria values of alternative k are as good as those of alternative i, and for at least one criterion, alternative k is better than i.

4.2.3 Non-Dominated Alternatives

An alternative that is not dominated by any other feasible alternative is called *non-dominated*, *Pareto optimal*, or *efficient* alternative. For a non-dominated alternative, an improvement in any one criterion is possible only at the expense of at least one other criterion.

4.2.4 Ideal Solution

It is the vector of the best values achievable for each criterion. In other words, if $f_j^* = \max f_{ij}$, then the ideal solution $= \left(f_1^*, f_2^* \ldots, f_p^* \right)$. Since the f_j^* values may correspond to different alternatives, the *ideal solution is not achievable* for MCSP. However, it provides good target values to compare against for a trade-off analysis.

4.3 Multiple-Criteria Ranking Methods

Weighted methods are commonly used to rank the alternatives under conflicting criteria. Based on the DM's preferences, a weight w_j is obtained for criterion j such that

$$w_j \geq 0 \text{ and } \sum_{j=1}^{p} w_j = 1. \quad (4.1)$$

Next, a weighted score of the criteria values is calculated for each alternative as follows:

$$\text{Score }(i) = \sum_{j=1}^{p} w_j f_{ij} \text{ for } i = 1, 2, \ldots n. \quad (4.2)$$

The alternatives are then ranked based on their scores. The alternative with the highest score is ranked at the top.

There are two common approaches for determining the criteria weights based on the DM's preferences, as discussed next.

4.3.1 Rating Method

Here, the DM is asked to provide a rating for each criterion on a scale of 1 to 10 (with 10 being the most important and 1 being the least important). The ratings are then normalized to determine the weights as follows:

$$w_j = r_j \bigg/ \sum_{j=1}^{p} r_j, \qquad (4.3)$$

where r_j is the rating assigned to criterion j for $j = 1, 2, \ldots, p$.

NOTE: $w_j \geq 0$ and $\sum_{j=1}^{p} w_j = 1$.

4.3.2 Borda Method

Under the ranking method devised by Jean Charles de Borda, (an eighteenth-century French physicist), the DM is asked to rank the p criteria from the most important (ranked first) to the least important (ranked last). The criterion that is ranked first gets p points, the criterion that is ranked second gets $(p - 1)$ points, and the last place criterion gets 1 point. The sum of all the points for the p criteria is given by

$$S = \frac{p(p+1)}{2}. \qquad (4.4)$$

The criteria weights are then calculated by dividing the points assigned to criterion j by the sum S, given by Equation 4.4.

Let us illustrate the basic definitions (dominated, non-dominated, and ideal solutions) and the two weighting methods with a numerical example.

Example 4.1 (Ravindran and Warsing 2013)

An Industrial Engineering department has interviewed five PhD candidates for a faculty position and has rated on a scale of 1–10 (10 being the best and 1 being the worst) on three key criteria—Research,

Evaluation of Service Systems

Teaching, and Service. The criteria values of the candidates are given in Table 4.2.

a. Determine the ideal solution to this problem. Is the ideal solution achievable?
b. Identify the dominated and non-dominated candidates.
c. Determine the ranking of the candidates using
   i. Rating method
   ii. Borda Count method

SOLUTION

a. Ideal solution represents the best values achievable for each criterion. Since all the criteria are to maximize, the ideal solution is given by (8, 8, 5). The ideal solution is not achievable since the criteria conflict with one another and no candidate has the ideal values.
b. Candidate A dominates E since A has higher values for all three criteria. Similarly, candidate C dominates B. On the other hand, candidates A, C, and D are non-dominated.
c.
   i. Rating method
      Assume that the ratings for Research, Teaching, and Service are 9, 7, and 4, respectively. Then, the weights for Research, Teaching, and Service are computed using Equation 4.3 as follows:

$$w_R = 9/(9+7+4) = 0.45$$
$$w_T = 7/20 = 0.35$$
$$w_S = 4/20 = 0.2$$

Note that the sum of the weights is equal to one. Using the criteria weights, the weighted score for candidate A is computed using Equation 4.2:

$$\text{Score (A)} = (0.45)8 + (0.35)4 + (0.2)3 = 5.6.$$

TABLE 4.2

Data for Example 4.1 (Faculty Recruiting)

	Criteria		
Candidate	Research	Teaching	Service
A	8	4	3
B	4	5	3
C	6	6	5
D	2	8	4
E	7	3	2

Similarly, the scores for the other candidates are computed and are given below:

Score (B) = 4.15
Score (C) = 5.8
Score (D) = 4.5
Score (E) = 4.6

The five candidates are then ranked, using their scores, from the highest to the lowest. Thus, candidate C is ranked first followed by candidates A, E, D, and B, respectively.

ii. Borda method

Assume that the three criteria are ranked as follows:

Rank	Criteria
1	Research
2	Teaching
3	Service

Thus, Research gets 3 points, Teaching 2 points, and Service 1 point. Their sum (S) is 6 and the weights are as follows:

$$w_R = 3/6 = 0.50$$
$$w_T = 2/6 = 0.33$$
$$w_S = 1/6 = 0.17$$

Using the above weights and Equation 2, the scores for candidates A, B, C, D, and E are 5.83, 4.17, 5.83, 4.33, and 4.83, respectively. Thus, both candidates A and C are tied for first place, followed by candidates E, D, and B. Note that the rankings are not exactly the same by the two methods. It does happen in practice.

4.3.3 Pairwise Comparison of Criteria

When there are many criteria, it would be difficult for a DM to rank order them precisely, as required by the Borda method. In practice, pairwise comparison of criteria is used to facilitate the criteria ranking. Here, the DM is asked to give the relative importance between two criteria C_i and C_j, whether C_i is preferred to C_j, C_j is preferred to C_i or both are equally important. When there are n criteria, the DM has to respond to $\dfrac{n(n-1)}{2}$ pairwise comparisons.

Evaluation of Service Systems

Based on the DM's response, the criteria rankings and their weights can be computed, following the steps given below:

Step 1: Based on the DM's response, a pairwise comparison matrix, $P_{(n \times n)}$, is constructed, whose elements p_{ij} are as given below:

$$p_{ii} = 1 \text{ for all } i = 1, 2, \ldots, n$$

$p_{ij} = 1, p_{ji} = 0$, if C_i is preferred to C_j ($C_i > C_j$)
$p_{ij} = 0, p_{ji} = 1$, if C_j is preferred to C_i ($C_i < C_j$)
$p_{ij} = p_{ji} = 1$ if C_i and C_j are equally important.

Step 2: Compute the row sums of the matrix P as, $t_i = \sum_j p_{ij}$, for $i = 1, 2, \ldots, n$

Step 3: Rank the criteria based on the t_i values and compute their weights, $W_j = \dfrac{t_j}{\sum_i t_i}, \forall j = 1, 2, \ldots n$

Example 4.2

Five criteria, A, B, C, D, and E, have to be ranked based on 10 pairwise comparisons given below.

- A > B, A > C, A > D, A > E
- B < C, B > D, B < E
- C > D, C < E
- D < E

SOLUTION

Step 1: Construct the pairwise comparison matrix P as shown in Table 4.3.

Step 2: Compute the row sums as $t_A = 5, t_B = 2, t_C = 3, t_D = 1,$ and $t_E = 4$.

TABLE 4.3
Pairwise Comparison Matrix (P) for Example 4.2

		A	B	C	D	E
	A	1	1	1	1	1
	B	0	1	0	1	0
$P_{(5 \times 5)} =$	C	0	1	1	1	0
	D	0	0	0	1	0
	E	0	1	1	1	1

Step 3: The ranking of the five criteria are A > E > C > B > D and their weights are as follows:

$$W_A = \frac{5}{15}, \ W_B = \frac{2}{15}, \ W_C = \frac{3}{15}, \ W_D = \frac{1}{15}, \text{ and } W_E = \frac{4}{15}.$$

4.3.4 Scaling Criteria Values

The major drawback of the ranking methods discussed so far is that they use criteria weights that require the criteria values to be scaled properly. For example, in Table 4.2, all the criteria values ranged between 2 and 8. In other words, they have been already scaled. In practice, the criteria are measured in different units. Some criteria values may be very large (e.g., cost), while others may be very small (e.g., quality, delivery time). If the criteria values are not scaled properly, the criteria with large magnitudes would simply dominate the final rankings, independent of the assigned weights. There are several methods available for scaling criteria values. They are discussed in Chapter 5, Section 5.3.7.

4.3.5 Analytic Hierarchy Process

The *Analytic Hierarchy Process* (AHP), developed by Saaty (1980), is another method used for ranking alternatives. AHP can accommodate both quantitative and qualitative criteria and does not require scaling of criteria values. It uses pairwise comparison of criteria and alternatives with strength of preference. Hence, its cognitive burden on the DM and computational requirements are much higher. A commercial software for AHP, called *Expert Choice*, is also available. For a detailed discussion of AHP and its applications, the reader is referred to Chapter 5, Section 5.3.6.4. Experiments done with real DMs have shown that the rankings by Borda method are as good as the AHP rankings and require less cognitive burden on the DM and fewer computations (Powdrell 2003; Ravindran et al. 2010; Velazquez et al. 2010).

4.4 MCMP Problems

Until now, our focus was on solving MCDM problems with a *finite* number of alternatives, where each alternative is measured by several conflicting criteria. These MCDM problems were called *Multiple-Criteria Selection Problems* (MCSP). The ranking methods we discussed earlier helped in identifying

Evaluation of Service Systems

the best alternative and rank order all the alternatives from the best to the worst.

In this and the subsequent sections, we will focus on MCDM problems with *infinite number of alternatives*. In other words, the feasible alternatives are not known *a priori* but are represented by a set of mathematical (linear/nonlinear) constraints. These MCDM problems are called *Multiple-Criteria Mathematical Programming* (MCMP) problems.

4.4.1 MCMP Problem

$$\text{Max } F(x) = \{f_1(x), f_1(x), \ldots, f_k(x),\} \\ \text{Subject to, } g_j(x) \leq 0 \quad \text{for } j = 1, \ldots, m \tag{4.5}$$

where x is an n-vector of *decision variables* and $f_i(x)$, $i = 1, \ldots, k$ are the k *criteria/objective functions*. All the objective functions are to maximize.

Let

$$S = \{x/g_j(x) \leq 0, \text{ for all } j\} \\ Y = \{y/F(x) = y, \text{ for some } x \in S\}$$

S is called the *decision space* and Y is called the *criteria or objective space* in MCMP.

A solution to MCMP is called a *superior solution* if it is feasible and maximizes all the objectives simultaneously. In most MCMP problems, superior solutions do not exist as the objectives conflict with one another.

4.4.2 Efficient, Non-Dominated, or Pareto Optimal Solution

A solution $x^o \in S$ to MCMP is said to be *efficient* if $f_k(x) > f_k(x^o)$ for some $x \in S$ implies that $f_j(x) < f_j(x^o)$ for at least one other index j. More simply stated, an efficient solution has the property that an improvement in any one objective is possible only at the expense of at least one other objective.

A *dominated solution* is a feasible solution that is not efficient.

Efficient set: A set of all efficient solutions is called the *efficient set* or *efficient frontier*.

NOTE: Even though the solution of MCMP reduces to finding the efficient set, it is not practical because there could be an infinite number of efficient solutions.

Example 4.3 (Ravindran and Warsing 2013)

Consider the following bi-criteria linear program:

$$\text{Max } Z_1 = 5x_1 + x_2$$

$$\text{Max } Z_2 = x_1 + 4x_2$$

Subject to

$$x_1 \leq 5$$

$$x_2 \leq 3$$

$$x_1 + x_2 \leq 6$$

$$x_1, x_2 \geq 0$$

SOLUTION

The *decision space* and the *objective space* are given in Figures 4.1 and 4.2, respectively. Corner points C and D are efficient solutions while corner points A, B, and E are dominated. The set of all efficient solutions is given by the line segment CD in both figures.

Ideal solution is the vector of individual optima obtained by optimizing each objective function separately ignoring all other objectives. In

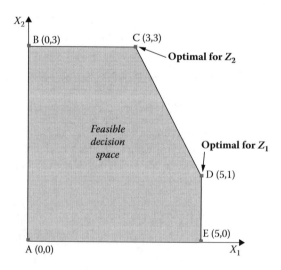

FIGURE 4.1
Decision space (Example 4.3).

Evaluation of Service Systems

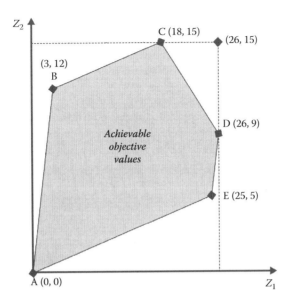

FIGURE 4.2
Objective space (Example 4.3).

Example 4.3, the maximum value of Z_1, ignoring Z_2, is 26 and occurs at point D. Similarly, maximum Z_2 of 15 is obtained at point C. Thus, the ideal solution is (26, 15) but is *not* feasible or achievable. One of the popular approaches to solving MCMP problems, called *compromise Programming*, is to find an efficient solution that comes "as close as possible" to the ideal solution. We will discuss this approach later in this Section 4.7.

4.4.3 Determining an Efficient Solution (Geoffrion 1968)

For the MCMP problem given by Equation 4.5, consider the following single-objective optimization problem, called the P_λ **problem**. The P_λ problem is also known as the *weighted objective problem*.

$$\text{Max } Z = \sum_{i=1}^{k} \lambda_i f_i(x) \tag{4.6}$$

Subject to: $x \in S$

$$\sum_{i=1}^{k} \lambda_i = 1$$

$$\lambda_i \geq 0$$

Theorem 4.1: (Sufficiency)

Let $\lambda_i > 0$ for all i be specified. If $\mathbf{x}^o$ is an optimal solution for the P_λ problem (Equation 4.6), then $\mathbf{x}^o$ is an efficient solution to the MCMP problem.

In Example 4.3, if we set $\lambda_1 = \lambda_2 = 0.5$ and solve the P_λ problem, the optimal solution will be at D, which is an efficient solution.

Warning: Theorem 4.1 is only a sufficient condition and is *not* necessary. For example, there could be efficient solutions to MCMP that could not be obtained as optimal solutions to the P_λ problem. Such situations occur when the objective space is not a convex set. However, for MCMP problems, where the objective functions and constraints are linear, Theorem 4.1 is both necessary and sufficient. ∎

4.4.4 Test for Efficiency

Given a feasible solution $\bar{x} \in S$ for MCMP, we can test whether or not it is efficient by solving the following single-objective problem:

$$\text{Max } W = \sum_{i=1}^{k} d_i$$

Subject to: $f_i(\mathbf{x}) \geq f_i(\bar{\mathbf{x}}) + d_i$ for $i = 1, 2, \ldots, k$

$$x \in S$$

$$d_i \geq 0$$

Theorem 4.2

i. If Max $W > 0$, then $\bar{x}$ is a dominated solution.
ii. If Max $W = 0$, then $\bar{x}$ is an efficient solution. ∎

NOTE: If Max $W > 0$, then at least one of the d_i's is positive. This implies that at least one objective can be improved without sacrificing the other objectives.

4.5 Classification of MCMP Methods

In MCMP problems, there are often an infinite number of efficient solutions and they are not comparable without the input from the DM. Hence, it is generally assumed that the DM has a real-valued *preference function* defined on the values of the objectives, but it is not known explicitly. With this assumption, the primary objective of the MCMP solution methods is to find the **best compromise solution**, which is an efficient solution that maximizes the DM's preference function.

In the last three decades, most MCDM research have been concerned with developing solution methods based on different assumptions and approaches to measure or derive the DM's preference function. Thus, the MCMP methods can be categorized by the basic assumptions made with respect to the DM's preference function as follows:

1. When *complete* information about the preference function is available from the DM.
2. When *no* information is available.
3. Where *partial* information is obtainable progressively from the DM.

In the following sections, we will discuss the MCMP methods—*Goal Programming*, *Compromise Programming*, and *Interactive Methods*, as examples of categories 1, 2, and 3 approaches, respectively.

4.6 Goal Programming (Ravindran et al. 2006)

One way to treat multiple criteria is to select one criterion as primary and the other criteria as secondary. The primary criterion is then used as the optimization objective function, while the secondary criteria are assigned acceptable minimum and maximum values and are treated as problem constraints. However, if careful considerations were not given while selecting the acceptable levels, a feasible design that satisfies all the constraints may not exist. This problem is overcome by *goal programming*, which has become a practical method for handling multiple criteria. Goal programming falls under the class of methods that use completely prespecified preferences of the DM in solving the MCMP problem.

In goal programming, all the objectives are assigned target levels for achievement and relative priority on achieving these levels. Goal programming treats these targets as *goals to aspire for* and not as absolute constraints. It then attempts to find an optimal solution that comes as "close as possible" to the targets in the order of specified priorities.

Before we discuss the formulation of goal programming models, we should discuss the difference between the terms *real constraints* and *goal constraints* (or simply *goals*) as used in goal programming models. The real constraints are absolute restrictions on the decision variables, while the goals are conditions one would like to achieve but are not mandatory. For instance, a real constraint given by

$$x_1 + x_2 = 3$$

requires all possible values of $x_1 + x_2$ to always equal 3. As opposed to this, a goal requiring $x_1 + x_2 = 3$ is not mandatory, and we can choose values of $x_1 + x_2 \geq 3$ as well as $x_1 + x_2 \leq 3$. In a goal constraint, positive and negative deviational variables are introduced as follows:

$$x_1 + x_2 + d_1^- - d_1^+ = 3 \quad d_1^+, d_1^- \geq 0.$$

Note that, if $d_1^- > 0$, then $x_1 + x_2 < 3$, and if $d_1^+ > 0$, then $x_1 + x_2 > 3$.

By assigning suitable weights w_1^- and w_1^+ on d_1^- and d_1^+ in the objective function, the model will try to achieve the sum $x_1 + x_2$ as close as possible to 3. If the goal were to satisfy $x_1 + x_2 \geq 3$, then only d_1^- is assigned a positive weight in the objective, while the weight on d_1^+ is set to zero.

4.6.1 Goal Programming Formulation

Consider the general MCMP problem given by Equation 4.5. The assumption that there exists an optimal solution to the MCMP problem involving multiple criteria implies the existence of some preference ordering of the criteria by the DM. The goal programming (GP) formulation of the MCMP problem requires the DM to specify an acceptable level of achievement (b_i) for each criterion f_i and specify a weight w_i (ordinal or cardinal) to be associated with the deviation between f_i and b_i. Thus, the GP model of an MCMP problem becomes

$$\text{Minimize } Z = \sum_{i=1}^{k} \left(w_i^+ d_i^+ + w_i^- d_i^- \right) \tag{4.7}$$

$$\text{Subject to } f_i(x) + d_i^- - d_i^+ = b_i \text{ for } i = 1,\ldots,k \tag{4.8}$$

$$g_j(x) \leq 0 \text{ for } j = 1,\ldots,m \tag{4.9}$$

$$x_j, d_i^-, d_i^+ \geq 0 \text{ for all } i \text{ and } j \tag{4.10}$$

Equation 4.7 represents the objective function of the GP model, which minimizes the weighted sum of the deviational variables. The system of equations (Equation 4.8) represents the *goal constraints* relating the multiple criteria to the goals/targets for those criteria. The variables, d_i^- and d_i^+, in Equation 4.8 are called *deviational variables*, representing *the underachievement* and *overachievement* of the *i*th goal. The set of weights (w_i^+ and w_i^-) may take two forms as given below:

1. Prespecified weights (cardinal)
2. Preemptive priorities (ordinal)

Under prespecified (cardinal) weights, specific values in a relative scale are assigned to w_i^+ and w_i^-, representing the DM's "trade-off" among the goals. Once w_i^+ and w_i^- are specified, the goal program represented by Equations 4.7 through 4.10 reduces to a single-objective optimization problem. The cardinal weights could be obtained from the DM using any of the methods discussed earlier, such as rating method and Borda count. However, in order for this method to work, the criteria values have to be scaled properly.

In reality, goals are usually *incompatible* (i.e., incommensurable) and some goals can be achieved only at the expense of some other goals. Hence, *preemptive goal programming*, which is more common in practice, uses *ordinal ranking* or *preemptive priorities* to the goals by assigning incommensurable goals to different priority levels and weights to goals at the same priority level. In this case, the objective function of the GP model (Equation 4.7) takes the form

$$\text{Minimize } Z = \sum_p P_p \sum_i \left(w_{ip}^+ d_i^+ + w_{ip}^- d_i^- \right), \qquad (4.11)$$

where P_p represents priority p with the assumption that P_p is much larger than P_{p+1} and w_{ip}^+ and w_{ip}^- are the weights assigned to the *i*th deviational variables at priority p. In this manner, lower-priority goals are considered only after attaining the higher-priority goals. Thus, *preemptive goal programming* is essentially a sequence of single-objective optimization problems, in which successive optimizations are carried out on the alternate optimal solutions of the previously optimized goals at higher priority.

In both preemptive and non-preemptive GP models, the DM has to specify the targets or goals for each objective. In addition, in the preemptive GP models, the DM specifies a preemptive priority ranking on the goal achievements. In the non-preemptive case, the DM has to specify relative weights for goal achievements.

To illustrate, consider the following bi-criteria linear program (BCLP):

Example 4.4 (BCLP)

$$\text{Max } f_1 = x_1 + x_2$$

$$\text{Max } f_2 = x_1$$

$$\text{Subject to } 4x_1 + 3x_2 \leq 12$$

$$x_1, x_2 \geq 0$$

Maximum f_1 occurs at $x = (0, 4)$ with $(f_1, f_2) = (4, 0)$. Maximum f_2 occurs at $x = (3, 0)$ with $(f_1, f_2) = (3, 3)$. Thus, the ideal values of f_1 and f_2 are 4 and 3, respectively, and the bounds on (f_1, f_2) on the efficient set will be

$$3 \leq f_1 \leq 4$$

$$0 \leq f_2 \leq 3$$

Let the DM set the goals for f_1 and f_2 as 3.5 and 2, respectively. Then, the GP model becomes

$$x_1 + x_2 + d_2^- - d_2^+ = 3.5 \tag{4.12}$$

$$x_1 + d_2^- - d_2^+ = 2 \tag{4.13}$$

$$4x_1 + 3x_2 \leq 12 \tag{4.14}$$

$$x_1, x_2, d_1^-, d_1^+, d_2^-, d_2^+ \geq 0 \tag{4.15}$$

Under the preemptive GP model, if the DM indicates that f_1 is much more important than f_2, then the objective function will be

$$\text{Min } Z = P_1 d_1^- + P_2 d_2^-$$

subject to the constraints (Equations 4.12 through 4.15), where P_1 is assumed to be much larger than P_2.

Under the non-preemptive GP model, the DM specifies relative weights on the goal achievements, say w_1 and w_2. Then, the objective function becomes

$$\text{Min } Z = w_1 d_1^- + w_2 d_2^-$$

subject to the same constraints (Equations 4.12 through 4.15).

4.6.2 Solution of Goal Programming Problems

4.6.2.1 Linear Goal Programs

Linear goal programming problems can be solved efficiently by the *partitioning algorithm* developed by Arthur and Ravindran (1978, 1980a). It is based on the fact that the definition of preemptive priorities implies that higher-order goals must be optimized before lower-order goals are even considered. Their procedure consists of solving a series of linear programming subproblems by using the solution of the higher-priority problem as the starting solution for the lower-priority problem. Care is taken that higher-priority achievements are not destroyed while improving lower-priority goals.

4.6.2.2 Integer Goal Programs

Arthur and Ravindran (1980b) show how the partitioning algorithm for linear GP problems can be extended with a modified branch and bound strategy to solve both pure and mixed integer GP problems. They demonstrate the applicability of the branch and bound algorithm by solving a multiple-objective nurse scheduling problem (Arthur and Ravindran 1981).

4.6.2.3 Nonlinear Goal Programs

Saber and Ravindran (1996) present an efficient and reliable method called the partitioning gradient-based (PGB) algorithm for solving nonlinear GP problems. The PGB algorithm uses the partitioning technique developed for linear GP problems and the generalized reduced gradient method to solve single-objective nonlinear programming problems. The authors also present numerical results by comparing the PGB algorithm against a modified pattern search method for solving several nonlinear GP problems. The PGB algorithm found the optimal solution for all test problems proving its robustness and reliability, while the pattern search method failed in more than half the test problems by converging to a nonoptimal point.

Kuriger and Ravindran (2005) have developed three intelligent search methods to solve nonlinear GP problems by adapting and extending the simplex search, complex search, and pattern search methods to account for multiple criteria. These modifications were largely accomplished by using partitioning concepts of goal programming. The paper also includes computational results with several test problems.

4.7 Method of Global Criterion and Compromise Programming

Method of Global Criterion (Hwang and Masud 1979) and *Compromise Programming* (Zeleny 1982) fall under the class of MCMP methods that do not require any preference information from the DM.

Consider the MCMP problem given by Equation 4.5. Let

$$S = \{x/g_j(x) \leq 0, \text{ for all } j\}$$

Let the ideal values of the objectives $f_1, f_2, \ldots, f_k$ be $f_1^*, f_2^*, \ldots, f_k^*$. The method of global criterion finds an efficient solution that is "closest" to the *ideal solution* in terms of the L_p distance metric. It also uses the ideal values to normalize the objective functions. Thus, the MCMP reduces to

$$\text{Minimize } Z = \sum_{i=1}^{k} \left(\frac{f_i^* - f_i}{f_i^*} \right)^p$$

Subject to $x \in S$.

The values of f_i^* are obtained by maximizing each objective f_i subject to the constraints $x \in S$, *but* ignoring the other objectives. The value of p can be 1, 2, 3, ..., ∞. Note that $p = 1$ implies equal importance to all deviations from the ideal. As p increases, larger deviations have more weight.

4.7.1 Compromise Programming

Compromise programming is similar in concept to the method of global criterion. It finds an efficient solution by minimizing the weighted L_p distance metric from the ideal point as given below.

$$\text{Min } L_p = \left[\sum_{i=1}^{k} \lambda_i^p \left(f_i^* - f_i \right)^p \right]^{1/p} \tag{4.16}$$

Subject to $x \in S$ and $p = 1, 2, \ldots, \infty$,

where λ_i's are weights that have to be specified or assessed subjectively. Note that λ_i could be set to $1/\left(f_i^* \right)$.

Theorem 4.3

Any point x* that minimizes L_p (Equation 4.16) for $\lambda_i > 0$ for all i, $\sum \lambda_i = 1$ and $1 \leq p < \infty$ is called a *compromise solution*. Zeleny (1982) has proved that these compromise solutions are non-dominated. As $p \to \infty$, Equation 4.16 becomes

$$\text{Min } L_\infty = \text{Min Max}_i \left[\lambda_i \left(f_i^* - f_i \right) \right]$$

and is known as the *Tchebycheff Metric*. ∎

4.8 Interactive Methods

Interactive methods for MCMP problems rely on the progressive articulation of preferences by the DM. These approaches can be characterized by the following procedure.

Step 1: Find a solution, preferably feasible and efficient.

Step 2: Interact with the DM to obtain his or her reaction or response to the obtained solution.

Step 3: Repeat steps 1 and 2 until satisfaction is achieved or until some other termination criterion is met.

When interactive algorithms are applied to real-world problems, the most critical factor is the functional restrictions placed on the objective functions, constraints, and the *unknown* preference function. Another important factor is *preference assessment styles* (hereafter called *interaction styles*). According to Shin and Ravindran (1991), the typical interaction styles are as follows:

a. *Binary pairwise comparison*—the DM must compare a pair of two-dimensional vectors at each interaction.

b. *Pairwise comparison*—the DM must compare a pair of p-dimensional vectors and specify a preference.

c. *Vector comparison*—the DM must compare a set of p-dimensional vectors and specify the best, the worst, or the order of preference (note that this can be done by a series of pairwise comparisons).

d. *Precise local trade-off ratio*—the DM must specify precise values of local trade-off ratios at a given point. It is the *marginal rate of substitution* between objectives f_i and f_j: in other words, trade-off ratio is how much the DM is willing to give up in objective j for a unit increase in objective i at a given efficient solution.

e. *Interval trade-off ratio*—the DM must specify an interval for each local trade-off ratio.

f. *Comparative trade-off ratio*—the DM must specify his preference for a given trade-off ratio.

g. *Index specification and value trade-off*—DM must list the indices of objectives to be improved or sacrificed, and specify the amount.

h. *Aspiration levels* (or reference point)—the DM must specify or adjust the values of the objectives that indicate his or her optimistic wish concerning the outcomes of the objectives.

Shin and Ravindran (1991) also provide a detailed survey of MCMP interactive methods. Their survey includes

- A classification scheme for all interactive methods
- A review of methods in each category based on functional assumptions, interaction style, progression of research papers from the first publication to all its extensions, solution approach, and published applications
- A rating of each category of methods in terms of the DM's cognitive burden, ease of use, effectiveness, and handling inconsistency

4.9 MCDM Applications

One of the most successful applications of multi-criteria decision making has been in the area of portfolio selection, an important problem faced by individual investors and financial analysts in investment companies. A portfolio specifies the amount invested in different securities, which may include bonds, common stocks, mutual funds, bank CD's, treasury notes, and others. Much of the earlier investment decisions were made by seat-of-the-pant approaches. Markowitz (1959) pioneered the development of the *modern portfolio theory*, which uses bi-criteria mathematical programming models to analyze the portfolio selection problem. By quantifying the trade-offs between risks and returns, he showed how an investor can diversify portfolios such that the portfolio risk can be reduced without sacrificing returns. Based on Markowitz's work, Sharpe (1963) introduced the concept of the *market risk* and developed a bi-criteria linear programming model for portfolio analysis. For their pioneering work in modern portfolio theory, both Markowitz and Sharpe shared the 1990 Nobel Prize in Economics. The Nobel award was the catalyst for the rapid use of modern portfolio theory by Wall Street firms in the 90s. A detailed discussion of the mean-variance model of Markowitz is given in Chapter 7 of this textbook.

Among the MCDM models and methods, the goal programming models have seen the most applications in industry and government. Chapter 4 of the textbook by Schniederjans (1995) contains an extensive bibliography (666 citations) on goal programming applications categorized by areas—accounting, agriculture, economics, engineering, finance, government, international, management, and marketing. Zanakis and Gupta (1995) also have a categorized bibliographic survey of goal programming applications. For an application of goal programming models in Supplier Selection, see Chapter 6, Section 6.4.8, of the textbook by Ravindran and Warsing (2013).

4.10 MCDM Software

One of the problems in applying MCDM methods in practice is the lack of commercially available software implementing these methods. There is some research software available. Two good resources for these are:

1. http://www.mcdmsociety.org/
2. http://www.sal.hut.fi/

The first is the web page of the International Society on Multiple-Criteria Decision Making. It has links to MCDM software and bibliography. A number of this software is available free for research and teaching use. The second link is to the research group at Helsinki University of Technology. It has links to some free software, again for research and instructional use.

4.11 Data Envelopment Analysis

Productivity is typically defined as the rate of output achieved per unit of input. One of the many challenges of measuring productivity of a service entity such as a bank or hospital or warehouse is that there are many inputs or resources that they manage and many output measures that are important. For example, a warehouse uses labor (e.g., order pickers), equipment such as forklifts and conveyor systems, information technology, warehouse space, and management. They perform a variety of functions including put-away, order picking, sortation, and packaging, and the warehouse manager will care about several outputs such as orders picked per unit time, number of picking errors per order, and value-added services provided. No single ratio of outputs to inputs, therefore, will measure the overall productivity of the warehouse.

We can illustrate this issue with a very simple example. Suppose there are three hospitals in a network. Each hospital has two inputs: number of

TABLE 4.4

Data for Hospital Network

Hospital	Healthcare Workers	Beds	Patients
H1	15	20	30
H2	16	21	30
H3	14	22	30

healthcare workers and number of hospital beds, and one output: average patients served per day. The values for each are shown in Table 4.4. In this example, we can think of productivity as patients served per healthcare worker or patients served per bed. H1 dominates H2 since it serves more patients per healthcare worker and more patients per bed. However, when comparing H1 to H3, we see that H1 has higher bed productivity but H3 has higher worker productivity. We cannot make an overall comparison of productivity based on this simple ratio idea, therefore.

DEA is an approach that determines an overall productivity measure called technical efficiency, which considers the whole portfolio of inputs and outputs. It is a nonparametric approach that can be formulated as a relatively simple optimization model. In this section, we will present the DEA framework along with examples and extensions. It is first necessary, however, to present some production economics fundamentals. The interested reader may find a more thorough coverage in the excellent text by Hackman (2008).

It is common in DEA to use the term decision-making unit (DMU). Each hospital in the previous discussion represents a DMU. DMU implies that there are decisions in determining how to best use inputs to get desired outputs. We will adopt this terminology for the remainder of this chapter.

4.11.1 Production Function

A *production function* $f(\cdot)$ describes how a DMU transforms a set of inputs $x = \{x_1, x_2, \ldots, x_n\}$ into an output y. We define the *input possibility set* $(L_{f(\cdot)}(u))$ of the production function as the set of inputs that achieve at least output level u. That is,

$$L_{f(\cdot)}(u) = \{x : f(x) \geq u\} \tag{4.17}$$

An *isoquant* $(Q_{f(\cdot)}(u))$ of the production function is the set of inputs that achieve exactly an output of u. That is,

$$Q_{f(\cdot)}(u) = \{x : f(x) = u\} \tag{4.18}$$

Production is called *efficient* if, for a given set of inputs, the maximum amount of output is produced (or, equivalently, the smallest levels of inputs

Evaluation of Service Systems

are used to achieve a given output). Further, the isoquant of efficient production may also be called the *efficient frontier*. We will distinguish between input and output efficiently later in this section.

There are several important characterizations of the production function. *Returns to scale* describes how the output changes when the input vector is multiplied by a scaler t. There are three classifications:

- If $f(tx) = tf(x)$ for all $t > 0$, then the production function exhibits *constant* returns to scale (CRS).
- If $f(tx) < tf(x)$ for all $t > 0$, then the production function exhibits *decreasing* returns to scale.
- If $f(tx) > tf(x)$ for all $t > 0$, then the production function exhibits *increasing* returns to scale.

Figure 4.3 illustrates the returns to scale graphically for a single input (x) and single output (y) example. The straight line shows the case for CRS and the segmented curve shows the case for variable returns to scale (VRS). For the VRS, there is increasing returns to scale as the input is increased to point A, beyond which there are decreasing returns to scale.

The *rate of technical substitution* (RTS) measures the rate at which one input (x_i) may be substituted for another input (x_j) to achieve the same output. It is defined by

$$RTS_{ij} = -\frac{\partial f(x)/\partial x_i}{\partial f(x)/\partial x_j}. \tag{4.19}$$

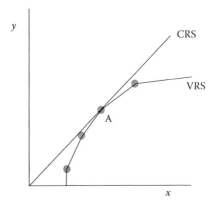

FIGURE 4.3
Illustration of returns to scale.

Output elasticity measures the percent change in output due to a small (1%) change in a single input (x_i). It is defined by

$$\varepsilon_i(x) = \frac{x_i}{f(x)} \frac{\partial f(x)}{\partial x_i}. \qquad (4.20)$$

Example 4.5

Consider the following production function (known as Cobb–Douglas):

$$f(x) = y = A x_1^\alpha x_2^\beta.$$

Figure 4.4 shows the input possibility set and isoquant when $\alpha = \beta = 0.5$, $A = 10$, and $y = 20$. We can determine the returns to scale as follows:

$$f(tx) = A(tx_1)^{0.5}(tx_2)^{0.5} = A t^{0.5+0.5} x_1^{0.5} x_2^{0.5} = t f(x).$$

Therefore, it exhibits CRS. Note that it is easy to see that, whenever $\alpha + \beta = 1$, Cobb–Douglas is CRS. The RTS when $\alpha = \beta = 0.5$ and $A = 10$ is equal to

$$RTS_{12} = -\frac{5 x_1^{-0.5} x_2^{0.5}}{5 x_1^{0.5} x_2^{-0.5}} = \frac{x_2}{x_1}.$$

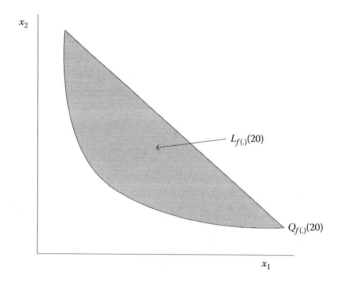

FIGURE 4.4
Cobb–Douglas input possibility set and isoquant ($\alpha = \beta = 0.5$, $A = 10$, $y = 20$).

Evaluation of Service Systems

If we evaluate this at $x_1 = 5$ and $x_2 = 10$, we get $RTS_{12} = 2$. The output elasticity for Cobb–Douglas with $\alpha = \beta = 0.5$ and $A = 10$ for input x_1 is given by

$$\varepsilon_i = \frac{x_1 \left(5 x_1^{-0.5} x_2^{0.5}\right)}{10 x_1^{0.5} x_2^{0.5}} = 0.5.$$

That is, a 1% change in x_1 will lead to a 0.5% change in output. Notice in this case that the output elasticity is constant. This holds for Cobb–Douglas only when $\alpha + \beta = 1$.

4.11.2 Technology

A DMU may have multiple inputs and outputs, and so a more general representation than the production function is needed. We will define *technology set* (T) as defining how the DMU transforms its inputs into outputs. For the single-output case, this is simply a different representation of the production function, which we can equivalently define as

$$T = \{(x, y) : f(x) \geq u\}. \tag{4.21}$$

For the multiple-input and -output case, we can define both input and output possibility sets. For each output vector y, the input possibility set is defined as

$$I(y) = \{x : (x, y) \in T\}, \tag{4.22}$$

and for each input vector x, the output possibility set is defined as

$$O(x) = \{y : (x, y) \in T\}. \tag{4.23}$$

We can therefore define the technology set as a collection of input possibility sets $\{I(y)\}$ or a collection of output possibility sets $\{O(x)\}$.

To this point, we have not discussed any underlying assumptions. However, there are several common properties that should hold for a technology to be well behaved. The most common ones are as follows:

1. A non-zero output vector can only be achieved with some positive input.
2. Each output possibility set is *closed* and *bounded*. Note that a subset S of real m-dimensional space $\mathcal{R}^m$ is closed if every point that is arbitrarily close to S belongs to S. It is bounded if it is contained in a sphere of finite radius.

3. Each input possibility set is *closed* and *convex*. Note that a set S in $\mathcal{R}^m$ is convex if for every $x_i, x_j \in S$ and for any real-valued α $(0 < \alpha < 1)$, the point $\alpha\, x_i + (1 - \alpha)\, x_j \in S$.

4. The technology set satisfies the condition of *free disposability*. Input free disposability means that if $(x, y) \in T$ and $x' \geq x$, then this implies $(x', y) \in T$. That is, adding additional inputs will not lead to a decrease in output. Similarly, output free disposability means that if $(x, y) \in T$ and $y' \leq y$, then this implies $(x, y') \in T$.

4.11.3 Nonparametric Models

The Cobb–Douglas model shown in Example 4.5 is an example of a parametric model, with parameters A, α, and β. In practice, if we know the form of production (e.g., is defined by Cobb–Douglas), then we can estimate the parameters from data. Parametric forms have the benefit of being differentiable, which helps facilitate analysis. In many cases, however, we do not know which parametric form to assume. In this section, we discuss how to estimate the technology using a nonparametric method.

Consider the simple case in Table 4.4 where an industry is made up of three DMUs (F1, F2, and F3), each using 2 inputs (x_1 and x_2) to produce an output y. F1 uses 2 units of x_1 and 4 units of x_2 to produce 3 units of output. F2 uses 5 units of x_1 and 2 units of x_2 to produce 3 units of output, and F3 uses 8 units of x_1 and 6 units of x_2 to produce 6 units of output. We can write the technology* as $T = \{(2,4,3), (5,2,3), (8,6,6)\}$. If we know that the industry exhibits CRS, then we can rewrite the technology so that all DMUs have the same output, namely, $T = \{(2,4,3), (5,2,3), (4,3,6)\}$. A plot of the input possibility set $I(3) = \{(2,4), (5,2), (4,3)\}$ is shown in Figure 4.5a. A question arises as to whether an input vector (e.g., (3,4)) is capable of producing an output of $y = 3$. This of course depends on the assumptions that are made about the technology set. If we assume free disposability, then this gives the input possibility set shown in Figure 4.5b. If we assume convexity and free disposability (as we did in Section 4.11.2), then this gives the input possibility set shown in Figure 4.5c.

It is interesting to note in Figure 4.5c that F3 is not on the isoquant of the input possibility set. This implies that under the assumption of convexity, the production of F3 is not efficient. That is, a firm with fewer inputs could achieve the same output as F3. We will discuss how to measure this "inefficiency" later in the section. First, we need to provide more formal treatment of nonparametric construction of the input technology set. We follow the approach developed in Hackman (2008).

We begin by presenting a conservative nonparametric construction developed by Hanoch and Rothschild (1972) from observed data on N DMUs,

---

* Note that many economics texts such as Varian (1992) use negative values for inputs and positive values for outputs.

Evaluation of Service Systems

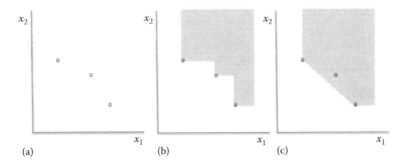

FIGURE 4.5
Plot of (a) input possibility set, (b) assuming free disposability, and (c) assuming convexity and free disposability.

where each DMU has multiple inputs and a single output. For a given output u, we define

$$\chi(u) = \{x_j\}_{j \in \Psi(u)}, \text{ where } \Psi(u) = \{j : u_j \geq j\} \tag{4.24}$$

Hanoch and Rothschild (H–R) define the input possibility set as the smallest convex and freely disposable set that contains the data. We can write this in equation form as

$$I^{H-R}(u) = \left\{ z : z \geq \sum_{j \in \psi(u)} \lambda_j x_j, \sum_{j \in \psi(u)} \lambda_j = 1, \lambda_j \geq 0 \ \forall j \right\}. \tag{4.25}$$

Note that this constructs the convex combinations from the data. Further, we are making no assumptions about the returns to scale. Finally, the method is conservative in the sense that it constructs the smallest set from the given data. Note that the smallest convex set on the given data is known as the *convex hull* and is constructed by a set of linear inequalities.

Example 4.6

Four DMUs are considered, each with 2 inputs (x_1 and x_2) and 1 output as shown in Figure 4.6. In the figure, output is given in parenthesis next to each point.

If we set the output to $u = 19$, then two DMUs will be in the input possibility set $I^{H-R}(19)$, which is shown in Figure 4.6a. Note that $I^{H-R}(19)$ does not differ from $I^{H-R}(20)$. If the output is reduced to 15, then all DMUs are in the set, which is shown in Figure 4.6b. It is clear from Figure 4.6a and b that the shaded areas are convex hulls and are defined by linear inequalities on the data.

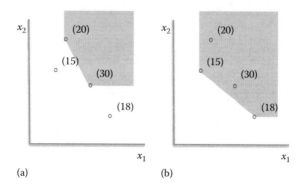

FIGURE 4.6
H–R input possibility sets for $u = 19$ (a) and $u = 15$ (b).

One of the key limitations of H–R is that it is only applicable for a single output model. Charnes et al. (1978) developed a less conservative method called DEA that assumes convexity on both the input and output sets. As such, the method is applicable to multiple inputs and outputs. The initial formulation assumed CRS. A VRS model was presented in Banker et al. (1984).

The technology set for the VRS model is the smallest convex and freely disposable set that contains all of the data. For N DMUs we have

$$T^{VRS} = \left\{ (x,y) : x \geq \sum_{i=1}^{N} \lambda_i x_i, \ y \leq \sum_{i=1}^{N} \lambda_i y_i, \ \sum_{i=1}^{N} \lambda_i = 1, \lambda_i \geq 0 \right\}. \quad (4.26)$$

All convex combinations of input and output bundles are feasible by definition. The first two terms are the convex combinations when we restrict $\sum_{i=1}^{N} \lambda_i = 1$. We therefore ensure that the technology is convex.*

The technology set for the CRS model is the smallest CRS technology that contains the VRS technology set. In this case,

$$T^{CRS} = \left\{ (x,y) : x \geq \sum_{i=1}^{N} \lambda_i x_i, \ y \leq \sum_{i=1}^{N} \lambda_i y_i, \ \lambda_i \geq 0 \right\}. \quad (4.27)$$

Notice that for CRS, the restriction on λ was removed. This can be explained as follows. Recall from the definition of CRS that for any scalar $t > 0$, (tx, ty) will also be in T. If we define $\mu_j = t\lambda_j$ and restrict $\sum_{j=1}^{N} \lambda_j = 1$, then we have $\sum_{j=1}^{N} \mu_j = t$ and is non-negative. Since t is not

---

* A technology set is convex when $(x_1, y_1) \in T$ implies $(\lambda x_1 + (1-\lambda)x_1, \lambda y_1 + (1-\lambda)y_1) \in T$, where $\lambda \in [0,1]$.

Evaluation of Service Systems

further restricted, then no further restriction need be applied to $\sum_{j=1}^{N}\mu_j$, and hence we can drop the restriction $\sum_{j=1}^{N}\lambda_j = 1$.

The construction of the input possibility set $I^{CRS}(y)$ is fairly straightforward to develop. Recall from the definition of CRS that $I^{CRS}(u) = uI^{CRS}(1)$. If we define $\hat{x}_i$ as the scaled vector x_i/y_i, then $\hat{x}_i \in I^{CRS}(1)$. We can then apply the H–R construction to the scaled data $(\hat{x}_i, 1)$. To determine this for other outputs, we simply apply the definition of CRS and scale from the origin. This is illustrated in the following example.

Example 4.7

Consider the case of four DMUs with two inputs (x_1 and x_2) to achieve a single output (y). The data are given in Table 4.5.

Note that the last two columns show the scaled data. Figure 4.7a shows a plot of the scaled data. The H–R construction on the scaled data forms the convex hull shown in Figure 4.7b, which also corresponds to $I^{CRS}(1)$. Finally, Figure 4.7c shows $I^{CRS}(2)$. It is constructed by multiplying each scaled input by 2 and forming the convex hull by H–R construction.

TABLE 4.5

DMU Data for Example 4.7

DMU	x_1	x_2	y	x_1/y	x_2/y
1	1	10	10	0.1	1.0
2	4	12	20	0.2	0.6
3	18	12	30	0.6	0.4
4	10	2	20	1.0	0.2

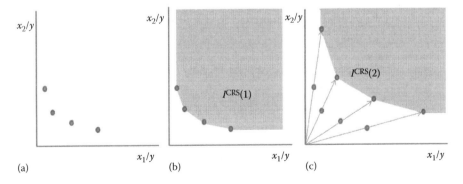

FIGURE 4.7
Plot of scaled data (a), input possibility set for an output of 1 (b), and input possibility set for an output of 2 (c).

Note that any input possibility set $I^{CRS}(t)$ may be determined by multiplying the scaled data by t and performing the H–R construction.

In Example 4.7, all points in Figure 4.7b lie on the boundary of the input possibility set. However, in Figure 4.5c, this is not the case; the data for F3 are in the interior. This implies that the output achieved by F3 could have been achieved with strictly less inputs. We can use the distance of the projection of F3 to the boundary as a measure of the *efficiency* of that DMU. Equivalently, we can consider the distance of the projection to the boundary of the output possibility set as a measure of efficiency.

Formally, we define the input ($TE^I(x,y)$) and output ($TE^O(x,y)$) efficiencies as follows:

$$TE^I(x,y) = \min\{\theta : (\theta x, y) \in T\} \tag{4.28}$$

$$TE^O(x,y) = \min\{\theta : (x, y/\theta) \in T\} \tag{4.29}$$

Since the technology set is made up of the original data (x, y), then for both technical efficiency measures, θ is in the interval $(0,1]$. Further, under the assumption of CRS, there is no difference between the two measures. One final observation is that in the measure of input technical efficiency, each input is scaled by the same amount. A value of $\theta = 0.9$ for a DMU means that the output obtained by the DMU could have been achieved using only 90% of each of the inputs.

There have been several different projections proposed, but we will use the equal proportions method throughout the rest of this chapter. In practice, we solve for technical efficiencies through the use of linear programming. We discuss this in the next section.

4.12 DEA Linear Programming Formulation

The equivalent DEA linear programs for Equations 4.28 and 4.29 are given below for DMU o for input vector x and output vector y:

Notation:

x_{ij} = value of input i for DMU j, for $i = 1,\ldots,n, j = 1,\ldots N$

y_{kj} = value of input k for DMU j, for $k = 1,\ldots,m, j = 1,\ldots N$

θ_O = technical efficiency for DMU o, $o \in \{1,\ldots N\}$

λ_j = intensity for DMU j, $j = 1,\ldots N$

Input-Oriented CRS DEA

$$\min \theta_o \tag{4.30}$$

Subject to

$$\sum_{j=1}^{N} \lambda_j x_{ij} \leq \theta_o x_{io}, \ \forall i \tag{4.31}$$

$$\sum_{j=1}^{N} \lambda_j y_{kj} \geq y_{ko}, \ \forall k \tag{4.32}$$

$$\lambda_j \geq 0, \ \forall j \tag{4.33}$$

Input-Oriented VRS DEA

$$\min \theta_o \tag{4.34}$$

Subject to

$$\sum_{j=1}^{N} \lambda_j x_{ij} \leq \theta_o x_{io}, \ \forall i \tag{4.35}$$

$$\sum_{j=1}^{N} \lambda_j y_{kj} \geq y_{ko}, \ \forall k \tag{4.36}$$

$$\sum_{j=1}^{N} \lambda_j = 1 \tag{4.37}$$

$$\lambda_j \geq 0, \ \forall j \tag{4.38}$$

We leave the output-oriented formulations as an exercise. Several comments are in order before giving an example. First, notice that the objective is specified for a particular DMU. Therefore, the linear program needs to be solved for each DMU. Second, it is important to remember that the technical efficiency is determined by an equal proportion projection to the boundary of T. We cannot, therefore, make direct comparisons between DMUs other than by a ranking. For example, if DMU i has $\theta_i = 0.9$ and DMU j has $\theta_j = 0.6$, all we can say is that i is more efficient at using its inputs than j. We cannot say that i is $100(0.9 - 0.6)/0.6 = 50\%$ more efficient than j. Finally, the decision variables are θ_o and λ_j ($j = 1, ..., N$). Any strictly positive λ_j implies that DMU j was used in the construction of the projection. This implies that DMU j lies on the boundary of the input possibility set and is hence efficient (i.e., it has a technical efficiency equal to 1).

Example 4.8

Consider five DMUs that use two inputs (x_1 and x_2) to obtain two outputs (y_1 and y_2). The data are given in Table 4.6. We want to find the input-oriented technical efficiency under CRS and VRS for DMU 3.

The input-oriented CRS formulation for DMU 3 is given by

$$\min \theta_3$$

Subject to

$$10\lambda_1 + 23\lambda_2 + 29\lambda_3 + 18\lambda_4 + 13\lambda_5 \leq 29\theta_3$$

$$15\lambda_1 + 19\lambda_2 + 20\lambda_3 + 12\lambda_4 + 20\lambda_5 \leq 20\theta_3$$

$$5\lambda_1 + 7\lambda_2 + 8\lambda_3 + 7\lambda_4 + 9\lambda_5 \geq 8$$

$$9\lambda_1 + 11\lambda_2 + 10\lambda_3 + 3\lambda_4 + 10\lambda_5 \geq 10$$

Solution of the linear program gives $\theta_3 = 0.948$, $\lambda_1 = \lambda_3 = \lambda_4 = 0$, $\lambda_2 = 0.345$, and $\lambda_5 = 0.621$. DMU 3 is therefore not efficient. We see from the values

TABLE 4.6

DMU Data for Example 4.8

DMU	x_1	x_2	y_1	y_2
1	10	15	5	9
2	23	19	7	11
3	29	20	8	10
4	18	12	7	3
5	13	20	9	10

of the intensities that DMUs 2 and 5 are both efficient. Finally, if we add the restriction on the intensities for the VRS model, then we add the constraint:

$$\lambda_1 + \lambda_2 + \lambda_3 + \lambda_4 + \lambda_5 = 1$$

If we solve this modified formulation, we obtain: $\theta_3 = 0.950$, $\lambda_3 = \lambda_4 = 0$, $\lambda_1 = 0.167$, $\lambda_2 = 0.167$, and $\lambda_5 = 0.666$.

4.13 Practical Considerations for DEA

There are several important practical considerations to consider for DEA. First, it is important to remember that the construction of the technology set is completely defined by the observed data. Consider the case where new DMUs are added to an existing set of DMUs D to form D'. For a DMU $i \in D$, it is the case that $\theta_i^D \leq \theta_i^{D'}$. Therefore, the technical efficiency for a DMU found by DEA is actually an upper bound. That is, the results only give information as to how a DMU is doing with respect to its "peers" and not to an absolute standard. A DMU with a technical efficiency of 1 may therefore have the false impression that they are doing everything as well as possible.

This naturally leads to the question of how many DMUs are required for the results to be meaningful. Unfortunately, there is not a simple answer. It depends on several factors including number of inputs, number of outputs, and the distribution of their values. Several researchers have recommended "rules of thumb" based on the number of inputs (n) and outputs (m). Examples include $N \geq 2(n + m)$ (Golany and Roll 1989), $N \geq 3(n + m)$ (Bowlin 1998), and $N \geq 2nm$ (Dyson et al. 2001).

A second issue with the use of the DEA model presented is that it is possible for a DMU to be technically efficient in the input-based DEA model and yet find some (x, y) that uses no more inputs than the DMU, with at least one of the inputs using strictly less input. This can be seen in the following simple case. Consider three DMUs with the data (x_1, x_2, y) equal to $\{(1, 5, 10), (1, 4, 10), (5, 2, 20)\}$. It is easy to see by observation that the results of the input-based CRS model are that all DMUs have a technical efficiency of 1. However, DMU 2 clearly dominates DMU 1. We will discuss the notion of allocative efficiency in Section 4.14, which helps alleviate this issue.

Third, and most limiting, is the assumption that the data used in the analysis contain no variability. In practice, the data will always have variability (and/or error). For example, the number of customers that a bank customer sees during an hour will vary based on the customer arrivals and the distribution of tasks she or he needs to perform for those customers. Because technical efficiency in DEA is completely based on the data, then even small perturbations can lead to large differences in efficiency. In Section 4.15, we will present a stochastic framework that helps address this.

Fourth, the proper selection of inputs and outputs is extremely important for a meaningful model. In addition, careful attention should be taken in defining the proper boundaries of the system of interest as well as the purpose of the analysis. Various statistical methods have been developed for variable selection. Nataraja and Johnson (2011) give an overview of the four most widely used approaches.

Finally, the DEA framework assumes that DMUs use their discretion in determining how to choose the mix of inputs and use them to achieve outputs. In some cases, however, the DM may not have influence on the input. For example, although floor space area may be an input in a DEA model on warehousing, the warehouse manager may not be able to change that space. Similarly, for the case of hospitals, although the number of beds would certainly be an input, the process of actually changing that number is a long and involved process (as discussed in Chapter 11) and hence the DM may not be able to adjust in the shorter term. We will call inputs that the DM cannot influence as *non-discretionary* inputs. We can therefore partition the input variables into n_D discretionary variables and n_{ND} non-discretionary variables, and modify the input constraints to

$$\sum_{j=1}^{N} \lambda_j x_{ij} \leq \theta_o x_{io}, \quad \text{for } i = 1, \ldots, n_D \tag{4.39}$$

$$\sum_{j=1}^{N} \lambda_j x_{ij} \leq x_{io}, \quad \text{for } i = 1, \ldots, n_{ND} \tag{4.40}$$

Notice that the non-discretionary constraints influence the intensities but do not directly influence the technical efficiency.

4.14 Allocative and Total Efficiency in DEA

The DEA model presented in Section 4.13 did not consider the costs of the inputs to achieve outputs. In many cases, it is not practical to estimate input costs, but when it is, further information can be determined about how well a DMU uses its inputs to achieve outputs. We first give a little background on the cost function. The basic definition of a cost function ($C(y,c)$) is

$$C(y,c) = \min\{c \cdot x : x \in I(y)\}. \tag{4.41}$$

Evaluation of Service Systems

Two standard assumptions are made on the cost function:

1. The cost function is *linearly homogeneous*. Note that a function $f(\cdot)$ is linearly homogeneous if $f(s \cdot x) = sf(x)$ for any scalar s ($s > 0$).
2. The cost function is *concave* in c for a given y. Note that a real-valued function is concave on a nonempty convex set in $\mathcal{R}^n$ if for every x and y in the set, $f(\lambda x + (1 - \lambda)y) \geq \lambda f(x) + (1 - \lambda) f(y)$. This comes directly from the definition of the cost function since the minimum of concave functions is concave.

We further assume that *all* DMUs face the same unit costs. We can illustrate the use of the cost function by returning to Example 4.5.

Example 4.9

Consider the Cobb–Douglas production function considered in Example 4.5. Suppose that a unit of x_1 is $10 and a unit of x_2 is $20. Further, we desire an output of 200 units. We get

$$C(y,c) = \min\left\{10x_1 + 20x_2 : 10x_1^{0.5}x_2^{0.5} \geq 20\right\}.$$

The solution to this function (given as an exercise) is $x_1 = 28.28$ and $x_2 = 14.14$. Figure 4.8 shows the boundary of input possibility set for $y = 200$ (straight line) and the line defining the minimum cost, which is tangent to the boundary (curved line). This line is commonly called the *isocost line*.

We can use the cost function to determine what is known as *allocative efficiency*. Allocative efficiency is a measure of the appropriateness of the

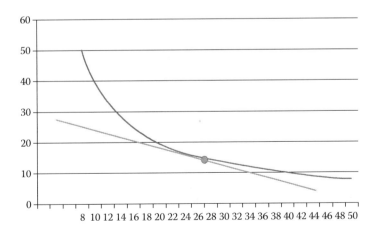

FIGURE 4.8
Boundary of input possibility set (curved line) and isocost line (straight line) for Example 4.9.

choice of inputs based on their costs. This can be illustrated graphically. Consider a two-input and one-output example as shown in Figure 4.9 for four DMUs, assuming CRS. The line through A and DMU 2 is the isocost line. Three of the DMUs are on the boundary of the input possibility set and hence have a technical efficiency of 1. DMU 3 (which we will call point C) is on the interior of the boundary, and hence has a technical efficiency strictly less than 1. Consider the ray drawn from the origin through DMU 3, and let $d(a, b)$ represent the distance from point a to b. The technical efficiency in this case will be $d(O,C)/d(O,B)$. The allocative efficiency is given by $d(O,A)/d(O,B)$. In this case, DMU 2 has an allocative efficiency of 1. Note that two firms (DMU 1 and DMU 4) are both technically efficient, but allocatively inefficient for the given costs. Finally, we can define an overall efficiency for DMU 3 as $d(O,A)/d(O,C)$, which is equivalent to the product of the technical and allocative efficiencies.

In terms of the costs, the overall efficiency for DMU j is defined as

$$OE(x_j, y_j, c_j) = \frac{C(y_j, c_j)}{c_j \cdot x_j} = \frac{\sum_{i=1}^{n} c_i x_i^*}{\sum_{j=1}^{n} c_j x_j}. \qquad (4.42)$$

We can determine the vector x^* for a given DMU o by solving the following linear program:

$$\min \sum_{i=1}^{n} c_i x_i \qquad (4.43)$$

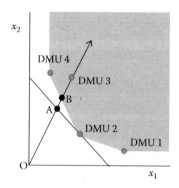

FIGURE 4.9
Illustration of technical, allocative, and overall efficiency.

Evaluation of Service Systems

Subject to

$$\sum_{j=1}^{N} \lambda_j x_{ij} \leq x_i, \quad \forall i \tag{4.44}$$

$$\sum_{j=1}^{N} \lambda_j y_{kj} \geq y_o, \quad \forall k \tag{4.45}$$

$$\lambda_j \geq 0, \quad \forall j \tag{4.46}$$

The decision variables are x_i ($i = 1, \ldots, n$) and λ_j ($j = 1, \ldots, N$). We illustrate with the following example.

Example 4.10

Consider the data given in Example 4.7. Suppose the costs are $c_1 = \$30$ and $c_2 = \$15$. Let us determine the allocative and overall efficiency for DMU 3. The linear programming model for costs is given by

$$\min 30x_1 + 15x_2$$

Subject to

$$10\lambda_1 + 23\lambda_2 + 29\lambda_3 + 18\lambda_4 + 13\lambda_5 \leq x_1$$

$$15\lambda_1 + 19\lambda_2 + 20\lambda_3 + 12\lambda_4 + 20\lambda_5 \leq x_2$$

$$5\lambda_1 + 7\lambda_2 + 8\lambda_3 + 7\lambda_4 + 9\lambda_5 \geq 8$$

$$9\lambda_1 + 11\lambda_2 + 10\lambda_3 + 3\lambda_4 + 10\lambda_5 \geq 10$$

The solution to this linear program is $x_1^* = 12.45$ and $x_2^* = 19.03$. The overall efficiency of DMU 3 is equal to

$$OE = \frac{30(12.45) + 15(19.03)}{30(20) + 15(8)} = 0.563.$$

We can find the allocative efficiency by the ratio of the overall efficiency to the technical efficiency (which we determined previously). In this case, we get AE = 0.563/0.950 = 0.593. It is clear that the biggest improvement for DMU 3 would come by considering a different input mix. Note that it is easy to extend the idea of costs on the inputs to prices on the outputs. We give this as an exercise.

4.15 Stochastic DEA

As mentioned previously, in practice, DMUs are typically subject to variability. DEA, however, assumes deterministic (and perfectly measured) inputs and outputs. In the presence of random variation, changes in the data can lead to significant changes in DEA results. As a result, there has been significant work in modeling variability in a DEA framework. One method for dealing with random variation is the use of stochastic DEA models. We present an extension to the previously presented DEA models that uses chance constraints (El-Demerdash et al. 2013; Ray 2004).

Ray (2004) developed an output-oriented stochastic DEA approach. In this model, inputs are deterministic and lead to outputs with random variation. It is assumed for DMU i that each output y_{ki} is normally distributed with mean μ_{ki} and standard deviation σ_{ki}. Further, Ray (2004) assumed that for each output variable, the output from DMU i is independent of the output from DMU j (i.e., the covariance, cov(y_{ki}, y_{kj}) = 0). If the covariance terms are not zero and are known, however, it is easy to include them in the model. We will do so here.

Because of the random nature of the outputs, it cannot be ensured with certainty that the output constraints won't be violated. We can develop *chance constrained* outputs that ensure that the probability that the inequality holds for a random sample of these variables does not fall below a certain level. We define chance output constraint as

$$P\left\{\sum_{j=1}^{N} \lambda_j y_{kj} \geq \theta_o y_{ko}\right\} = (1 - \alpha_k), \quad \forall k. \tag{4.47}$$

In this case, we are setting the probability ($P\{\cdot\}$) of the constraint holding for output k to significance level $(1 - \alpha_k)$. We define random variable v as

$$v_{ko} = \sum_{j=1}^{N} \lambda_j y_{kj} - \theta_o y_{ko}. \tag{4.48}$$

Evaluation of Service Systems

Using the definition of expectation, we get

$$E[v_{ko}] = \sum_{j=1}^{N} \lambda_j \mu_{kj} - \theta_o \mu_{ko} \tag{4.49}$$

$$\text{var}[v_{ko}] = \sum_{\substack{j=1 \\ j \neq o}}^{N} \lambda_j^2 \sigma_{kj}^2 + (\lambda_o - \theta_o)^2 \sigma_{ko}^2 + 2\,\text{cov}(y_{kj}, y_{ko}). \tag{4.50}$$

Because the output is normally distributed, then we have

$$P\left\{\sum_{j=1}^{N} \lambda_j y_{kj} \geq \theta_o y_{ko}\right\} = P\{v \leq 0\} = P\left\{z \leq \frac{v - \mu_{ko}}{\sigma_{ko}}\right\}, \tag{4.51}$$

where z is the standard normal variable. From the standard normal,

$$P\{v \leq 0\} = P\left\{z \leq \frac{-\mu_{ko}}{\sigma_{ko}}\right\} = 1 - \Phi\left(\frac{\mu_{ko}}{\sigma_{ko}}\right), \tag{4.52}$$

where $\Phi(\cdot)$ is the cumulative density for the standard normal. If we set this to the significance level $(1 - \alpha_k)$, this gives

$$1 - \Phi\left(\frac{\mu_{ko}}{\sigma_{ko}}\right) \geq (1 - \alpha_k) \Rightarrow \mu_{ko} \leq Z_{\alpha_k} \sigma_{ko}. \tag{4.53}$$

We now have the components we need to modify the output constraints. The resulting output-oriented stochastic DEA formulation (VRS) is as follows:

$$\max \theta_o \tag{4.54}$$

Subject to

$$\sum_{j=1}^{N} \lambda_j x_{ij} \leq x_{io}, \quad \forall i \tag{4.55}$$

$$\sum_{j=1}^{N} \lambda_j \mu_{kj} \geq \theta_o \mu_{ko}$$

$$+ Z_{\alpha_k} \sqrt{\sum_{\substack{j=1 \\ j \neq o}}^{N} \lambda_j^2 \sigma_{kj}^2 + (\lambda_o - \theta_o)^2 \sigma_{ko}^2 + 2\text{cov}(y_{kj}, y_{ko})}, \quad \forall k \qquad (4.56)$$

$$\sum_{j=1}^{N} \lambda_j = 1 \qquad (4.57)$$

$$\lambda_j \geq 0, \quad \forall j \qquad (4.58)$$

If only a subset of the outputs is assumed to be stochastic, the output constraints can be split into chance-constrained and deterministic. In addition, a very similar analysis can be done for the case where outputs are assumed to be deterministic and inputs are subject to random variation. We illustrate this in the following example.

Example 4.11

Consider the DMU data (Table 4.6) from Example 4.8. We wish to find the technical efficiency for DMU 3. It turns out that rather than being deterministic, x_2 is now normally distributed as given in Table 4.7. In addition, the covariance between DMU 3 and the other DMUs for this output is given in Table 4.8. All other data are the same.

TABLE 4.7

Distribution of Input Variable for Example 4.11

DMU	1	2	3	4	5
μ_{2i}	15	19	20	12	20
σ_{2i}^2	3	4	4	1	4

TABLE 4.8

Covariance Data for Example 4.11

	DMU 1	DMU 2	DMU 4	DMU 5
DMU 3	0.4	0.6	0.7	0.5

Evaluation of Service Systems

If we want to be 90% confident of satisfying the chance constraint (i.e., $Z_{0.90} = 1.28$), the input-oriented VRS stochastic DEA model to find the technical efficiency for DMU 3 is

$$\min \theta_3$$

Subject to

$$10\lambda_1 + 23\lambda_2 + 29\lambda_3 + 18\lambda_4 + 13\lambda_5 \leq 29\theta_3$$

$$15\lambda_1 + 19\lambda_2 + 20\lambda_3 + 12\lambda_4 + 20\lambda_5 - 20\theta_3$$
$$\leq 1.28\Big[\big(3\lambda_1 + 4(\lambda_1 - \theta_3)^2 + 2(0.4)\big)$$
$$+ \big(4\lambda_2 + 4(\lambda_1 - \theta_3)^2 + 2(0.6)\big)$$
$$+ \big(1\lambda_4 + 4(\lambda_1 - \theta_3)^2 + 2(0.7)\big)$$
$$+ \big(3\lambda_5 + 4(\lambda_1 - \theta_3)^2 + 2(0.5)\big)\Big]^{0.5}$$

$$5\lambda_1 + 7\lambda_2 + 8\lambda_3 + 7\lambda_4 + 9\lambda_5 \geq 8$$

$$9\lambda_1 + 11\lambda_2 + 10\lambda_3 + 3\lambda_4 + 10\lambda_5 \geq 10$$

$$\lambda_1 + \lambda_2 + \lambda_3 + \lambda_4 + \lambda_5 = 1$$

It is important to note that the chance constraint is nonlinear. We therefore need to use a nonlinear solver. The solution to the formulation is: $\theta_3 = 0.735$, $\lambda_3 = \lambda_4 = 0$, $\lambda_1 = 0.167$, $\lambda_2 = 0.167$, and $\lambda_5 = 0.666$. Note that if we increase the significance level to 95% (i.e., $Z_{0.95} = 1.96$), the technical efficiency becomes $\theta_3 = 0.639$.

The nonlinearity of the chance constraint can be computationally difficult to solve for large problems. In addition, the assumption of random variation being normally distributed is limiting. Stochastic DEA is a very active area of research and many other models exist. The interested reader can find several models in Zhu (2015).

4.16 DEA Software

There are several commercially available DEA solvers available. However, there are also many open-source versions. Although the authors of this text do not recommend one version over any other, some examples include the following:

1. http://www.opensourcedea.org/index.php?title=Open_Source_DEA
2. https://www.deaos.com/login.aspx?ReturnUrl=%2f
3. https://cran.r-project.org/web/packages/nonparaeff/nonparaeff.pdf

The first link is for open-source solver software that works in both Windows and Linux. The site also includes a solver library that can be used with Java. The second link is for a web-based program that can solve limited size instances. Finally, the third link is to a package written in R-cran, which is open-source software. Several other packages are available on this site.

Exercises

4.1 List three or more equally important criteria that might be considered in each of the following decision situations:
   a. Choosing an apartment for rent.
   b. Purchase of a car.
   c. Selecting a university to pursue a PhD degree.
   d. Choosing a partner in marriage.

   Also indicate whether each of the criteria should be treated as a goal or as an objective and list some attributes to measure them.

4.2 Discuss the differences between the following terms in MCDM:
   a. Efficient and ideal solutions
   b. Decision space and objective space
   c. Dominated and Pareto optimal solutions
   d. Multiple-criteria selection problem and multiple-criteria mathematical programming problem
   e. Goals and constraints
   f. Preemptive and non-preemptive goal programs

4.3 Discuss how elasticity meaningfully differs from the derivative in DEA.

4.4 Give a practical example for when you believe constant returns to scale, increasing returns to scale, and decreasing returns to scale are appropriate.

4.5 Consider a DEA model for a university system. What do you believe are the five most important inputs? What are the three most important outputs? Discuss how you would measure them.

4.6 You have been asked to rank a set of universities using technical efficiency found from DEA. Discuss the main advantages and disadvantages of this approach.

4.7 Discuss the most limiting assumptions on allocative efficiency in DEA.

4.8 Prove which of the following problems is an MCDM problem. Justify your conclusion.

i. Max $Z_1 = x_1 + x_2$; Max $Z_2 = 2x_1 + x_2$

Subject to

$$0 \leq x_1 \leq 7$$

$$0 \leq x_2 \leq 10$$

ii. Max $Z_1 = x_1 + x_2$; Max $Z_2 = x_1 - x_2$

Subject to

$$0 \leq x_1 \leq 5$$

$$0 \leq x_2 \leq 8$$

4.9 Consider the following bi-criteria LP model:

Maximize

$$z_1 = 5x_1 + x_2$$

Minimize

$$z_2 = x_1 + x_2$$

Subject to:

$$x_1 + x_2 \leq 6$$

$$x_1 \leq 5$$

$$x_2 \leq 3$$

$$x_1, x_2 \geq 0$$

a. Show graphically the decision space and the objective space.
b. Show graphically the set of efficient solutions in both the decision space and the objective space.
c. Find two dominated and non-dominated solutions.
d. What is the ideal solution? Is it achievable?
e. Determine the best lower and upper bounds for z_1 and z_2.

4.10 XYZ company has an MCDM problem with six criteria—C1, C2, C3, C4, C5, and C6.

The company, wishing to determine the order of importance of these criteria, uses the method of paired comparisons and Borda method to obtain a ranking. The company's judgments for the paired comparisons are as follows:

$C^1 > C^2 \quad C^2 < C^3 \quad C^3 < C^4 \quad C^4 > C^5 \quad C^5 > C^6$
$C^1 > C^3 \quad C^2 < C^4 \quad C^3 > C^5 \quad C^4 > C^6$
$C^1 < C^4 \quad C^2 > C^5 \quad C^3 > C^6$
$C^1 > C^5 \quad C^2 > C^6$
$C^1 > C^6$

For the above data, derive the Preference Matrix and compute the ordinal ranking of the criteria for the company in terms of order of importance and the criteria weights. Show all steps.

4.11 You just completed your PhD in Industrial Engineering, have job offers from six different universities, and are faced with the problem of choosing the right university to work for. You have collected the necessary data about each university, as given in Table 4.9.

Acting as the new PhD, use the following methods to determine the criteria weights and the ranking of the universities. [*Notes:* (i) You will be maximizing criteria 1, 3, 4, and 5 and minimizing criteria 2 and 6. (ii) Scale the salaries by dividing by $10,000.]

a. Rating method
b. Borda method, using pairwise comparison of criteria

Do you find any rank reversals?

TABLE 4.9
Data for Exercise 4.11

University Criteria	1	2	3	4	5	6
1. 9-month salary (in $)	$94,000	$93,000	$96,000	$97,500	$92,000	$90,000
2. Average teaching load (hours/semester)	9	8	10	12	6	11
3. Geographic location[a]	2	3	5	2	4	1
4. Summer support[a]	7	6	3	8	10	4
5. Research support[a]	4	4	3	1	5	2
6. Number of different preparations/year	3	2	4	5	2	3

[a] Scaled—where larger numbers represent more desirable outcomes.

4.12 Consider the MCLP problem:

Max
$$f_1(x) = -x_1 + x_2$$

Max
$$f_2(x) = 10x_1$$

Min
$$f_3(x) = 2x_1 - 3x_2$$

Subject to:
$$-x_1 + x_2 \leq 2$$

$$x_1 \leq 7$$

$$x_1 + x_2 \leq 8$$

$$x_1, x_2 \geq 0$$

Determine at least two properly efficient points to the above MCLP using Geoffrion's theorem. (*Note:* You may use graphical means to solve the LP problems.)

4.13 Consider the MCLP problem:

Max
$$f_1 = x_1 + x_2$$

Max
$$f_2 = 2x_1 - x_2$$

Subject to:
$$x_1 + x_2 \geq 1$$
$$x_2 \leq 4$$
$$x_1, x_2 \geq 0$$

a. Show graphically the decision space and the objective space.
b. Determine all efficient solutions to the MCLP problem.
c. Explain what happens when you try to solve a P_λ problem with
$$\lambda_1 > 0 \text{ and } \lambda_2 > 0.$$

4.14 Consider the following bi-criteria problem:

Maximize
$$f_1 = 2x_1 - 3x_2 + 5x_3 - x_4 + x_5$$

Minimize
$$f_2 = 5x_1 - 3x_2 - x_3 + 6x_4 + x_5$$

Subject to:
$$4x_1 + x_2 + 3x_3 + x_4 = 24$$
$$3x_1 + x_2 + 2x_3 - x_5 = 4$$
$$\text{All } x_i \geq 0, \text{ for } i = 1, \ldots, 5.$$

Prove whether or not the following solutions are efficient:
a. $X^{(1)} = (0,0,2,18,0)$
b. $X^{(2)} = (0,0,8,0,12)$
c. $X^{(3)} = (0,4,0,20,0)$

Hint: Use Theorem 4.2, Section 4.4.4.

4.15 The following bi-criteria problem is to be solved using compromise programming:

Maximize
$$f_1(x) = -4x_1 + 3x_2$$

Maximize
$$f_2(x) = 7x_1 + 5x_2$$

Subject to:
$$x_1 + x_2 \geq 3$$
$$-2x_1 + 3x_2 \leq 12$$
$$6x_1 + x_2 \leq 42$$
$$x_2 \leq 6$$
$$x_1, x_2 \geq 0$$

a. Determine the ideal solution f_1^* and f_2^*. You can solve it graphically or use any LP software.
b. Using the distance measure

$$L_p = \left\{ \sum_{i=1}^{2} \left[\frac{f_i^* - f_i(x)}{f_i^*} \right]^p \right\}^{1/p} \quad \text{for } p = 1, 2, \ldots, \infty.$$

Write down the objective functions for L_1, L_2, and L_∞ for the given example.

4.16 Write the output-based VRS DEA model.

4.17 For the Cobb–Douglas production function given in Example 4.5, use $A = 20$, $\alpha = 0.7$, and $\beta = 0.5$. Answer the following:
   a. What can you say about the returns to scale?
   b. What is the output and elasticity if both inputs equal 20?
   c. What is the rate of technical substitute at that same point?

4.18 Solve for the optimal inputs for Example 4.9.

4.19 Expand the model given in Section 4.14 to include prices on the outputs. What key assumptions do you need to make?

4.20 Consider the data for six DMUs given in Table 4.10.
   a. Draw the input possibility set using the H–R construction for an output of 15.
   b. Repeat for an output of 8.
   c. Repeat for an output of 5.

4.21 Using the same data as given in Exercise 4.20:
   a. Scale the inputs by the output and apply the H–R construction.
   b. Draw the input possibility set for an output of 10 [you can directly use the results from part (a)].

TABLE 4.10

DMU Data for Exercise 4.20

DMU	x_1	x_2	y_1
1	8	2	10
2	6	3	8
3	8	4	16
4	3	3	5
5	2	4	6
6	9	4	4

TABLE 4.11

DMU Data for Exercise 4.22

DMU	x_1	x_2	y_1	y_2
1	23	12	19	25
2	16	30	8	40
3	18	40	20	18
4	30	30	15	15
5	28	34	26	19
6	9	42	24	14

4.22 Consider the data for six DMUs given in Table 4.11.
   a. Formulate and solve as a VRS input-based DEA model for DMU 4.
   b. Formulate and solve as a VRS output-based DEA model for DMU 4.
   c. Will these always be the same? Why or why not?

4.23 For Exercise 4.20, suppose the cost for input 1 is $15 per unit and that for input 2 is $10 per unit.
   a. Determine the overall efficiency for DMU 4.
   b. Determine the allocative efficiency for DMU 4.

4.24 For Exercise 4.20, suppose output 2 is stochastic. The mean is equal to the deterministic value, and the variance is equal to 10 for the first three DMUs and 20 for the next three. Determine the technical efficiency for DMU 4.

References

Arthur, J. L. and A. Ravindran. 1978. An efficient goal programming algorithm using constraint partitioning and variable elimination. *Management Science*. 24: 867–868.

Arthur, J. L. and A. Ravindran. 1980a. PAGP: An efficient algorithm for Linear Goal Programming problems. *ACM Transactions on Mathematical Software*. 6: 378–386.

Arthur, J. L. and A. Ravindran. 1980b. A branch and bound algorithm with constraint partitioning for integer goal programs. *European Journal of Operational Research*. 4: 421–425.

Arthur, J. L. and A. Ravindran. 1981. A multiple objective nurse scheduling model. *Institute of Industrial Engineers Transactions*. 13: 55–60.

Banker, R. D., A. Charnes, and W. W. Cooper. 1984. Models for estimation of technical and scale inefficiencies in data envelopment analysis. *Management Science*. 30: 1078–1092.

Bowlin, W. F. 1998. Measuring performance: An introduction to data envelopment analysis (DEA). *Journal of Cost Analysis*. 7: 3–27.

Charnes, A., W. W. Cooper, and E. Rhodes. 1978. Measuring the efficiency of decision making units. *European Journal of Operational Research*. 2: 429–444.

Dyson, R. G., R. Allen, A. S. Camanho, V. V. Podinovski, C. S. Sarrico, and E. A. Shale. 2001. Pitfalls and protocols in DEA. *European Journal of Operational Research*. 132: 245–259.

El-Demerdash, B. E., I. A. El-Khodary, and A. A. Tharwat. 2013. Developing a stochastic input oriented data envelopment analysis (SIODEA) model. *International Journal of Advanced Computer Science and Applications*. 4: 40–44.

Geoffrion, A. 1968. Proper efficiency and theory of vector maximum. *Journal of Mathematical Analysis and Applications*. 22: 618–630.

Golany, B. and Y. Roll. 1989. An application procedure for DEA. *Omega*. 17: 237–250.

Hackman, S. T. 2008. *Production Economics: Integrating the Microeconomic and Engineering Perspectives*. Berlin: Springer-Verlag.

Hanoch, G. and M. Rothschild. 1972. Testing the assumptions of production theory: A nonparametric approach. *Journal of Political Economy*. 80: 256–275.

Hwang, C. L. and A. Masud. 1979. *Multiple Objective Decision Making-Methods and Applications*. New York: Springer-Verlag.

Kuriger, G. and A. Ravindran. 2005. Intelligent search methods for nonlinear goal programs. *Information Systems and Operational Research*. 43: 79–92.

Markowitz, H. 1959. *Portfolio Selection: Efficient Diversification of Investments*. NY: Wiley.

Masud, A. S. M. and A. Ravindran. 2008. Multiple criteria decision making. In *Operations Research and Management Science Handbook*. ed. A. R. Ravindran, Chapter 5. Boca Raton, FL: CRC Press.

Masud, A. S. M. and A. Ravindran. 2009. Multiple criteria decision making. In *Operation Research Methodologies*. ed. A. R. Ravindran, Chapter 5. Boca Raton, FL: CRC Press.

Nataraja, N. R. and A. L. Johnson. 2011. Guidelines for using variable selection techniques in data envelopment analysis. *European Journal of Operational Research*. 215: 662–669.

Powdrell, B. J. 2003. Comparison of MCDM algorithms for discrete alternatives. MS Thesis, Department of Industrial Engineering, Pennsylvania State University.

Ravindran, A., K. M. Ragsdell, and G. V. Reklaitis. 2006. *Engineering Optimization: Methods and Applications*, 2nd Edition, Chapter 11. Hoboken, NJ: John Wiley.

Ravindran, A. R., U. Bilsel, V. Wadhwa, and T. Yang. 2010. Risk adjusted multicriteria supplier selection models with applications. *International Journal of Production Research*. 48: 405–424.

Ravindran, A. R. and D. P. Warsing. 2013. *Supply Chain Engineering: Models and Applications*. Boca Raton, FL: CRC Press.

Ray, S. R. 2004. *Data Envelopment Analysis: Theory and Techniques for Economics and Operations Research*. Cambridge University Press.

Saaty, T. L. 1980. *The Analytic Hierarchy Process*. New York: McGraw-Hill.

Saber, H. M. and A. Ravindran. 1996. A Partitioning Gradient Based (PGB) algorithm for solving nonlinear goal programming problem. *Computers and Operations Research*. 23: 141–152.

Schniederjans, M. 1995. *Goal Programming: Methodology and Applications*. Boston, MA: Kluwer Academic Publishers.

Sharpe, W. F. 1963. A simplified model for portfolio analysis. *Management Science*. 9: 277–293.

Shin, W. S. and A. Ravindran. 1991. Interactive multi objective optimization: Survey I—continuous case. *Computers and Operations Research*. 18: 97–114.

Varian, H. R. 1992. *Microeconomic Analysis*, 3rd Edition. New York: W.W. Norton and Company.

Velazquez, M. A., D. Claudio, and A. R. Ravindran. 2010. Experiments in multiple criteria selection problems with multiple decision makers. *International Journal of Operational Research*. 7(4): 413–428.

Zanakis, S. H. and S. K. Gupta. 1995. A categorized bibliographic survey of goal programming. *Omega: International Journal of Management Science*. 13: 211–222.

Zeleny, M. 1982. *Multiple Criteria Decision Making*. New York: McGraw-Hill.

Zhu, J. 2015. *Data Envelopment Analysis: A Handbook of Models and Methods*. Boston: Springer.

5

*Supply Chain Engineering**

Our focus in this chapter is on the design of the *supply chain system*, which involves connecting many production and distribution systems, often across wide geographic distances, in such a way that the businesses involved can ultimately satisfy consumer demand as efficiently as possible, resulting in maximum financial returns to those businesses connected to that supply chain system.

In this chapter, we begin with the meaning of supply chain engineering (SCE) and describe, at a high level, the types of decisions made in managing a supply chain. We introduce a variety of supply chain performance measures and show how they relate to a company's financial measures. We then discuss the importance of SCE and the distinguishing characteristics of those firms recognized as leaders in supply chain management (SCM). Next, we focus on location and distribution strategies for designing and operating supply chain networks. Integer programming models introduced in Chapter 3 will be used to solve supply chain network optimization problems. Next, we discuss outsourcing decisions and supplier selection. Multiple-criteria ranking methods, discussed in Chapter 4, will be used for selecting the best suppliers. We end this chapter with a discussion of supply chain logistics, risk pooling strategies, and contracting in supply chains.

5.1 Supply Chain Decisions and Design Metrics

Before we formally define *supply chain engineering*, we begin with the definition of what a *supply chain* is.

A supply chain consists of the following:

1. A series of *stages* (e.g., suppliers, manufacturers, distributors, retailers, and customers) that are physically distinct and geographically separated at which inventory is either stored or converted in form and/or in value.

---

* Adapted with permission from Ravindran, A. R. and D. P. Warsing. 2013. *Supply Chain Engineering: Models and Applications.* Chapters 1, 5, and 6. Boca Raton, FL: CRC Press.

2. A coordinated set of *activities* concerned with the procurement of raw materials, production of intermediate and finished products, and the distribution of these products to customers within and external to the chain.

Thus, a supply chain includes all the partners involved in fulfilling customer demands and all the activities performed in fulfilling those demands. Figure 5.1 illustrates a typical supply chain.

It is important to recognize that the different stages of the supply chain (suppliers, plants, distribution centers [DCs] and retailers) may be located in different countries for a multinational company with a global supply chain network. It is also possible that a firm may employ fewer supply chain stages than those represented in Figure 5.1, or perhaps more. Indeed, some business researchers (e.g., Fine 2000) argue that the supply chains in various industries follow historical cycles that move from periods of significant vertical integration to periods of significantly less integration, where firms in the supply chain rely more on partnerships than on owning substantial portions of the value chain within a single firm. In vertical integration periods, the supply chain may employ only a few stages from raw material extraction to final production, owned primarily or exclusively by a single firm. A good example of this would be the early days of Ford Motor Company in the 1920s. In the less integrated periods, it is the horizontal, across-firm relationships that are prominent. A good example would be in the 1990s and early 2000s, when Dell led the global market for personal computers with a highly decentralized supply chain in which they served only as the final assembler and direct distributor. In this latter case, Dell not only relied heavily on its suppliers to independently manage the production and supply of components but also simply bypassed independent distributors and retailers and dealt with the final consumers directly, without the "middlemen." Interestingly, Fine's (2000) hypothesis regarding cycles of change in various industries may be

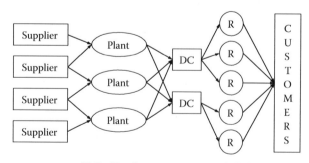

(DC = Distribution center, R = Retailer)

FIGURE 5.1
Typical supply chain network.

coming to light in the PC industry, since Dell has recently added the retail middlemen back to its supply chain.

5.1.1 Flows in Supply Chains

Following Chopra and Meindl (2001), the key flows in a supply chain are the following:

- *Products:* Includes raw materials, work-in-progress (WIP), subassemblies, and finished goods
- *Funds:* Includes invoices, payments, and credits
- *Information:* Includes orders, deliveries, marketing promotions, plant capacities, and inventory

Thus, the flows in the supply chain are not just "goods." Tracking flows from the suppliers to the customers is called *moving downstream* in the supply chain. Tracking flows from the customers to the suppliers is called *moving upstream* in the supply chain.

5.1.2 Meaning of SCE

According to Ravindran and Warsing (2013), SCE encompasses the following key activities for the effective management of a supply chain:

1. Design of the supply chain network, namely, the location of plants, DCs, warehouses, and so on
2. Procurement of raw materials and parts from suppliers to the manufacturing plants
3. Management of the production and inventory of finished goods to meet customer demands
4. Management of the transportation and logistics network to deliver the final products to the warehouses and retailers
5. Managing the integrity of the supply chain network by mitigating supply chain disruptions at all levels.

Most of the above activities involved in SCE also come under the rubric of *supply chain management* (SCM). Chopra and Meindl (2001) define SCM as "the management of flows between and among supply chain stages to maximize supply chain profitability." A more complete definition of SCM by Simchi-Levi et al. (2003) states, "SCM is a set of approaches utilized to efficiently integrate suppliers, manufacturers, warehouses and stores, so that merchandise is produced and distributed at the right quantities to the right locations, and at the right time in order to minimize system-wide costs while satisfying service level requirements." Their definition brings out all the key aspects of SCM including the two conflicting objectives in SCM—minimizing supply chain costs while simultaneously maximizing customer service.

According to Ravindran and Warsing (2013), an important distinction between SCE and SCM is the *emphasis in SCE on the design of the supply chain network and the use of mathematical models and methods to determine the optimal strategies for managing the supply chain.*

Without doubt, commerce has become increasingly global in scope over the past several decades. This trend toward *globalization* has resulted in supply chains whose footprint is often huge, spanning multiple countries and continents. Since products and funds now regularly flow across international boundaries, engineering a global supply chain becomes impractical—or at least ill-advised—without the use of sophisticated mathematical models. The use of the methods of Operations Research and applied mathematics for SCE decisions will be the primary focus of this chapter.

5.1.3 Supply Chain Decisions

The various decisions in SCE can be broadly grouped into three types—*strategic, tactical,* and *operational*.

1. *Strategic Decisions*

    Strategic decisions deal primarily with the design of the supply chain network and the selection of partners. These decisions not only are made over a relatively long time period (usually spanning several years), and have greater impact in terms of the company's resources, but also are subject to significant uncertainty in the operating environment over this lengthy span of time. Examples of strategic decisions include the following:
    - Network design—where to locate and at what capacity?
        - Number and location of plants and warehouses
        - Plant and warehouse capacity levels
    - Production and sourcing—make or buy?
        - Produce internally or outsource
        - Choice of suppliers, subcontractors, and other partners
    - Information technology—how to coordinate the supply chain?
        - Develop software internally or purchase commercially available packages—for example, SAP, Oracle.

2. *Tactical Decisions*

    Tactical decisions are primarily *supply chain planning decisions* and are made in a time horizon of moderate length, generally as monthly or quarterly decisions, covering a planning horizon of 1 or 2 years.

Thus, these decisions are typically made in an environment characterized by less uncertainty relative to strategic decisions, but where the effects of uncertainty still are not inconsequential. Examples of tactical decisions include the following:

- Purchasing decisions—for example, *how much to buy and when?*
- Production planning decisions—for example, *how much to produce and when?*
- Inventory management decisions—for example, *how much and when to hold to balance the costs of resupply with the risks of shortages?*
- Transportation decisions—for example, *which modes to choose and how frequently to ship on them?*
- Distribution decisions—for example, *how to coordinate DC replenishment with production schedules?*

3. *Operational Decisions*

Operational decisions are short-term decisions made on a daily/weekly basis, at which point much of the operational uncertainty that existed when the strategic and tactical decisions were made has been resolved. In addition, since the time scale is so short, most of these decisions involve a significantly lower expenditure of funds. Examples include the following:

- Setting delivery schedules for shipments from suppliers
- Setting due dates for customer orders
- Generating weekly or daily production schedules
- Allocating limited supply (e.g., between backorders and new customer demand)

It is important to recognize that the three types of supply chain decisions—strategic, tactical, and operational—are interrelated. For example, the number and locations of plants affect the choice of suppliers and modes of transport for raw materials. Moreover, the number and locations of plants and warehouses also affect the inventory levels required at the warehouses and the delivery times of products to customers. Aggregate production planning decisions affect product movement decisions and customer fulfillment.

5.1.4 Enablers and Drivers of Supply Chain Performance

5.1.4.1 Supply Chain Enablers

Enablers make things happen and, in the case of SCM, are considered essential for a supply chain to perform effectively. Without the necessary enablers, the supply chain will not function smoothly. Simply having the necessary enablers, however, does not guarantee a successful supply chain performance.

Based on a survey of supply chain managers, Marien (2000) describes four enablers of effective management of the supply chain. In order of their ranked importance by the survey respondents, they are *organizational infrastructure, technology, strategic alliances,* and *human resource management.*

- *Organizational infrastructure:* The essential issue in this case is whether SCM activities internal to the firm and across firms in the supply chain are organized in more of a vertical orientation, or with greater decentralization. The fact that this enabler ranked first among practicing supply chain managers is clearly consistent with much that has been written in the trade press regarding the fact that intrafirm SCM processes must be in place and operating effectively before there is any hope that interfirm collaboration on SCM activities is to be successful.
- *Technology:* Two types of technology are critical to success in designing and managing supply chains effectively: information technology and manufacturing technology. Although the emphasis of many consultants and software providers is often on information technology, product design can often have more impact on whether supply chain efficiencies can ultimately be achieved. Prominently, product design should account for manufacturability—for example, using modular components that utilize common interfaces with multiple final products—and allow for efficiencies in managing inventories and distribution processes.
- *Alliances:* The effectiveness of alliances is particularly important in supply chains that are more decentralized, wherein more authority is given to suppliers. In some cases, these suppliers may take on roles that go beyond just supplying components, becoming instead outsourcing partners who assume significant responsibility for product design and manufacturing.
- *Human resources:* Two categories of employees are critical to effective SCM. First, technical employees assume an important role in designing networks that minimize costs while simultaneously achieving high levels of customer service performance. These employees must have a solid understanding of the types of mathematical tools we discuss in this book. Second, managerial employees must similarly have a solid conceptual grasp on the key issues addressed by the models and tools of the technical staff and must clearly understand how such tools can be applied to ultimately allow the firm to achieve its strategic goals.

5.1.4.2 Supply Chain Drivers

Supply chain drivers represent the critical areas of decision making in SCE, those that ultimately generate the outcomes that affect the supply chain performance. Thus, they appear as the decision/design variables in the optimization models used in SCE decision making. The key drivers of supply chain performance are described below.

1. *Inventory*

    Companies maintain inventory of raw materials, WIP, and finished goods to protect against unpredictable demand and unreliable supply. Inventory is considered an *idle asset* to the company and is one of the major portions of supply chain costs. Maintaining large inventories increases supply chain costs but provides a higher level of customer service. The key decision variables here are *what items to hold in inventory* (raw materials, WIP, and finished products), *how much and when to order* (inventory policies), and *where to hold inventory* (locations).

2. *Transportation*

    Transportation is concerned with the movement of items between the supply chain stages—suppliers, plants, DCs, and retailers. Use of faster transportation modes such as air and roadways incurs higher transportation costs but reduces delivery times and increases reliability. The key decision variables here are as follows:

    i. Whether to outsource transportation decision making and execution to a *third-party logistics* (3PL) provider
    ii. What transportation mode(s) to use (air, sea, road, etc.) and for what items (raw materials, WIP, and finished goods)
    iii. Distribution options for finished goods (either shipped to customers directly or through intermediate DCs)

3. *Facilities*

    Facilities (plants and DCs) play a key role in SCE. They are generally considered strategic decisions and directly affect the performance of the supply chain. The key decision variables under facilities are the following:

    i. Number of plants and their locations
    ii. Plant capacities and product mix allocated to plants
    iii. Number of DCs and their locations
    iv. Distribution strategies

4. *Suppliers*

    Raw material cost accounts for 40%–60% of the production cost in most manufacturing industries. In fact, for the automotive industry, the cost of components and parts from outside suppliers may exceed 50% of sales (Wadhwa and Ravindran 2007). For technology firms, component cost, as a fraction of sales, could be as high as 80%.

Hence, the selection of suppliers for raw materials and intermediate components is considered a critical area of strategic decision making in SCE.

5.1.5 Assessing and Managing Supply Chain Performance

The idea that there are key drivers of supply chain performance is useful in thinking about another theme emphasized by many authors and first proposed by Fisher (1997) in an important article that advanced the notion that "one size fits all" is not an effective approach to managing supply chains. Fisher (1997) cogently lays out a matrix that matches product characteristics—what he describes as a dichotomy between *innovative* products like technology-based products and *functional* products like toothpaste or other staple goods—and supply chain characteristics—another dichotomy between *efficient* (cost-focused) supply chains and *responsive* (customer service-focused) supply chains. Chopra and Meindl (2001) take this conceptual model a step further, first by pointing out that Fisher's product characteristics and supply chain strategies are really continuous spectra and then by superimposing the Fisher model, as it were, on a frontier that represents the natural trade-off between responsiveness and efficiency. Clearly, it stands to reason that a firm, or a supply chain, cannot maximize cost efficiency and customer responsiveness simultaneously. Some aspects of each of these objectives necessarily work at cross purposes. A combined version of Chopra and Meindl's frontier and Fisher's product dichotomy is presented in Figure 5.2.

The value of this perspective is that it clearly identifies a market-driven basis for strategic choices regarding the supply chain drivers: Should our inventory management decisions be focused more on efficiency—for example, minimizing inventory levels—or on responsiveness—for example, maximizing

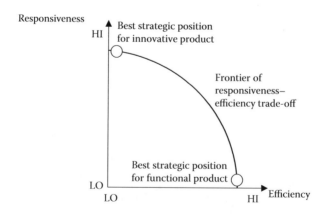

FIGURE 5.2
Responsiveness–efficiency trade-off frontier.

product availability? Should our transportation choices be focused more on efficiency—for example, minimizing transportation costs, perhaps through more extensive economies of scale—or on responsiveness—for example, minimizing delivery lead times and maximizing reliability? Should our facilities (network design) decisions be focused more on efficiency—for example, minimizing the number of locations and maximizing their size and scale—or on responsiveness—for example, seeking high levels of customer service by choosing many, focused locations closer to customers?

Below, we discuss efficiency and responsiveness in more detail, and we also introduce *supply chain risk* as an additional criterion to consider in designing the supply chain network and its associated operating policies.

5.1.5.1 Supply Chain Efficiency

Generally, efficiency is measured by a ratio of the level of output generated to the level of input consumed to generate that output. This concept can be applied to both physical systems—for example, an automobile engine that converts the energy stored in the fuel consumed by the engine into horsepower generated by the engine to drive the wheels of the vehicle—and businesses—for example, the conversion of dollar-valued inputs (labor, materials, and the costs of owning and/or operating physical assets like plants and warehouses) into sales revenue. Therefore, the efficiency of a given supply chain focuses on how well resources are utilized across the chain in fulfilling customer demand.

In the conceptual framework of Fisher (1997), discussed above, efficient supply chains are more focused on cost minimization, the idea being that a supply chain that requires less cost input to generate the same amount of sales revenue output is more efficient. Therefore, efficiency measures in SCE are often focused on costs and include the following:

- Raw materials cost
- Manufacturing cost
- Distribution cost
- Inventory holding cost
- Facility operating costs
- Freight transportation costs
- Shortage costs

In addition, other measures that may influence the costs listed above include the following:

- Product cycle time—this is the time that elapses from the start of production of the item up to its conversion into a product that can be shipped to the customer. Clearly, longer cycle times can result in larger costs—for example, labor costs and/or inventory holding costs.

- Inventory levels—again, higher levels of inventory can result in a number of associated costs, beyond just the cost of tying up the firm's cash in currently idle assets. Higher inventory levels generate greater needs for storage space and for labor hours and/or employee levels in order to manage these inventories as they reside in and flow between storage facilities. Later in this chapter, we will present two important measures of inventory levels, namely, *inventory turns* and *days of inventory*.

Typically, supply chain optimization models focus on minimizing costs, since the decisions of supply chain managers often involve choices that directly influence costs, while revenue may often be outside the scope of the supply chain manager's decisions. Some SCE models, however, may appropriately involve maximizing profit, to the extent that it is clear that the decision at hand has both cost and revenue implications.

5.1.5.2 Supply Chain Responsiveness

Responsiveness refers to the extent to which customer needs and expectations are met, and also the extent to which the supply chain can flexibly accommodate changes in these needs and expectations. Thus, in the efficiency–responsiveness trade-off introduced by the Fisher (1997) framework discussed above, firms whose supply chains are focused on responsiveness are willing to accept higher levels of cost (i.e., lower cost efficiency) in order to improve their ability to meet and flexibly accommodate customer requirements (i.e., higher responsiveness). Common measures of responsiveness are as follows:

- Reliability and accuracy of fulfilling customer orders
- Delivery time
- Product variety
- Time to process special or unique customer requests (customization)
- Percentage of customer demand filled from finished goods inventory versus built to order from raw materials or component inventories

5.1.5.3 Supply Chain Risk

According to Ravindran and Warsing (2013), a third supply chain criterion has gained attention in recent years. The September 11 terrorist attacks in the United States in 2001 obviously had broad and lasting impacts on society in general. From the standpoint of managing supply chains, the disruption in material flows over the days and weeks after September 11 caused companies to realize that a singular emphasis on the cost efficiency of the supply chain can actually make the chain brittle and much more susceptible to the risk of disruptions. This includes not only catastrophic disruptions like large-scale terrorist attacks, but even mundane, commonly occurring disruptions like a labor strike at a supplier. Thus, effective SCM no longer just involves moving products efficiently along the supply chain, but it also includes mitigating risks along the way. Supply chain risk can be broadly classified into two types:

1. *Hazard risks*—these are disruptions to the supply chain that arise from large-scale events with broad geographic impacts, such as natural disasters (e.g., hurricanes, floods, and blizzards), terrorist attacks, and major political actions like wars or border closings.
2. *Operational risks*—these are more commonly occurring disruptions whose impacts are localized (e.g., affecting only a single supplier) and resolved over a relatively short period of time. Examples include information technology disruptions (e.g., a server crash due to a computer virus infection), supplier quality problems, and temporary logistics failures (e.g., temporarily "lost" shipments).

Chapter 7 of the textbook by Ravindran and Warsing (2013) is devoted entirely to supply chain risk management. Multi-criteria optimization models that consider profitability, customer responsiveness, and supply chain risk are discussed in this context.

5.1.5.4 Conflicting Criteria in Supply Chain Optimization

It is important to recognize that efficiency and responsiveness are conflicting criteria in managing supply chains. For example, customer responsiveness can be increased by having a larger inventory of several different products, but this increases inventory costs and thereby reduces efficiency. Similarly, using fewer DCs reduces facility costs and, as we show in Section 5.5, can also reduce inventory levels across the network through "pooling" effects. The downside, however, is that such a network design increases delivery time and thereby reduces responsiveness, and it also increases supply chain risk by concentrating the risk of distribution failure in fewer facilities. Thus, supply chain optimization problems are generally multiple-criteria optimization models. They can be solved using the multiple-criteria optimization methods discussed in Chapter 4 (Section 4.1), Chapter 5 (Section 5.3.6), and Chapter 7 (Sections 7.3 and 7.4).

5.1.6 Relationship between Supply Chain Metrics and Financial Metrics

We will demonstrate in this section that improvement in some selected supply chain metrics also result in improvements in some important financial metrics of the firm, which should, of course, be closely correlated with its overall business performance. To illustrate this relationship, let us consider several interrelated inventory measures—*inventory turns, days of inventory,* and *inventory capital*—and how they affect some important financial measures—*return on assets, working capital,* and *cash-to-cash cycle.*

5.1.6.1 Supply Chain Metrics

1. *Inventory turns*

    Inventory turns is a measure of how quickly inventory is turned over from production to sales, specifically

    $$\text{Inventory turns} = \frac{\text{Annual sales}}{\text{Average inventory}}. \quad (5.1)$$

    For example, if the annual sales are 1200 units and the average inventory is 100 units, then the inventory turns would be 1200/100 = 12. In other words, on the average, goods are stored in inventory for 1 month before they are sold. From a financial standpoint, companies prefer a higher value for inventory turns so that products reach the end customer as soon as possible after production and the funds tied up in inventory can be freed up and converted into cash (or accounts receivable) more quickly. For many years, Dell's inventory turns exceeded 100, indicating a quick turnover of computers after assembly.

2. *Days of inventory*

    Days of inventory refers to how many days of customer demand is carried in inventory, specifically

    $$\text{Days of inventory} = \frac{\text{Average inventory}}{\text{Daily sales}}. \quad (5.2)$$

    Using Equations 5.1 and 5.2, *days of inventory* can be written as follows:

    $$\begin{aligned}
    \text{Days of inventory} &= \frac{\text{Average inventory}}{\left(\dfrac{\text{Annual sales}}{365}\right)} \\
    &= \frac{365}{\left(\dfrac{\text{Annual sales}}{\text{Average inventory}}\right)} \quad (5.3) \\
    &= \frac{365}{\text{Inventory turns}}.
    \end{aligned}$$

Supply Chain Engineering

For Dell, with inventory turns equal to 100, the days of inventory will be $\frac{365}{100} = 3.65$ days. In other words, Dell carries less than 4 days of inventory, on average. However, Dell requires its suppliers to carry 10 days of inventory (Dell 2004).

A Harvard Business School study in 2007 reported that the consumer goods industry carries, on average, 11 weeks (77 days) of inventory and retailers carry 7 weeks (49 days) of inventory. Despite that, the stockout rate in the retail industry averages nearly 10%! Note that, increasing the days of inventory is one means of attempting to increase customer responsiveness, but this increase would come at the expense of additional supply chain cost.

3. *Inventory capital*

    Inventory capital refers to the total investment in inventory, specifically

$$\text{Inventory capital} = \sum_{k=1}^{N} I_k V_k, \quad (5.4)$$

where
I_k = average inventory of item k
V_k = value of item k per unit
N = total number of items held in inventory

Using Equations 5.1 through 5.4, we can state that *increasing the inventory turns* will affect the other inventory measures as follows (assuming the same annual demand):

- Average inventory decreases.
- Days of inventory decreases.
- Inventory capital decreases.

5.1.6.2 Business Financial Metrics

4. *Return on assets*

    Return on assets refers to the ratio of company's net income to its total assets, computed as

$$\text{Return on Assets (ROA)} = \frac{\text{Annual income (\$)}}{\text{Total assets (\$)}}. \quad (5.5)$$

ROA provides a general proxy for the overall operational efficiency of a company, that is, its ability to utilize assets (input) to generate profits (output). Since inventory capital is included in the total assets, reducing inventory capital may increase ROA for the same annual income, as long as it decreases assets in total (see "Working capital" below for details).

5. *Working capital*

   Working capital is the difference between a company's short-term assets (e.g., cash, inventories, and accounts receivable) and its short-term liabilities (e.g., accounts payable, interest payments, and short-term debt). Thus, working capital, particularly the portion reflected by cash, represents the amount of flexible funds available to the company to invest in research and development (R&D) and other projects. Increasing inventory turns, therefore, can shift working capital toward cash, thereby freeing up funds for immediate use in profit-generating activities and projects.

6. *Cash-to-cash cycle*

   Cash-to-cash cycle refers to the difference in the length of time it takes for a company's *accounts receivable* to be converted into cash inflows and the length of time it takes for the company's *accounts payable* to be converted into cash outflows. Historically, it was often the case that a company paid for its raw materials, production, and distribution of its products on a cycle that was shorter than the cycle by which it received payments from its customers after sales. Thus, for most companies, the cash-to-cash cycle has historically been positive. Companies would, however, like to decrease their cash-to-cash cycle in order to improve their profitability. For example, when a company increases its inventory turns, the finished goods reach the end customer sooner. Thus, the company converts purchased components into receivables more quickly, and the cash-to-cash cycle decreases.

In summary, there is a direct and significant relationship between supply chain inventory metrics and a company's financial metrics. An *increase in inventory turns* has a cascading effect on the other inventory and financial measures as follows:

- Days of inventory decreases.
- Inventory capital decreases.
- ROA may increase.*
- Working capital may increase.*
- Cash-to-cash cycle decreases.

---

* Recall that while inventory is part of working capital, so is cash. Thus, if an increase in inventory turns reduces inventory capital by shifting it to cash, there is no net change in assets or working capital.

5.1.7 Importance of SCM

As shown in Section 5.1.6, SCM can have a significant impact on business performance. Based on a 2003 Accenture study, done in conjunction with Stanford University, Mulani (2005) reported the following:

- Nearly 90% of the companies surveyed said that SCM is critical or very important.
- 51% said that the importance of SCM had increased significantly in the 5 years leading up to the survey.
- SCM accounted for nearly 70% of the companies' operating costs and comprised at least half of all the typical company's assets.

Mulani (2005) also reported that a significant percentage of promised synergies for many company mergers and acquisitions came from SCM. For example, during the Hewlett–Packard/Compaq merger, it was estimated that the merger would result in a savings of $2.5 billion, of which $1.8 billion would be due to supply chain efficiency.

Moreover, failure to excel in SCM can negatively affect a company's stock price. Hendricks and Singhal (2005) found this to be true in a study of 885 supply chain disruptions reported by publicly traded companies during 1989–2000. The list of companies included small, medium, and large companies with respect to market capitalization and covered both manufacturing and IT industries. Hendricks and Singhal found that companies suffering a supply chain disruption experienced the following effects, on average:

- A loss of more than $250 million in shareholder value per disruption
- 10% reduction in stock price
- 92% reduction in ROA
- 7% lower sales
- 11% increase in cost of doing business
- 14% increase in inventory

SCM has become sufficiently important to business performance to warrant a mantra, of sorts, namely, that "companies do not compete with each other, but their supply chains do." While some might debate whether entire supply chains could literally compete with each other, there is no doubt that efficient management of the supply chain has become a competitive differentiator for many companies. In the next section, we discuss the top 25 supply chains and what supply chain characteristics make them industry leaders.

5.1.7.1 Supply Chain Top 25

Since 2004, AMR Research (now owned by Gartner Group) has annually ranked the "Supply Chain Top 25," identifying the supply chain leaders and highlighting their best practices. The rankings are based broadly on both qualitative and quantitative measures. Half of the ranking weight (50%) is based on the company's performance measures, specifically ROA (20%), inventory turns (10%), revenue growth (10%), and corporate social responsibility (10%). The data are obtained from publicly available company information. The other half of the ranking weight (50%) is based on voting by 38 Gartner analysts (25%) and a peer panel of 185 senior-level supply chain executives (25%).

In addition to the ranking of *Top 25 supply chain companies*, Gartner also lists *Supply Chain Masters*, a new category introduced in 2015. The Masters category recognizes companies with sustained supply chain leadership over the years, namely, companies ranked in the top five, the last 7 out of 10 years. The three companies recognized as Supply Chain Masters in 2017 include Apple, Amazon, and Procter & Gamble. The top 10 companies from the 2017 supply chain rankings are listed in Table 5.1.

A complete list of the Supply Chain Top 25 for 2017 and the data used in their rankings are available in the Gartner report (www.gartner.com/doc/3728317). Based on the study of the top 25 companies over the years, the supply chain leaders as a group exhibit the following attributes compared to their competitors:

- Carry 15% less inventory
- Are 60% faster to market
- Complete 17% more "perfect" orders
- Have 35% shorter cash-to-cash cycles
- Have 5% higher profit margins

TABLE 5.1

Top 10 Supply Chains (2017)

Rank	Company
1	Unilever
2	McDonald's
3	Inditex
4	Cisco Systems
5	H&M
6	Intel
7	Nestle
8	Nike
9	Colgate–Palmolive
10	Starbucks

In addition, these top 25 companies have also outperformed the S&P 500 index in terms of average stock price growth. They have agile supply chain networks that can respond quickly to changes in customer demand and supply chain network disruptions. The goal of Gartner rankings is to raise the importance of SCM and its potential to positively impact companies' profitability. By identifying the leaders and their best practices, the hope is that other companies can learn from them and improve their supply chain performance.

5.2 Supply Chain Network Design and Distribution

The problems of managing an existing supply chain network are far more pervasive and frequently encountered in practice compared to designing a new network or redesigning an existing one. One might justifiably ask at this point, "How did the decision makers actually decide the location of the facilities in the network and the assignment of those facilities to the customers they serve?" Therefore, in this section, we focus our attention to "location and distribution" strategies or designing and operating the supply chain network. The key questions to answer in this effort are the following:

- How do we design a good (or perhaps the *best*) network?
- What are the key objectives for the network?
- What are the primary decision variables? Are they strategic or tactical decisions?
- What are the key constraints of the network?

The key decision variables in the supply chain optimization models will be the number and location of various types of facilities—spanning as far as supplier sites, manufacturing sites, and distribution sites—and the quantities shipped from upstream sites to the downstream sites and ultimately out to customers. We will apply the integer programming models with binary variables, discussed in Chapter 3, for location and distribution decisions in SCM.

5.2.1 Supply Chain Distribution Planning

The selection of optimal sites for warehouse location is a *strategic decision*. We shall first consider the *tactical decision* of distributing products to retail outlets from a given set of warehouses. The distribution problem is basically a *transportation problem* that is discussed in most operations research textbooks (Ravindran et al. 1987). We shall illustrate this with a numerical example.

Example 5.1 (Ravindran and Warsing 2013)

Consider a distribution planning problem with 3 warehouses and 12 retailers. Tables 5.2 and 5.3 give the supply available at the warehouses and the retailer demands, respectively. The unit cost of shipping the product from a given warehouse to each retailer is given in Table 5.4. The problem is to determine the optimal distribution plan that will minimize the total cost.

SOLUTION

The total supply available at the warehouses is 13,000 units, while the total retailer demand is 10,600. Since any warehouse can supply any

TABLE 5.2

Warehouse Supplies (Example 5.1)

	Warehouses		
	W_1	W_2	W_3
Supply	5000	5000	3000

TABLE 5.3

Retailer Demands (Example 5.1)

	Retailers											
	R_1	R_2	R_3	R_4	R_5	R_6	R_7	R_8	R_9	R_{10}	R_{11}	R_{12}
Demand	650	400	850	1900	3100	250	350	400	500	400	1350	450

TABLE 5.4

Unit Shipping Cost in Dollars (Example 5.1)

	Warehouse		
Retailer	W_1	W_2	W_3
R_1	3.50	4.30	3.00
R_2	3.60	4.90	3.90
R_3	4.30	3.80	3.30
R_4	4.80	3.70	4.20
R_5	3.10	4.40	3.40
R_6	3.70	3.90	3.10
R_7	3.80	4.10	3.50
R_8	3.80	4.00	2.80
R_9	3.70	4.30	3.10
R_{10}	3.80	4.75	3.50
R_{11}	4.50	3.00	3.80
R_{12}	3.90	4.90	4.10

Supply Chain Engineering

retailer, it is feasible to meet all the retailer demands with the available supply. To determine the least cost distribution plan, we formulate the following transportation problem:

Decision variables
X_{ij} = amount shipped to retailer i (R_i) from warehouse j (W_j); $i = 1, 2, ..., 12$ and $j = 1, 2, 3$. Thus, we have 36 decision variables.

Supply constraints at each warehouse
Each warehouse cannot supply more than its capacity. Thus, we get

$$\sum_{i=1}^{12} X_{i1} \leq 5000 \qquad \text{(Warehouse 1)}$$

$$\sum_{i=1}^{12} X_{i2} \leq 5000 \qquad \text{(Warehouse 2)}$$

$$\sum_{i=1}^{12} X_{i3} \leq 3000 \qquad \text{(Warehouse 3)}$$

Demand constraints for each retailer
The total amount shipped to a retailer from the three warehouses should be equal to the retailer's demand.

$$X_{11} + X_{12} + X_{13} = 650 \qquad \text{(R1-demand)}$$

$$X_{21} + X_{22} + X_{23} = 400 \qquad \text{(R2-demand)}$$

$$X_{31} + X_{32} + X_{33} = 850 \qquad \text{(R3-demand)}$$

$$X_{41} + X_{42} + X_{43} = 1900 \qquad \text{(R4-demand)}$$

$$X_{51} + X_{52} + X_{53} = 3100 \qquad \text{(R5-demand)}$$

$$X_{61} + X_{62} + X_{63} = 250 \qquad \text{(R6-demand)}$$

$$X_{71} + X_{72} + X_{73} = 350 \qquad \text{(R7-demand)}$$

$$X_{81} + X_{82} + X_{83} = 400 \qquad \text{(R8-demand)}$$

TABLE 5.5

Optimal Distribution Plan (Example 5.1)

Retailer	W_1	W_2	W_3
R_1	0	0	650
R_2	400	0	0
R_3	0	0	850
R_4	0	1900	0
R_5	3100	0	0
R_6	0	0	250
R_7	0	0	350
R_8	0	0	400
R_9	0	0	500
R_{10}	400	0	0
R_{11}	0	1350	0
R_{12}	450	0	0

(Warehouse column header spans W_1, W_2, W_3.)

$$X_{91} + X_{92} + X_{93} = 500 \quad \text{(R9-demand)}$$

$$X_{10,1} + X_{10,2} + X_{10,3} = 400 \quad \text{(R10-demand)}$$

$$X_{11,1} + X_{11,2} + X_{11,3} = 1350 \quad \text{(R11-demand)}$$

$$X_{12,1} + X_{12,2} + X_{12,3} = 450 \quad \text{(R12-demand)}$$

Objective function

Minimize total cost of shipping given by

$$Z = (3.5X_{11} + 4.3X_{12} + 3X_{13}) + (3.6X_{21} + 4.9X_{22} + 3.9X_{23}) + \ldots\ldots\ldots$$
$$+ (4.5X_{11,1} + 3X_{11,2} + 3.8X_{11,3}) + (3.9X_{12,1} + 4.9X_{12,2} + 4.1X_{12,3}).$$

Thus, the transportation problem has 36 variables and 15 constraints. Solving it in Microsoft's Excel Solver, we get the optimal solution as shown in Table 5.5. The minimum shipping cost is $34,830.

5.2.2 Location–Distribution Problem

Let us now consider an integrated example, where both the location decisions and distribution decisions have to be made simultaneously.

Supply Chain Engineering

Example 5.2 (Ravindran et al. 1987)

A retail firm is planning to expand its activities in an area by opening two new warehouses. Three possible sites are under consideration as shown in Figure 5.3. Four customers have to be supplied whose annual demands are D_1, D_2, D_3, and D_4.

Assume that any two sites can supply all the demands but site 1 can supply customers 1, 2, and 4 only; site 3 can supply customers 2, 3, and 4; while site 2 can supply all the customers. The unit transportation cost from site i to customer j is C_{ij}. For each warehouse, Table 5.6 gives the data on capacity, annual investment, and operating costs. The optimization problem is to select the proper sites for the two warehouses that will minimize the total costs of investment, operation, and transportation.

SOLUTION

Each warehouse site has a fixed capital cost independent of the quantity stored and a variable cost proportional to the quantity shipped. Thus, the total cost of opening and operating a warehouse is a nonlinear function of the quantity stored. Through the use of binary integer variables,

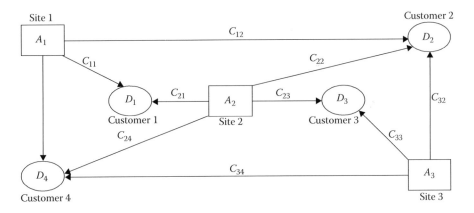

FIGURE 5.3
Supply chain network for Example 5.2.

TABLE 5.6
Warehouse Data (Example 5.2)

Site	Capacity	Initial Capital Investment ($)	Unit Operating Cost ($)
1	A_1	K_1	P_1
2	A_2	K_2	P_2
3	A_3	K_3	P_3

the warehouse location–distribution problem can be formulated as an integer program.

Let the binary integer variable δ_i denote the decision to select site i. In other words,

$$\delta_i = \begin{cases} 1, & \text{if site } i \text{ is selected} \\ 0, & \text{otherwise} \end{cases}.$$

Let X_{ij} denote the quantity shipped from site i to customer j.

The supply constraint for site 1 is given by

$$X_{11} + X_{12} + X_{14} \leq A_1 \delta_1. \qquad \text{(Site 1)}$$

When $\delta_1 = 1$, site 1 is selected with capacity A_1 and the quantity shipped from site 1 cannot exceed A_1. When $\delta_1 = 0$, the nonnegative variables X_{11}, X_{12}, and X_{14} will automatically become zero, implying no possible shipment from site 1.

Similarly, for sites 2 and 3, we obtain

$$X_{21} + X_{22} + X_{23} + X_{24} \leq A_2 \delta_2 \qquad \text{(Site 2)}$$

$$X_{32} + X_{33} + X_{34} \leq A_3 \delta_3 \qquad \text{(Site 3)}$$

To select exactly two sites, we need the following constraint:

$$\delta_1 + \delta_2 + \delta_3 = 2.$$

Since the δ_i's can assume values of 0 or 1 only, the new constraint will force two of the δ_i's to be one.

The demand constraints can be written as

$$X_{11} + X_{21} = D_1 \qquad \text{(Customer 1)}$$

$$X_{12} + X_{22} + X_{32} = D_2 \qquad \text{(Customer 2)}$$

$$X_{23} + X_{33} = D_3 \qquad \text{(Customer 3)}$$

$$X_{14} + X_{24} + X_{34} = D_4 \qquad \text{(Customer 4)}$$

To write the objective functions, we note that the total cost of investment, operation, and transportation for site 1 is

$$K_1 \delta_1 + P_1(X_{11} + X_{12} + X_{14}) + C_{11} X_{11} + C_{12} X_{12} + C_{14} X_{14}.$$

Supply Chain Engineering

When site 1 is not selected, δ_1 will be zero. This will force X_{11}, X_{12}, and X_{14} to become zero. Similarly, the cost functions for sites 2 and 3 can be written. Thus, the complete formulation of the warehouse location problem reduces to the following mixed integer program:

Minimize

$$Z = K_1\delta_1 + P_1(X_{11} + X_{12} + X_{14}) + C_{11}X_{11} + C_{12}X_{12} + C_{14}X_{14} \\ + K_2\delta_2 + P_2(X_{21} + X_{22} + X_{23} + X_{24}) + C_{21}X_{21} + C_{22}X_{22} + C_{23}X_{23} + C_{24}X_{24} \\ + K_3\delta_3 + P_3(X_{32} + X_{33} + X_{34}) + C_{32}X_{32} + C_{33}X_{33} + C_{34}X_{34}$$

Subject to

$$X_{11} + X_{12} + X_{14} \le A_1\delta_1$$

$$X_{21} + X_{22} + X_{23} + X_{24} \le A_2\delta_2$$

$$X_{32} + X_{33} + X_{34} \le A_3\delta_3$$

$$\delta_1 + \delta_2 + \delta_3 = 2$$

$$X_{11} + X_{21} = D_1$$

$$X_{12} + X_{22} + X_{32} = D_2$$

$$X_{23} + X_{33} = D_3$$

$$X_{14} + X_{24} + X_{34} = D_4$$

$$\delta_i \in (0,1) \quad \text{for } i = 1,2,3$$

$$X_{ij} \ge 0 \quad \text{for all } (i,j)$$

5.2.3 Location–Distribution with Dedicated Warehouses (Srinivasan 2010)

In Example 5.2, we allowed multiple deliveries to a customer. In other words, a customer can receive his demand from more than one warehouse. Suppose the customers demand single deliveries. In this case, each customer has to be supplied by one warehouse only, even though a warehouse may supply more than one customer. We call these as "dedicated warehouse" problems. This condition can be easily modeled with a minor change in the definition of variables X_{ij}. In Example 5.2, X_{ij} was a continuous variable denoting the quantity shipped from site i to customer j. To incorporate dedicated warehouses, we define X_{ij} as a binary variable.

$$X_{ij} = \begin{cases} 1, & \text{if site } i \text{ supplies customer } j; i=1,2,3 \; j=1,2,3,4 \\ 0, & \text{otherwise} \end{cases}$$

To guarantee that customer 1 receives supply from one of the sites only, we write the constraint:

$$X_{11} + X_{21} = 1.$$

The supply constraint at site 1 will become

$$D_1 X_{11} + D_2 X_{12} + D_4 X_{14} \leq A_1 \delta_1.$$

The complete formulation is given below:

$$\begin{aligned}
\text{Minimize } Z = & K_1 \delta_1 + P_1(D_1 X_{11} + D_2 X_{12} + D_4 X_{14}) + C_{11} D_1 X_{11} + C_{12} D_2 X_{12} \\
& + C_{14} D_4 X_{14} + K_2 \delta_2 + P_2(D_1 X_{21} + D_2 X_{22} + D_3 X_{23} + D_4 X_{24}) \\
& + C_{21} D_1 X_{21} + C_{22} D_2 X_{22} + C_{23} D_3 X_{23} + C_{24} D_4 X_{24} + K_3 \delta_3 \\
& + P_3(D_2 X_{32} + D_3 X_{33} + D_4 X_{34}) + C_{32} D_2 X_{32} + C_{33} D_3 X_{33} + C_{34} D_4 X_{34}
\end{aligned}$$

Subject to

$$D_1 X_{11} + D_2 X_{12} + D_4 X_{14} \leq A_1 \delta_1$$

$$D_1 X_{21} + D_2 X_{22} + D_3 X_{23} + D_4 X_{24} \leq A_2 \delta_2$$

$$D_2 X_{32} + D_3 X_{33} + D_4 X_{34} \leq A_3 \delta_3$$

$$\delta_1 + \delta_2 + \delta_3 = 2$$

$$X_{11} + X_{21} = 1$$

Supply Chain Engineering

$$X_{12} + X_{22} + X_{32} = 1$$

$$X_{23} + X_{33} = 1$$

$$X_{14} + X_{24} + X_{34} = 1$$

$$\delta_i \in (0,1) \quad \text{for } i = 1,2,3$$

$$X_{ij} \in (0,1) \quad \text{for all } (i,j)$$

5.2.4 Supply Chain Network Design

Since warehouses can be built of different capacities and cost, we shall consider a multi-state supply chain network design problem, which determines not only the locations of the warehouses but their right capacities to meet the customer demand.

Example 5.3 (Ravindran and Warsing 2013)

XYZ Company is looking at improving its distribution system. XYZ's three-stage supply chain—a factory, two potential warehouse sites, and four retailers—is shown in Figure 5.4. At each of the potential sites,

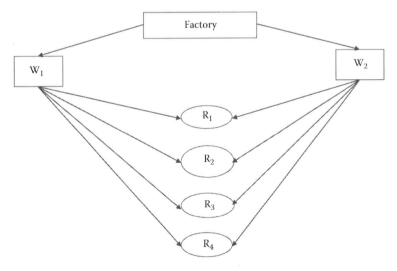

FIGURE 5.4
XYZ's three-stage supply chain (Example 5.3).

the company can build a warehouse in three different sizes—small, medium, or large. The investment cost and capacities at the two warehouse sites are given in Table 5.7. The company can build a warehouse at either location or both. However, it cannot build more than one warehouse at each location. In other words, the company cannot build both a large and a medium-size warehouse at the same location. The customer demands must be met and are given in Table 5.8. The unit cost of shipping XYZ's product from the factory to the warehouse sites and from each warehouse site to the customer is given in Table 5.9.

Note that the construction costs are lower at site 2, but its distribution costs are higher. The company has a total investment limit of $30,000 for construction. The objective is to minimize the total cost of investment and transportation such that all customers' demands are met.

TABLE 5.7

Warehouse Capacities and Investment Cost (Example 5.3)

Size	Capacity	Investment Cost (Annualized) Site 1 (W_1)	Site 2 (W_2)
Small	3000 units	$10,000	$8,000
Medium	5000 units	$15,000	$12,000
Large	10,000 units	$20,000	$17,000

TABLE 5.8

Annual Demand (Example 5.3)

	Retailer Demands		
R_1	R_2	R_3	R_4
2000	2500	1000	3500

TABLE 5.9

Unit Shipping Cost (Example 5.3)

i. Factory to Warehouse

	W_1	W_2
Factory	$10	$15

ii. Warehouse to Retailers

Warehouse	Retailer			
	R_1	R_2	R_3	R_4
W_1	$2	$3	$5	$4
W_2	$5	$4	$6	$7

Supply Chain Engineering

SOLUTION

Binary Variables

$$\delta_{iS} = \begin{cases} 1, & \text{if small warehouse is built at site } i \\ 0, & \text{otherwise} \end{cases}, \text{ for } i = 1, 2.$$

Similarly, we define additional binary variables for δ_{iM} and δ_{iL} for building medium and large warehouses at site i.

Continuous Variables

X_{Fi} = amount shipped from the factory to warehouse i ($i = 1, 2$)
X_{ij} = amount shipped from warehouse i to retailer j ($i = 1, 2$ and $j = 1, 2, 3, 4$)

Thus, we have 16 decision variables (6 binary and 10 continuous).

Constraints

1. At each site, no more than one warehouse can be built.

$$\text{Site 1: } \delta_{1S} + \delta_{1M} + \delta_{1L} \leq 1 \tag{5.6}$$

$$\text{Site 2: } \delta_{2S} + \delta_{2M} + \delta_{2L} \leq 1 \tag{5.7}$$

2. Amount shipped from the factory to the warehouse cannot exceed that warehouse's capacity.

$$\text{Site 1: } X_{F1} \leq 3000\delta_{1S} + 5000\delta_{1M} + 10,000\delta_{1L} \tag{5.8}$$

$$\text{Site 2: } X_{F2} \leq 3000\delta_{2S} + 5000\delta_{2M} + 10,000\delta_{2L} \tag{5.9}$$

(Recall the formulation of constraints with multiple right-hand side constants discussed in Chapter 3, Section 3.1.3.)

3. A warehouse cannot ship more than what it receives from the factory.

$$\text{Site 1: } X_{11} + X_{12} + X_{13} + X_{14} \leq X_{F1} \tag{5.10}$$

$$\text{Site 1: } X_{21} + X_{22} + X_{23} + X_{24} \leq X_{F2} \tag{5.11}$$

4. Retailer demands must be met.

$$\text{Retailer 1: } X_{11} + X_{21} = 2000 \tag{5.12}$$

$$\text{Retailer 2: } X_{12} + X_{22} = 2500 \tag{5.13}$$

$$\text{Retailer 3: } X_{13} + X_{23} = 1000 \tag{5.14}$$

$$\text{Retailer 4: } X_{14} + X_{24} = 3500 \tag{5.15}$$

5. Total investment for construction cannot exceed $30,000.

$$\begin{aligned} &10,000\delta_{1S} + 15,000\delta_{1M} + 20,000\delta_{1L} + 8000\delta_{2S} \\ &+ 12,000\delta_{2M} + 17,000\delta_{2L} \le 30,000 \end{aligned} \tag{5.16}$$

Thus, we have 11 constraints in the model.

Objective Function

The total cost of investment and transportation is to be minimized.

1. *Investment Cost (IC)*

$$IC = 10,000\delta_{1S} + 15,000\delta_{1M} + 20,000\delta_{1L} + 8000\delta_{2S} + 12,000\delta_{2M} + 17,000\delta_{2L}$$

2. *Shipping Cost (SC)*

$$SC = 10X_{F1} + 15X_{F2} + 2X_{11} + 3X_{12} + 5X_{13} + 4X_{14} + 5X_{21} + 4X_{22} + 6X_{23} + 7X_{24}$$

$$\text{Minimize Total Cost} = TC = IC + SC.$$

NOTES
1. In Equation 5.8, the right-hand side values can only be equal to 0, 3000, 5000, or 10,000, since no more than one binary variable (δ_{1S}, δ_{1M}, δ_{1L}) can be one for site 1 due to Equation 5.6.
2. If no warehouse is built at site 1, then all the three binary variables δ_{1S}, δ_{1M}, and δ_{1L} will be zero. In such a case, the right-hand side value of Equation 5.8 will zero and it will force $X_{F1} = 0$. When $X_{F1} = 0$, all the shipping variables from site 1 to the four retailers (X_{11}, X_{12}, X_{13}, X_{14}) will be forced to zero due to Equation 5.10. Then, Equations 5.12 through 5.15 will force all the shipping to the retailers from site 2 only.

Optimal Solution

The optimal solution to this integer program was obtained by using Microsoft's Excel Solver software. Table 5.10 gives the optimal solution to Example 5.3. From Table 5.10, it can be inferred that one large warehouse is built at site 1. No other warehouses are built. All shipments to and from the second site are zero, which again ensures that all retailer demands are met from the large warehouse at site 1 alone.

TABLE 5.10

Optimal Solution for Example 5.3

i. Warehouse Locations and Capacities

Binary Variable	Value
δ_{1S}	0
δ_{1M}	0
δ_{1L}	1
δ_{2S}	0
δ_{2M}	0
δ_{2L}	0

ii. Factory to Warehouse Shipments

	Warehouses	
	W_1	W_2
Factory	9000	0

iii. Warehouse to Retailer Shipments

	Retailers			
	R_1	R_2	R_3	R_4
W_1	2000	2500	1000	3500
W_2	0	0	0	0

5.2.5 Real-World Applications

There are several published results of real-world applications using integer programming models for supply chain network design and distribution problems. In this section, we discuss briefly a few of the applications in practice. For interested readers, the cited references will provide more details on the case studies.

5.2.5.1 Multinational Consumer Products Company

Here, we discuss two real-world applications illustrating the use of Integer Programming (IP) for location and distribution problems. Both applications are related to a leading global health and hygiene company listed in the Fortune 500, selling paper and personal care products. The company's sales are close to $20 billion a year. Its global brands are sold in 150 countries holding first or second market positions.

5.2.5.1.1 Case 1: Supply Chain Network Design

The first application, described in Portillo (2016), was related to the company's largest international division; selling products in 22 countries, across one continent with 21 manufacturing plants, 45 DCs, 100 customer zones, and 22 brands. The company's supply chain was formed by different types of customers, from multinational chains and large distributors to thousands of small "mom and pop" stores. A Mixed Integer Linear Programming (MILP) model was developed for a complete strategic and tactical optimization of the manufacturing and distribution network based on customer demand projections for a 5-year horizon. The MILP model, with the objective of maximizing profits, consisted of three stages:

Stage 1: Determine the ability of the company to fulfill present and projected sales based on the current supply chain design.

Stage 2: Evaluate how the company's ability would be improved, considering the potential expansions of plants and DCs already in the horizon.

Stage 3: Optimize the global supply chain design that would deliver the best results for the entire 5-year time horizon.

The MILP model had 7500 variables, of which 300 were binary and 7000 constraints. The mathematical model was coded in ILOG and solved using the CPLEX solver (www.ILOG.com). The solver reached optimality in 2 minutes.

The stage 1 analysis showed a very close to full capacity utilization of the existing facilities under current demand levels. When considering future demand, the overall results showed that the existing supply chain could fulfill only 75% of the total projected demand, primarily restricted by production and distribution capacity.

The stage 2 analysis included capacity expansions in production and distribution facilities already considered by the management. Close to a dozen new production lines were planned within a 2-year horizon. Although management had already decided on their locations, multiple options were allowed in the model to confirm their choices. The results showed that the majority of the chosen locations were optimal. Although the locations were the best to maximize profit, the supply chain network was capable of meeting only 85% of the projected demand.

During the stage 3 analysis, additional levels of facility expansions were considered. The results provided production, distribution, and sales levels to maximize profits. However, the optimal demand fulfillment ratio was only 96%, highlighting specific product–market combinations that were not profitable.

Additional details on the MILP model are available in Portillo (2016) as a case study for global supply chain network design.

5.2.5.1.2 Case 2: Supply Chain Distribution Planning

Cintron et al. (2010) describe a real-world application of a multiple-criteria integer programming model to determine the best distribution network in one of the countries for the same consumer product company discussed in the first application. The country under study had 4 manufacturing plants, 66 retailers, 5 independent distributors, and 2 DCs (one company-owned and one leased). Multiple products were made at the plants to meet customer demand. No product was made in more than one plant. The company was the global competitor in that country with sales of more than $86 million annually. The company was looking to reduce their distribution costs by improving the region's distribution network design. Specifically, the company was interested in designing the flow of products from the manufacturing plants to the customers.

The case study considered four distribution options for the customers to receive their products. Products can be supplied from

i. The manufacturing plant directly
ii. The company's DC
iii. An independent distributor, who is supplied directly by the plant
iv. An independent distributor, who is supplied by the company's DC

A MILP model was developed to select the best option for each customer to maximize the profitability and customer responsiveness among other things. The MILP model had 2790 variables, of which 2500 were binary and 900 constraints. The model was solved using the GAMS software in less than 30 seconds. The optimal network recommended by the MILP model eliminated the need for the leased DC for the company and increased the direct shipment from plants to customers from 33% of all demands (existing policy) to 83%. This resulted in the reduction of supply chain distribution cost from 12% to 3% of net sales, a savings of $7 million annually. The complete details of the model and the results are available in Cintron (2016).

5.2.5.2 Procter and Gamble

Procter and Gamble (P&G) sells more than 300 brands of consumer products around the world. In 1993, P&G's Operations Research team undertook a major study, called *Strengthening Global Effectiveness*, to restructure P&G's global supply chain. A major part of the study was to examine the North American supply chain, which had 60 plants, 15 DCs, and more than 1000 customer zones. To be globally competitive, P&G decided to consolidate manufacturing plants to reduce cost and improve speed to market. The Operations Research team decomposed the overall supply chain problem into two subproblems: one dealing with the *location of the DCs* and another dealing with *product sourcing*, one for each product category. Thus, the DC locations were chosen independent of the plant locations. This was justified based on the fact that only 10%–20% of product volume goes through DC, the fact that the manufacturing costs are much larger than the distribution cost, and the need to locate DCs closer to the customer zones to provide good customer service.

The *DC location model* was formulated as an uncapacitated facility location model, as discussed in Section 5.2.2. The solution to this integer program determined the location of the DCs and the customer zones assigned to each DC. The *product sourcing model* determined the location of the manufacturing plants, the products each plant makes, and the distribution of the products either directly to the customer zone or through the DCs. Instead of formulating this as a mixed integer programming model, the Operations Research team developed several scenarios for plant locations and their products. Thus, the product sourcing model was essentially reduced to solving a series of transportation problems, similar to the one we discussed in Example 5.1.

The Operations Research study was completed in 1994 and was implemented in mid-1996. It resulted in closing 20% of the manufacturing plants at 12 sites and a savings of more than $200 million annually. For more details on the study, readers are referred to Camm et al. (1997).

5.2.5.3 Hewlett–Packard

The supply chain team at Hewlett–Packard (HP), called SPaM (Strategic Planning and Modeling), has been using quantitative methods to solve its global supply chain design problems since 1994 (Lee and Billington 1995). The SPaM team's approach is to generate alternate scenarios based on intuition and expert knowledge for HP's global supply chain network design problems and use Operations Research for analyzing the scenarios. SPaM's initial success was solving the problems of spiraling inventory and declining customer satisfaction in the early 90s. Their initial project dealt with the personal computer and desk jet printer divisions. The development of the *Worldwide Inventory Network Optimizer* (WINO) became the building block of the complete SCM models at HP. Customer fill rate and finished goods inventory were the two conflicting objectives of WINO. WINO achieved inventory reduction of 10% to 30% and increased customer satisfaction simultaneously. From inventory modeling, SPaM moved to supply chain design strategies in manufacturing and distribution.

In a follow-up case study (Laval et al. 2005), the SPaM team reported on how scenario analysis and optimization were combined to design the global supply chain for HP's Imaging and Printing Group. The main objective of the group was to reduce the number of its contract manufacturing partners worldwide. Using expert knowledge, the SPaM team selected scenarios that included closing several existing sites and opening some new ones centrally. The scenario-based modeling approach helped reduce the number of binary variables and constraints of the integer programming model. Each scenario defined a set of contract manufacturing locations, the products handled at those locations, and the demand areas they served. Uncertainties in demands and lead times were also included in the scenarios. The optimization model

Supply Chain Engineering

was a mixed integer programming model. The objective function was to minimize the total supply chain cost that included outbound and inbound transportation cost and manufacturing cost (fixed production cost plus variable labor cost). The model's recommendations resulted in cost savings of more than $10 million in supply chain cost, without sacrificing the existing service levels.

5.2.5.4 BMW

BMW produces cars in eight plants located in Germany, the United Kingdom, the United States, and South Africa. Its engines are manufactured at four other sites. It also has six assembly plants. This case study reported on the development of a strategic planning model to design BMW's global supply chain for a 12-year time horizon (Fleischmann et al. 2006). The optimization model addressed the allocation of car models to its global production facilities, supply of materials, and the distribution of cars to the global markets. The global market was aggregated into 10 sales regions.

The basic supply chain model was a multi-period MILP. The binary variables identified whether a certain product was produced at a given plant. The continuous variables represented the annual volume in supply, production, and distribution. The objective function included fixed capital expenditure as well as variable costs for supply, production, and distribution. An example model in the case study had six plants, 36 products, and eight sales regions for a 12-year planning horizon. The MILP model had 60,000 variables (2000 binary) and 145,000 constraints. The model was solved using the ILOG/CPLEX solver on a 1.6-GHz processor in just 4 minutes!

The use of the MILP model improved BMW's long-term strategic planning process. It made the planning process more transparent and reduced the planning effort. It also resulted in the reduction of 5.7% in investment and the cost of materials, production, and distribution.

5.2.5.5 AT&T

AT&T provides telecommunication equipment and long-distance services to the telemarketing industry. To help its telemarketing customers for their site location decisions, AT&T developed a Decision Support System (Spencer et al. 1990). When the "800-service" was introduced in 1967, the cost of 800-service determined their locations, since the labor cost was minimal. The Midwest, particularly Omaha (Nebraska), became the "800-captial" of the world. However, in the 80s, labor cost started becoming very significant. AT&T provided the *Decision Support System* (DSS) as a value-added service to its customers.

The DSS was basically a MILP model, similar to the capacitated facility location model we discussed in Example 5.2. The MILP model determined the best locations for the telemarketing centers from a set of candidate locations, and the volume of customer traffic from different regions handled by each center. The objective was to minimize the total cost of labor, fixed facility, and communication. For example, in 1988, the model helped 46 AT&T telemarketing customers with their site location decisions. AT&T secured business worth $375 million in annual communication revenues and $31 million in equipment sales.

5.3 Outsourcing Decisions and Supplier Selection

5.3.1 Importance of Supplier Selection

The contribution of the purchasing function to the profitability of the supply chain has assumed greater proportions in recent years; one of the most critical functions of purchasing is selection of suppliers. For most manufacturing firms, the purchasing of raw material and component parts from suppliers constitutes a major expense. Raw material cost accounts for 40%–60% of production costs for most U.S. manufacturers. In fact, for the automotive industry, the cost of components and parts from outside suppliers may exceed 50% of sales (Wadhwa and Ravindran 2007). For technology firms, purchased materials and services account for 80% of the total production cost. It is vital to the competitiveness of most firms to be able to keep the purchasing cost to a minimum. In today's competitive operating environment, it is impossible to successfully produce low-cost, high-quality products without good suppliers. A study carried out by the Aberdeen Group (2004) found that more than 83% of the organizations engaged in outsourcing achieved significant reduction in purchasing cost, more than 73% achieved reduction in transaction cost, and more than 60% were able to shrink sourcing cycles.

Supplier selection process is difficult because the criteria for selecting suppliers could be conflicting. Figure 5.5 illustrates the various factors that could affect the supplier selection process (Sonmez 2006). Supplier selection is a multiple-criteria optimization problem that requires trade-off among different qualitative and quantitative factors to find the best set of suppliers. For example, the supplier with the lowest unit price may also have the lowest quality. The problem is also complicated by the fact that several conflicting criteria must be considered in the decision-making process.

Most of the time, buyers have to choose among a set of suppliers by using some predetermined criteria such as quality, reliability, technical capability, lead times, and so on, even before building long-term relationships. To accomplish these goals, two basic and interrelated decisions must be made by a firm. The firm must decide which suppliers to do business with and how much to order from each supplier. Weber et al. (1991) refer to this pair of decisions as the supplier selection problem.

Supply Chain Engineering

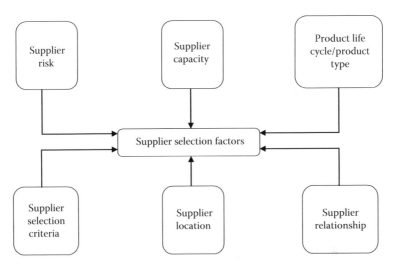

FIGURE 5.5
Supplier selection factors.

5.3.2 Supplier Selection Process

Figure 5.6 illustrates the steps in the supplier selection process. The first step is to determine whether to *make or buy* the item. Most organizations buy those parts that are not core to the business or not cost-effective if produced in-house. The next step is to define the various criteria for selecting the suppliers. The criteria for selecting a supplier of critical product may not be the same as a supplier of maintenance, repair, and operating items. Once a decision to buy the item is taken, the most critical step is selecting the right supplier. Once the suppliers are chosen, the organization has to negotiate terms of contract and monitor their performance. Finally, the suppliers have to be constantly evaluated and the feedback should be provided to the suppliers.

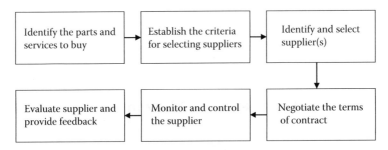

FIGURE 5.6
Supplier selection steps.

5.3.3 In-House or Outsource

As illustrated in Figure 5.6, the important sourcing decision is to decide whether to make the part or perform the service in-house or to outsource the activity. In the past, companies used to be vertically integrated in such a way that all the necessary parts were produced internally and support services, such as transportation, procurement, payroll, and so on, were provided by the companies themselves. However, in recent years, companies have started outsourcing their non-core functions. For example, many manufacturing companies outsource their shipping to 3PL providers, such as, UPS, FedEx, Penske, and so on. This generally reduces cost and provides better service by taking advantage of the 3PL's expertise in transportation. However, the company loses control and flexibility. When manufacturing of critical parts is outsourced, the company may also lose intellectual property. Thus, Step 1 in Figure 5.6, whether to do it in-house or outsource, is an important strategic decision for critical parts and services.

In addition to the qualitative factors, a company can also use a cost criterion to decide whether to make or buy a certain part. This process is illustrated in Example 5.4.

Example 5.4 (Ravindran and Warsing 2013)

A manufacturing company is currently producing a part required in its final product internally. It uses about 3000 parts annually. The setup of a production run costs $600 and the part costs $2. The company's inventory carrying cost is 20% per year. The company is considering an option to purchase this part from a local supplier since the fixed cost of ordering will only be $25. However, the unit cost of each part will increase from $2 to $2.50. The problem is to determine whether the company should buy the parts or produce them internally.

SOLUTION

a. In-house production option
   The optimal number of parts per production run P can be calculated using the Economic Order Quantity formula given in Section 5.4.3.

$$P = \sqrt{\frac{(2)(600)(3000)}{(0.2)(2)}} = 3000 \text{ units}$$

$$\text{Total yearly cost} = \sqrt{(2)(600)(3000)(0.4)} + (2)(3000)$$

$$= \$1200 + \$6000 = \$7200$$

Supply Chain Engineering

b. Outsourcing option

$$\text{Optimal Order Quantity} = \sqrt{\frac{(2)(25)(3000)}{(0.2)(2.5)}} = 548 \text{ units.}$$

Total yearly cost = $\sqrt{(2)(25)(3000)(0.5)} + (2.5)(3000) = \$274 + \$7500 = \$7774$.

Hence, based on minimum cost, the company should continue to make the parts internally.

5.3.4 Supplier Selection Methods

As mentioned earlier, the supplier selection activity plays a key role in cost reduction and is one of the most important functions of the purchasing department. Different mathematical, statistical, and game theoretical models have been proposed to solve the problem. Weber et al. (1991), Aissaoui et al. (2007), De Boer et al. (2001), and Dickson (1966) provide an overview of supplier selection methods. De Boer et al. (2001) stated that that supplier selection is made up of several decision-making steps as shown in Figure 5.7.

5.3.4.1 Sourcing Strategy

It is important for a company to recognize the importance of the sourcing strategy. It can be either strategic or tactical depending on the type of item

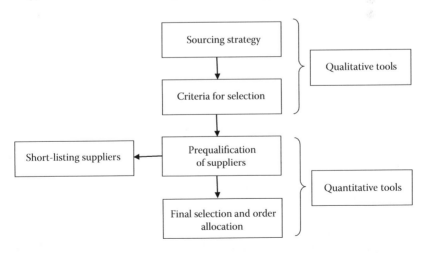

FIGURE 5.7
Supplier selection problem.

being purchased. For example, if the item being purchased is a critical component or the most expensive part of the final product, then the supplier selection becomes a strategic decision. The company would most likely want to develop a long-term relationship with the supplier. In this case, subjective criteria such as trust and keeping abreast of emerging technologies would be more important than lead time or even base price. On the other hand, when an item is a commodity that can be easily found in the market from several suppliers, then the supplier selection is a tactical decision. Quantitative criteria such as cost, lead time, delivery, and geographic location play a key role.

5.3.4.2 Criteria for Selection

Depending on the buying situation, different sets of criteria may have to be employed. Criteria for supplier selection have been studied extensively since the 60s. Dickson's (1966) study was the earliest to review the supplier selection criteria. He identified 23 selection criteria with varying degrees of importance for the supplier selection process. Dickson's study was based on a survey of purchasing agents and managers. A follow-up study was done by Weber et al. (1991). They reviewed 74 articles published in the 70s and 80s and categorized them based on Dickson's 23 selection criteria. In their studies, net price, quality, delivery, and service were identified as the key selection criteria. They also found that under Just-in-Time manufacturing, quality and delivery became more important than net price. Zhang et al. (2003) conducted a similar study and found that price, quality, and delivery are the top three key criteria for supplier selection. Table 5.11 compares the relative importance of the various supplier selection criteria identified in the three studies. Supplier selection criteria may also change over time. Wilson (1994) examined studies conducted in 1974, 1982, and 1993 on the relative importance of the selection criteria. She found that quality and service criteria began to dominate price and delivery. She concluded that globalization of the marketplace and an increasingly competitive atmosphere had contributed to the shift. Table 5.12 shows the top 15 supplier selection criteria as given in Weber et al. (1991).

5.3.4.3 Prequalification of Suppliers

Prequalification (Figure 5.7) is defined as the process of reducing a large set of potential suppliers to a smaller manageable number by ranking the suppliers under a predefined set of criteria. The primary benefits of prequalification of suppliers are as follows (Holt 1998):

1. The possibility of rejecting good suppliers at an early stage is reduced.
2. Resource commitment of the buyer toward purchasing process is optimized.
3. With the application of preselected criteria, the prequalification process is rationalized.

TABLE 5.11
Importance of Supplier Selection Criteria

Criterion	Dickson (1966) Ranking[a]	Rating[b]	Weber et al. (1991) Ranking[a]	%	Zhang et al. (2003) Ranking[a]	%
Net price	6	A	1	80	1	87
Quality	1	A+	3	53	2	82
Delivery	2	A	2	58	3	73
Production facilities and capacity	5	A	4	30	4	44
Technical capacity	7	A	6	20	5	33
Financial position	8	A	9	9	6	29
Geographical location	20	B	5	21	7	11
Management and organization	13	B	7	13	7	11
Performance history	3	A	9	9	7	11
Operating controls	14	B	13	4	7	11
Communication systems	10	B	15	3	7	11
Reputation and position in industry	11	B	8	11	12	7
Repair service	15	B	9	9	13	4
Packaging ability	18	B	13	4	13	4
Training aids	22	B	15	3	13	4

(*Continued*)

TABLE 5.11 (CONTINUED)
Importance of Supplier Selection Criteria

Criterion	Dickson (1966) Ranking[a]	Rating[b]	Weber et al. (1991) Ranking[a]	%	Zhang et al. (2003) Ranking[a]	%
Procedural compliance	9	B	15	3	13	4
Labor relations record	19	B	15	3	13	4
Warranties and claims policies	4	A	—	0	13	4
Attitude	16	B	12	8	19	2
Reciprocal arrangements	23	C	15	3	19	2
Impression	17	B	15	3	21	0
Desire for business	12	B	21	1	21	0
Amount of past business	21	B	21	1	21	0

[a] Ranking is based on the frequency that the criterion is discussed in reviewed papers.
[b] A+, extremely important; A, considerably important; B, averagely important; C, slightly important.

TABLE 5.12
Key Criteria for Supplier Selection

Rank	Criteria
1	Net price
2	Delivery
3	Quality
4	Production facilities and capabilities
5	Geographical location
6	Technical capability
7	Management and organization
8	Reputation and position in industry
9	Financial position
10	Performance history
11	Repair service
12	Attitude
13	Packaging ability
14	Operational control
15	Training aids

Source: Weber, C. A., J. R. Current, and W. C. Benton. 1991. Supplier selection criteria and methods. *European Journal of Operational Research.* 50: 2–18.

Prequalification is an orderly process of ranking various suppliers under conflicting criteria. It is a multiple-criteria ranking problem that requires the buyer to make trade-off among the conflicting criteria, some of which may be qualitative. There are a number of multiple-criteria ranking methods available for prequalification of suppliers. The various ranking methods we discussed in Chapter 4, Section 4.1.2 are applicable for ranking suppliers. We will illustrate them with a case study in Section 5.3.6.

5.3.4.4 Final Selection

Most of the publications in the area of supplier selection have focused on final selection. In the final selection step (Figure 5.7), the buyer identifies the suppliers to do business with and allocates order quantities among the chosen supplier(s). In reviewing the literature, there are basically two kinds of supplier selection problem, as stated by Ghodsypour and O'Brien (2001):

- **Single sourcing**, which implies that any one of the suppliers can satisfy the buyer's requirements of demand, quality, delivery, and so on
- **Multiple sourcing**, which implies that there are some limitations in suppliers' capacity, quality, and so on and multiple suppliers have to be used

The primary objective of a single sourcing strategy is to minimize the procurement costs. A secondary objective is to establish a long-term relationship with a single supplier. However, the single sourcing strategy exposes a company to a greater risk of supply chain disruption. Toyota's 1977 brake valve crisis is an example of such a strategy (Mendoza 2007). Toyota was using a single supplier (Aisin Seiki) for brake valves used in all its cars (Nishiguchi and Beaudet 1998). The supplier's plant was shut down due to a fire, which resulted in the closure of Toyota's assembly plants for several days. It affected the production of 70,000 cars, at an estimated cost of $195 million. Thereafter, Toyota decided to use at least two suppliers for every part (Treece 1997).

A multiple sourcing strategy results in obtaining raw materials and parts from more than one supplier and provides greater flexibility to a company. It introduces competition among the suppliers, which helps drive the procurement costs down and improve quality of parts (Jayaraman et al. 1999). It also protects the company from major supply disruptions. However, the overhead costs will increase as the company has to deal with multiple suppliers for each part. Bilsel and Ravindran (2011) discuss a hybrid sourcing strategy, wherein the company uses one main supplier but has several "backup" suppliers who can be used in case of supply disruption at the main supplier.

5.3.5 Single Sourcing Methods for Supplier Selection

Single sourcing is a possibility when a relatively small number of parts are procured. In this section, we will discuss two most commonly used methods in single sourcing

1. *Linear Weighted Point (LWP) method:* The LWP method is the most widely used approach for single sourcing. The approach uses a simple scoring method that heavily depends on human judgment. Some of the references that discuss this approach include Wind and Robinson (1968) and Zenz (1981). The multiple-criteria ranking methods discussed in Section 5.3.6 can also be used for single sourcing for selecting the top-ranked supplier.

   To illustrate the LWP method, consider a simple supplier selection problem with two suppliers and five criteria (quality, service, capacity, price, and risk) as shown in Table 5.13. Each criterion is measured on a scale of 1 to 10, with higher numbers preferred over lower ones. Note that Supplier A has higher quality, service, and capacity but costs more and has higher risk of supply disruptions. The *ideal values* in Table 5.13 represent the best values for each criterion. However, the ideal values are not achievable simultaneously for all the criteria, because the criteria conflict with one another.

TABLE 5.13
Supplier Selection by the LWP Method

Criteria	Weights	Supplier A	Supplier B	Ideal Value
Quality	15	10	9	10
Service	20	8	7	8
Capacity	10	10	8	10
Price	45	7	10	10
Risk	10	6	9	9
Total	100%	785	895	

In the LWP method, the purchasing manager assigns a weight to each criterion using a scale of 1 to 100. In Table 5.13, price has the highest weight (45%) followed by service and quality. Capacity and risk are tied for the last place. A total weighted score is then calculated by summing the product of each criterion weight and its value (see Table 5.13). Based on the total score, Supplier B would be selected even though Supplier A was better in three out of the five criteria.

The LWP method has several drawbacks. It is generally difficult to come up with exact criteria weights. The method also requires all the criteria values to be scaled properly if the measurement units are very different. For example, quality may be measured in "percent defectives," service may be measured by lead time in days, capacity may be measured in hundreds of units, and price may be measured in dollars. The use of linear additive value function violates the normal economic principle of "Diminishing Marginal Utility." In Sections 5.3.6 and 5.3.7, we will discuss several methods for ranking suppliers that overcome these drawbacks.

2. *Total Cost of Ownership (TCO):* TCO is another supplier selection method used in practice for single sourcing. TCO looks beyond the unit price of the item to include costs of other factors, such as quality, delivery, supply disruption, safety stock, and so on. Unlike the LWP method, TCO assigns a cost to each criterion and computes the total cost of ownership with respect to each supplier. Finally, the business is awarded to the supplier with the lowest total cost. General Electric Wiring Devices have developed a total cost supplier selection method that takes into account risk factors, business desirable factors, and measurable cost factors (Smytka and Clemens 1993). The TCO approach has also been used by Ellram (1995), Degraeve and Roodhoft (1999, 2000), and Degraeve et al. (2004). Example 5.5 illustrates the TCO approach for a simple supplier selection problem.

Example 5.5

A company is considering two potential suppliers, one domestic and one foreign, for one of its parts needed in production. The relevant supplier data are given in Table 5.14.

The daily usage of this part is 20 units. The company follows an inventory policy of maintaining a safety stock of 50% of the lead time demand and uses an inventory carrying charge of 25% per year (i.e., $0.25/dollar/year). Defective parts encountered in manufacturing costs the company $5 per unit. The company wishes to choose one of the suppliers using the TCO approach.

The total cost of ownership includes the following costs:

- Procurement cost
- Inventory holding costs (cycle inventory + safety stock)
- Cost of quality

SOLUTION

We will compute the annual cost of ownership for each supplier.

Supplier A (domestic):
Annual procurement cost = $(10) (20) (365) = $73,000
Lead time demand = (10) (20) = 200 units
Safety stock = (50%) (200) = 100 units
Average cycle inventory = $\frac{500}{2}$ = 250 units
Annual inventory holding cost = (250 + 100) (0.25) ($10) = $875
Annual cost of quality = ($5) (7300) (5%) = $1825
TCO for Supplier A = $73,000 + $875 + $1825 = $75,700 per year

Supplier B (foreign):
Annual procurement cost = $ (9) (20) (365) = $65,700
Lead time demand = (90) (20) = 1800 units
Safety stock = (50%) (1800) = 900 units
Average cycle inventory = $\frac{3000}{2}$ = 1500 units
Annual inventory holding cost = (1500 + 900) (0.25) ($9) = $5400
Annual cost of quality = ($5) (7300) (20%) = $7300
Total cost of ownership for Supplier B = $65,700 + $5400 + $7300 = $78,400 per year

TABLE 5.14

Supplier Data for Example 5.5

	Supplier A (Domestic)	Supplier B (Foreign)
Item cost/unit	$10	$9
Minimum order	500	3000
Lead time	10 days	90 days
Quality (% defective)	5%	20%

Supply Chain Engineering

Even though the per-unit cost of the item is lower with the foreign supplier, the total cost of ownership is less with the domestic supplier when we take into account the cost of inventory and cost of quality.

5.3.6 Multi-Criteria Ranking Methods for Supplier Selection

Many organizations have a large pool of suppliers to select from. The supplier selection problem can be solved in two phases. The first phase reduces the large number of candidate suppliers to a manageable size (prequalification of suppliers). In Phase 2, an optimization model is used to allocate order quantities among the shortlisted suppliers (order allocation). Phase 2 will be discussed in Section 5.3.10 under Multiple Sourcing. Figure 5.8 shows the steps involved in a two-phase supplier selection model used by Ravindran et al. (2010) for a case study.

The problem of ranking suppliers (prequalification) represents a class of multiple-criteria optimization problems that deal with the ranking of a finite number of alternatives, where each alternative is measured by several conflicting criteria. In this section, we illustrate several multiple-criteria ranking approaches for the supplier selection problem, namely, the prequalification of suppliers (Phase 1), with the help of an actual case study.

In the prequalification process, readily available qualitative and quantitative data are collected for the various suppliers. These data can be obtained from trade journals, Internet, and past transactions to name a few sources. Once these data are gathered, these suppliers are evaluated using multiple-criteria ranking methods. The decision maker (DM) then selects a portion of the suppliers for extensive evaluation in Phase 2.

5.3.6.1 Case Study 1: Ranking of Suppliers (Source: sRavindran and Wadhwa 2009)

The first step in prequalification is defining the selection criteria. For this case study, we have used the following 14 prequalification criteria as an

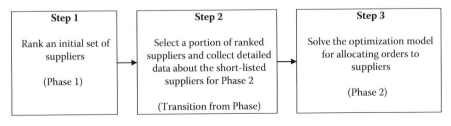

FIGURE 5.8
Two-phase supplier selection model.

illustration. The prequalification criteria have been split into various categories, such as organizational criteria, experience criteria, and so on. The various prequalification criteria are described below:

- Organizational Criteria:
  - Size of company (C1): Size of the company can be either its number of employees or its market capitalization.
  - Age of company (C2): Age of the company is the number of years that the company has been in business.
  - R&D activities (C3): Investment in R&D.
- Experience Criteria:
  - Project type (C4): Specific types of projects completed in the past.
  - Project size (C5): Specific sizes of projects completed in the past.
- Performance Criteria:
  - Cost overruns (C6): Cost overruns in the past.
  - Capacity (C7): Capacity of the supplier to fulfill orders.
  - Lead time (C8): Meeting promised delivery time.
- Quality Criteria:
  - Responsiveness (C9): If there is an issue concerning quality, how fast the supplier reacts to correct the problem.
  - Acceptance rate (C10): Perfect orders received within acceptable quality.
- Cost Criteria:
  - Order change and cancellation charges (C11): Fee associated with modifying or changing orders after they have been placed.
  - Cost savings (C12): Overall *reduction* in procurement cost.
- Miscellaneous Criteria:
  - Labor relations (C13): Number of strikes or any other labor problems encountered in the past.
  - Procedural Compliances (C14): Conformance to national/international standards (e.g., ISO 9000).

In this case study, we assume that there are 20 suppliers during prequalification. The 14 supplier criteria values for the initial set of 20 suppliers are given in Table 5.15. Smaller values are preferred for criteria C6, C11, and C13; larger values are preferred for other criteria. Next, we apply several multiple-criteria ranking methods for short-listing the suppliers and illustrate them

TABLE 5.15
Supplier Criteria Values for Case Study 1

	C1	C2	C3	C4	C5	C6	C7	C8	C9	C10	C11	C12	C13	C14
S1	0.75	1	0.46	1	0.92	0.9	1	0	0.13	0.18	0.18	0.01	0.26	0.79
S2	0.22	0	0.33	1	0.94	0.35	0.9	0.13	0.02	0	0.38	0.95	0.88	0.72
S3	0.53	0	0.74	0	0.03	0.89	0.1	0.12	0	0.3	0.66	0.08	0.86	0.22
S4	0.28	1	0.8	0	0.54	0.75	0.85	1	1	0.87	0.33	0.5	0.78	0.12
S5	0.3	0	0.79	1	0.6	0.49	0.8	0.15	0.97	0.79	0.83	0.13	0.46	0.15
S6	0.5	0	0.27	0	0.43	0.52	0.12	0	0	0.25	0.9	0.07	0.26	0
S7	0.25	1	0.6	1	0.1	0.18	0	0.13	1	0.85	0.51	0.59	0.12	1
S8	0.76	1	0.68	1	0.55	0.87	0	0.14	0	1	0.98	0.19	0.86	0.99
S9	0.25	1	0.5	1	0.26	0.92	0.94	0.03	0.15	1	0.7	0.41	0.95	1
S10	0.16	1	0.7	0	0.46	0.62	0.9	0	0.03	0	0.3	0.68	0.61	1
S11	0.31	0	0.3	0	0.09	0.73	1	1	1	0	0.87	0.3	0.98	0
S12	0.34	1	0.39	1	0.75	0.94	0.78	0.3	0	0.85	0.94	0.61	0.46	0.3
S13	0.08	0	0.27	0	0.14	0.42	1	0.91	0	0.82	0.45	0.42	0.81	1
S14	0.62	1	0.02	1	0.15	0.97	0.15	0.01	0.18	0.92	0.55	0.23	0.12	0.97
S15	0.49	0	0.98	0	0.52	0.68	0	0.24	0.06	0	0.52	0.84	0.05	0.76
S16	0.1	1	0.32	1	0.67	0.21	1	0.85	0.16	0.29	0.49	0.41	0.29	0.27
S17	0.08	0	0.19	1	0.24	0.87	0	0.72	0.26	1	0.84	0.99	0.64	0.04
S18	0.86	0	0.28	1	0.95	0.08	1	0.12	0.2	0	0.4	0.76	0.66	1
S19	0.72	0	0.88	0	0.15	0.93	0.97	1	1	1	0.75	0.64	0.26	1
S20	0.15	1	0.92	1	0.77	0.63	0	0	0.3	0.22	0.22	0.94	0.93	0.26

using the case study data. Each method has advantages and limitations. The methods that we will discuss are as follows:

1. L_p metric method
2. Rating method
3. Borda method
4. Analytic hierarchy process (AHP)

For a more detailed discussion of multi-criteria ranking methods, the reader is referred to Masud and Ravindran (2008, 2009).

5.3.6.2 Use of the L_p Metric for Ranking Suppliers

Mathematically, the L_p metric represents the distance between two vectors **x** and **y**, where $\mathbf{x}, \mathbf{y} \in R^n$, and is given by

$$\|\mathbf{x}-\mathbf{y}\|_p = \left[\sum_{j=1}^{n} |x_j - y_j|^p\right]^{1/p}. \quad (5.17)$$

One of the most commonly used L_p metrics is the L_2 metric ($p = 2$), which measures the Euclidean distance between two vectors. The ranking of suppliers is done by calculating the L_p metric between the ideal solution (H) and each vector representing the supplier's ratings for the criteria. The ideal solution represents the best values possible for each criterion from the initial list of suppliers. Since no supplier will have the best values for all criteria (e.g., a supplier with minimum cost may have poor quality and delivery time), the ideal solution is an artificial target and cannot be achieved. The L_p metric approach computes the distance of each supplier's attributes from the ideal solution and ranks the supplier's based on that distance (smaller the better). We shall illustrate the steps of the L_2 metric method using the supplier data in Table 5.15.

5.3.6.2.1 Steps of the L_2 metric method

Step 1: Determine the ideal solution. The ideal values (H) for the 14 criteria of Table 5.15 are given in Table 5.16.

Step 2: Use the L_2 metric to measure the closeness of supplier to the ideal values. The L_2 metric for supplier k is given by

$$L_2(k) = \sqrt{\sum_{j=1}^{n} (H_j - Y_{jk})^2}, \quad (5.18)$$

where H_j is the ideal value for criterion j and Y_{jk} is the jth criterion value for supplier k.

TABLE 5.16
Ideal Values (H) for Case Study 1

Criteria	Ideal Value	Criteria	Ideal Value
C1	0.86	C8	1
C2	1	C9	1
C3	0.98	C10	1
C4	1	C11	0.18
C5	0.95	C12	0.99
C6	0.08	C13	0.05
C7	1	C14	1

TABLE 5.17
Supplier Ranking Using L_2 Metric (Case Study 1)

Supplier	L_2 Value	Rank	Supplier	L_2 Value	Rank
Supplier 1	2.105	7	Supplier 11	2.782	18
Supplier 2	2.332	11	Supplier 12	2.083	5
Supplier 3	3.011	20	Supplier 13	2.429	15
Supplier 4	1.896	3	Supplier 14	2.347	13
Supplier 5	2.121	8	Supplier 15	2.517	16
Supplier 6	2.800	19	Supplier 16	1.834	2
Supplier 7	1.817	1	Supplier 17	2.586	17
Supplier 8	2.357	4	Supplier 18	2.092	6
Supplier 9	2.206	9	Supplier 19	1.970	4
Supplier 10	2.339	12	Supplier 20	2.295	10

Step 3: Rank the suppliers using the L_2 metric. The supplier with the smallest L_2 value is ranked first, followed by the next smallest L_2 value, and so on. Table 5.17 gives the L_2 distance from the ideal for each supplier and the resulting supplier rankings.

5.3.6.3 Rating (Scoring) Method

Rating is one of the simplest and most widely used ranking methods under conflicting criteria. This is similar to the LWP method discussed in Section 5.3.5. First, an appropriate rating scale is agreed to (e.g., from 1 to 10, where 10 is the most important and 1 is the least important selection criteria). The scale should be clearly understood by the DM to be used properly. Next, using the selected scale, the DM provides a rating r_j for each criterion, C_j. The same rating can be given to more than one criterion. The ratings are then normalized to determine the weights of the criteria j.

Assuming n criteria:

$$W_j = \frac{r_j}{\sum_{j=1}^{j=n} r_j} \quad \text{for } j = 1, 2, \ldots n \qquad (5.19)$$

NOTE: $\sum_{j=1}^{n} W_j = 1$

Next, a weighted score of the attributes is calculated for each supplier as follows:

$$S_i = \sum_{j=1}^{n} W_j f_{ij}, \quad \forall i = 1 \ldots K,$$

where f_{ij}'s are the criteria values for supplier i. The suppliers are then ranked based on their scores. The supplier with the highest score is ranked first. Rating method requires relatively little cognitive burden on the DM.

Table 5.18 illustrates the ratings and the corresponding weights for the 14 criteria. Since criteria C6, C11, and C13 are to minimize, their respective weights have been set as negative values. Table 5.19 shows the final scores for different suppliers and their respective rankings under the rating method.

TABLE 5.18

Criteria Weights Using Rating Method (Case Study 1)

Criterion	Rating	Weight	Criterion	Rating	Weight
C1	6	0.073	C8	1	0.012
C2	7	0.085	C9	8	0.098
C3	5	0.061	C10	7	0.085
C4	9	0.110	C11	(−) 6	−0.073
C5	10	0.122	C12	7	0.085
C6	(−) 2	−0.024	C13	(−) 4	−0.049
C7	3	0.037	C14	7	0.085

TABLE 5.19

Supplier Ranking Using Rating Method (Case Study 1)

Supplier	Total Score	Rank	Supplier	Total Score	Rank
Supplier 1	0.475	2	Supplier 11	0.095	19
Supplier 2	0.360	13	Supplier 12	0.404	9
Supplier 3	0.032	20	Supplier 13	0.196	17
Supplier 4	0.408	7	Supplier 14	0.394	10
Supplier 5	0.375	12	Supplier 15	0.247	16
Supplier 6	0.131	18	Supplier 16	0.394	11
Supplier 7	0.522	1	Supplier 17	0.250	15
Supplier 8	0.412	5	Supplier 18	0.450	3
Supplier 9	0.411	6	Supplier 19	0.405	8
Supplier 10	0.308	14	Supplier 20	0.430	4

5.3.6.4 Borda Method

Named after Jean Charles de Borda, an eighteenth-century French physicist, the method works as follows:

- The n criteria are ranked 1 (most important) to n (least important)
    - Criterion ranked 1 gets n points, the second rank gets $n - 1$ points, and the last place criterion gets 1 point.
- Weights for the criteria are calculated as follows:
    - Criterion ranked $1 = \dfrac{n}{s}$
    - Criterion ranked $2 = \dfrac{n-1}{s}$
    - Last criterion $= \dfrac{1}{s}$

where s is the sum of all the points $= \dfrac{n(n+1)}{2}$

Table 5.20 illustrates the calculations of criteria weights using the Borda method. For example, criterion 3 is ranked first among the 14 criteria and gets 14 points. Criterion 11 is ranked last and gets 1 point. Thus, the weight for criterion $3 = \dfrac{14}{105} = 0.133$. Using these criteria weights, the supplier scores are calculated as before for ranking as shown in Table 5.21.

TABLE 5.20

Criteria Weights Using Borda Count (Case Study 1)

Criterion	Ranking Points	Weight	Criterion	Ranking Points	Weight
C1	9	0.086	C8	13	0.124
C2	7	0.067	C9	2	0.019
C3	14	0.133	C10	8	0.076
C4	6	0.057	C11	(−) 1	−0.010
C5	5	0.048	C12	12	0.114
C6	(−) 10	−0.095	C13	(−) 4	−0.038
C7	11	0.105	C14	3	0.029

TABLE 5.21

Supplier Ranking Using Borda Count (Case Study 1)

Supplier	Total Score	Rank	Supplier	Total Score	Rank
Supplier 1	0.342	9	Supplier 11	0.238	17
Supplier 2	0.335	11	Supplier 12	0.387	6
Supplier 3	0.085	20	Supplier 13	0.331	12
Supplier 4	0.478	2	Supplier 14	0.230	18
Supplier 5	0.345	8	Supplier 15	0.274	15
Supplier 6	0.138	19	Supplier 16	0.462	3
Supplier 7	0.399	5	Supplier 17	0.270	16
Supplier 8	0.325	13	Supplier 18	0.416	4
Supplier 9	0.351	7	Supplier 19	0.503	1
Supplier 10	0.312	14	Supplier 20	0.336	10

5.3.6.4.1 Pairwise Comparison of Criteria

When there are many criteria, it would be difficult for a DM to rank order them precisely. In practice, pairwise comparison of criteria is used to facilitate the criteria ranking required by the Borda method. Here, the DM is asked to give the relative importance between two criteria C_i and C_j, whether C_i is preferred to C_j, C_j is preferred to C_i, or both are equally important. When there are n criteria, the DM has to respond to $\frac{n(n-1)}{2}$ pairwise comparisons. Based on the DM's responses, the criteria rankings and their weights can be computed. Refer to Chapter 4, Section 4.1.2.3, for a discussion of the detailed steps of getting the criteria rankings and weights based on the pairwise comparison responses, illustrated by a numerical example.

Supply Chain Engineering   253

5.3.6.5 Analytic Hierarchy Process

The AHP, developed by Saaty (1980), is a multi-criteria decision making method for ranking alternatives. Using AHP, the DM can assess not only quantitative but also various qualitative factors, such as financial stability and feeling of trust in the supplier selection process. The buyer establishes a set of evaluation criteria and AHP uses these criteria to rank the different suppliers. AHP can enable the DM to represent the interaction of multiple factors in complex and unstructured situations. AHP does not require the scaling of criteria values.

5.3.6.5.1 Basic Principles of AHP

- Design a hierarchy: Top vertex is the main objective and bottom vertices are the alternatives. Intermediate vertices are criteria/subcriteria (which are more and more aggregated as you go up in the hierarchy).
- At each level of the hierarchy, a pairwise comparison of the vertices criteria/subcriteria is performed from the point of view of their "contribution (weights)" to each of the higher-level vertices to which they are linked.
- Uses both the rating method and the pairwise comparison method. A numerical scale 1–9 (1, equal importance; 9, most important) is used.
- Uses pairwise comparison of alternatives with respect to each criterion (subcriterion) and gets a numerical score for each alternative on every criterion (subcriterion).
- Computes the total weighted score for each alternative and ranks the alternatives accordingly.

To design the hierarchy for Case Study 1, the 14 supplier criteria are grouped into 6 major criteria and several subcriteria as shown in Figure 5.9.

5.3.6.5.2 Steps of the AHP Model

Step 1: In the first step, carry out a pairwise comparison of criteria using the 1–9 degree of importance scale shown in Table 5.22.

If there are n criteria to evaluate, then the pairwise comparison matrix for the criteria is given by $A_{(n \times n)} = [a_{ij}]$, where a_{ij} represents the relative importance of criterion i with respect to criterion j. Set $a_{ii} = 1$ and $a_{ji} = \dfrac{1}{a_{ij}}$. The pairwise comparisons, with the degree of importance, for the six major criteria in Case Study 1, are shown in Table 5.23.

254 Service Systems Engineering and Management

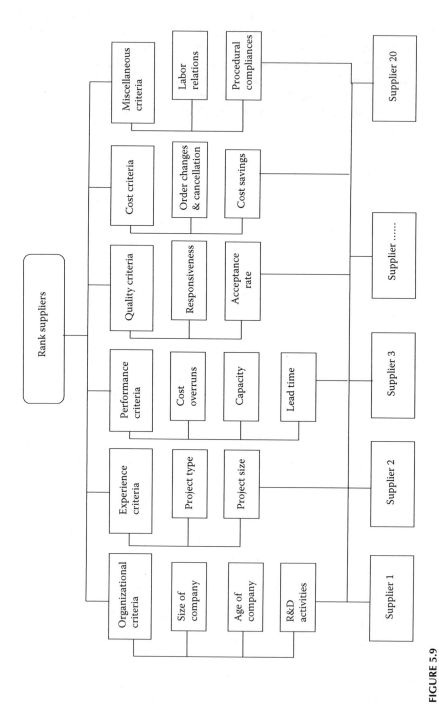

FIGURE 5.9
AHP hierarchy for supplier criteria (Case Study 1).

TABLE 5.22

Degree of Importance Scale in AHP

Degree of Importance	Definition
1	Equal importance
3	Weak importance of one over other
5	Essential or strong importance
7	Demonstrated importance
9	Absolute importance
2, 4, 6, 8	Intermediate values between two adjacent judgments

Step 2: Compute the normalized weights for the main criteria. We obtain the weights using L_1 norm. The two-step process for calculating the weights is as follows:

- Normalize each column of A matrix using L_1 norm.

$$r_{ij} = \frac{a_{ij}}{\sum_{i=1}^{n} a_{ij}}$$

- Average the normalized values across each row to get the criteria weights.

$$w_i = \frac{\sum_{j=1}^{n} r_{ij}}{n}$$

Table 5.24 shows the criteria weights for Case Study 1 obtained as a result of Step 2.

Step 3: In this step, we check for consistency of the pairwise comparison matrix using Eigen value theory as follows (Saaty 1980).

1. Using the pairwise comparison matrix A (Table 5.23) and the weights W (Table 5.24), compute the vector AW. Let the vector $X = (X_1, X_2, X_3, ..., X_n)$ denote the values of AW.
2. Compute

$$\lambda_{max} = \text{Average}\left[\frac{X_1}{W_1}, \frac{X_2}{W_2}, \frac{X_3}{W_3} .... \frac{X_n}{W_n}\right].$$

TABLE 5.23
Pairwise Comparison of Criteria (Case Study 1)

	Organizational	Experience	Performance	Quality	Cost	Miscellaneous
Organizational	1	0.2	0.143	0.33	0.33	1
Experience	5	1	0.5	2	2	5
Performance	7	2	1	5	4	7
Quality	3	0.5	0.2	1	1	3
Cost	3	0.5	0.25	1	1	3
Miscellaneous	1	0.2	0.143	0.33	0.33	1

TABLE 5.24
Final Criteria Weights using AHP (Case Study 1)

Criteria	Weight
Organizational	0.047
Experience	0.231
Performance	0.430
Quality	0.120
Cost	0.124
Miscellaneous	0.047

3. Consistency index (CI) is given by

$$CI = \frac{\lambda_{max} - n}{n - 1}.$$

Saaty (1980) generated a number of random positive reciprocal matrices with $a_{ij} \in (1, 9)$ for different sizes and computed their average CI values, denoted by RI, as given below:

N	1	2	3	4	5	6	7	8	9	10
RI	0	0	0.52	0.89	1.11	1.25	1.35	1.4	1.45	1.49

Saaty defines the consistency ratio (CR) as $CR = \frac{CI}{RI}$. If $CR < 0.15$, then accept the pairwise comparison matrix as consistent. Using the above steps, CR is found to be 0.009 for our example problem. Since the CR is less than 0.15, the pairwise comparison matrix is assumed to be consistent.

Step 4: In the next step, we compute the relative importance of the subcriteria in the same way as done for the main criteria. Step 2 and Step 3 are carried out for every pair of subcriteria with respect to their main criterion. The final weights of the subcriteria are the product of the weights along the corresponding branch. Table 5.25 illustrates the final weights of the various criteria and subcriteria for Case Study 1.

Step 5: Repeat Steps 1, 2, and 3 and obtain

a. Pairwise comparison of alternatives with respect to each criterion using the same ratio scale (1–9).

b. Normalized scores of all alternatives with respect to each criterion. Here, an $(m \times m)$ matrix S is obtained, where S_{ij} = normalized score for alternative i with respect to criterion j and m is the number of alternatives.

TABLE 5.25

AHP Subcriteria Weights for Case Study 1

Criteria (Criteria Weight)	Subcriteria	Subcriteria Weight	Global Weight (Criteria Weight × Subcriteria Weight)
Organizational (0.047)	Size of company	0.143	0.006
	Age of company	0.429	0.020
	R&D activities	0.429	0.020
Experience (0.231)	Project type	0.875	0.202
	Project size	0.125	0.028
Performance (0.430)	Cost overruns	0.714	0.307
	Capacity	0.143	0.061
	Lead time	0.143	0.061
Quality (0.120)	Responsiveness	0.833	0.099
	Acceptance rate	0.167	0.020
Cost (0.124)	Order change	0.833	0.103
	Cost savings	0.167	0.020
Miscellaneous (0.047)	Labor relations	0.125	0.005
	Procedural compliances	0.875	0.041

Step 6: Compute the total score (TS) for each alternative as follows $TS_{(m\times1)} = S_{(m\times n)}W_{(n\times1)}$, where W is the weight vector obtained after Step 4. Using the total scores, the alternatives are ranked. The total scores for all the suppliers obtained by AHP for Case Study 1 are given in Table 5.26.

NOTE: There is commercially available software for AHP called Expert Choice. Interested readers can refer to www.expertchoice.com for additional information.

5.3.7 Scaling Criteria Values

The major drawback of most ranking methods (L_p metric, rating, and Borda methods) is that they use criteria weights that require the criteria values to be scaled properly. For example, in Table 5.15, all the supplier criteria values ranged from 0 to 1. In other words, they have been scaled already. In practice, supplier criteria are measured in different units. Some criteria values may be very large (e.g., cost) while others may be very small (e.g., quality, delivery time). If the criteria values are not scaled properly, the criteria with large magnitudes would simply dominate the final rankings, independent of the weights assigned to them. In this section, we shall discuss some common approaches to scaling criteria values. It is important to note that AHP does

TABLE 5.26
Supplier Ranking Using AHP (Case Study 1)

Supplier	Total Score	Rank
Supplier 1	0.119	20
Supplier 2	0.247	14
Supplier 3	0.325	12
Supplier 4	0.191	18
Supplier 5	0.210	16
Supplier 6	0.120	19
Supplier 7	0.249	13
Supplier 8	0.328	11
Supplier 9	0.192	17
Supplier 10	0.212	15
Supplier 11	0.427	7
Supplier 12	0.661	1
Supplier 13	0.431	6
Supplier 14	0.524	5
Supplier 15	0.422	8
Supplier 16	0.539	4
Supplier 17	0.637	2
Supplier 18	0.421	9
Supplier 19	0.412	10
Supplier 20	0.543	3

not use the actual criteria values. Using the pairwise comparison of criteria, the criteria values are converted to strength of preference values using a scale of 1 through 9 (see Tables 5.22 and 5.23). Hence, AHP does not require the scaling of criteria values.

To describe the various scaling methods, consider a supplier selection problem with m suppliers and n criteria, where f_{ij} denotes the value of criterion j for supplier i. Let F denote the supplier–criteria matrix.

$$F_{(m \times n)} = [f_{ij}]$$

Determine $H_j = \max_i f_{ij}$ and $L_j = \min_i f_{ij}$.

H_j will be the ideal value, if criterion j is maximizing and L_j is its ideal value, if it is minimizing.

5.3.7.1 Simple Scaling

In simple scaling, the criteria values are multiplied by 10^K where K is a positive or negative integer including zero. This is the most common scaling method used in practice. If a criterion is to be minimized, its values should be multiplied by (−1) before computing the weighted score.

5.3.7.2 Ideal Value Method

In this method, criteria values are scaled using their ideal values as given below:

$$\text{For "max" criterion: } r_{ij} = \frac{f_{ij}}{H_j} \quad (5.20)$$

$$\text{For "min" criterion: } r_{ij} = \frac{L_j}{f_{ij}} \quad (5.21)$$

Note that the scaled criteria values will always be ≤1, and all the criteria have been changed to maximization. The best value of each criterion is 1, but the worst value need not necessarily be zero. In the next approach, all criteria values will be scaled between 0 and 1, with 1 for the best value and 0 for the worst.

5.3.7.3 Simple Linearization (Linear Normalization)

Here, the criteria values are scaled as given below.

$$\text{For "max" criterion: } r_{ij} = \frac{f_{ij} - L_j}{H_j - L_j} \quad (5.22)$$

$$\text{For "min" criterion: } r_{ij} = \frac{H_j - f_{ij}}{H_j - L_j} \quad (5.23)$$

Here, all the scaled criteria value will be between 0 and 1 and all the criteria are to be maximized after scaling.

5.3.7.4 Use of L_p Norm (Vector Scaling)

The L_p norm of a vector $X \in R^n$ is given by L_p norm $= \left[\sum_{j=1}^{n} |X_j|^p\right]^{\frac{1}{p}}$, for $p = 1, 2, \ldots \infty$.

The most common values of p are $p = 1, 2,$ and ∞.

$$\text{For } p = 1, L_1 \text{ norm} = \sum_{j=1}^{n} |X_j|. \quad (5.24)$$

Supply Chain Engineering

For $p = 2$, L_2 norm $= \left[\sum_{j=1}^{n} |X_j|^2\right]^{\frac{1}{2}}$ (length of vector X). (5.25)

For $p = \infty$, L_∞ norm $= \max\left[|X_j|\right]$ (Tchebycheff norm). (5.26)

In this method, scaling is done by dividing the criteria values by their respective L_p norms. After scaling, the L_p norm of each criterion will be one.

We shall illustrate the above scaling methods with a numerical example.

5.3.7.5 Illustrative Example of Scaling Criteria Values

Example 5.6 (Scaling criteria values)

Consider a supplier selection problem with three suppliers A, B, and C and three selection criteria—TCO, service, and experience. The criteria values are given in Table 5.27. TCO has to be minimized, while the service and experience criteria have to be maximized.

SOLUTION

Note that the high-cost supplier (A) gives the best service and has the most experience, while supplier C has the lowest cost and experience and gives poor service. The criteria values are not scaled properly, particularly cost measured in dollars. If the values are not scaled, the TCO criterion will dominate the selection process irrespective of its assigned weight. If we assume that the criteria weights are equal $\left(\frac{1}{3}\right)$, then the weighted score for each supplier would be

$$S_A = \frac{(-125{,}000 + 10 + 9)}{3} = -41{,}660$$

$$S_B = \frac{(-95{,}000 + 5 + 6)}{3} = -31{,}663$$

TABLE 5.27

Supplier Criteria Values for Example 5.6

	TCO (Min)	Service (Max)	Experience (Max)
Supplier A	$125,000	10	9
Supplier B	$95,000	5	6
Supplier C	$65,000	3	3

$$S_C = \frac{(-65{,}000 + 3 + 3)}{3} = -21.665$$

Note that the cost criterion (TCO) has been multiplied by (−1) to convert it to a maximization criterion, before computing the weighted score. Supplier C has the maximum weighted score and the rankings will be Supplier C > Supplier B > Supplier A in that order. In fact, even if the weight for TCO is reduced, cost will continue to dominate, as long as it is not scaled properly. Let us look at the rankings after scaling the criteria values by the methods given in this section.

5.3.7.6 Simple Scaling Illustration

Dividing TCO values by 10,000, we get the scaled values as 12.5, 9.5, and 6.5, which are comparable in magnitude with the criteria values for service and experience. Assuming equal weights again for the criteria, the new weighted scores of the suppliers are as follows:

$$S_A = \frac{(-12.5 + 10 + 9)}{3} = 2.17$$

$$S_B = \frac{(-9.5 + 5 + 6)}{3} = 0.5$$

$$S_C = \frac{(-6.5 + 3 + 3)}{3} = -0.17$$

Now, Supplier A is the best followed by Suppliers B and C.

5.3.7.7 Illustration of Scaling by Ideal Value

For Example 5.6, the maximum and minimum criteria values are as given below:

C1—TCO: $H_1 = 125{,}000$, $L_1 = 65{,}000$
C2—Service: $H_2 = 10$, $L_2 = 3$
C3—Experience: $H_3 = 9$, $L_3 = 3$

The ideal values for the three criteria are 65,000, 10, and 9 respectively. Of course, the ideal solution is not achievable.

The scaled criteria values, using the ideal value method, are computed using Equation 5.21 for TCO and Equation 5.20 for service and experience. They are given in Table 5.28.

TABLE 5.28
Scaled Criteria Values by the Ideal Value Method (Example 5.6)

	TCO	Service	Experience
Supplier A	0.52	1	1
Supplier B	0.68	0.5	0.67
Supplier C	1	0.3	0.33

Note that the scaled values are such that all criteria (including TCO) have to be maximized. Thus, the new weighted scores are

$$S_A = \frac{(0.52 + 1 + 1)}{3} = 0.84$$

$$S_B = \frac{(0.68 + 0.5 + 0.67)}{3} = 0.62$$

$$S_C = \frac{(1 + 0.3 + 0.33)}{3} = 0.54$$

The final rankings are Suppliers A, B, and C have the same ranking obtained using simple scaling.

5.3.7.8 Simple Linearization (Linear Normalization) Illustration

Under this method, the scaled values are computed using Equations 5.22 and 5.23 and are given in Table 5.29.

Note that the best and worst values of each criterion are 1 and 0, respectively, and all the criteria values are now to be maximized. The revised weighted sums are

$$S_A = \frac{(0 + 1 + 1)}{3} = 0.67$$

TABLE 5.29
Scaled Criteria Values by Simple Linearization (Example 5.6)

	TCO	Service	Experience
Supplier A	0	1	1
Supplier B	0.5	0.29	0.5
Supplier C	1	0	0

$$S_B = \frac{(0.5 + 0.29 + 0.5)}{3} = 0.43$$

$$S_C = \frac{(1 + 0 + 0)}{3} = 0.33$$

The rankings are unchanged with Supplier A as the best, followed by Suppliers B and C.

5.3.7.9 Illustration of Scaling by L_p Norm

We shall illustrate using L_∞ norm for scaling. The L_∞ norms for the three criteria are computed using Equation 5.26.

L_∞ norm for TCO = Max (125,000, 95,000, 65,000) = 125,000
L_∞ norm for service = Max (10, 5, 3) = 10
L_∞ norm for experience = Max (9, 6, 3) = 9

The criteria values are then scaled by dividing them by their respective L_∞ norms and are given in Table 5.30.

Note that the scaling by L_p norm did not convert the minimization criterion (TCO) to maximization as the previous two methods (ideal value and simple linearization) did. Hence, the TCO values have to be multiplied by (–1) before computing the weighted score. Note also that the L_∞ norm of each criterion (column) in Table 5.30 is always 1.

The new weighted scores are as follows:

$$S_A = \frac{(-1 + 1 + 1)}{3} = 0.33$$

$$S_B = \frac{(-0.76 + 0.5 + 0.67)}{3} = 0.14$$

$$S_C = \frac{(-0.52 + 0.3 + 0.33)}{3} = 0.04$$

TABLE 5.30

Scaled Criteria Values using L_∞ Norm (Example 5.6)

	TCO	Service	Experience
Supplier A	1	1	1
Supplier B	0.76	0.5	0.67
Supplier C	0.52	0.3	0.33

Supply Chain Engineering

Once again, the rankings are the same, namely, Supplier A, followed by Suppliers B and C.

It should be noted that even though the scaled values, using different scaling methods, were different, the final rankings were always the same. Occasionally, it is possible for rank reversals to occur.

5.3.8 Group Decision Making

Most purchasing decisions, including the ranking and selection of suppliers, involve the participation of multiple DMs and the ultimate decision is based on the aggregation of DMs' individual judgments to arrive at a group decision. The rating method, Borda method, and AHP discussed in this section can be extended to group decision making as described below:

1. **Rating Method:** Ratings of each DM for every criterion are averaged. The average ratings are then normalized to obtain the group criteria weights.
2. **Borda Method:** Points are assigned based on the number of DMs that assign a particular rank for a criterion. These points are then totaled for each criterion and normalized to get criteria weights. (This is similar to how the college polls are done to get the top 25 football or basketball teams.)
3. **AHP:** There are two methods to get the group rankings using AHP.
    a. **Method 1:** Strength of preference scores assigned by individual DMs are aggregated using geometric means and then used in the AHP calculations.
    b. **Method 2:** First, all the alternatives are ranked by each DM using AHP. The individual rankings are then aggregated to a group ranking using the Borda method.

5.3.9 Comparison of Ranking Methods

Different ranking methods can provide different solutions resulting in rank reversals. In extensive empirical studies with human subjects, it has been found (Powdrell 2003, Patel 2007) that the Borda method (with pairwise comparison of criteria) rankings are generally in line with AHP rankings. Given the increased cognitive burden and expensive calculations required for AHP, the Borda method is an appropriate and efficient method for supplier rankings. Even though the rating method is easy to use, it could lead to several ties in the final rankings, thereby making the results less useful.

Velazquez et al. (2010) studied the best combination of weighting and scaling methods for single and multiple DMs. The scaling methods considered were ideal value, linear normalization, and vector normalization using L_p norm for $p = 1, 2, 3$, and ∞. The weighting methods included in the study were rating, Borda method, and AHP. The L_p metric method for $p = 1, 2, 3,$ and ∞ were also included for ranking. Experiments were done with real DMs. They found that the best scaling method was influenced by the chosen weighting method. The best combination was scaling by L_∞ norm and ranking by the Borda method. The worst combination was scaling by L_∞ norm and ranking by L_∞ metric. The conclusions were the same for both single and multiple DMs.

5.3.10 Multiple Sourcing Methods for Supplier Selection and Order Allocation

In multiple sourcing, a buyer purchases the same item from more than one supplier. Multiple sourcing can offset the risk of supply disruptions. Mathematical programming models are the most appropriate methods for multiple sourcing decisions. It allows the buyer to consider different constraints including capacity, delivery time, quality, and so on, while choosing the suppliers and their order allocations. Two types of mathematical models are found in the literature, single-objective and multiple-objective models. The single-objective models invariably use cost as the main objective function. We shall illustrate the single-objective model in this section using Case Study 2. Multiple-objective models are discussed in detail in Ravindran and Wadhwa (2009) and Ravindran and Warsing (2013). In addition to cost, multiple-objective models consider quality, delivery, and supplier risk as additional objectives.

5.3.10.1 Case Study 2 (Ravindran and Warsing 2013)

Consider a supplier selection and order allocation problem with two products, two buyers (multiple buyers represent situations when different divisions of a company buy through one central purchasing department), two suppliers, and each supplier offering "incremental" price discounts to each buyer (not "all unit discount"). The objective is to minimize *total cost*, which consists of fixed cost and the variable cost. Fixed cost is a one-time cost that is incurred if a supplier is used for any product, irrespective of the number of units bought from that supplier.

The constraints in the model include the following:

1. Capacity constraints of the suppliers
2. Buyer's demand constraints
3. Buyer's quality constraints
4. Buyer's lead time constraints
5. Price break constraints for products

Supply Chain Engineering

The customer demand data are given in Table 5.31. Data regarding supplier capacities, lead time, and product quality are given in Tables 5.32 through 5.34.

The quantity discounts and price breaks offered by the suppliers are given in Table 5.35. The Level 1 break points represent the quantity at which price discounts apply. Level 2 break points represent the maximum quantity of a particular product a supplier can provide to that buyer. For example, for Product 1, Buyer 1, and Supplier 1, the first 85 units will cost $180/unit and the next 65 units (i.e., 150 − 85) will cost $165/unit; no more than 150 units of Product 1 can be purchased from Supplier 1 by Buyer 1.

The lead time requirements of the buyers for the two products are given in Table 5.36.

TABLE 5.31

Product Demand

Product	Buyer	Demand
1	1	150
1	2	175
2	1	200
2	2	180

TABLE 5.32

Supplier Capacities

Product	Supplier	Capacity
1	1	400
1	2	450
2	1	480
2	2	460

TABLE 5.33

Lead Time of Products in Days

Product	Buyer	Supplier	Lead Time
1	1	1	8
1	1	2	17
1	2	1	14
1	2	2	24
2	1	1	28
2	1	2	8
2	2	1	16
2	2	2	12

TABLE 5.34

Quality of Product (Measured by Percentage of Rejects)

Product	Supplier	Quality
1	1	3%
1	2	9%
2	1	6%
2	2	2%

TABLE 5.35

Product Prices (Including Shipping Cost) of Suppliers with Quantity Discounts and Their Corresponding Price Break Points

Product	Buyer	Supplier	Level	Unit Price	Break Point
1	1	1	1	180	85
1	1	1	2	165	150
1	1	2	1	178	95
1	1	2	2	166	175
1	2	1	1	180	85
1	2	1	2	165	150
1	2	2	1	178	95
1	2	2	2	166	175
2	1	1	1	80	120
2	1	1	2	70	170
2	1	2	1	83	125
2	1	2	2	69	160
2	2	1	1	80	120
2	2	1	2	70	170
2	2	2	1	83	125
2	2	2	2	69	160

TABLE 5.36

Average Lead Time Requirements of the Buyers

Product	Buyer	Lead
1	1	12.5
1	2	19
2	1	18
2	2	14

Buyers 1 and 2 limit the average reject levels to 6% and 4%, respectively, from all suppliers. The fixed costs of using Suppliers 1 and 2 are $3500 and $3600, respectively.

5.3.10.2 Solution of Case Study 2

Decision Variables

x_{ijkm}—Quantity of product i that buyer j buys from supplier k, at price level m

$$\delta_k - \begin{cases} 1, & \text{if supplier } k \text{ is used} \\ 0, & \text{Otherwise} \end{cases}$$

$$\beta_{ijkm} - \begin{cases} 1, & \text{if buyer } j \text{ buys product } i \text{ from supplier } k \text{ at price level } m \\ 0, & \text{Otherwise} \end{cases}$$

Product, $i = 1, 2$
Buyer, $j = 1, 2$
Supplier, $k = 1, 2$
Price level, $m = 1, 2$

Thus, there are 18 binary variables and 16 continuous variables.

Objective Function

Minimize cost = Fixed cost + Variable cost
$$= 3500\,\delta_1 + 3600\,\delta_2 + 180\,x_{1111} + 165\,x_{1112}$$
$$+ 178\,x_{1121} + 166\,x_{1122} + 180\,x_{1211} + 165\,x_{1212}$$
$$+ 178\,x_{1221} + 166\,x_{1222} + 80\,x_{2111} + 70\,x_{2112} + 83\,x_{2121}$$
$$+ 69\,x_{2122} + 80\,x_{2211} + 70\,x_{2212} + 83\,x_{2221} + 69\,x_{2222}$$

The objective function is to minimize the sum of fixed and variable costs.

Constraints

1. Buyers' demand constraints

$x_{1111} + x_{1112} + x_{1121} + x_{1122} = 150$	(Product 1, Buyer 1)
$x_{1211} + x_{1212} + x_{1221} + x_{1222} = 175$	(Product 1, Buyer 2)
$x_{2111} + x_{2112} + x_{2121} + x_{2122} = 200$	(Product 2, Buyer 1)
$x_{2211} + x_{2212} + x_{2221} + x_{2222} = 180$	(Product 2, Buyer 2)

2. Suppliers' capacity constraints

$x_{1111} + x_{1112} + x_{1211} + x_{1212} \leq 400\,\delta_1$	(Product 1, Supplier 1)
$x_{1121} + x_{1122} + x_{1221} + x_{1222} \leq 450\,\delta_2$	(Product 1, Supplier 2)
$x_{2111} + x_{2112} + x_{2211} + x_{2212} \leq 480\,\delta_1$	(Product 2, Supplier 1)
$x_{2121} + x_{2122} + x_{2221} + x_{2222} \leq 460\,\delta_2$	(Product 2, Supplier 2)

3. Buyers' quality constraints

   Buyer 1: The weighted average rejects of both products received from the two suppliers (measured in percent defectives) cannot exceed 6% for Buyer 1.

$$\frac{0.03(x_{1111} + x_{1112}) + 0.09(x_{1121} + x_{1122}) + 0.06(x_{2111} + x_{2112}) + 0.02(x_{2121} + x_{2122})}{150 + 200} \leq 0.06$$

   Buyer 2: Similarly for Buyer 2, it is limited to 4%.

$$\frac{0.03(x_{1211} + x_{1212}) + 0.09(x_{1221} + x_{1222}) + 0.06(x_{2211} + x_{2212}) + 0.02(x_{2221} + x_{2222})}{175 + 180} \leq 0.04$$

4. Buyer's lead time constraints

   Buyer 1: The weighted average lead time is 12.5 days for Product 1 and 18 days for Product 2 as follows:

$$\frac{8(x_{1111} + x_{1112}) + 17(x_{1121} + x_{1122})}{150} \leq 12.5$$

$$\frac{28(x_{2111} + x_{2112}) + 8(x_{2121} + x_{2122})}{200} \leq 18$$

   Buyer 2: Similarly, the lead time constraints for Buyer 2 are as follows:

$$\frac{14(x_{1211} + x_{1212}) + 24(x_{1221} + x_{1222})}{175} \leq 19$$

$$\frac{16(x_{2211} + x_{2212}) + 12(x_{2221} + x_{2222})}{180} \leq 14$$

Supply Chain Engineering

5. Price Break Constraints for Products

    (Recall the formulation of price break constraints under quantity discounts in Chapter 3, Section 3.1.4.)

    i. Product 1, Buyer 1, Supplier 1

    $$0 \leq x_{1111} \leq 85\beta_{1111}$$

    $$x_{1111} \geq 85\beta_{1112}$$

    $$0 \leq x_{1112} \leq 65\beta_{1112}$$

    Note that if β_{1112} is one, then it would force β_{1111} to be one and $x_{1111} = 85$ units. In other words, Buyer 1 cannot buy Product 1 from Supplier 1 at the level 2 (lower) price unless he has ordered at least 85 units at the level 1 (higher) price. Similarly, we get the other price break constraints for the other supplier, product, and buyer as given below:

    ii. Product 1, Buyer 1, Supplier 2

    $$0 \leq x_{1121} \leq 95\beta_{1121}$$

    $$x_{1121} \geq 95\beta_{1122}$$

    $$0 \leq x_{1122} \leq 80\beta_{1122}$$

    iii. Product 1, Buyer 2, Supplier 1

    $$0 \leq x_{1211} \leq 85\beta_{1121}$$

    $$x_{1211} \geq 85\beta_{1212}$$

    $$0 \leq x_{1212} \leq 65\beta_{1212}$$

    iv. Product 1, Buyer 2, Supplier 2

    $$0 \leq x_{1221} \leq 95\beta_{1221}$$

    $$x_{1221} \geq 95\beta_{1222}$$

    $$0 \leq x_{1222} \leq 80\beta_{1222}$$

v. Product 2, Buyer 1, Supplier 1

$$0 \leq x_{2111} \leq 120\beta_{2111}$$

$$x_{2111} \geq 120\beta_{2112}$$

$$0 \leq x_{2112} \leq 50\beta_{2112}$$

vi. Product 2, Buyer 1, Supplier 2

$$0 \leq x_{2121} \leq 125\beta_{2121}$$

$$x_{2121} \geq 125\beta_{2122}$$

$$0 \leq x_{2122} \leq 35\beta_{2122}$$

vii. Product 2, Buyer 2, Supplier 1

$$0 \leq x_{2211} \leq 120\beta_{2211}$$

$$x_{2211} \geq 120\beta_{2212}$$

$$0 \leq x_{2212} \leq 50\beta_{2212}$$

viii. Product 2, Buyer 2, Supplier 2

$$0 \leq x_{2221} \leq 125\beta_{2221}$$

$$x_{2221} \geq 125\beta_{2222}$$

$$0 \leq x_{2222} \leq 35\beta_{2222}$$

ix. Binary and Non-negativity Constraints

$$\delta_k \in (0,1) \forall k = 1, 2$$

$$\beta_{ijkm} \in (0,1) \forall i, j, k, m$$

$$x_{ijkm} \geq 0$$

TABLE 5.37

Optimal Order Allocation (Units)

		Supplier 1		Supplier 2	
		Level 1	Level 2	Level 1	Level 2
Buyer 1	Product 1	85	65	0	0
	Product 2	40	0	125	35
Buyer 2	Product 1	85	65	25	0
	Product 2	20	0	125	35

The above mixed integer program with 18 binary variables, 16 continuous variables, and 38 constraints was solved using Microsoft's Excel Solver software. The optimal order allocation is given in Table 5.37. Both suppliers are used for purchases. The policy results in a total cost of $93,980.

5.4 Supply Chain Logistics

Logistics decisions are those that affect the process of the distribution of goods across the components of the supply chain. Warehousing is the most important intrafacility component of logistics, and is discussed in Chapter 6. We focus on the interfacility decisions here.

Logistics costs make up a significant component of a country's gross domestic product (GDP). In the United States, these costs were 8.3% of GDP in 2015 (down from 9.9% of GDP in 2009, when oil prices were significantly higher and trucking capacities much tighter) (Wilson 2015). Trucking makes up the largest component of these costs, totaling more than $700 billion in 2014. The other key modes are rail ($80 billion), water ($40 billion), and air ($28 billion). Total carrying costs were estimated to be $476 billion in 2014 (Wilson 2015).

The effective distribution of goods between locations involves two key decisions: (i) the choice of the mode of transportation to use and, in the case when delivery is across a set of customers, (ii) the routing for those customers. In the next sections, we discuss each of these decisions. It is important to remember that both of these decisions are dependent on the network design, as discussed in Section 5.2. Inventory modeling also influences and is influenced by logistics decisions. As a result, we present a basic introduction in Section 5.4.3.

5.4.1 Mode Selection

As mentioned previously, the four key distribution modes are air, rail, truck, and water. Trucking can be broken down into two main categories: full truckload

(FTL) and less than truckload (LTL). As the name implies, FTL is when the shipper will use the entire truck capacity to go from an origin to a destination. LTL occurs when the shipper may only need a part of the capacity. In LTL, there are multiple stops across the set of customers (i.e., there is an associated route). Note that pickup and delivery may both be done on the same LTL route.

Often, however, a good may require several modes of transportation to go from the origin to the destination. For example, an item shipped from Mascoutah, Illinois to Hong Kong might use the following sequence: (i) local trucking from Mascoutah, Illinois to St. Louis, Missouri, (ii) rail from St. Louis, Missouri to Long Beach, California, (iii) ocean carrier from Long Beach, California to Hong Kong, and (iv) local delivery by truck in Hong Kong. An alternative would be to replace (ii) and (iii) by air shipment from St. Louis, Missouri to Hong Kong.

In general, the mode selection problem is a matter of costing all alternatives between the origin and destination and picking the cheapest. One complication is that the lead times may differ between alternatives, which would affect both inventory and customer service. A second complication is if a firm manages its own fleet of truck and does multi-stop routes, then the LTL cost will be a function of the route. The second complication is rather challenging to solve, and will not be considered here, though the interested reader can see Fazi (2014).

5.4.2 Vehicle Routing

The simplest form of vehicle routing problem consists of finding the shortest tour (which could be measured in time) that covers a set of customers. This is called the traveling salesperson problem (TSP) and has been well studied in the area of combinatorial optimization and belongs to a class of problems known as NP-hard. Practically, this implies that the even relatively small instances of the problem can take a significant amount of time to solve. An excellent discussion of the TSP and its history is found in Cook (2011).

Assume that we have N locations that we wish to visit. If we let d_{ij} be the distance between two locations and x_{ij} equal 1 if we travel directly from location i to location j, then the following formulation will determine the "optimal" tour. The formulation is given by

$$\min \sum_{i=1}^{N} \sum_{j=1}^{N} d_{ij} x_{ij} \qquad (5.27)$$

Subject to

$$\sum_{i=1}^{N} x_{ij} = 1 \quad \forall i \qquad (5.28)$$

$$\sum_{j=1}^{N} x_{ij} = 1 \quad \forall j \tag{5.29}$$

$$u_1 = 1 \tag{5.30}$$

$$2 \leq u_i \leq N \quad \forall i \neq 1 \tag{5.31}$$

$$u_i - u_j + 1 \leq N(1 - x_{ij}) \quad \forall i, j \neq 1 \tag{5.32}$$

$$x_{ij} \in \{0,1\} \quad \forall i, j \tag{5.33}$$

In the above formulation, Constraints 5.28 and 5.29 ensure that each location is visited exactly once. Constraints 5.30 through 5.32 ensure that there are no subtours generated. Note that in the Excel Solver, the subtour elimination constraints can be accomplished by simply adding the constraint "alldifferent."[*]

Although the above formulation can be solved for small to medium instances, there are often several complications that also need to be considered. These included truck capacity constraints, pickup and delivery time windows, and the use of multiple trucks to service a set of customers. For this reason, heuristic techniques are often employed. Common heuristics include construction heuristics such as nearest-neighbor (go to the closest unvisited location) and improvement heuristics such as two-opting (if any two arcs cross on a route, reorder them so they don't). A two-opting improvement is shown in Figure 5.10.

For the case of capacitated multi-vehicle routing, there are two decisions to be made: assignment of locations to a route and routing of assigned locations. A simple (and hence popular) heuristic used for multivehicle routing is the Savings Method (also known as the Clarke–Wright Savings Algorithm). This approach works by initially assigning each location to its own route. Routes are combined in a "greedy" way that doesn't violate capacity constraints. The process is repeated until no more combinations are possible. The specific algorithm is given below. Let t_i be the time (or cost) of route i starting and ending at location T (T, ..., i,T) and t_j be the cost of route (T,j, ..., T). The "savings" of combing these two routes is given by

$$s_{ij} = (t_i + d_{iT} + t_j + d_{Tj}) - (t_i + d_{ij} + c_j) = d_{iT} + d_{jT} - d_{ij}. \tag{5.34}$$

---

[*] van Hoeve, W.-J. (2001). The Alldifferent Constraint: A Survey. In Proceedings of the Sixth Annual Workshop of the ERCIM Working Group on Constraints. http://www.arxiv.org/html/cs/0110012

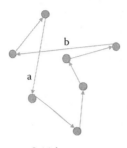

Initial route   Improved route from two-opt

FIGURE 5.10
Illustration of route improvement by two-opt on arcs a and b.

The method is given by

Step 1. *Initialize*

- Compute s_{ij} for all $i \neq j$.
- Create n initial routes (T,i,T) for all i.
- Form a savings list by ordering the s_{ij} values in a non-increasing fashion.

Step 2. *Route Merge Step*

- Starting from the top of the savings list, execute the following:
  - For the given s_{ij}, determine whether the route staring with (T,j) can be merged with the route ending with (i,T).
  - If the merger is feasible, combine the two routes by deleting (T,j) and (i,T) and introducing (i,j).
  - Delete s_{ij} from the list.

Example 5.7

This simple daily routing example illustrates Savings Method. A depot (T) is located at $(5,5)$. The depot serves four customers (A to D) with the following locations and daily demands:

Customer	X-location	Y-location	Daily Demand
C_1	2	8	12
C_2	7	3	11
C_3	9	9	6
C_4	3	8	8

Supply Chain Engineering

Cost per unit distance of travel is $100. Truck capacity is 20 units, and distance is measured using the Euclidean metric. The distance between each location pair is given by

	C_1	C_2	C_3	C_4
T	4.24	2.83	5.66	3.61
C_1		7.07	7.97	1.00
C_2			6.32	6.40
C_3				6.08

This leads to the following savings matrix, using Equation 5.34:

s_{ij}	C_2	C_3	C_4
C_1	0.00	2.83	6.85
C_2		2.17	0.04
C_3			3.19

The algorithm would say that based on the savings matrix to first combine routes (T, 1, T) and (T, 4, T) to form (T, 1, 4, T). We would then try to combine (T, 3, T) with (T, 1, 4, T), but this would violate the capacity constraint. Similarly, adding (T, 1, T) to this route is infeasible. Finally, we would combine routes (T, 2, T) and (T, 3, T) to form (T, 2, 3, T). There are no more possible combinations, and so we would finish with these two routes.

5.4.3 Inventory Modeling

The key inventory decisions are the determination of when to order, how much to order, and how much to hold as a buffer. When an order is placed, there is typically a time that elapses until the order arrives, which is the replenishment lead time. Inventory is typically classified into three types: cycle stock, safety stock, and pipeline stock. Cycle stock is inventory held to handle average demand between orders. Safety stock is to be held as a buffer to uncertainty in demand. Pipeline stock is the inventory that is in transit. In this section, we discuss three basic inventory models: (i) economic order quantity, (ii) continuous review (Q, R) model, and (iii) newsvendor model.

The *economic order quantity* (EOQ) model is used when annual demand (d) is constant, the replenishment lead time is zero, and stockouts are not allowed. We assume that the cost per item is c, the cost to place an order is A, and the cost to hold an item for one year is h. The holding cost arises from several sources: the cost (interest) of tying up money in inventory, space costs, taxes, and theft. It is common to apply a rate i (in %), so that $h = ic$. Common values for i range from 15% to 33%. We can determine the optimal order quantity Q, but considering the components of total cost ($TC(Q)$): annual order cost, annual holding cost, and total purchase cost.

$$TC(Q) = A\frac{d}{Q} + h\frac{Q}{2} + cd \tag{5.35}$$

Note that in the first term, d/Q is the number of orders placed over the year. In the second term, $Q/2$ represents the average inventory (since it goes from Q to 0). The third term is the total purchase cost. Since it isn't a function of Q, the purchase cost doesn't impact the decision. By taking the derivative of the total cost and setting it to 0, we get the EOQ formula:

$$Q = \sqrt{\frac{2Ad}{h}}. \tag{5.36}$$

It is easy to extend the EOQ formula for several cases such as allowing backorders if there are stockouts, adding a lead time, and so on. These will be developed in the exercises at the end of this chapter.

In most cases, demand is not actually known. If demand can be characterized according to a stationary distribution (f_d), where d is now a random variable (we will assume that the demand is normally distributed with annual demand μ and standard deviation σ), then a common approach is to use the *continuous review method* (Q, R). In this case, the inventory position (which consists of on hand plus on order minus backordered inventory) is tracked over time. Once it hits or falls below level equal to R (the reorder point), then an order of size Q is placed. We will assume that the replenishment lead time (L) is constant. The decision variables then are to determine the appropriate values of Q and R. A popular approach to set Q is to use the expected annual demand in the EOQ formula. The value of R is based on a "service level" $(1 - \alpha)$, which is defined as the probability that a stockout does not occur between orders. That is, $R = 1 - P$ {inventory position < 0}. Given a service level, then the value of R is defined as the mean demand over lead time plus the safety stock. That is:

$$R = \mu_L + Z_{1-\alpha}\sigma_L \tag{5.37}$$

where $\mu_L = \mu/L$ and $\sigma_L = \sqrt{1/L}\sigma$. The Z-value comes from the standard normal distribution. We illustrate both the EOQ and (Q, R) models with the following example.

Example 5.8

A retail outlet orders item A1 from a regional DC. The item costs $100 per unit, the order cost is $1000 per order, and the discount rate used for holding is 25%. In addition, it is desired that the service level is 95%. Annual demand for the item is normally distributed with a mean of 4000 and standard deviation of 1200. Replenishment lead time is 1 week. The outlet wishes to develop the (Q, R) model using the EOQ. In this example, the EOQ gives

$$Q = \sqrt{\frac{2(1000)(4000)}{0.25(100)}} = 566 \text{ units}.$$

In addition, $\mu_L = 4000/12 = 333.3$ and $\sigma_L = \sqrt{\frac{1}{12}}(1200) = 346.4$. For $\alpha = 0.05$, we get from the standard normal table (or the Excel command normsinv(0.95)) $Z_{0.95} = 1.645$. Therefore, we have $R = 333.3 + 1.645(346.4) = 901$ units. When the inventory position falls at or below 901, the retailer should order 566 units. Note that the safety stock for this case is given by the second component in the R equation (i.e., 1.645(346.4)) or 570 units.

One last inventory item to discuss is the *newsvendor model*. In this case, it is assumed that items are a "fashion good." That is, there is a selling period over which the item can be sold. At the end of the period, they can be salvaged for some value (which may be $0) below cost. Newspapers clearly fit this assumption since they have value on the day they are sold, but whose value is much diminished the following day (yesterday's news is no longer interesting). Other items that fit this are fashionable clothing and perishable goods (e.g., milk). We assume that only one order can be placed at the beginning of the selling period; leftover items are salvaged at a value (S) less than costs (C), and stocking out is equivalent to lost profit (i.e., each unmet demand corresponds to a marginal profit loss of selling price P minus cost C). The fundamental trade-off is the cost of having too many versus the cost of having too few.

We can determine the optimal order quantity in this case by performing a marginal analysis. Assume that we have ordered Q. The marginal benefit of ordering one more unit is equal to $P - C$ if the item is sold (i.e., if $P\{d > Q\}$, which is equal to $1 - F_d(Q)$). The marginal cost is equal to $C - S$, since we would salvage an unsold unit. Setting the marginal cost equal to the marginal benefit (which would occur at optimality) gives the critical ratio

$$F_d(Q) = \frac{P - C}{P - S}. \tag{5.38}$$

We want to find the value of Q for which this ratio holds. Since demand is discrete, then we should find the smallest value of Q for which the left-hand side of the above expression is greater than the right-hand side. An example of using the newsvendor model is given in Section 5.5.2.

5.5 Supply Chain Pooling and Contracting

Two difficult challenges to deal with in a supply chain are uncertainty/variability and risk. Uncertainty/variability in a supply chain leads to increases in inventory and/or decreases in customer service. This becomes even more of an issue as replenishment lead times or delays in information (such as demand data) increase. Risk can differ by component in supply chain and may lead to local decisions that hurt overall supply chain performance. In this section, we discuss pooling strategies that help reduce variability and hence inventory as well as contracting strategies that can help better share risk across components and improve overall profitability.

5.5.1 An Introduction to Pooling

Inventory pooling is based on the notion of variability reduction through consolidation. At its simplest form, an organization serves multiple markets from the same inventory. Consolidation can occur through two main approaches: pooling by location and pooling by product (or "SKU rationalization"). In both cases, the reduction in inventory comes from the statistical averaging of demand. To see this, consider the following example where two systems are compared to serve N identical customers with independent and identically normally* distributed demands. In the first case, let us assume that each customer orders their own inventory using a (Q, R) model and desires a service level of $1 - \alpha$. The safety stock at each customer is then given by $Z_{1-\alpha}\sigma_L$, and hence, total safety stock is $NZ_{1-\alpha}\sigma_L$. If, instead, all of the customers were to pool their inventory at one location, then the distribution of demand at the pooled location would be normally distributed with a mean of $N\mu$ and a standard deviation of

$$\sigma = \left(\sum_{i=1}^{N} \sigma_i^2\right)^{0.5} = (N\sigma^2)^{0.5} = \sqrt{N}\sigma. \tag{5.39}$$

Therefore, the pooled safety stock would be equal to $\sqrt{N}Z_{1-\alpha}\sigma_L$. This illustrates the square root law that "average safety stock increases proportionally to the square root of the number of locations that inventory is held." Note also that if the EOQ is used, that cycle stock inventory also decreases through pooling, and the square root law also holds. We see this from the basic EOQ formula. For N independent orders, cycle stock equals $1/2N\sqrt{2Ad/h}$, where for the pooled system, the cycle stock is $1/2\sqrt{2adN/h} = 1/2\sqrt{N}\sqrt{2Ad/h}$. Note that inventory pooling does not affect pipeline stock.

---

* Note that it is not a requirement that the demand be normally distributed for pooling to work.

5.5.1.1 Location Pooling

In location pooling, product is held at a local warehouse or DC as opposed to each individual location. The key advantage is the corresponding reduction in safety and cycle stock for the same service level. The key disadvantage, however, is that product is pulled farther away from the customer. In addition, there are typically additional material handling operations that need to be performed, since product is now moving both through the regional warehouse and through the final location.

5.5.1.2 Postponement

Postponement is the practice of waiting until the last possible moment to differentiate product for a customer. An example is Benetton and Hewlett–Packard. For the case of Benetton, they delayed the dyeing process of shirts and sweaters by producing a "generic" undyed garment and then dyeing it at local markets. This allowed them to better respond to uncertainty in demand by pooling the variability across the colors. In addition to inventory savings, unsold stock went down from more than 30% to less than 10%.

Example 5.9

Nittany Hair Products (NHP) sells brushes to Target, CVS, and Walmart, each requiring unique packaging. The hair brushes are manufactured in Malaysia and sent to a DC in State College, Pennsylvania, with a lead time of 12 weeks. Annual demand from each customer is 300,000, with a standard deviation of 80,000. NHP uses a (Q, R) model for inventory with a 96% service level. In this case, total safety stock is

$$SS = Z_{0.96}\sqrt{\frac{12}{52}}(240,000) = 201,841.$$

NHP decides instead to wait to uniquely package the brushes at the DC. This process adds 3 days to the lead time. Under this scenario, total safety stock reduces to

$$SS = Z_{0.96}\sqrt{\frac{12}{52}}\left(\sqrt{3}(80,000)\right) + Z_{0.96}\sqrt{\frac{3}{365}}(240,000) = 154,624.$$

Note that this solution would require packaging capabilities at the DC. This may add new constraints that would have to be considered in the analysis.

5.5.1.3 Product Pooling

Product pooling is the process of using a "universal design" for a set of products to meet multiple demands. For example, instead of using a unique car horn in each type of car that Ford makes, a common horn for all cars could be used. The universal design leads to a reduction in both safety and cycle stocks.

Example 5.10

Ford Motor Company currently uses unique horns in the Mustang and Focus. U.S. annual demand for the Mustang is normally distributed with a mean of 87,000 and a standard deviation of 18,000. U.S. demand for the Focus is normally distributed with a mean of 198,000 and a standard deviation of 32,000. Ford desires a service level of 98% and uses a (Q, R) model for inventory. Assume lead time is 1 week and that demands for the automobiles are independent. If each car horn is stocked, then the total safety stock (SS) is

$$SS = Z_{0.98}\sqrt{\frac{1}{12}(18,000+32,000)} = 29,643.$$

If instead a universal horn is used for both cars, then the safety stock reduces to

$$SS = Z_{0.98}\sqrt{\frac{1}{12}(18,000^2+32,000^2)^{0.5}} = 21,767.$$

This benefit can also be accomplished through the process of SKU rationalization. In this case, a firm determines if all of the different SKUs are actually needed to satisfy demand. For example, Tide detergent may offer 15 different package types for Tide HE Liquid. In this case, the supply chain needs to manage each of these SKUs and maintain appropriate inventory for each. It is possible that the same demand could be adequately met with 10 package types, thus saving supply chain costs. This would lead to increased profitability if the total demand did not decrease from certain customers only buying a package type that was removed. A well-publicized example is the process that Clorox used to drastically reduce the number of SKUs it manages (Pegels and Van Hoek 2006).

5.5.1.4 Correlation and Pooling

To this point, we have assumed that the demand distributions are independent and identically distributed. If, however, demands are correlated, then this affects the effectiveness of correlation. Given random variables $x_1, x_2, \ldots, x_n$, let x be defined as the sum of the random variables. The variance of x is given by

$$\text{Var}[x] = \sum_{j=1}^{n}\text{Var}(x_i) + 2\sum_{i<j}\text{COV}(x_i,x_j). \qquad (5.40)$$

The COV(x_i, x_j) is the covariance of random variables x_i and x_j, which we can define by the correlation coefficient ρ_{ij}:

$$\rho_{ij} = \frac{\text{COV}(x_i, x_j)}{\sigma_i \sigma_j}. \tag{5.41}$$

Note that if the demands for two goods are positively correlated, then they move in the same direction and would be "complimentary goods." Examples would be milk and cereal, printers and ink cartridges, and automobiles and gas. If demands are negatively correlated, then they move in opposite directions and are known as "substitute goods." Examples would be butter and margarine, Coca-Cola and Pepsi,* and movie tickets and video streaming downloads.

Suppose in Example 5.10 that demands are correlated ($\rho_{12} = -0.72$). Note that if demands are positively correlated, the benefit of pooling is decreased, but if demands are negatively correlated, the benefit goes up. In this case, the total safety stock becomes

$$SS = Z_{0.98} \sqrt{\frac{1}{12}} \left(18{,}000^2 + 32{,}000^2 - 2(0.72)(18{,}000)(32{,}000)\right)^{0.5} = 13{,}501.$$

When using product pooling, therefore, it is always important to consider correlation (when appropriate) and search for opportunities from products with negatively correlated demands.

5.5.2 Supply Chain Contracting

In a supply chain, it is often the case that each component is responsible for many of its own local decisions. This may include decisions on how much inventory to hold, what orders to place (and their quantities), and delivery schedules. Clearly independent and local decision making can lead to overall poor supply chain performance. One reason for this is that certain components in the supply chain may have more uncertainty and risk than others. For example, consider the simple supply chain with two echelons in Figure 5.11, where a distributor orders product from a supplier that satisfies the newsvendor assumptions.

In this case, the distributor has to deal with all of the demand uncertainty. That is, they face all of the downside risk. For this reason, if the cost of having excess supply is large, they will be conservative in how many items they order. For this reason, the whole supply chain is likely to be made worse off.

---

* It must be pointed out that one of the authors lived in Atlanta, Georgia, where Pepsi would certainly not be considered a substitute for Coca Cola! It can therefore depend on conditions.

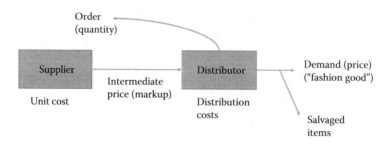

FIGURE 5.11
Two-echelon supply chain for a fashion good.

Supply chain contracting is one way to help spread the risk throughout the supply chain and thereby improve performance for all components. We discuss two such contracts in the next section: buy-back contracts and wholesale-price contracts. Many other types of contracts have been developed and studied, but these two will illustrate the benefit of such an approach. A rich literature exists on this topic. Some recommended readings that contain further details of supply chain contracting may be found in Cachon and Lariviere (2005), Fayezi et al. (2012), Narayanan and Raman (2004), Tayur et al. (2012), and Tsay et al. (1999).

Consider a supply chain where a manufacturer (M) supplies a distributor (D) with a fashion good (winter coats). In this case, it costs M $100 per unit in production costs. Further, M is responsible for shipping the product to D at a cost of $2 per unit, independent of the order quantity. M sells the product at $150 per unit to D. The market price for D is $200 per unit and there is a material handling cost of $1 per unit that D incurs. The demand x for the coat over the selling season is uniformly distributed between 500 and 1000 units. D determines the order quantity (Q) and can only order from M at the beginning of the selling season, so that the newsvendor assumptions hold. Further, we assume that M is a reliable manufacturer and will always deliver the quantity desired on time. Any coats left over after the selling season are salvaged for $25 through a discount outlet.

In this case, we can determine Q using the newsvendor model. We need to find the smallest Q that satisfies the ratio

$$F(Q) \geq \frac{P-C}{P-S} = \frac{200-151}{200-25} = 0.28.$$

Using the uniform distribution, we solve

$$\frac{Q-500}{1000-500} = 0.286,$$

which gives $Q = 640$. We can determine the profit for each party. For M, the expected profit is given by

$$\pi_M = 640(150 - 100 - 2) = \$30{,}720.$$

The expected profit for D depends on demand and is given by

$$\pi_D = \int_{500}^{640} \left((200 - 151)x + 25(640 - x)\right) \frac{1}{500} dx + \int_{640}^{1000} (200 - 151) \frac{640}{500} dx.$$

The solution gives an expected profit of \$30,890. The total expected profit for the supply chain in this case is \$61,610.

One important aspect to notice is that M is charging a markup over their costs in order to make profit. The higher the markup, the less that D will order because the cost of having excess inventory relative to the cost of not having enough demand increases. That is, D is incurring all of the risk of the demand uncertainty.

We could obtain the "optimal" expected profit for the total if the supply chain if it were to become centralized. In this case, there is no immediate markup, and so the newsvendor critical ratio becomes

$$F(Q) \geq \frac{P - C}{P - S} = \frac{200 - 103}{200 - 25} = 0.554.$$

Using the uniform distribution as before gives $Q = 777$. The total profit for the supply chain in this case would be given by

$$\pi_T = \int_{500}^{777} \left((200 - 103)x + 25(777 - x)\right) \frac{1}{500} dx + \int_{777}^{1000} (200 - 103) \frac{777}{500} dx.$$

The solution is \$69,845, which is roughly 13.4% higher than in the original decentralized case. The question is if we can structure some mechanism by which the decentralized supply chain can (i) achieve the same profit as the centralized one and (ii) make both M and D better off than they were in the decentralized case. This last condition is important because the mechanism will not work in a decentralized setting unless both parties agree to it. If we can find a "contract" that satisfies both conditions, we call it a *coordinating mechanism*.

One way to encourage D to order more is to reduce their cost of having excess items at the end of the selling period. M could offer to "buy back" the items at a value B that is strictly greater than the salvage value S. If B is high enough, this will encourage D to order the same amount as the centralized supply chain. We can determine this amount by finding B that satisfies the critical ratio of the centralized case. That is,

$$F(Q) \geq \frac{P-C}{P-S} = \frac{200-151}{200-B} = 0.554.$$

Solving for the buy-back value gives $B = \$111.55$. In this case, D will order 777 (the same as if centralized) and the expected profit for D is given by

$$\pi_T = \int_{500}^{777} \left((200-151)x + 111.55(777-x)\right)\frac{1}{500}dx + \int_{777}^{1000}(200-103)\frac{777}{500}dx.$$

The solution is \$59,502.58. The expected profit for M depends on the quantity of items that they buy back. We assume in this case that M buys back at \$111.55 per coat and then salvages them at \$25 per coat. The expected profit is then given by

$$\pi_M = 777(150-100-2) + \int_{500}^{777}(25-111.55)(777-x)\frac{1}{500}dx,$$

which equals \$10,342.42. Note that the sum of the expected profits for M and D equals the expected profit for the centralized case. However, M is made worse off compared to the decentralized case. In order to make this buy-back contract an effective coordinating mechanism, D will need to share some of its profit with M. If D gives M between \$20,378 and \$28,612, then both M and D will be strictly better off than without the contract, and the decentralized supply chain will make the maximum amount possible.

An alternative contract is the *wholesale price contract*. In this case, M sells to D at cost (\$102). This will lead D to purchase 777 coats. Since M sells at cost, then D will make all of the expected profit (\$69,845) and M will make no profit. In this case, if D were to give M between \$30,721 and \$38,954, then it would be coordinating.

A few comments are in order. It is important to remember that the contracts were based on analysis using expected profits. Therefore, on average, each party will be strictly better off, but there will be times when (due to changing conditions) they make less money. Therefore, these coordinating mechanisms are more effective if the contract is over a long period of time (e.g., repeated selling seasons). Second, the key benefit of the solution is that a simple mechanism can coordinate a more complicated system. Just knowing the best way to control a system doesn't mean that the agents

Supply Chain Engineering

in the system will behave as they should. Through contracting, however, we are able to align the individual incentives of the agents with the overall goal of the supply chain. Finally, although the method works well to coordinate different groups within a company, there can be legal issues that need to be considered when separate companies engage in buy-back or wholesale price contracts together. We illustrate with one of the more popular examples of contracting in a supply chain, namely, Blockbuster video (Cachon and Lariviere 2001).

In the Blockbuster video rental example (this was before the days of downloading videos through Netflix), customers were complaining that rentals for new releases were consistently stocked out in Blockbuster stores. The reason is that Blockbuster had to pay high prices for the videos, which took many rentals before they paid off. Since demand for new releases is only high for a short period of time, Blockbuster could not afford to stock to the peak demand. In order for Blockbuster to purchase more videos, a revenue sharing arrangement (similar to the wholesale price contract) was established. In this case, the distributor would lower the price for the video if Blockbuster would give a portion of its revenue to the distributor. The arrangement was very successful. However, relatively soon after implementation, Disney sued Blockbuster for $120 million,[*] believing that Blockbuster withheld some of the revenue amount they were supposed to share. Shortly thereafter, Blockbuster was accused of violating antitrust law from the arrangement (Warren and Peers 2002).

Exercises

5.1 Discuss at least two differences among the strategic, tactical, and operational decisions in a supply chain. Give two examples of each type of decision.

5.2 Discuss the differences between the two key criteria of supply chain performance—efficiency and responsiveness. Identify three measures for each criterion.

5.3 Describe the four major drivers of supply chain performance. For each driver, identify one efficiency measure and one responsiveness measure.

5.4 What is supply chain risk? Discuss the differences between *hazard risks* and *operational risks*.

---

[*] Bloomberg News (2003). Disney Sues Blockbuster Over Contract. New York Times, January 4, 2013. Available at: http://www.nytimes.com/2003/01/04/business/disney-sues-blockbuster-over-contract.html

5.5 A typical supply chain manager has to make the decisions listed below. Categorize each decision as strategic, tactical, or operational and briefly justify your choice.
   a. Number of warehouses needed and where to locate them
   b. Assignment of customer orders to inventory or production
   c. Choosing vendors to supply critical raw materials
   d. Selecting a transportation provider for shipments to customers
   e. Given the factory capacity, determining the quarterly production schedule
   f. Choosing to produce a part internally or to outsource that part
   g. Setting inventory policies at a retail location
   h. Choice of transportation mode for shipping
   i. Reordering materials to build inventory
   j. Allocating inventory to customer backorders
   k. Creating the distribution plan to move finished goods inventory from warehouses to retail locations

5.6 Sales of cars at a local auto dealer total 3000 cars per year. The dealer reports inventory turns of 30 per year. Compute
   a. Average inventory in the dealer's lot
   b. The length of time the average car sits in the dealer lot before it is sold

   Suppose the dealer wants to increase his customer responsiveness. What actions should he take? How will these actions affect inventory turns, average inventory, days of inventory, and supply chain cost?

5.7 XYZ company's annual sales for its products is 5000 units and its average inventory is 500 units.
   a. Compute the company's inventory turns.
   b. If XYZ decides to *increase* its inventory turns, how will this affect (increase or decrease) the following?
      i. Response time to customers
      ii. Inventory capital
      iii. Days of inventory
      iv. ROA
      v. Cash-to-cash cycle
      vi. Working capital

Supply Chain Engineering

5.8 State whether each of the following statements is true or false, or whether the answer ultimately depends on some other factors (which you should identify).
   a. Speed of delivery is an efficiency measure for the transportation driver.
   b. Large inventory is a warning sign for poor supply chain efficiency.
   c. Location of plants and DCs are tactical decisions.
   d. Selection of suppliers is an operational decision.
   e. Increasing product variety improves supply chain responsiveness.
   f. Increasing inventory turns increases working capital.
   g. Increasing days of inventory increases ROA.
   h. To decrease supply chain cost, inventory turns should be increased.
   i. The Japanese earthquake and tsunami of 2011 were a hazard risk to the supply chain.
   j. Failure of IT systems is considered an operational risk to the supply chain.
   k. Operational risks have a more significant impact on supply chain performance than hazard risks.
   l. Hazard risks are uncontrollable rare events that disrupt supply chain performance.
   m. Functional products are best matched with responsive supply chains.

5.9 Discuss the pros and cons of single sourcing and multiple sourcing.

5.10 Why is scaling of criteria values necessary in supplier ranking?

5.11 Discuss the similarities and differences between L_p metric and AHP for ranking suppliers.

5.12 (Adapted from Srinivasan 2010) Consider a supply chain network with three potential sites for warehouses and eight retailer regions. The fixed costs of locating warehouses at the three sites are given below:

Site 1: $100,000
Site 2: $80,000
Site 3: $110,000

The capacities of the three sites are 100,000, 80,000 and 125,000, respectively. The retailer demands are 20,000 for the first four retailers and 25,000 for the remaining.

The unit transportation costs ($) are given in Table 5.38.

a. Formulate a mixed integer linear program to determine the optimal location and distribution plan that will minimize the total cost. You must define your variables clearly, write out the constraints, explaining briefly the significance of each, and write the objective function. Assume that the retailers can receive supply from multiple sites. Solve using any optimization software. Write down the optimal solution.

b. Reformulate the optimization problem as a linear integer program, assuming dedicated warehouses, that is, each retailer has to be supplied by *exactly* one warehouse. Solve the integer programming model. What is the new optimal solution?

c. Compare the two optimal solutions and comment on their distribution plans.

5.13 Consider the following design proposals and financial performance measures for Mighty Manufacturing given in Table 5.39. Both sets of numbers are for 2017 with all possible markets open and plants open in both Denver and Covington. Scenario A has DCs in Denver, Chicago, Pittsburgh, and Atlanta. Scenario B has DCs in Denver,

TABLE 5.38

Data for Exercise 5.12

	R_1	R_2	R_3	R_4	R_5	R_6	R_7	R_8
Site 1	4	5	5	4	4	4.2	3.3	5
Site 2	2.5	3.5	4.5	3	2.2	4	2.6	5
Site 3	2	4	5	2.5	2.6	3.8	2.9	3.5

TABLE 5.39

Data for Mighty Manufacturing (Exercise 5.13)

	Scenario A	Scenario B
Total sales ($000)	$76,537	$75,092
Return on assets	62.03%	57.93%
Average time through supply chain (days)	42.68	40.03
Order cycle time to customer (days)	4.38	3.42
Achieved service level	98.95%	97.08%
Average production capacity utilization	89.22%	82.47%

Los Angeles, Portland, Dallas, Chicago, Boston, Pittsburgh, and Atlanta.
a. Which scenario offers better customer service? Why?
b. Which scenario would you suspect has higher transportation costs? Why?
c. Under which scenario is Mighty Manufacturing in a better position if forecasts for new markets are too high? Why?
d. Under which scenario is Mighty Manufacturing in a better position if forecasts for new markets are too low? Why?

5.14 A company needs two parts, A and B, for its product. It can either buy them from another company, or can make them in its own plant, or do both. The costs of each alternative and the in-house production rates are given in Table 5.40.

The company must have at least 100 units of part A and 200 units of part B each week. There are 40 hours of production time per week and idle time on the machine costs $3.00/hour. Furthermore, no more than 60 units of A can be made each week, and no more than 120 units of B can be made each week. Also, no more than 150 units of B can be bought per week. The company wants to determine an optimal plan that will minimize the total costs per week. Formulate this as a linear programming problem and solve. What is the optimal make or buy plan for the company?

5.15 Recall Example 5.6 discussed in Section 5.3.7. Determine the ranking of the suppliers using the following methods:
a. L_1 norm to scale the criteria values and L_1 metric to rank
b. L_2 norm to scale the criteria values and L_2 metric to rank

Do you find any rank reversals including the rankings obtained in Section 5.3.7? Explain any differences.

5.16 Consider a supplier selection problem where the five most important criteria are identified as follows.

C1—Risk

C2—Delivery time

C3—Quality

TABLE 5.40

Data for Exercise 5.14

	Part A	Part B
Make	$1.00/unit	$2.00/unit
Buy	$1.20/unit	$1.50/unit
In-house production rate	3 units/hour	5 units/hour

C4—Price

C5—Business Performance

a. Using the Borda method and your own preferences, rank the criteria and compute their weights.

b. Suppose 5% of purchasing managers were interviewed and they were asked to rank the above criteria. The summary of their responses are given in Table 5.41.

Determine the weights of the criteria and their rankings based on the sample survey of the purchasing managers, using the Borda method for multiple decision makers discussed in Section 5.3.8. How does their ranking compare with yours?

5.17 Suppose you are planning to use single sourcing and have narrowed down the choices to four suppliers A, B, C and D. Your criteria for selection are price (min), company size in market capitalization in millions of dollars (max), and quality on a scale of 1 to 100 (max). The relevant data are given in Table 5.42.

a. Scale the data using the Ideal value method.

b. Use the L_∞ metric method to rank the four suppliers.

5.18 Consider an order allocation problem under multiple sourcing, where it is required to buy 2000 units of a certain product from three different suppliers. The fixed setup cost (independent of the order

TABLE 5.41

Criteria Rankings for Exercise 5.16

Rank / Criteria	1	2	3	4	5
C_1	9	4	4	7	6
C_2	3	13	4	8	2
C_3	6	3	11	9	1
C_4	8	7	9	2	4
C_5	4	3	2	4	17

TABLE 5.42

Supplier Data for Exercise 5.17

	Price ($) (Min)	Size (Millions of Dollars) (Max)	Quality (Max)
A	180,000	2800	75
B	160,000	3200	85
C	140,000	2600	80
D	190,000	3600	65

Supply Chain Engineering

TABLE 5.43

Supplier Data for Exercise 5.18

Supplier	Fixed Cost	Capacity	Unit Price
1	$100	600 units	$10/unit for the first 300 units
			$7/unit for the remaining 300 units
2	$500	800 units	$2/unit for all 800 units
3	$300	1200 units	$6/unit for the first 500 units
			$4/unit for the remaining 700 units

quantity), variable cost (unit price), and the maximum capacity of each supplier are given in Table 5.43 (two suppliers offer quantity discounts).

The objective is to minimize the total cost of purchasing (fixed plus variable cost). Formulate this as a linear integer programming problem. You must define all your variables clearly, write out the constraints to be satisfied with a brief explanation of each, and develop the objective function.

5.19 **Case Study 3 (Mendoza et al. 2008)**

In this case study, you will be working on a supplier ranking problem. It is an actual application for a manufacturing company located in Tijuana, Mexico. Because of confidentiality issues, the data given here have been disguised. The supplier selection criteria and subcriteria have been defined by the Purchasing Manager as shown in Figure 5.12. Note that Flexibility, Process Capability (C_{pk} Index), and Service have to be maximized, while the other criteria have to be minimized. The company is considering 21 potential suppliers and the supplier data with respect to the criteria are given in Table 5.44.

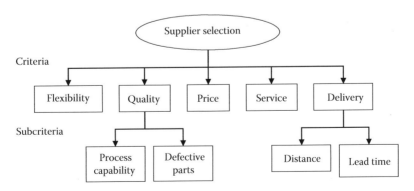

FIGURE 5.12
Supplier selection criteria and subcriteria (Exercise 5.19).

TABLE 5.44
Supplier Criteria Data (Exercise 5.19)

			Criteria Values				
Supplier	Price ($)	C_{pk} (Index)	Defective Parts (ppm)	Flexibility (%)	Service (%)	Distance (km)	Lead Time (hours/part)
1	50	0.95	105,650	10	75	500	0.25
2	80	2.00	340	0	100	1500	0.60
3	45	0.83	158,650	25	65	50	0.20
4	60	1.00	66,800	15	85	5000	0.80
5	40	1.17	22,750	18	90	9500	0.95
6	60	1.50	1350	5	99	7250	0.50
7	65	1.33	6200	0	100	10	0.10
8	70	1.50	1350	0	50	15,000	1.50
9	45	1.00	66,800	5	80	7500	1.75
10	70	1.25	12,225	10	85	12,500	2.00
11	75	0.83	158,650	15	75	1345	1.25
12	65	1.00	66,800	0	80	6680	1.15
13	80	1.33	6200	0	85	5000	1.00
14	75	1.15	22,750	2	87	16,000	0.90
15	70	1.33	6200	5	86	17,000	0.95
16	70	1.05	44,500	0	65	1860	1.50
17	85	1.25	12,225	5	70	1789	1.45
18	65	0.95	105,650	0	77	1775	0.90
19	55	0.83	158,650	10	89	2500	0.75
20	80	1.25	12,225	10	85	12,500	1.50
21	85	0.83	158,650	0	50	17,500	2.00

Questions:
a. Scale the supplier criteria values using linear normalization (simple linearization).
b. Using the scaled values from part (a), apply the L_2 metric method to rank the suppliers.
c. Determine the criteria/subcriteria weights using the following methods (use your own judgment):
   i. Rating method.
   ii. Pairwise comparison of criteria/subcriteria and Borda method.
   iii. AHP. Test the consistency of all pairwise comparison matrices.
d. Using the weights obtained in part (c) and the scaled supplier data in part (a), determine the supplier rankings by all three methods.

[NOTE: For AHP, do not perform pairwise comparison of alternatives (suppliers) with each criterion/subcriterion. Instead, use directly the scaled supplier data from part (a) as the S matrix.]

e. Compare the supplier rankings obtained by L_2 metric, rating method, Borda method, and AHP. Discuss any rank reversals.

5.20 Suppose backorders are allowed in the EOQ model that was discussed in Section 5.4.3. Each item backordered results in some cost of b per item per year backordered. Derive the new EOQ formula.

5.21 In Example 5.8, suppose the lead time is reduced from 1 week to 4 days. What R should be used? What would the resulting safety stock be?

5.22 In Example 5.8, sketch a plot of safety stock versus service level. Use α values of 0.01, 0.02, 0.03, 0.05, 0.075, 0.10, 0.125, and 0.150. What can you conclude from the graph?

5.23 Demand is normally distributed with a mean of 500 and a standard deviation of 200. Newsvendor assumptions hold. The selling price per unit is $400, cost per unit is $200, and salvage per unit is $40. Determine the optimal order quantity. How would your answer change if the salvage value were $0?

5.24 Suppose a distributor faces an annual demand for a car horn for small cars that is normally distributed with a mean of 2300 items and a standard deviation of 300. The cost of placing an order is $150 and the holding cost is $10/item/year. The replenishment lead time is 9 days (assume 352 days in a year). Answer the following:
a. Compute the economic order quantity (EOQ).
b. If a service level of 5% is desired, what should R be in the (Q, R) model?
c. For your answer to part (b), what is the average cycle, safety, and pipeline stock equal to?

5.25 The bookstore orders Penn State Volleyball shirts once for a selling season from a manufacturer in Malaysia. They estimate that the demand is according to a triangular distribution with $a = 1500$, $b = 3000$, and $c = 2000$. It costs the bookstore $5 per shirt, and they sell the shirt for $20. If at the end of the volleyball season there are any shirts left over, they have a salvage value of $3. Determine the optimal order quantity for the bookstore.

5.26 Suppose for Exercise 5.24, there is a second car horn for medium-sized cars with demand that is normally distributed with a mean of 1500 and a standard deviation of 200. All other information from Exercise 5.24 is the same. The car manufacturer is considering using only one car horn type rather than the two they carry. Answer the following:

a. Determine the (Q, R) parameters for the second horn.
b. Determine the total safety, cycle, and pipeline stock for both horns.
c. Repeat parts (a) and (b) for a combined horn (i.e., one that serves both small and medium cars).

5.27 Techtronic operates a single DC in St. Louis. Annual demand for a product they carry is normally distributed with a mean of 80.6 and a standard deviation of 58.81. There is a 3-week lead time and the target in-stock probability is 99.9%. For this case, we won't worry about Q, but just the safety stock needed to guarantee the service level. Answer the following:

a. Determine the safety stock required for a single DC.
b. Techtronic is considering splitting into multiple DCs (assume that they do it in a way that demand is split evenly between them). Determine the safety stock if there are 2, 4, and 8 DCs.

5.28 Demands over time for two products (D1 and D2) are given in Table 5.45. Determine the correlation between them.

5.29 For Exercise 5.26 (c), what would the answer be if the correlation between the two horns is −0.6? What would it be if the correlation is 0? What would it be if the correlation is 0.4?

5.30 Demand for an item over the next five periods is forecasted to be 10, 20, 15, 16, and 18. There are currently 20 items on hand. The order cost

TABLE 5.45
Data for Exercise 5.28

Time	1	2	3	4	5	6	7	8
D1	100	130	90	150	170	140	110	80
D2	100	80	100	110	70	75	90	120

is $100 per order, and each item costs $50. The time value of money per period is 0.1. Write the formulation to determine the optimal order quantities. Use Excel Solver to get the solution. Approximate by determining the average demand and solve using EOQ. Compare.

5.31 Distance from a DC to two customers is 130 and 70 miles, respectively. Demands from each are 150 and 260 cases, respectively. Find the best mode of delivery from among the three choices:

  a. Contract Carrier: DHL charges $5.80 per case. All customers are in the same delivery zone and so the price is the same for each.
  b. LTL: The rate is $2.45 per mile traveled plus a handling fee of $2.50 per case.
  c. FTL: The rate is $1.40 per mile traveled plus a handling fee of $400. The handling fee is paid regardless of the number of cases. The capacity of a truck is 200 cases.

5.32 In Example 5.7, consider the addition of two more locations: (12, 12) with a demand of 13, and (3, 11) with a demand of 5.

  a. Ignore capacity (and demands) and solve the TSP using the nearest neighbor (from and to the DC).
  b. From your answer in (a), identify an opportunity of a two-opt (if any) and perform a local improvement.
  c. Formulate the TSP problem as an optimization problem in Excel (use the alldifferent constraint) and solve.

5.33 For Exercise 5.32, use the Savings Method to solve (using the truck capacities given in Example 5.7). Comment on your solution.

5.34 The Gap sells seven T-shirts of different colors. U.S. demand for each color is independent and normally distributed with a mean of 100,000 per year and a standard deviation of 34,000. They want a 95% service level on each color. Lead time from the supplier (located in Malaysia) is 3.6 months. Do the following:

  a. Compute the total safety stock.
  b. Consider the case where the Gap dyes their shirts in the United States. The supply chain now has a lead time from the manufacturer (who ships white shirts) to the dyeing operation of 3.5 months. Lead time from the dyeing operation to the DC is 0.2 months. Determine the total safety stock in this case.

5.35 In and Out burger in Los Angeles orders their beef from a Canadian supplier. The supplier ships directly to the 12 different restaurants in the LA area. In and Out wants to determine if it makes sense to have a pooling DC. Currently, it takes 2 weeks to ship to each store. Demand at each location is independent and normally distributed with a mean of 25,000 patties per year and a standard deviation of

10,000. They want a 99% service level. In the pooling case, the time from the pooling DC to a location is 2 days. Compare the total safety stock in each case.

5.36 Consider a two-stage supply chain (A supplies B that supplies customer demand). Customer demand is uniformly distributed between 500 and 1000. The cost to make a unit at A is $100. They sell the item to B for $150 who sells to a customer for $200. The transportation cost between A to B is $2 per unit, and the material handling cost at B is $1 per unit. Newsvendor assumptions hold. If there is any leftover stock anywhere, it has a salvage value of $30. Answer the following:

   a. Determine the order quantities and expected profits (for each party) if A and B act independently.
   b. Determine the optimal order quantities if the supply chain behaves in a centralized way.
   c. See if you can determine a buy-back contract that "coordinates" the supply chain.
   d. See if you can determine a revenue sharing contract that "coordinates" the supply chain.
   e. Compare your solutions to parts (c) and (d).

References

Aberdeen Group. 2004. Outsourcing portions of procurement now a Core Strategy. *Supplier Selection and Management Report.* 04(07): 4.

Aissaoui, N., M. Haouari, and E. Hassini. 2007. Supplier selection and order lot sizing modeling: A review. *Computers & Operations Research.* 34(12): 3516–3540.

Bilsel, R. U. and A. Ravindran. 2011. A multi-objective chance constrained programming model for supplier selection. *Transportation Research Part B.* 45(8): 1284–1300.

Bloomberg News, Disney Sues Blockbuster Over Contract. Jan. 4, 2003. *New York Times* Available at: http://www.nytimes.com/2003/01/04/business/disney-sues-blockbuster-over-contract.html.

Cachon, G. P. and M. A. Lariviere. 2001. Turning the supply chain into a revenue chain. *Harvard Business Review.* March.

Cachon, G. P. and M. A. Lariviere, M. A. 2005. Supply chain coordination with revenue-sharing contracts: Strengths and limitations. *Management Science.* 51(1): 30–44.

Camm, J. D., T. E. Chorman, F. A. Dill, J. R. Evans, D. J. Sweeney, and G. W. Wegryn. 1997. Blending OR/MS, judgement and GIS: Restructuring P&G's supply chain. *Interfaces.* 27: 128–142.

Chopra S. and P. Meindl. 2001. *Supply Chain Management: Strategy, Planning and Operation.* Upper Saddle River, NJ: Prentice Hall.

Cintron, A. 2016. Multi-criteria distribution planning model for a consumer products company. In *Multiple Criteria Decision Making in Supply Chain Management.* ed. A. R. Ravindran, Chapter 5. Boca Raton, FL: CRC Press.

Cintron, A., A. Ravindran, and J. A. Ventura. 2010. Multi-criteria mathematical models for designing the distribution network of a consumer products company. *Computers and Industrial Engineering.* 58: 584–593.

Cook, W. J. 2011. *In Pursuit of the Traveling Salesman: Mathematics at the Limits of Computation.* Princeton University Press.

De Boer, L., E. Labro, and P. Morlacchi. 2001. A review of methods supporting supplier selection. *European Journal of Purchasing and Supply Management.* 7: 75–89.

Degraeve, Z. and F. Roodhooft. 1999. Improving the efficiency of the purchasing process using total cost of ownership information: The case of heating electrodes at CockerillSambre S.A. *European Journal of Operational Research.* 112: 42–53.

Degraeve, Z. and F. Roodhooft. 2000. A mathematical programming approach for procurement using activity based costing. *Journal of Business Finance and Accounting.* 27:69–98.

Degraeve, Z., E. Labro, and F. Roodhooft. 2004. Total cost of ownership purchasing of a service: The case of airline selection at Alcatel Bell. *European Journal of Operational Research.* 156: 23–40.

Dell, M. September 3, 2004. Conversations with Dell. *Presentation at the Pennsylvania State University.*

Dickson, G. W. 1966. An analysis of vendor selection systems and decisions. *Journal of Purchasing.* 2(1): 5–17.

Ellram, L. M. 1995. Total cost of ownership, an analysis approach for purchasing. *International Journal of Physical Distribution and Logistics Management.* 25(8): 4–23.

Fayezi, S., A. O'Loughlin, and A. Zutshi. 2012. Agency theory and supply chain management: A structured literature review. *Supply chain Management: An International Journal.* 17(5): 556–570.

Fazi, S. 2014. Model Selection, Routing and Scheduling for Inland Container Transport. PhD Thesis, Eindhoven University of Technology. Available at https://pure.tue.nl/ws/files/3947824/777919.pdf

Fine, C. H. 2000. Clock speed based strategies for supply chain design. *Production and Operations Management.* 9(3): 213–221.

Fisher, M. 1997. What is the right supply chain for your product? *Harvard Business Review.* March–April issue. 105–116.

Fleischmann, B., S. Ferber, and P. Henrich. 2006. Strategic planning of BMW's global production network. *Interfaces.* 36(3): 194–208.

Ghodsypour, S. H. and C. O'Brien. 2001. The total cost of logistics in supplier selection, under conditions of multiple sourcing, multiple criteria and capacity constraint. *International Journal of Production Economics.* 73(1): 15–27.

Hendricks, K. B. and V. R. Singhal. 2005. Association between supply chain glitches and operating performance. *Management Science.* 51(5): 695–711.

Holt, G. D. 1998. Which contractor selection methodology? *International Journal of Project Management.* 16(3): 153–164.

Jayaraman, V., R. Srivastava, and W. C. Benton. 1999. Supplier selection and order quantity allocation: A comprehensive model. *Journal of Supply Chain Management.* 35(2): 50–58.

Laval, C., M. Feyhland, and S. Kakaouros. 2005. Hewlett–Packard combined OR and expert knowledge to design its supply chains. *Interfaces.* 35(3): 238–247.

Lee, H. L. and C. Billington. 1995. The evolution of supply chain management models and practice at Hewlett–Packard Company. *Interfaces*. 25(5): 42–46.0.

Marien, E. J. 2000. The four supply chain enablers. *Supply Chain Management Review*. March–April issue. 60–68.

Masud, A. S. M. and A. Ravindran. 2008. Multiple criteria decision making. In *Operations Research and Management Science Handbook*. ed. A. R. Ravindran, Chapter 5. Boca Raton, FL: CRC Press.

Masud, A. S. M. and A. Ravindran. 2009. Multiple criteria decision making. In *Operation Research Methodologies*. ed. A. R. Ravindran, Chapter 5. Boca Raton, FL: CRC Press.

Mendoza, A. 2007. Effective methodologies for supplier selection and order quantity allocation. PhD dissertation, Department of Industrial Engineering, Pennsylvania State University.

Mendoza, A., A. Ravindran, and E. Santiago. 2008. A three phase multi-criteria method to the supplier selection problem. *International Journal of Industrial Engineering*. 15(2): 195–210.

Mulani, N. 2005. High performance supply chains. *Production and Operations Management Society (POMS) Conference*. Chicago, IL.

Narayanan, V. G. and A. Raman. 2004. Aligning incentives in supply chains. *Harvard Business Review*. 82(11): 94–102.

Nishiguchi, T. and A. Beaudet. 1998. Case study: The Toyota group and the Aisin fire. *Sloan Management Review*. 40(1): 49–59.

Patel, U. R. 2007. Experiments in group decision making in the Analytic Hierarchy Process. MS Thesis, Department of Industrial Engineering, Pennsylvania State University.

Pegels, K. and R. Van Hoek. 2006. Growing by cutting SKUs at Clorox. *Harvard Business Review*. 84(4): 22.

Portillo, R. C. 2016. Designing resilient global supply chain networks. In *Multiple Criteria Decision Making in Supply Chain Management*. ed. A. R. Ravindran, Chapter 3. Boca Raton, FL: CRC Press.

Powdrell, B. J. 2003. Comparison of MCDM algorithms for discrete alternatives. MS Thesis, Department of Industrial Engineering, Pennsylvania State University.

Ravindran, A. R., R. U. Bilsel, V. Wadhwa, and T. Yang. 2010. Risk adjusted multicriteria supplier selection models with applications. *International Journal of Production Research*. 48(2): 405–424.

Ravindran, A., D. T. Philips, and J. Solberg. 1987. *Operations Research: Principles and Practice*. 2nd Edition. NY: John Wiley & Sons, Inc.

Ravindran, A. R. and V. Wadhwa. 2009. Multiple criteria optimization models for supplier selection. In *Handbook of Military Industrial Engineering*, eds. A. Badiru and M. U. Thomas, Chapter 4. Boca Raton, FL: CRC Press.

Ravindran, A. R. and D. P. Warsing. 2013. *Supply Chain Engineering: Models and Applications*. Boca Raton, FL: CRC Press.

Saaty, T. L. 1980. *The Analytic Hierarchy Process*. New York: McGraw Hill.

Simchi-Levi, D., P. Kaminisky, and E. Simchi-Levi. 2003. *Designing and Managing the Supply Chain*. 2nd Edition. New York: McGraw-Hill.

Smytka, D. L. and M. W. Clemens. 1993. Total cost supplier selection model: A case study. *International Journal of Purchasing and Materials Management*. 29(1): 42–49.

Sonmez, M. 2006. A review and critique of Supplier Selection Process and Practices. Loughborough University Business School.

Spencer, T., A. J. Brigani, D. R. Dargon, and M. J. Sheehan. 1990. AT&T's telemarketing site selection system offers customer support. *Interfaces.* 20(1): 83–96.

Srinivasan, G. 2010. *Quantitative Models in Operations and Supply Management.* New Delhi: Prentice Hall.

Tayur, S., R. Ganeshan, and M. Magazine, eds. 2012. *Quantitative Models for Supply Chain Management.* Vol. 17. Springer Science & Business Media.

Treece, J. 1997. Just-too-much single-sourcing spurs Toyota purchasing review: Maker seeks at least 2 suppliers for each part. *Automotive News.* 3: 3.

Tsay, A. A., S. Nahmias, and N. Agrawal. 1999. Modeling supply chain contracts: A review. In *Quantitative Models for Supply Chain Management.* 299–336. Springer US.

van Hoeve, W.-J. 2001. The All different constraint: A survey. In *Proceedings of the Sixth Annual Workshop of the ERCIM Working Group on Constraints.* http://www.arxiv.org/html/cs/0110012.

Velazquez, M. A., D. Claudio, and A. R. Ravindran. 2010. Experiments in multiple criteria selection problems with multiple decision makers. *International Journal of Operational Research.* 7(4): 413–428.

Wadhwa, V. and A. Ravindran. 2007. Vendor selection in outsourcing. *Computers and Operations Research.* 34: 3725–3737.

Warren, A. and M. Peers. June 13, 2002, B10. Video Retailers Have Day in Court—Plaintiffs Say Supply Deals between Blockbuster Inc. and Studios Violates Laws. *Wall Street Journal*, New York.

Weber, C. A., J. R. Current, and W. C. Benton. 1991. Supplier selection criteria and methods. *European Journal of Operational Research.* 50: 2–18.

Wilson, E. J. 1994. The relative importance of supplier selection. *International Journal of Purchasing and Materials Management.* 30: 34–41.

Wilson, R. 2015. 26th Annual State of Logistics Report, Council of Supply Chain Management Professionals. Chicago, IL.

Wind, Y. and P. J. Robinson. 1968. The determinants of vendor selection: Evaluation function approach. *Journal of Purchasing and Materials Management.* 4(8): 29–46.

Zenz, G. 1981. *Purchasing and the Management of Materials.* New York: Wiley.

Zhang, Z., J. Lei, N. Cao, K. To, and K. Ng. 2003. Evolution of Supplier Selection Criteria and Methods. E-article retrieved from http://www.pbsrg.com/overview/downloads/Zhiming%20Zhang_Evolution%20of%20Supplier%20Selection%20Criteria%20and%20Methods.pdf on June 8, 2010.

6

Warehousing and Distribution

Warehouses facilitate the intermediate storage of products in the supply chain. They serve the purpose of meeting customer demand more effectively by acting as a buffer to variability and by strategically positioning products with respect to the customers. In addition, warehouse operations can help reduce transportation costs through consolidation across a set of suppliers to satisfy customer orders.

Total logistics costs made up close to 8% of U.S. gross domestic product in 2016. Warehousing made up more than 30% of these logistics costs, or $420 billion (Council of Supply Chain Management Professionals 2017). Much of warehousing operations is labor based, and as such, it is important to design and support these operations to be efficient.

In this chapter, we give an introduction to the topic of warehousing and distribution. We first describe the basic functions of a warehouse, including warehouse type. We then discuss various equipment required to support these strategies. Next, we provide some basics of warehouse design and describe effective layouts for a warehouse and the equipment required to support the operations. Next, we discuss three aspects of item retrieval in a warehouse, namely, order picking, slotting, and design of a forward picking area. This is followed by a discussion of line balancing and a self-organizing strategy called bucket brigades. We conclude this chapter with a discussion on cross dock design and operations.

6.1 Warehouse Functions

Warehousing is concerned with the efficient storage and handling of materials. As such, warehouses make up an important component of the supply chain. Note that the words "warehouse" and "distribution center" (DC) are often used interchangeably (as we do in this chapter), but technically a warehouse is only concerned with the intermediate storage of materials whereas DCs have the purpose of order fulfillment, which means that items may enter and exit the DC in different forms. The key functions of a warehouse are shown in Figure 6.1.

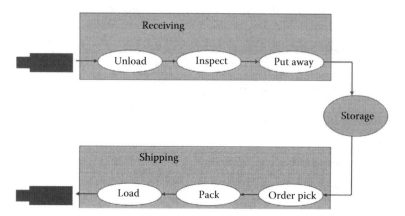

FIGURE 6.1
Basic warehouse functions.

We define these functions as follows:

- Unload—the part of receiving where products are taken off of the truck and brought into the facility.
- Inspect—ensures that the items are in good shape and that what has been received matches what was requested.
- Put away—the process of moving items into their respective storage area.
- Storage—the process of holding the products until requested; this may involve various types of equipment (defined later in this chapter).
- Order pick—the process of removing products from storage to fulfill an order (or set of orders).
- Pack—picked items are sorted by order and then packed into the requested unit; appropriate order labels including barcodes are placed on the package.
- Load—the packed orders are put onto trucks for shipment.

When considering the flow of product through a warehouse, it is important to consider that there are different units of handling and different types of requests. Products are typically identified by a stock keeping unit (SKU), which is a unique code. For example, if we carry cotton T-shirts in a warehouse that comes in three colors (white, blue, and red), four sizes (small, medium, large, and x-large), and either has a front pocket or not, then we would have $3 \times 4 \times 2 = 24$ different SKUs for this style of T-shirt. We define an order as a set of lines for SKUs, where a line defines a quantity for the defined SKU. Consider the case where you are shopping at a grocery store. Your order is

- Two 13-ounce boxes of Triscuit crackers (original)
- One 21.6-ounce Dawn Ultra dishwashing liquid soap (original scent)
- One 6-count Bounty paper towels, prints, big roll
- Two 12-count 12-ounce Diet Coke fridge pack cans

In this case, the order is made up of four lines, each with an associated SKU, and six total items retrieved. We will use the term "pick" to define a specific line retrieval. Note that your roommate might give you their order that they would like you to get them since you are going to the store anyway. Her order is

- One 6-count Bounty paper towels, prints, big roll
- One 12-count 12-ounce Diet Coke fridge pack cans
- One Charmin ultra strong toilet paper, 6 mega rolls

Notice that two of the lines in this order match your order. We could consolidate these into a single order of five lines (and nine total pieces to pick). As we will see later in this chapter, consolidation can be a useful strategy to reduce picking effort.

Although each SKU represents a unique item, it can come in several different-sized handling units. Notice that in the order above, one of the SKUs was a fridge pack of Diet Cokes. Cans of Diet Coke can come in different forms in the store, for example, 6 packs, 12 packs, and cases of 24 cans. We will use the following handling units, defined from smallest to largest: piece, carton, case, and pallet. These are not universal units, however. For example, if we consider the 12-ounce can of Diet Coke, there may be three units of handling in a warehouse: piece—a single can; case—a set of 24 cans; pallet—a set of 144 cases. On the other hand, if we consider ball point pens, there may be five types of handling units: item—a single pen; pack—a set of 12 pens; multipack—a set of 5 packs; case—a set of 50 multipacks; pallet—a set of 8 cases.

A particular SKU may be requested in different forms within the same order, as shown in Figure 6.2. If, for example, an order consisted of 12 multipacks of pens (broken case) and two cases of pens (full case), these would be treated as separate lines. This is because the different handling units may be stored in different areas of the warehouse, and so two separate picks would be required to retrieve them. It is therefore important to keep track of both the SKU and also the handling unit in an order.

We can classify different types of warehouses by their hierarchy in the supply chain. This is illustrated in a retail store example (Figure 6.3). Cross docks are used to consolidate orders as described in Section 6.8. Central DCs can be quite large in size, ranging from 1 million to 3 million square feet, and serve a set of regional distribution centers (RDCs), though some products

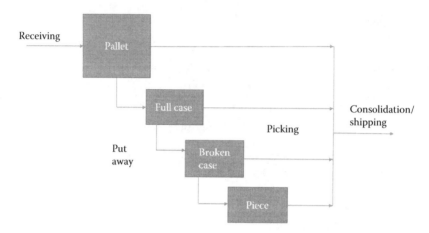

FIGURE 6.2
Different types of handling units.

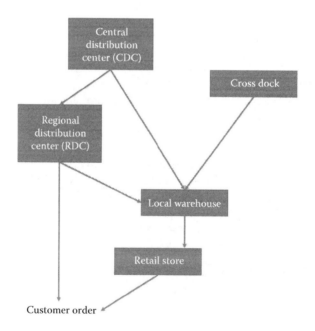

FIGURE 6.3
Warehouse hierarchy.

may be directly sent to the local warehouse. RDCs typically run from 200,000 to 400,000 square feet. They serve the local warehouse for the store, though they may also directly supply a customer through an online ordering process, as demonstrated in Figure 6.4. Note that for this case, an RDC will set aside a distinct area to handle online orders separately from store orders.

Warehousing and Distribution

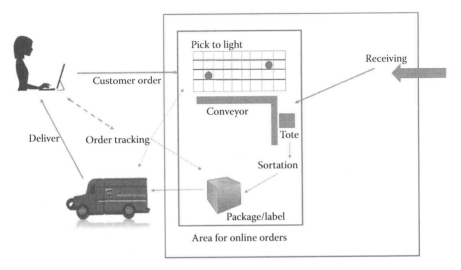

FIGURE 6.4
Warehouse with online fulfillment capabilities.

There are several specialty types of warehouses based on the characteristics of the type of product being stored. For example, several types of products such as produce or types of pharmaceuticals (e.g., vaccines) need refrigeration. The process to support the distribution of products of this type is called a "cold chain." Other types of climate control may include humidity (e.g., for flowers) or dust free for sensitive electronics. The different requirements and different types of warehouses can make measuring the performance of a warehouse challenging (Hackman et al. 2001).

6.2 Warehouse Equipment

Although warehouses have a significant labor component, there is also equipment that is needed to support the operations. This equipment is typically categorized into four areas: moving product, storing product, sorting product, and tracking product. We discuss each of these types briefly. Note that the College Industry Council on Material Handling Education maintains a website that contains images of the vast majority of warehouse equipment (http://www4.ncsu.edu/~kay/mhetax/). The reader is encouraged to visit this site for the material in this section, as well as the excellent images provided in Kay (2015; http://www4.ncsu.edu/~kay/Warehousing.pdf). A sampling of these images (with permission) is provided here.

6.2.1 Moving Product

Products are generally moved in one of three ways: conveyors, cranes, or industrial trucks (both manual and automated). Conveyors move product over a fixed path and work by gravity (e.g., a chute conveyor), manually (e.g., manual roller conveyor), or by power (e.g., a powered roller conveyor [Figure 6.5] or belt conveyor). The product is often placed in a bin or tote when moved by the conveyor.

Cranes are typically used to move heavy objects or to move loads when there isn't enough volume to justify the investment in a conveyor system. The movement is typically limited to a smaller area.

Industrial trucks move products over variable paths. Pallets are normally moved by this type of equipment. The simplest type of industrial truck is a hand truck. This is used manually and moves cases or stacks of cases. Pallet jacks have forks that allow for the insertion in a pallet to lift it and move it. They can come in manual or powered form. A lift truck (Figure 6.6) is a vehicle with a set of forks to move pallets. Not only can they be driven through the warehouse, but they have the ability to lift the pallet to either stack or place in shelving. The most automated version of an industrial truck is an automated guided vehicle. These can travel along a fixed guide path (e.g., along a buried wire or magnetic tape network) or can be guided

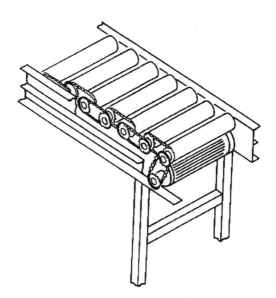

FIGURE 6.5
Powered roller conveyor. (College Industry Council on Material Handling Equipment. 1999. Material Handling Equipment. Available at http://www4.ncsu.edu/~kay/mhetax/. Last accessed Feb. 23, 2018. Out of courtesy.)

Warehousing and Distribution 309

FIGURE 6.6
Lift truck (also known as a forklift). (College Industry Council on Material Handling Equipment. 1999. Material Handling Equipment. Available at http://www4.ncsu.edu/~kay/mhetax/. Last accessed Feb. 23, 2018. Out of courtesy.)

through a navigation system such as laser guidance or magnetic spot guidance system, which uses reference points that are used to determine the position through triangulation.

6.2.2 Storing Product

Storage equipment depends on the handling unit. Often, equipment is used to form what is known as a "unit load" in order to standardize the handling size. Examples of unit loads are pallets (wood, plastic, or paper-based frames of 48 inches by 40 inches on which items are stacked) and totes (reusable plastic containers whose size depends on the application).

If pallet storage is required, they are typically block stacked or held in a pallet rack. In block stacking, pallets are placed directly on top of each other, and no special type of storage equipment is required. The height of the stack depends on the material since stacking too high would actually crush the product. For example, block stacking pallets of 12-ounce cans of soft drinks ranges from one to four pallets high in warehouses. The block stacks tend to be arranged in aisles so that lift trucks can have easy access to them. However, space can be saved by stacking them closer together as shown in Figure 6.7. In this figure, we see a top view of 24 pallets (stacked 2-high) in two arrangements. In the left figure, aisles are set up so that each pallet can be accessed by the lift truck through aisles. In the right figure, much less space is required, but each pallet is not directly

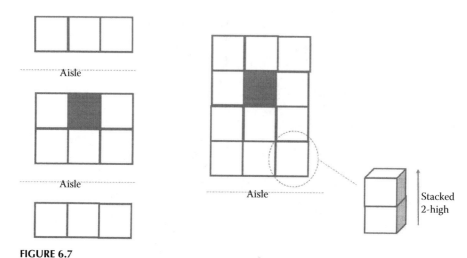

FIGURE 6.7
Block stacking pallets.

accessible. Block stacks are defined by three measures: number of lanes (which would be three in the image on the right in Figure 6.7), lane depth (defined as the number of stacks along an aisle; or four in the figure), and lane height (which would be two in this figure).

There are different policies for utilizing space in block stacking. Many warehouses employ a last-in–first-out method for block stacking. In this case, once pallets are stored in a lane, additional pallets are not added to a lane until all of the pallets in the lane are retrieved. This can lead to underutilized space, called "honeycombing," as illustrated in Figure 6.8.

An alternative would be to "dig out" specific pallets, which can require multiple movements of the same pallet. For example, in Figure 6.7, to retrieve the shaded pallet on the case on the right, we would need to move four (if on the top stack) or five (if on the bottom stack) pallets to access it.

Pallets can be stored in rack systems as well. This allows them to be stacked without the risk of damaging the product since the rack supports the weight. The pallet racks can be single or double deep (i.e., deep reach).

Case and broken case storage can also be stored in racks. This can be in standard racks or what is known as flow-through (or gravity) racks (Figure 6.9). In flow-through racks, items are loaded at the rear of the rack (i.e., high end). As cases are picked at the front end, the cases behind it "flow" toward the front. These types of shelving help to more efficiently use the footprint space of the warehouse.

There are also several types of automated systems that can be used to move cases to the picking operation. One example is the carousel system, as illustrated in Figure 6.10. In this case, each shelf (or set of shelves, depending on the carousel type) can rotate in either direction. This brings the product

Warehousing and Distribution

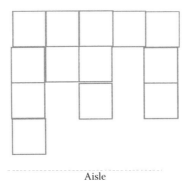

FIGURE 6.8
Honeycombing in block stacking.

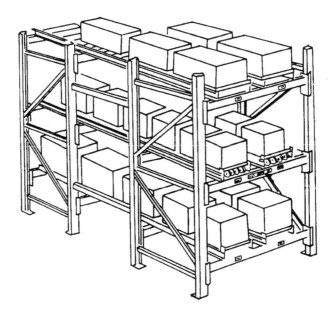

FIGURE 6.9
Flow-through (or gravity) rack. (College Industry Council on Material Handling Equipment. 1999. Material Handling Equipment. Available at http://www4.ncsu.edu/~kay/mhetax/. Last accessed Feb. 23, 2018. Out of courtesy.)

to the individual that picks it. The picker therefore only needs to move along the carousels to pick all of the items requested, thus saving walking time.

Automated storage systems are often combined with "pick to light." In this case, an LED (light-emitting diode) will turn on to signal the operator which item is to be picked. This can greatly increase picking efficiency by reducing search time.

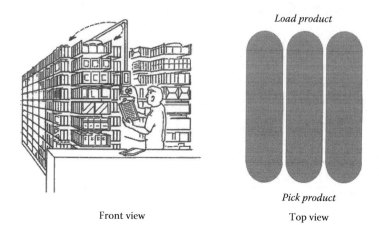

FIGURE 6.10
Carousel system. (Left: College Industry Council on Material Handling Equipment. 1999. Material Handling Equipment. Available at http://www4.ncsu.edu/~kay/mhetax/. Last accessed Feb. 23, 2018. Out of courtesy.)

6.2.3 Sorting Product

Equipment for sortation is typically automated and integrated with the conveyor network. A key benefit is that it improves the efficiency of picking operations. For example, consider three orders made up of two lines each:

- Order 1: 25 of A and 15 of C
- Order 2: 19 of A and 22 of B
- Order 3: 11 of A, 16 of B, and 20 of C

Instead of having a picker for each order, one picker could pick the 54 items of A, another the 38 items of B, and the other the 35 items of C (Figure 6.11). This would reduce the total travel time. The sortation equipment would then ensure that the orders are constructed as requested. For example, for order 2, the number of requested A's and B's would be sent down the second order-specific conveyor for packing.

6.2.4 Tracking Product

Warehouses can have hundreds of thousands of SKUs stored in different handling units. It is therefore very important to know where an individual product is, as well as where items are put together in orders. Tracking product is essential for inventory control and accuracy (eliminating human error), understanding the current state, providing immediate feedback about the

Warehousing and Distribution

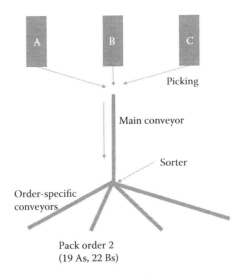

FIGURE 6.11
Sortation system for a three-order example.

FIGURE 6.12
Barcode label examples (1D and 2D).

product, and providing feedback on the status of fulfillment. The most common approach for tracking product is through the use of barcodes.

Barcode technology uses a paper label in an optical and machine-readable format. Both 1D and 2D versions of barcodes are available (Figure 6.12) and are read by a handheld barcode scanner. The barcode system is made up of a network of hardware and software to automate the data collection and label generation. The data are tied into the warehouse management system for operational planning and control, including planning for picking, stocking, and restocking. Although more expensive, radio frequency identification (RFID) is also used for asset tracking. RFIDs use electromagnetic fields to identify and track. In addition to product tracking, RFID tags can also be used to track assets such as warehouse equipment to know where these assets are at all times.

6.3 Warehouse Design

Warehouse design is concerned with the staffing, equipment selection, and layout to support warehouse functions in order to best meet the organizational objectives. Since warehouses don't generate revenue, there are often multiple and competing objectives. Example warehouse objectives include the following:

- Effectively utilize the labor staffing (labor efficiency, cross-training, employee satisfaction)
- Maintain a high level of customer service (timeliness and accuracy)
- Effectively use capital investment (space and equipment efficiency)
- Effectively manage the inventory (efficiency, damage/theft reduction)

Design depends on the requirements of the warehouse. Fundamentally, the trade-off is between resources (building and equipment costs) and handling costs. However, there are several principles that generally lead to effective design. These include the following:

- Minimize the amount of product handling and unnecessary movement/travel—handling of product is labor intensive, and the less it is done, the lower those costs and the less likely it is for product to get damaged or misplaced.
- Effectively use "all" space—it is important to consider not only the footprint of the warehouse but also the height. In some instances, making full use of the space (i.e., cube) can greatly improve performance and/or reduce costs.
- Maintain flexibility when possible—things change over time; having an inflexible layout will lead to significant problems when those changes occur. Holding space buffers can also help address variability.
- Consider horizontal movement by conveyors—although conveyors can be both capital intensive and inflexible, they greatly improve the efficiency of movement between warehouse areas.
- Consider a warehouse within the warehouse—in many situations, it is useful to think of having separate warehouse functional areas. Examples include a forward area for picking (discussed in Section 6.6), a separate area to handle online orders, and a separate area to consider process returns. This helps focus the unique capabilities that are required for a particular function.

Warehousing and Distribution 315

- Organize the warehouse to reduce travel time—as mentioned previously, labor travel times make up a significant amount of warehousing cost. As such, the design should be developed to reduce travel effort.
- Organize the warehouse to support smooth flow—the layout should be organized in order to minimize the high areas of congestion where various flows require the same area. Further, the layout should support the movement to take place in a logical order. Organizing areas so that product and people flow as a U-shape or S-shape can help support this. Figure 6.13 shows an example layout that has a U-shaped flow. The product comes in the upper left and exits on the lower left. In this layout, there are built-in buffers (e.g., the overflow area) and a low number of intersection points. The highest congestion area in this warehouse would likely be in the center, and so additional space helps reduce the potential congestion.
- Maintain adequate space—aisle sizes that are too narrow do not allow space for material handling equipment such as forklifts to "turn around" in. Aisles that are too wide waste valuable footprint. Further, there should be adequate buffer space between warehouse operations.

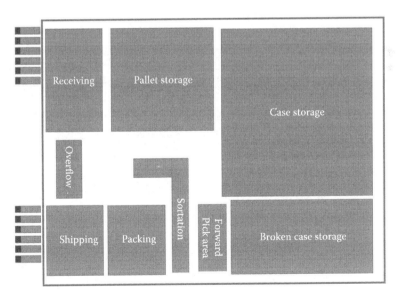

FIGURE 6.13
Simple warehouse layout.

6.3.1 Material Flow and Staffing

Bartholdi and Hackman (2016) describe the warehouse as a queuing system where SKUs may be thought of as the arriving customers to the receiving dock (λ) and the queue corresponds to storage in the warehouse with waiting time W. The relationship with number of items (e.g., pallets) in the warehouse (L) is given by Little's law:

$$L = \lambda W.$$

There must be sufficient labor to handle at least at a rate of λ.

To illustrate, suppose there are, on average, 25,000 pallets in the warehouse, each requiring equivalent material handling labor effort. The turnover ratio (annual volume/average inventory) for pallets is 5 per year. We can use Little's law to estimate the arrival rate:

$$\lambda = 25,000(5) = 125,000.$$

The labor must be sufficient to handle, then, at least 125,000 pallets per year.

As discussed in Section 3.3.4, Little's law applies at any scale. It can, therefore, be applied across each area in the warehouse or across the various set of customers, and the total staffing requirements simply added up. This is illustrated in the exercises at the end of this chapter.

6.3.2 Storage Strategies

When considering how to store items in a warehouse, there is a fundamental trade-off between the space requirements and the ability to specify a specific area for a particular item. This can be seen by considering the following three storage strategies:

- Dedicated storage policy—an item is assigned to a specific location (or set of locations).
- Random storage policy—an item may be placed in any open location.
- Class-based storage policy—the warehouse space is broken down into classes (each with a defined set of items); an item may be placed in any open location in their class.

This trade-off can be seen in the following example. Consider the seven items, each with daily storage requirements (in slots) that are normally distributed, as defined in Table 6.1. For the sake of simplicity, let us assume that each slot has the same size.

Let us determine the amount of dedicated space that would be needed for the three different storage policies. For class-based storage, assume that

Warehousing and Distribution

TABLE 6.1

Mean and Standard Deviation of Daily Storage Requirements for Seven Items

SKU	Mean "Slots"	Standard Deviation
1	55	10
2	70	25
3	83	20
4	60	20
5	40	16
6	35	10
7	35	13

there are two classes, with class 1 having SKUs 1, 2, and 3 assigned to it and class 2 having the remaining SKUs. In addition, assume we desire a 90% service level (i.e., 90% of the time, there will be sufficient space for an SKU).

For dedicated storage, we can determine the slots required for SKU 1 by finding the value of W that satisfies

$$P\{\text{number of slots required} \leq W\} = 0.9.$$

For the first SKU, using the standard normal distribution, the solution of this gives $W = 55 + 10(z_{0.9}) = 68$. Note that we rounded up from 67.82. If we perform this calculation for each of the SKUs and sum it up, we have 525 slots required.

For randomized storage, space can be shared between all SKUs. Therefore, we can take the convolution of the seven normally distributed random variables. For the normal distribution, the mean and variance are additive. This means that the total space requirement is normally distributed with a mean of 378 slots and a standard deviation of 45.3 slots. The required number of slots for the 90% service level is $W = 378 + 45.3(z_{0.9}) = 436$ slots, which is a reduction of 17%.

For class-based storage, space can be shared between SKUs within a class. Performing a similar calculation as above, we find that the space requirements are 251 slots for Class 1 and 209 slots for Class 2, for a total of 460 slots. As expected, this falls between dedicated and randomized storage requirements.

Although randomized storage uses less space than dedicated storage, there can be two significant drawbacks. First, with dedicated storage, we can ensure that the faster-moving SKUs are placed closer to the entry/exit point so that travel time is reduced. Second, in practice, when randomized storage is used, it is often the closest open spot that is chosen rather than a random spot. This means that a slow-moving SKU can be placed in a prime slot that is close to the entry/exit. It is easy to see that, over time, the slowest-moving items will all be closest, which leads to poor performance.

6.3.3 Warehouse Layout

Warehouse layout is concerned with the spatial organization of warehouse departments within the warehouse space. Determining the best layout can depend on a wide range of objectives that may conflict with each other, including the following:

- Minimize total travel time
- Minimize handling of material
- Minimize work in process
- Maximize safety
- Minimize congestion
- Maximize flexibility

Although satisfying all of these objectives is quite difficult, Thompkins et al. (2010) and Muther and Hales (2015) define a simple and practical systematic layout planning (SLP) process. The approach (initially developed by Muther in the 1970s) is described extremely well in Muther and Hales (2015) and is based on three fundamentals:

- Relationships—a measure of importance of "closeness" between two departments
- Space—the amount of space required as well as specific shape requirements
- Adjustment—the arrangement of departments into a realistic best fit

Given a set of activities (i.e., departments or work groups) to be placed in the facility along with the building shape (which is already existing), the SLP process consists of six steps. To illustrate each step, we will use a very simple example of five departments (D1 to D5), which are to be placed in a rectangular facility.

1. Chart the relationships—define how important it is for departments to be close to each other, and why. Note that, in some cases, it is important for certain departments to be located far apart. One way to accomplish this is through a relationship chart shown in Figure 6.14. In this figure, a letter is given based on the importance of proximity (A, E, I, O, U) in decreasing importance. Note that X is used to represent that it is important that these departments be placed far apart.

2. Establish space requirements—estimate the space requirements for each of the departments. It is important to consider factors such as working areas or access aisles in these calculations. Basic shape

Warehousing and Distribution

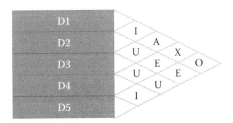

FIGURE 6.14
Relationship chart example for five departments.

constraints should also be noted. Let us assume for our example that space requirements are 25,000 square feet for departments D1 to D4 and 50,000 square feet for department D5. The building footprint is 300 feet by 500 feet.

3. Diagram activity relationships—a graph expressing the relationship can be a useful way to accomplish this, where the nodes represent departments and the edges represent the relationships between them. This is demonstrated in Figure 6.15. Two different graphs are demonstrated. The graph on the right better satisfies the given relationships, though it can be further improved.

4. Draw space relationship layouts—next, we use the relationship layout developed in step 3 to guide the placement of departments in the facility. This is also called a "block" layout. Specific shape constraints should be considered in the construction. Figure 6.16 illustrates a block layout for the two graphs given in Figure 6.15.

5. Evaluate alternative arrangements—layouts generated in step 4 need to be evaluated. This is typically done based on either adjacencies or

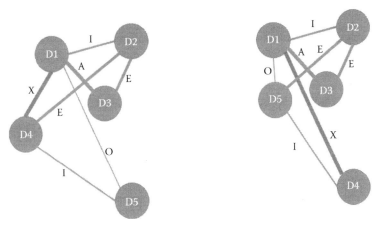

FIGURE 6.15
Examples of activity relationship layouts.

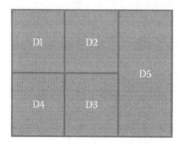
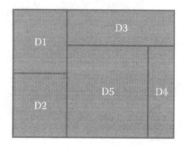

FIGURE 6.16
Examples of block layouts.

distances between pairs of departments. For example, department pairs with an "A" relationship should be adjacent. Several metric scores can be used. One example would be $\max Z = \sum_{i=1}^{m-1}\sum_{j=1+1}^{m} w_{ij}X_{ij}$, where X_{ij} equals 1 if department i and j are adjacent in the layout and 0 otherwise, and w_{ij} is the value of the relationship. An example would be to use A = 5, E = 4, I = 3, O = 1, U = 0, X = −10 for these weight values.

6. Detail the selected layout plan—once the block layout from step 5 is chosen, then the specific details of that layout are specified, including aisle locations, equipment locations, restroom facilities, and work spaces.

A large number of "computerized" layout software packages have been developed for SLP, including Craft, Corelap, and LayOpt. Although there are significant differences between them, the basic idea is to minimize the sum of weighted distances between departments subject to space/shape constraints. For example:

$$\min z = \sum_{i=1}^{m-1}\sum_{j=i+1}^{m} w_{ij}D_{ij},$$

where w_{ij} is the "importance" weight of the closeness of departments i and j and D_{ij} is the distance between the centroids of the two departments in the layout. As enforcing the space/shape constraints is a challenging problem mathematically, heuristics are typically employed. One common approach is to start with an initial layout and then construct alternatives using a local improvement search such as 2-opting:

- *2-opting*—Pick two departments i and j and switch their location in the layout if feasible; keep the switch if it improves the objective.

This could also be done with three departments (3-opting) or using some other local improvement approach. Our view, however, is that although SLP is a very useful framework for developing and evaluating layouts, due in large part to the structured discipline it imposes, computerized software tends to not be very useful in practice. The exception to this is that visualization with simulated flows can be very helpful in identifying potential layout issues such as congestion.

6.3.4 Aisle Layout

Since large sections of warehouses consist of aisles (often for pallets or other types of unit loads), it is important to consider their layout explicitly. General practice is for aisles to be straight and perpendicular to cross lanes. Figure 6.17 shows two examples. In the layout on the left of this figure, there are long aisles with cross lanes at each end. In the layout on the right, a cross lane is added across the middle. The addition of these lanes does take away some storage space, but has the potential benefit of reducing pick path lengths since it supports crossing aisles rather than having to travel all the way the end to cross.

Gue and Meller (2009) developed a different type of configuration for parallel pick aisle where the cross lanes are not perpendicular to the aisles. They considered the problem of minimizing the expected travel distance assuming randomized storage with a single cross lane (in addition to the top and bottom cross lanes). They found that a reduction of roughly 20% in travel time could be achieved if the cross lane was placed in what was essentially amounted to a "V" shape; called a "Flying-V" cross lane. Making the V shape linear along the diagonal, as shown in Figure 6.18 (also known as the fishbone aisle layout), yielded roughly the same performance. Although still not common, this layout is starting to be employed by several organizations.

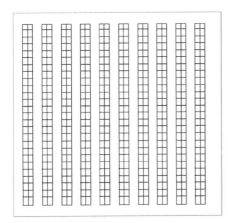 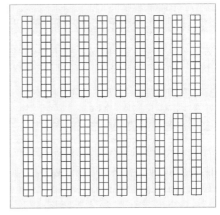

FIGURE 6.17
Aisle layouts with differing number of cross lanes.

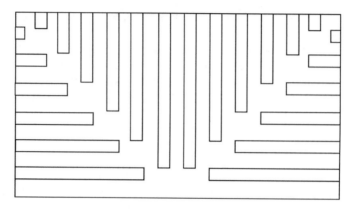

FIGURE 6.18
Fishbone aisle layout.

6.4 Order Picking

Order picking is the process of collecting SKUs (in the quantity and handling unit requested) from a warehouse in order to meet the set of requested customer orders. For most warehouses, order picking is heavily labor intensive and makes up the greatest expense, roughly 45%–60% of total operating expenses, more than half of which is traveling time. We can classify order picking into four categories (Frazelle 2001):

1. Picker to part—pickers go into warehouse storage (which may typically include designated picking areas) to retrieve items to form an order.
2. Part to picker—equipment is used to move the part from the storage area to the picking area, typically into picking bays, where pickers then remove items from the bay to fill an order.
3. Sorting system—automated material handling systems are used to retrieve and sort items into requested orders.
4. Pick to box—in this case, there are designated picking areas, and the pickers fill partial orders to a "box," which are then organized through connected conveyors to form an order.

In this section, we will focus on picker-to-part order picking. For this case, an example pick path in a pick area made up of rows of shelving for a picker is shown in Figure 6.19. Note that a picker may be assigned individual orders that she or he would pick one line at time. This is known as "discrete order

Warehousing and Distribution

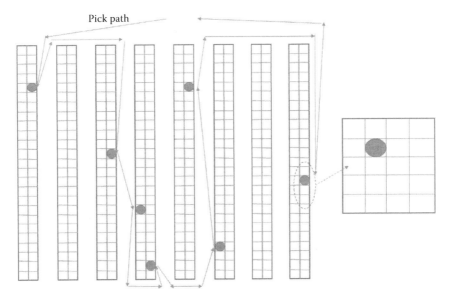

FIGURE 6.19
Example pick path for a picker in a warehouse.

picking." However, efficiency can be gained by using one of the following variations:

- Batch picking—assigning more than one order (i.e., a batch) to a picker at one time. The orders are picked one line at a time. This can be done with a single tote or with a tote for each unique order (sometimes called "cluster" picking).
- Zone picking—assigning the picker to a specific area in the warehouse. In this case, the picker would pick a group of lines from multiple orders. A sortation/consolidation process would be needed to convert the "picked lines" into the customer orders after the picking is completed.
- Wave picking—this is essentially a type of discrete order picking, but where a scheduling window is used to pick orders at a specific time of the day.

If a set of SKUs has been assigned to a picker, regardless of the pick strategy used, the problem becomes sequencing the SKUs so that they can be picked as efficiently as possible. In most cases, this is equivalent to minimizing the total length of the pick path (which is equivalent to minimizing the time to pick the set of SKUs). Ratliff and Rosenthal (1983) showed that, when picking in a rectangular warehouse (as in Figure 6.19), that problem is the same as the well-known traveling salesperson problem:

- Given a start/end point and set of locations to visit once and only once, find the shortest tour.

This was discussed in Section 5.4.2, and the same approaches that were presented in that section are directly applicable here.

There have been several specific heuristics developed for order picking in a rectangular warehouse. Some of the more popular are called the S-shape, largest gap, and aisle-by-aisle heuristics (Roodbergen and de Koster 2000). The S-shape heuristic, which is extremely easy to implement in practice is made up of three steps:

S-Shape Heuristic

- Aisles that are visited are totally traversed.
- Aisles with no items to pick are skipped.
- After the last item is picked, the picker returns to the front of the aisle.

This is illustrated in Figure 6.20.

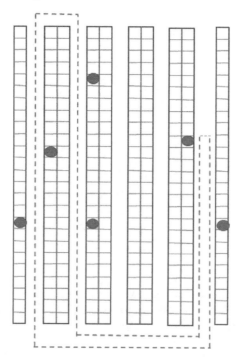

FIGURE 6.20
Example pick path using S-shape heuristic.

Determining the best batch size for an order is a rather complicated problem to model. However, Gong and Koster (2009) show that for a parallel-aisle rectangular warehouse (as shown in Figure 6.19), the objective is convex in batch size. Therefore, a simple local search strategy on the objective will determine the optimal size.

6.5 Slotting

Order picking efficiency depends in large part on how the product (normally in handling units of cases) is arranged within the facility. The determination of where to place individual cases within the warehouse is called "slotting." Effective slotting can lead to several benefits. These include the following:

- Reducing picking and replenishment labor requirement
- Reducing the time to pick an order (i.e., better response)
- Reduced capital expenditure due to effective space utilization
- Better accuracy in picking (since slots are well defined)

The task of slotting in general depends on two things:

- How to prioritize SKUs
- How to assign SKUs to locations based on the prioritization

In general, standard practice is to place fast-moving items near the shipping area in order to reduce the travel time (i.e., increase pick efficiency). However, there are several complicating factors that need to be considered. These include demand variability, heterogeneity of SKU factors such as size and weight, and seasonality.

6.5.1 SKU Profiling

Examining past order/line data (i.e., SKU profiling) can provide valuable information for order picking operations. Some of the key factors to estimate for each SKU include the following:

- Total line volume—this provides information for prioritizing the SKU.
- Lines per order—this gives information about the amount of picking effort that will be required for an order.
- Lines per item—this defines the popularity of the item.

- Items per order—this provides information about the amount of picking effort that will be required for the SKU.
- Cube per order—this helps select the picking method.
- Weight per order—this helps select the picking method.
- Seasonality—this provides information about whether the SKU will need to be repositioned.
- Correlation—this gives information on how often one SKU is picked with another (and hence how important it is that they are co-located). The correlation coefficient for two SKUs X and Y for n observations is defined by

$$\rho_{XY} = \frac{\sum_{i=1}^{n}(x_i - \bar{x})(y_i - \bar{y})}{(n-1)s_x s_y},$$

where x_i is the number of items to pick in line i for SKU X, $\bar{x}$ is the average for the n observations for SKU X, and s_x is the standard deviation of these observations. The equivalent notation is for SKU Y. Note that $-1 \le \rho_{xy} \le 1$. However, it is more common to use the percentage of orders for which both SKUs appear.

We can illustrate the usefulness of these measures with the following simple example.

Example 6.1

Assume that we have 10 SKUs (S1 to S10) in the warehouse and have collected a representative sample of 14 orders (O1 to O14) over a week. The order information is given in Table 6.2, along with weight (in pounds) and cube (in cubic inches) for an individual SKU item on the last two rows.

We summarize the profiling measures in Tables 6.3 and 6.4. Note that in calculating these measures, we do not include SKUs if they are not part of an order. For example, SKU S2 was part of two different orders, and so the average items per order is (14 + 16)/2 = 15.

If we compute demand correlation (as a percentage of similar SKUs between two orders), the two extremes are S1 and S2, which has a correlation of 0%, and S3 and S6, which has a correlation of 67%. There would be no reason to locate S1 and S2 near each other, but we should try to locate S3 and S6 near each other in order to save travel time.

There are a few other observations that we can make from the analysis. First, S3 is the most popular SKU (in terms of number of times picked) and should be located close to the front (origin/destination) point. Further, S10 is a large and heavy item with a reasonably high item volume and should be located near the front in an easy-to-reach place to help with material handling effort. We also see that there is a wide variation in both cube per order

Warehousing and Distribution

TABLE 6.2

Sample Data for 14 Orders (Example 6.1)

Order	SKU (# Requested in the Order)									
	S1	S2	S3	S4	S5	S6	S7	S8	S9	S10
O1			15			7				33
O2	26		22	7						
O3			8			9		3		
O4		14			25					
O5	26								19	
O6							40			
O7			13			26				
O8			17						9	
O9	25						40			
O10							40			
O11		16		22						33
O12			21			19				
O13	27									33
O14			22					40	14	
Weight (pounds)	1.5	2.0	3.9	1.5	2.3	7.8	1.3	4.1	5.0	11.3
Cube (cubic inches)	250	78	60	40	100	25	43	45	90	300

TABLE 6.3

Order-Based Metrics (Example 6.1)

Order	Line Volume	Lines per Order	Volume per Order	Weight per Order
O1	55	3	10,975	486
O2	55	3	8100	135.3
O3	20	3	840	113.7
O4	39	2	3592	85.5
O5	45	2	8210	134
O6	40	1	1720	52
O7	39	2	1430	253.5
O8	26	2	1830	111.3
O9	65	2	7970	89.5
O10	40	1	1720	52
O11	71	3	12,028	437.9
O12	40	3	1735	230.1
O13	60	2	16,650	413.4
O14	76	3	4300	207.8

TABLE 6.4

SKU-Based Metrics (Example 6.1)

SKU	Items per Order	Lines per Item	Item Volume
S1	26	4	104
S2	15	2	30
S3	16.9	7	118
S4	14.5	2	29
S5	25	1	25
S6	15.3	4	61
S7	40	4	160
S8	3	1	3
S9	14	3	42
S10	33	3	99

and weight per order, which implies that the picking process will be rather complicated and likely need multiple types of equipment to pick. Finally, it is worth noting that S7 and S10 have the same item quantity each time in an order (e.g., S7 is always picked in quantities of 40). Storing these SKUs in that size handling unit may be beneficial. In addition, if there are some differences (e.g., S1 is picked in quantities from 25 to 27), encouraging ordering in a standard quantity would also help simplify the picking process.

6.5.2 Prioritizing SKUs

A common method for prioritizing SKUs for dedicated storage is based on cube per order (COI), which bases the location of an SKU based on the volume of the pick space required for the SKU (cube) and how frequently it is picked. For example, we can use the following process:

Prioritizing SKUs for Dedicated Slotting (COI)
- For each SKU i to be slotted, determine the flow f_i, maximum number of slots required S_i, and item volume v_i.
- Order SKUs from most important to least based on the COI ratio: $(v_i S_i)/f_i$, that is $(v_1 S_1)/f_1 \leq (v_2 S_2)/f_2 \leq (v_3 S_3)/f_3 \leq \ldots$
- Assign SKU with priority 1 to the "least cost" (typically closest) location down to SKU n to the highest cost location.

The simple policy works reasonably well if the handling costs are equivalent. As an example, we will apply this prioritization to the data from Table 6.2 (Example 6.1). The rankings and slot requirements are shown in Table 6.5. Note that the flow values are the item volumes for the week. In order to determine the S_i

TABLE 6.5

SKU Prioritization Result for Example 6.1

SKU	f_i	v_i	s_i	Ratio	Rank
S1	104	250	150	360.58	9
S2	30	78	45	117.00	7
S3	118	60	130	66.10	4
S4	29	40	33	45.52	2
S5	25	100	30	120.00	8
S6	61	25	100	40.98	1
S7	160	43	175	47.03	3
S8	3	45	5	75.00	5
S9	42	90	50	107.14	6
S10	99	300	130	393.94	10

values provided in this table, we determine the max weekly flow for each SKU (data used for this calculation not shown).

One of the drawbacks of this solution is that weight of the items wasn't considered. Note that SKU S10 is both large and heavy. Since heavier items typically take more material handling effort, it is often desirable to locate them near the front. Using general ratios like COI can give a good initial assignment, but it is highly recommended that the slotting performance is evaluated through an approach such as simulation.

For aisle storage, Figure 6.21 shows a general strategy for assigning SKUs to shelves. Note that what is termed the "golden zone" (shown as the middle shaded region) runs roughly down the middle third of the area (pick direction running from left to right). In this area, the items are at a height that is easiest to reach for the picker and they don't need to bend down or reach up

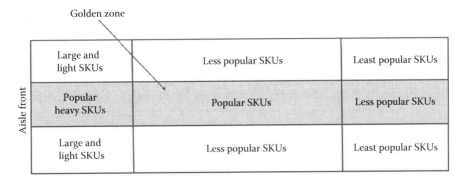

FIGURE 6.21
Slotting layout for an aisle.

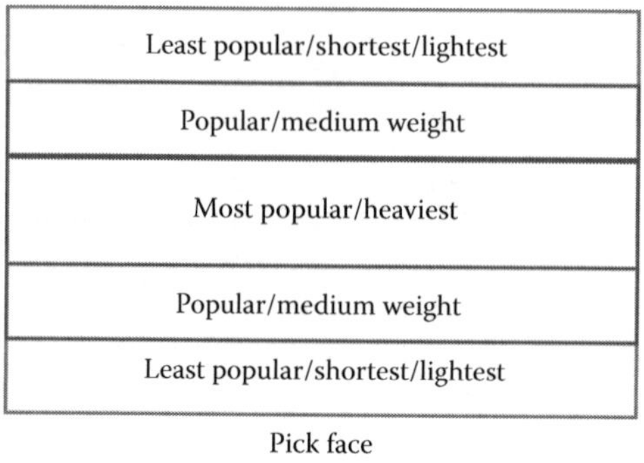

Least popular/shortest/lightest
Popular/medium weight
Most popular/heaviest
Popular/medium weight
Least popular/shortest/lightest

Pick face

FIGURE 6.22
General slotting arrangement for the pick face of flow racks.

to retrieve them. It is, therefore, best to put the most popular SKUs in this region. As you move farther down the aisle, the real estate is less valuable, and hence less popular SKUs are placed there.

A similar slotting strategy may be applied to a pick face of gravity flow racks. This is illustrated in Figure 6.22.

6.5.3 Packing SKUs

After the slotting locations have been determined, the remaining problem is packing the item cartons into the space allocated. Note that a case can be stacked in one of six orientations. Consider a two-foot by one-foot by six-inch (2' × 1' × 6") carton. The six stack orientations are shown (top view and side view) in Figure 6.23, with horizontal stripes representing the top, vertical stripes representing the long side, and diagonal stripes representing the short side. The shelf boundaries, which are fixed, are shown in the figure as well.

As presented in Bartholdi and Hackman (2016), a basic packing algorithm has the following form:

Packing Algorithm
- Construct the sorted SKU list (as described in Section 6.5.2).
- Iteratively move each SKU from the list and place it in the most suitable location based on a packing strategy (e.g., next fit).

A simple packing strategy, called "next fit," works as follows:

- For a given shelf, start with the highest rated SKU.
- If there is space to hold the SKU, add it to the shelf, if not, use the next SKU on the list.
- Remove the shelved SKU from the list.

Additional strategies are discussed in Bartholdi and Hackman (2016).

Warehousing and Distribution

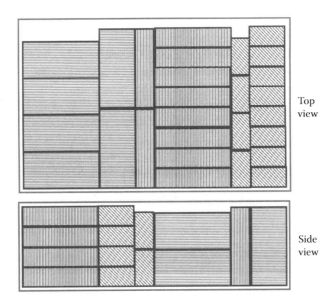

FIGURE 6.23
Six stack orientations for a carton.

6.6 Forward–Reserve

One of the most labor-intensive functions in a warehouse is piece picking. As a result, many warehouses will use a forward (or "fast pick") area. This is illustrated in Figure 6.24. In this case, the items stored in the forward area are relatively easy (in terms of labor effort) to pick. Those SKUs not stored in the forward area may be picked in the reserve area but are relatively more costly

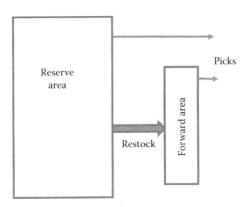

FIGURE 6.24
Forward and reserve areas.

to pick. The forward area is restocked from the reserve area periodically. The basic problem is to determine which SKUs to stock in the forward area and how much forward area volume to dedicate to each SKU.

The basic trade-off is the cost of restocking versus the ease of picking from the forward area. Hackman et al. (1990) formulated the problem as follows:

Notation:

s is the "savings" per fulfillment request for SKU i if it is stored in the forward area

c_i is the cost per internal replenishment for SKU i

R_i is the number of fulfillment requests for SKU i

D_i is the demand per unit time (in units of volume) for SKU i

N is the number of SKUs in the warehouse

V is the volume of the forward area

z_i is the volume of the forward area dedicated to SKU i

x_i equals 1 if SKU i is assigned to the forward area and 0 otherwise

$$\max \sum_{i=1}^{N} f_i(z_i) \qquad (6.1)$$

Subject to

$$\sum_{i=1}^{N} z_i \leq V$$

$$z_i \geq 0 \quad \forall i,$$

where

$$f_i(z_i) = \begin{cases} sR_i - \dfrac{c_i D_i}{z_i} & \text{if } z_i > 0 \\ 0 & \text{if } z_i = 0 \end{cases}.$$

Note that this formulation has a knapsack structure with the added complication of a nonlinear and noncontinuous (at zero) objective function. Bartholdi and Hackman (2008) also considered this problem and developed several interesting insights.

Let us first consider which SKUs to put into the forward area. Let S be the set of SKUs that should be assigned to the forward area (which we don't

Warehousing and Distribution 333

know). Further, let s be the savings per pick from the forward area compared to the reserve area, c_r be the restock cost, p_i be the number of picks for SKU i for the planning horizon, and f_i be the annual volume per year for SKU i. We can define the "bang for the buck" of placing SKU i in the forward area by

$$\left(\frac{p_i}{\sqrt{f_i}}\right)\left(s\sum_{j\in S}\sqrt{f_j}\right) - c_r\left(\sum_{j\in S}\sqrt{f_j}\right)^2. \tag{6.2}$$

Although we don't know S, Bartholdi and Hackman (2008) show that ranking SKUs by "bang for the buck" is independent of S. This observation allows for the following equivalent measure (called "labor efficiency") for sorting:

$$\frac{p_i}{\sqrt{f_i}}. \tag{6.3}$$

The following procedure based on this measure is near optimal in performance.

Procedure: What to Put in the Forward Area
1. Sort SKUs in order of labor efficiency (greatest to least).
2. Successively evaluate the total net cost of putting no SKUs in the forward area; putting only the first SKU in the forward area; only the first two SKUs in the forward area; and so on.
3. Find the SKU j in step 2 where the cost is minimum. Assign all SKUs $i \leq j$ to the forward area.

Next, we can determine how much of each SKU to store in the forward area. Bartholdi and Hackman (2008) show the following useful observation:

- The "Law of Uniform Restocking"—Under optimal allocations, each unit of storage space is restocked at the same frequency.

This leads to the following simple procedure.

Procedure: How to Determine the Volume to Allocation for SKU i in the Forward Area
1. For each SKU $i \in S$ assigned to the forward area, determine

$$z_i = \frac{\sqrt{f_i}}{\sum_j \sqrt{f_j}}.$$

2. Assign z_i amount of space from V to SKU i.

We can demonstrate what to put in the forward area as well as how much to put in the forward area with a simple example.

Example 6.2

Consider the case that the forward area holds 500 cubic feet. There are a total of 11 SKUs (S1 to S11) that are being considered to be placed in the forward area. Data for the SKUs are provided in Table 6.6. Assume that the cost per restock is the same for all SKUs and is $20. Similarly, assume the cost per pick in the reserve area is the same for all SKUs and is $0.20/pick.

The rankings are shown in Table 6.6 based on the labor efficiency ratio from Equation 6.3. We can now compute the total costs, consisting of pick plus restock costs, in order of the rankings. Note that the portion of 500 square feet assigned to the SKU is determined from z_i. As an example, consider the case where the first highest two ranked SKUs (S3 and S1) are assigned to the forward area. We then have

$$z_1 = \frac{\sqrt{1200}}{\sqrt{1200} + \sqrt{1600}} = 0.46,$$

and hence, $z_2 = 1 - 0.46 = 0.54$. The number of restocks for S1 is given by

$$\frac{1200}{0.46(500)} = 5.2 \text{ restocks/year},$$

TABLE 6.6
Data for Forward Reserve (Example 6.2)

SKU	Cost per Pick (c_i) in Forward	Annual Number of Picks (p_i) (1)	Volume per Item (square feet) (2)	Annual Flow Volume (f_i) (3) = (1)*(2)	Ratio (1)/sqrt(3)	Rank
S1	0.13	800	2	1600	20.00	2
S2	0.13	500	3	1500	12.91	6
S3	0.12	1200	1	1200	34.64	1
S4	0.15	400	4	1600	10.00	7
S5	0.18	600	2	1200	17.32	4
S6	0.17	300	5	1500	7.75	8
S7	0.13	900	4	3600	15.00	5
S8	0.10	700	2	1400	18.71	3
S9	0.14	200	5	1000	6.32	10
S10	0.18	200	7	1400	5.35	11
S11	0.17	350	8	2800	6.61	9

Warehousing and Distribution 335

TABLE 6.7

Calculations for Forward Reserve (Example 6.2)

SKUs Assigned to Forward Area	Portion of Forward Area Assigned to Each SKU	Number of Restocks per Year	Total Cost
None	N/A	N/A	6150 (0.20) = $1230
S3	$Z_3 = 1.00$	2.4 for S3	4950 (0.2) + 2.4 (20) + 1200 (0.12) = $1182
S3 and S1	$Z_3 = 0.46$ $Z_1 = 0.54$	5.2 for S3 5.9 for S1	$1532

and similarly for S2, we get 1600/((0.54)(500)) = 5.9 restocks per year. The total cost, then, is

$$(5.2 + 5.9)(20) + 1200(0.12) + 800(0.13) + (6150 - 1200 - 800)(0.20) = \$1532$$

The calculations for each step are shown in Table 6.7.

In this case, we would only place S3 in the forward area. The reason is the high restock cost ($20/restock) compared to the relatively much less savings that result from picking in the forward area. In the exercises, the restock cost is greatly reduced and you are asked to resolve this example. Several additional SKUs are placed in the forward area.

Note that several extensions are provided in Bartholdi and Hackman (2016). These include accounting for safety stock, consideration of product families, and labor capacity limitations on restocks or total picks.

6.7 Line Balancing and Bucket Brigades

In assembly operations, the workers each perform a set of tasks to complete an order. This can be illustrated by the fast-food restaurant Chipotle. When a customer enters the restaurant to order a burrito, there are two workers that assemble the ingredients to make the desired order, typically a burrito. This includes choosing the tortilla, adding the ingredients, adding condiments, rolling and wrapping the burrito, and checking the customer out. In order to effectively use the workers, the work should effectively be divided equally between them. The process of doing this is called line balancing.

More formally, consider a set of n tasks $S = \{s_1, \ldots, s_n\}$, each with a task time t_i ($i = 1, \ldots, n$). In addition, there may be a set of precedence relations A in the form that before task i can be performed, task j must be performed first (e.g., before a burrito is wrapped, it must first be rolled up). Let F_i represent the tasks that must follow task i (we could equivalently have used the set of predecessors). In addition, consider that there are m workers. The goal is to assign tasks to workers in a way that doesn't violate the precedence constraints and where the maximum time for a worker to perform the set of tasks (i.e., bottleneck of the operation, which we will call cycle time [CT]) is minimized. We can formulate this as a binary program, where $x_{ij} = 1$ if task i is assigned to worker j, and 0 otherwise.

$$\min z = CT \qquad (6.4)$$

Subject to

$$\sum_{i=1}^{n} x_{ij} t_i \leq CT \quad \forall j$$

$$\sum_{j=1}^{m} x_{ij} = 1 \quad \forall i$$

$$x_{jl} \leq \sum_{k \leq l} x_{ik} \quad \forall (i,j) \in A$$

$$x_{ij} \in \{0,1\} \quad \forall i,j$$

In this formulation, the first constraint determines the bottleneck of the line, the second constraint ensures that each task is assigned to a single worker, and the third constraint ensures that the precedence relations are satisfied.

Several heuristics have been developed for the line balancing problem. Weiss (2013) presents an Excel implementation of the following popular heuristic for a given cycle time.

Line Balancing Heuristic
    Step 1. Determine feasible tasks that
        i. Have had their precedence constraints met
        ii. Have not yet been scheduled
        iii. Do not require more time than the time remaining at the worker
    Step 2. If there does not exist a feasible task, then
        i. Start a new worker (increment the station count)
        ii. Allocate a complete cycle time to that station
    Step 3. Schedule the feasible task with the longest task time.

Warehousing and Distribution

Note that in step 3, we used longest task time as a priority rule. Other priority rules could be used in its place including picking a random feasible task. We can illustrate this heuristic with the following example.

Example 6.3

Consider the case where seven tasks are to be assigned, with the task times and precedence constraints listed in Table 6.8. We want the total cycle time for a picker to be no more than 2 minutes.

For the first step of the procedure, the feasible tasks are {T1, T2}, and we would choose T1 since it has the longest task time. The steps are illustrated in Table 6.9 (note that ties for feasible tasks were broken arbitrarily).

The resulting solution is that P1 performs tasks T1, T2, and T3 with 10 seconds of idle time; P2 performs tasks T4 and T5 with 45 seconds of idle time; P3 performs tasks T6 and T7 with 18 seconds of idle time. Note that

TABLE 6.8

Data for Line Balancing (Example 6.3)

Task	Time (seconds)	Required Precedence Tasks
T1	35	
T2	30	
T3	45	T1
T4	25	T1
T5	50	T2, T4
T6	47	T2, T4
T7	55	T4

TABLE 6.9

Solution to the Line Balancing Problem (Example 6.3)

Step	Picker	Time Remaining (seconds)	Feasible Tasks	Task Chosen
1	P1	120	T1, T2	T1
2	P1	85	T2	T2
3	P1	55	T3, T4	T3
4	P1	10	None (go to next picker)	
5	P2	120	T4	T4
6	P2	95	T5, T6, T7	T5
7	P2	45	None (go to next picker)	
8	P3	120	T6, T7	T7
9	P3	65	T6	T6

there is quite an imbalance between pickers in terms of their resulting idle time.

An additional limitation of line balancing is that it does not consider several factors, including:

- Task times can be variable (e.g., workers get tired, a problem occurs on the line).
- The number of workers can be variable due to breaks or absences.
- There is variation between workers (e.g., workers with more experience may be faster).

Bartholdi and Eisenstein (1999) developed an approach where the assembly line balances itself; that is, it is self-organizing. The approach is called "bucket brigades." In this approach, it is assumed that the tasks are in an assembly line. In addition to the Chipotle example mentioned earlier, some other examples include

- Manufacturing assembly lines
- Small-parts zone where product is pulled from carousels on a flow rack
- Aisles of flow rack where pickers complete orders and place them on a carousel

The basic approach for bucket brigades is to use the following principle: "Each worker carries a product towards completion; when the last worker finishes her/his product she/he walks back upstream to take over the work of their predecessor, who walks back and takes over the work of her/his predecessor and so on, until, after relinquishing his product, the first worker walks back to the start to begin a new product" (Bartholdi and Eisenstein 1999). In this case, workers should be sequenced from slowest to fastest and they are not allowed to pass each other. In addition to the excellent performance in practice, the beauty of the approach is in the simplicity of implementation. Bartholdi and Eisenstein have implemented this in numerous locations and have observed a significant performance improvement as a result.

6.8 Cross Docking

Cross docks are a useful warehousing strategy for consolidating orders from multiple suppliers for a set of customers. This process is illustrated in Figure 6.25. In the figure on the left, each supplier must deliver to each of the stores. In the figure on the right, an intermediate location is established that consolidates product from suppliers into orders for customers, thus

Warehousing and Distribution

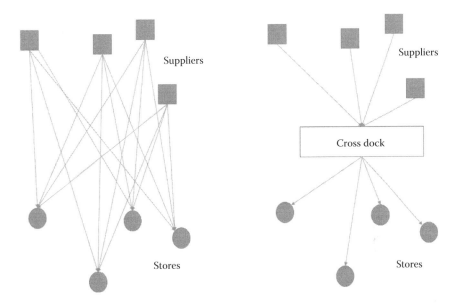

FIGURE 6.25
Consolidation benefit through cross docking.

potentially reducing transportation costs. If inbound goods are pre-tagged for specific customers (i.e., customer assignment is made before shipment), this is known as pre-distribution cross docking. If, on the other hand, the allocation of goods to stores is determined at the cross dock, this is known as post-distribution cross docking.

A pure cross dock facility holds no inventory beyond that required for the consolidation, as shown in Figure 6.26. Typically, the product will stay no more than 24 hours in the facility. In this case, the cross dock is only made up of receiving, sorting, and shipping areas.

In a cross dock, the material is commonly moved by forklift or pallet jack from inbound trailers at a receiving (or "strip") door to the appropriate trailer at the shipping (or "stack") door. The following are the two types of primary staging methods:

- Sort-at-shipping—loads are taken off of the trailer at receiving and placed in a lane that is aligned with the receiving door. At the shipping end, workers pull loads from the appropriate lane and put them on the shipping trailer (i.e., sortation occurs during the shipping stage).
- Sort-at-receiving—loads are taken off of the trailer at receiving and placed in the lane for the shipping trailer. At the shipping end, loads are pulled directly from their shipping lane (i.e., sortation occurs during the receiving stage).

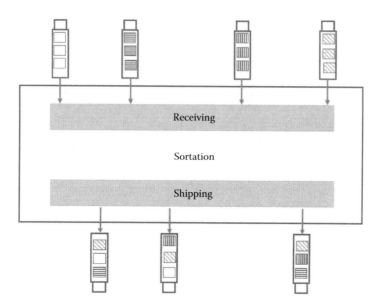

FIGURE 6.26
Cross dock layout.

Managing the dynamics of the receiving and shipping process is the key operational challenge. This includes determining which door to assign to an arriving trailer.

This process can be simplified by adding an intermediate stage. In this case, there are receiving and shipping lanes and an intermediate space between them. The sortation is done in the intermediate area. Organizing in this way simplifies the door assignment since fixed positions don't need to be known in advance. It does, however, require additional handling and additional space. A cross dock of this type is known as a two-stage cross dock.

Effective cross docking requires good communications and coordination. For example, if the loads from the shippers do not arrive at close to the same time, the products that have arrived have to be held in intermediate storage (or the trailer needs to wait in the parking lot). In fact, all of the different loads must arrive before they can be completely sorted, loaded onto the outbound trailers, and sent out. This is similar to the "wait to match" times that are common in assembly systems (Hopp and Spearman 2008). For this reason, if the wait-to-match times are relatively long, holding intermediate inventory can help improve delivery performance to the customer.

The shape, or geometry, of cross docks plays an important role in effectiveness. Most cross docks are long rectangles, or "I" shaped, as shown in Figure 6.27. However, there can be problems with this design, particularly as the number of doors grows large. In this case, there can be significant congestion in the middle of the cross dock. In addition, the travel distance between door pairs increases significantly.

Warehousing and Distribution

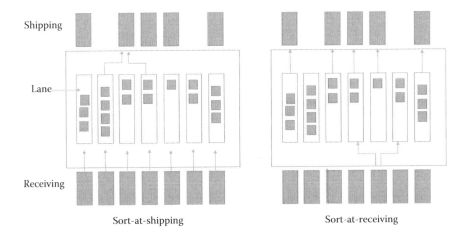

FIGURE 6.27
Staging in a cross dock (sort-at-shipping and sort-at-receiving).

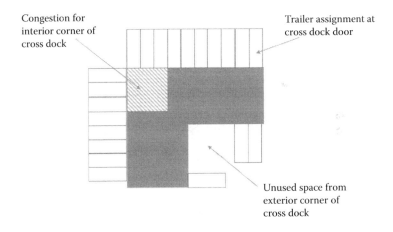

FIGURE 6.28
The impact of interior and exterior corners for cross docks.

Examples of shapes found in practice that are not I-shaped include "T," "X," and "L" shapes. Many factors influence what is the best shape, including the land plot on which the cross dock is located. One of the key factors to consider with regard to shape is corners. This is illustrated in Figure 6.28 for an L-shaped cross dock. Note that for an interior corner (shown in diagonal stripe), there is less space assigned to each door, which leads to congestion. For an exterior corner, as shown on the opposite side, there is unused space since the trailers aren't able to pull into it. A very nice detailed discussion of the impact of cross dock shape on performance may be found in Bartholdi and Gue (2004).

Operationally, a key task is assigning both outbound and inbound trailers to doors. Several approaches have been developed for this. For example, Bartholdi and Gue (2000) use a local search strategy to minimize the labor costs. Further, they found that door assignment has a significant effect on overall productivity.

We can formulate the door assignment through an optimization model (Zhu et al. 2009), which makes assignments in order to minimize the transportation costs within the cross dock. Consider the following notation:

- Set of inbound (receiving) trailers $m \in M$
- Set of outbound (shipping) trailers $n \in N$
- Set of inbound (strip) doors $i \in I$
- Set of outbound (stack) doors $j \in J$
- Distance between door pair d_{ij}
- Amount of goods to unload from inbound trailer s_m
- Amount of goods to go onto outbound trailer r_n
- Flow of goods between inbound and outbound trailers w_{mn}
- Capacity of strip door S_i
- Capacity of stack door R_j
- Assignment of inbound trailer to strip door (binary) x_{mi}
- Assignment of outbound trailer to stack door (binary) y_{nj}

The formulation is given by

$$\min z = \sum_{m=1}^{M}\sum_{i=1}^{I}\sum_{n=1}^{N}\sum_{j=1}^{J} w_{mn} d_{ij} x_{mi} y_{nj}. \qquad (6.5)$$

Subject to

$$\sum_{i=1}^{I} x_{mi} = 1 \quad \forall m$$

$$\sum_{j=1}^{J} y_{nj} = 1 \quad \forall n$$

$$\sum_{m=1}^{M} s_m x_{mi} \leq S_i \quad \forall i$$

$$\sum_{m=1}^{M} r_n x_{nj} \le R_j \quad \forall j$$

$$x_{mi}, y_{nj} \in \{0,1\}$$

The objective minimizes the total distance required to move all of the loads through the cross dock. The first two constraints ensure that an inbound trailer is assigned exactly one door and an outbound trailer is assigned exactly one door. The next two constraints ensure that the strip and stack door capacities aren't exceeded.

The formulation given above is static and assumes that the assignments occur for one unload/load cross dock cycle; that is, there is a time window over which all of the inbound trailers are unloaded and all of the outbound trailers loaded. In many cases, however, loading and unloading don't occur within a cycle. For this scenario, a dynamic policy is more appropriate. Bartholdi and Gue (2000) present a simple policy that they show works well based on the freight mix. The basic question is to determine which inbound trailer should be processed next at a free strip door:

- If the freight mix is uniform across all inbound trailers, then use first in–first out.
- If the freight mix is not uniform, choose the trailer from those waiting with the smallest processing time.

6.9 Warehouse Location and Inventory

The resources required for warehouse operations depend in large part on the inventory that needs to be held in it. Inventory is a function of (among many other things) the location of the set of warehouses with respect to suppliers and customers. Warehouse location is part of the supply chain network design problem as discussed in Chapter 5. We refer the reader back to that chapter for more detailed coverage of location, but present a simple formulation here for discussion. Consider the case where we wish to locate m warehouses with respect to a set of customer zones, where each zone may be served by multiple warehouses. We use the following notation:

- Set of potential facility locations $i \in I$
- Set of customer zones $j \in J$
- Fixed cost (annualized) of a facility F_i

- Total cost of satisfying all of customer zone j from warehouse i is c_{ij}
- Warehouse assignment to location i (binary) y_i
- Portion of customer zone j demand satisfied by warehouse i is x_{ij}

The resulting formulation is

$$\min z = \sum_{i=1}^{N}\sum_{j=1}^{M} c_{ij} x_{ij} + \sum_{i=1}^{N} F_i y_i. \tag{6.6}$$

Subject to

$$\sum_{i=1}^{N} x_{ij} = 1 \quad \forall j$$

$$x_{ij} \leq y_i \quad \forall j$$

$$x_{ij} \geq 0; \quad y_i \in \{0,1\} \quad \forall i,j$$

The objective minimizes total fixed and variable costs. The first constraint ensures that customer demand is satisfied and the second (a linking constraint) ensures that a customer can't be supplied by a location unless the warehouse is located there (and hence fixed costs paid). This formulation does not assume limited capacity, but this constraint can be easily added (and is an exercise at the end of this chapter).

As we mentioned early in this chapter, the distribution network typically has multiple layers. As such, multiple stages would need to be included in the formulation. However, an additional important issue arises. In the above formulation, we have not explicitly considered inventory. For the single-stage case, we can use a service level to estimate the required inventory, which gets incorporated into the c_{ij} costs. When considering multiple stages, however, if an overall service level s desired (e.g., 95% of the time, inventory will be available for the full replenishment cycle), it is not sufficient to simply set s at each stage. Consider the stages in Figure 6.29. In this case, setting the RDC service level to 95% will not guarantee an overall customer service level of 95% since it assumes there would be no stock outs at the CDC. The total service level would therefore be less than 95%.

There are multiple ways that a service level could be met. We could hold more inventory at the CDC and less at the RDC or more at the RDC and less at the CDC, both achieving a service level of s. The distribution of total inventory, however, is not arbitrary as is shown in Figure 6.30.

Warehousing and Distribution

FIGURE 6.29
Multiple stages in a warehouse network.

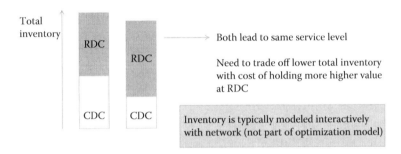

FIGURE 6.30
Two possible ways to meet service level s in Figure 6.29.

In this figure, both alternatives lead to the same service level. However, in this case, we would hold less total inventory (and hence have lower costs) if we held more at the RDC. This is case specific, and which stage has more or less inventory depends on several factors such as replenishment lead times. The key point is that the inventory decisions can be modeled interactively to determine the best allocation. In practice, this is usually done independently of the location optimization model.

6.10 Further Readings

Because of space limitations, there are some important topics in warehousing and distribution that were not covered here. In particular, detailed warehouse layout, advanced order picking, the impact on CO_2 emissions, warehouse performance, warehouse risk, financial issues of warehousing, and warehouse management systems were left out. There are several excellent sources of material for these topics that we would recommend. First and foremost, we recommend *Warehouse Science* by Bartholdi and Hackman (2016); it is a well-written text by two of the leaders in the field and has the added benefit of being open source. Other good reference sources include Frazelle (2001), Gu et al. (2007), and Richards (2014). For the topic of cross docking and other warehousing topics, Kevin Gue hosts a very nice blog (https://kevingue.wordpress.com/) that we recommend.

Exercises

6.1 Search the Web to find estimates of the breakdown of warehouse costs into its components (e.g., labor, equipment, space). Make sure to define your source(s).

6.2 Choose an item that you regularly buy at the grocery store. Find what the different handling units are for that item.

6.3 Pick a company from the S&P500. Find all of their DC/warehouse locations. Comment on what you find.

6.4 Consider a 100% cotton T-shirt sold by Walmart. From the cotton farm to the final customer, list all of the types of warehouses and DCs that would be used.

6.5 Go online and pick a men's clothing store (e.g., Brooks Brothers) and select a style of shirt. List all the SKUs that would exist for that style.

6.6 Consider the following items to be stored in a unit load (pallet) storage area, stacked 1-high. Determine the number of pallet locations required for dedicated, randomized, and class-based storage for a 90% and then 95% service level. Assume that for class-based, there are three classes (A = {I1, I2}, B = {I3}, C = {I4, I5, I6}). Daily storage requirements provided in the following table are normally distributed.

Item	Mean	Standard Deviation
I1	300	50
I2	400	120
I3	800	300
I4	250	100
I5	400	120
I6	300	145

6.7 How would you estimate daily requirements in practice (i.e., how would you get the data for Exercise 6.6)?

6.8 Develop several other SKU priority rules. Discuss the advantages and disadvantages of each one.

6.9 Re-solve the forward reserve example in Section 6.6. In this case, assume that the restocking cost is $2 rather than $20.

6.10 Compare the benefits and drawbacks to pre-distribution and post-distribution cross docks. In addition, what information requirements would be needed for each strategy?

6.11 Modify the formulation in Equation 6.6 to include warehouse capacity.

6.12 Modify the formulation in Equation 6.6 to include two stages. That is, assume that a supplier supplies a set of DCs that supply a set of local warehouses that supply a set of customers.

6.13 Solve the line balancing problem given in the following table with a cycle time of 100 seconds:

Task	Time (seconds)	Required Precedence Tasks
T1	35	
T2	30	T1
T3	45	T1
T4	25	T1
T5	50	T4
T6	47	T2, T4
T7	55	T3

6.14 Why in bucket brigades are pickers ordered from fastest to slowest (right to left)? What would happen if the ordering were done the other way?

6.15 Consider a carton that is 3 feet by 2 feet by 1 foot. This item handling unit will be placed on a shelf that is 5 feet wide, 4 feet high, and 3 feet deep. For the six possible carton orientations, determine the number of packages that can be placed on the shelf if they are all placed in the same orientation. Assume no part of a carton can overhang a shelf. Determine the space efficiency (utilization) for each case.

6.16 For Figure 6.19, determine the pick path length using the S-shape and nearest-neighbor heuristics. Next, try to determine the best pick path visually. Compare your results.

6.17 Find at least one other order picking heuristic for a rectangular warehouse and describe how it works. Apply it to Figure 6.19.

References

Bartholdi III, J. J. and D. D. Eisenstein. 1999. A production line that balances itself. *Operations Research.* 44: 21–34.

Bartholdi III, J. J. and K. R. Gue. 2000. Reducing labor costs in an LTL cross-docking terminal. *Operations Research.* 48: 823–832.

Bartholdi III, J. J. and K. R. Gue. 2004. The best shape for a crossdock. *Transportation Science.* 38: 235–244.

Bartholdi III, J. J. and S. T. Hackman. 2008. Allocating space in a forward pick area of a distribution center for small parts. *IIE Transactions.* 40: 1046–1053.

Bartholdi III, J. J. and S. T. Hackman. 2016. *Warehouse & Distribution Science*. Available at http://www.warehouse-science.com/

Council of Supply Chain Management Professionals (CSCMP). 2017. *28th Annual State of Logistics Report*. Lombard, IL.

Frazelle, E. 2001. *World Class Warehousing and Material Handling*. New York: McGraw-Hill Education.

Gong, Y. and R. de Koster. 2009. Approximate optimal order batch sizes in a parallel-aisle warehouse. In *Innovations in Distribution Logistics*, eds. L. Bertazzi et al. Berlin: Springer-Verlag.

Gu, J., M. Goetschalckx, and L. F. McGinnis. 2007. Research on warehouse operations: A comprehensive review. *European Journal of Operational Research*. 177: 1–21.

Gue, K. R. and R. D. Meller. 2009. Aisle configurations for unit-load warehouses. *IIE Transactions*. 41: 171–182.

Hackman, S. T., E. H. Frazelle, P. M. Griffin, S. O. Griffin, and D. A. Vlatsa. 2001. Benchmarking warehousing and distribution operations: An input–output approach. *Journal of Productivity Analysis*. 16: 241–261.

Hackman, S. T., M. J. Rosenblatt, and J. M. Olin. 1990. Allocation items to an automated storage and retrieval system. *IIE Transactions*. 22: 7–14.

Hopp, W. J. and M. L. Spearman. 2008. *Factory Physics*, 3rd Edition. Long Grove, IL: Waveland Press, Inc.

Kay, M. G. 2015. *Warehousing*. Available at http://www4.ncsu.edu/~kay/Warehousing.pdf

Muther, R. and L. Hales. 2015. *Systematic Layout Planning*, 4th Edition. Marietta, GA: Management & Industrial Research Publications.

Ratliff, H. D. and A. S. Rosenthal. 1983. Order picking in a rectangular warehouse: A solvable case of the traveling salesman problem. *Operations Research*. 31: 507–521.

Richards, G. 2014. *Warehouse Management: A Complete Guide to Improving Efficiency and Minimizing Costs in the Modern Warehouse*, 2nd Edition. Philadelphia PA: Kogan Page.

Roodbergen, K. J. and R. de Koster. 2000. Routing methods for warehouses with multiple cross aisles. *International Journal of Production Research*. 39: 1865–1883.

Thompkins, J. A., J. A. White, Y. A. Bozer, and J. M. A. Tanchoco. 2010. *Facilities Planning*, 4th Edition. Hoboken, NJ: John Wiley & Sons.

Weiss, H. J. 2013. Teaching note: Implementing line balancing heuristics in spreadsheets. *INFORMS Transactions on Education*. 13: 114–125.

Zhu, Y. R., P. M. Hahn, Y. Liu, and M. Guignard. 2009. New approach for the cross dock assignment problem. In *Proceedings of the XLI Brazilian Symposium of Operations Research*.

7

Financial Engineering

An important problem faced by individual investors and financial analysts in investment firms (banks, mutual funds, and insurance companies) is the determination of an optimal investment portfolio. A portfolio specifies the amount of money invested in different securities, which may include bank accounts, bonds, common stocks, real estate, U.S. Treasury notes, and others. Because of the economic significance of the investment problem and its complexity, a number of mathematical models have been proposed for analyzing the portfolio selection problem. The pioneering work, known as the modern portfolio theory, was done by Harry Markowitz in the 1950s. In this chapter, we begin with the basic concepts of investing and then develop Markowitz's mean-variance quadratic programming model (Markowitz 1952, 1959) and its applications in selecting a diversified investment portfolio. We will also discuss extensions by William Sharpe, who developed a bi-criteria linear programming model for portfolio optimization (Sharpe 1963). For their contributions to the modern portfolio theory, both Markowitz and Sharpe shared the Nobel Prize in Economics in 1990. We end this chapter with a discussion of investing in bonds and prudent strategies for investing in general.

7.1 Basic Concepts in Investing

The basic questions in investment are *where to invest and how much to invest*. An investor has thousands of securities available for investment. Each investment security has to be measured in terms of three conflicting criteria—liquidity, return, and risk. Liquidity refers to how easily accessible is your investment. Return represents how much profit you can expect over time from your investment. Finally, risk refers to the safety of your capital, namely, will your investment be still there when you need it?

The investment securities can be broadly classified into three major categories as follows:

- Cash
- Bonds
- Stocks

We shall describe each investment category in detail and how they stack up under the conflicting criteria of liquidity, return, and risk. A good investment strategy is basically a trade-off between return and risk.

7.1.1 Investing in Cash

Cash refers to securities such as checking, savings, and money market accounts in commercial banks and credit unions. Funds invested in these securities are highly liquid (e.g., 24-hour access through ATMs), give practically no return on investment, but are very safe (insured up to $250,000 by the U.S. government). Generally, funds needed in an emergency should be kept in these types of securities.

7.1.2 Investing in Bonds

Bonds are also known as income-producing securities and are basically loans that investors make to corporation and governments (local, state, or federal). They are primarily "IOUs" issued by companies and governments, backed strictly by their reputations. Every bond has a stated principal amount (par value), yearly interest payment (coupon rate), and the length of the loan (maturity). Bonds are bought and sold through security dealers and hence are not very liquid. However, their returns on investment are higher than cash. Their risk (safety of capital and payment of interest) depend on the credit worthiness of the bond issuers. For example, treasury bonds issued by the U.S. government are considered to be the safest investment securities in the world. Hence, the demand for U.S. Treasury securities is very high and naturally they pay comparatively low interest. On the other hand, a bond issued by a start-up company has to pay a much higher interest in order to attract investors because it is considered to be very risky.

In addition to the credit worthiness of the bond issuers, bond prices also vary from its par value depending on the prevailing interest rates set by the government. When interest rates are increased, bond prices fall and their returns become less attractive. However, when the interest rates fall, prices of older bonds increase, resulting in an overall increase in their return on investment. Section 7.5 explains investing in bonds in more detail.

7.1.3 Investing in Stocks

Known also as *equities*, stocks basically represent ownership of a publicly held company. When a company has 1 million outstanding shares and you buy 10,000 shares, you basically own 1% of the company! Shares are bought and sold through security dealers at different stock exchanges (e.g., New York Stock Exchange and NASDAQ). Hence, stock investments are not easily

TABLE 7.1

Comparison of Cash, Bond, and Stock Investments

	Liquidity	Return	Risk
Cash	High	Low	Low
Bond	Medium	Medium	Medium
Stock	Low	High	High

accessible on liquidity criterion. Returns from stock investments can take two forms:

1. Cash dividends
2. Price appreciation

Cash dividends are declared periodically by companies to their shareholders. These represent a portion of company's revenue returned to the shareholders. When a company does well in the economy and continues to grow (e.g., Amazon, Google, and Apple), its share prices also increase. The increase in share price is also considered part of the return on stock investment. Since a company's fortune can change over time, its annual returns can also vary widely. Hence, stocks are considered risky investments but have a high potential for larger returns. For example, annual returns of Google (now Alphabet Inc.) stock have averaged nearly 20% since it went public in 2004. There are more than 7000 stocks available for investment in the U.S. Stock Market.

With globalization of the economy, investors have also opportunities to buy international stocks of companies in Asia, Africa, and Europe. When investing internationally, currency risk may also affect the overall return. In other words, the exchange rates between the U.S. dollar and other international currencies may fluctuate, affecting the real returns from those stocks.

Table 7.1 compares the investments in cash, bond, and stock with respect to liquidity, return, and risk.

7.1.4 Asset Allocation

A prudent investor should invest in all three categories of investment. Immediate needs, including emergency funds, should be kept in cash. The remaining funds should be invested in bonds and stocks to maximize return and minimize risk. Asset allocation refers to the amount of funds invested in cash, bonds, and stocks. Academic research has shown that more than 90% of the performance of an investment portfolio is directly correlated to the asset allocation percentages.

Asset allocation varies depending on how much return an investor wants and how much risk he or she is willing to accept. For example, a young investor at the beginning of her career may be willing to take more risk and expect a higher return of her portfolio. On the other hand, a retiree, who needs funds for his day-to-day living, may invest more in cash and bonds, compared to stocks.

In addition to the broad classification, asset allocation is also used to decide how the stock and bond investments are diversified. For example, a stock portfolio may include U.S. stocks (small, mid-size, or large companies) and international stocks (developed countries in Europe and emerging countries such as India and China). Similarly, an asset allocation strategy for a bond portfolio may include corporate bonds (big and small companies) and government bonds (local, state, and federal).

7.1.5 Mutual Funds

Mutual funds are investment companies that have become attractive for investment in recent years. They are popular with both individual and professional investors. A stock mutual fund receives funds from investors by selling shares of its company. These funds are then invested in a variety of stocks and their stock prices basically determine the net asset value of the mutual fund share. Because of the availability of the large pool of investment money, a mutual fund has the ability to buy a lot of shares in several different companies and diversify its portfolio. Of course, the mutual fund charges its shareholders an annual fee for its investment management cost. The management fee usually varies between 0.2% and 3% of the shareholder's investment.

There are more than 5000 mutual funds available for investment in the United States. Some of the largest mutual fund families include Vanguard, Fidelity, and T. Rowe Price. Each fund family may offer a variety of mutual funds to cater to individual investor needs. They can be classified broadly as stock funds, bond funds, and balanced funds. Balanced funds invest in a mix of stocks and bonds. We will discuss investing in mutual funds more in detail later in Section 7.6.3.

7.2 A Simple Model for Portfolio Selection

Assume we have N securities for possible investment and are interested in determining the portion of available capital C that should be invested in each

Financial Engineering

of the securities for the next investment period. Let the decision variables be denoted by x_j, $j = 1,2..., N$, which represent the dollar amount invested in security j. We have the following system of constraints on the decision variables:

$$x_1 + x_2 + \ldots + x_N \leq C$$

$$x_j \geq 0 \quad j = 1,2,\ldots,N$$

Generally, we have historical data on the performance of each security for the past several years, which give the price fluctuations and dividend payments. We can then estimate the return on investment for each security from the past data. Let $r_j(t)$ denote the total return per dollar invested in security j during year t. Then,

$$r_j(t) = \frac{p_j(t+1) - p_j(t) + d_j(t)}{p_j(t)}, \tag{7.1}$$

where $p_j(t)$ is the price of security j at the beginning of year t and $d_j(t)$ is the total dividends received in year t, for $t = 1, 2, 3, \ldots, T$.

Note that the values of $r_j(t)$ are not constants and can fluctuate widely from year to year. In addition, $r_j(t)$ may be positive, negative, or zero. Hence, to assess the investment potential of security j, we can compute the *average* or *expected return* from security j per dollar invested, denoted by μ_j as follows:

$$\mu_j = \frac{1}{T}\sum_{t=1}^{T} r_j(t) \tag{7.2}$$

The total expected return from the investment portfolio is then given by

$$E = \sum_{j=1}^{N} \mu_j x_j = \mu^T x$$

where $\mu^T = (\mu_1, \mu_2, \ldots, \mu_N)$ and $x = (x_1, x_2, \ldots, x_N)^T$.

7.2.1 Linear Programming Model

A simple optimization model for the investor's problem is to maximize the total expected return subject to the constraints on investment goals as follows:

$$\text{Maximize} \quad Z = \sum_{j=1}^{N} \mu_j x_j \tag{7.3}$$

$$\text{Subject to} \quad \sum_{j=1}^{N} x_j \leq C \qquad x_j \geq 0 \tag{7.4}$$

A number of policy constraints may also be imposed on the portfolio. Most investment companies limit the amount that can be invested in common stocks, whose returns are subject to wide variations. This can be expressed as

$$\sum_{j \in J_1} x_j \leq b_1 \tag{7.5}$$

where J_1 represents the common stock securities and b_1 is the maximum investment ($) in common stocks.

Many investment companies also require a certain amount in ready cash or fluid state to meet withdrawal requests from customers. This may be expressed as

$$\sum_{j \in J_2} x_j \geq b_2 \tag{7.6}$$

where J_2 represents the securities in fluid state (e.g., savings accounts, checking accounts) and b_2 is the minimum cash reserve required. Similarly, one can introduce several such policy constraints on the investment portfolio. We shall illustrate the linear programming (LP) model for portfolio selection with Example 7.1.

Example 7.1

Mary has $100,000 to invest in mutual funds. She has identified 10 potential funds for investment as shown in Table 7.2. The table provides the average annual returns over a 10-year period. Growth funds invest in companies with high potential for increases in stock prices but generally

Financial Engineering

TABLE 7.2
Performance Data of Funds for Example 7.1

Type of Fund	Average Annual Return (%)
Growth funds	
Fund 1	30
Fund 2	20
Fund 3	15
Fund 4	25
Growth and income funds	
Fund 5	16
Fund 6	17
Fund 7	12
Income funds	
Fund 8	8
Fund 9	6
Money market fund	
Fund 10	4

do not pay any dividends. Examples of such companies include Apple, Amazon, and Google. Growth and income funds invest in companies, who pay steady dividends with potential for some price appreciation. Examples include AT&T, General Electric, IBM, and General Motors. Income funds invest primarily in government and corporate bonds. Money market funds represent cash.

Even though Mary is interested in the "Growth" of her portfolio, she has the following investment policy constraints:

i. All the money should be invested among the 10 funds.
ii. No more than $20,000 should be invested in any one fund.
iii. Between 20% and 40% should be invested in growth funds.
iv. At least $5000 should be available as ready cash in the money market fund.
v. At least 10% should be invested in income funds.
vi. At least 15% should be invested in growth and income funds.

Mary's objective is to maximize the total expected return of the portfolio.

LP MODEL

Define X_j as the dollar amount invested in mutual fund j, where $j = 1, 2, \ldots, 10$. The objective function is to maximize the total expected return in dollars given by

$$\text{Maximize} \quad Z = 0.3X_1 + 0.2X_2 + 0.15X_3 + 0.25X_4 + 0.16X_5$$
$$+ 0.17X_6 + 0.12X_7 + 0.08X_8 + 0.06X_9 + 0.04X_{10}$$

(7.7)

The constraints of the LP model are as follows.

i. Invest the entire $100,000 on the 10 funds

$$X_1 + X_2 + \ldots + X_{10} = 100,000 \qquad (7.8)$$

ii. No more than $20000 should be invested in each fund

$$X_j \leq 20,000 \quad \text{for } j = 1, 2, \ldots, 10 \qquad (7.9)$$

iii. Invest 20% and 40% of $100,000 in growth mutual funds

$$20,000 \leq X_1 + X_2 + X_3 + X_4 \leq 40,000 \qquad (7.10)$$

iv. Maintain at least $5000 in cash

$$X_{10} \geq 5,000 \qquad (7.11)$$

v. At least 10% should be invested in income funds

$$X_8 + X_9 \geq 10,000 \qquad (7.12)$$

vi. Invest at least 15% in growth and income funds

$$X_5 + X_6 + X_7 \geq 15,000 \qquad (7.13)$$

In addition, the fund investments can be positive or zero only

$$X_j \geq 0, \quad j = 1, 2 \ldots, 10 \qquad (7.14)$$

The LP model is solved using Microsoft's Excel Solver and the optimal solution is given in Table 7.3.

The maximum value of the objective function, which represents the total expected return in dollars, is given below:

Maximum $Z = (0.3)(20,000) + (0.25)(20,000) + (0.16)(20,000)$

$\qquad + (0.17)(20,000) + (0.12)(5000) + (0.08)(10,000) + (0.04)(5000)$

$\qquad = \$19,200 \quad \text{or} \quad 19.2\%$

Financial Engineering

TABLE 7.3

Optimal Solution to Example 7.1

Mutual Fund Type	Decision Variable	Value ($)
Growth	X1	$20,000
	X2	0
	X3	0
	X4	$20,000
Growth and income	X5	$20,000
	X6	$20,000
	X7	$5000
Income	X8	$10,000
	X9	0
Money market	X10	$5000
Total portfolio		$100,000

7.2.2 Drawbacks of the LP Model

A major drawback of the simple LP model is that it completely ignores the risk associated with investments. Hence, it can lead to an investment portfolio, which may have the potential for a very high average return over the years, but also has a very high risk associated with it. Because of the higher risk, the annual returns would fluctuate widely and the actual return for any given year may be far less than the average value!

By examining the LP optimal solution given by Table 7.3 for Example 7.1, it is clear that the simple LP model invests as much as possible in mutual funds with higher average annual returns. If Mary had not imposed any policy constraints on her investments, then the entire portfolio of $100,000 would be invested in Fund 1, with the maximum return of $30,000 (30%). However, such an investment will be a very risky portfolio, which will go against the proverbial advice, "Don't put all your eggs in one basket." Even though such a portfolio may have a very high average return, the year-to-year fluctuations of this portfolio's returns will be very large. Thus, the actual return for any given year may even be negative! For example, Google stock's average return between 2005 and 2013 was 33% per year. However, the annual returns widely fluctuated during this period, as high as 110% in 2005 and as low as −56% in 2008!

Table 7.4 gives the 20-year returns for four major categories of securities during the period 1997–2016. Even though U.S. stocks gave the largest average annual return of 7.7%, it also had the largest variability in returns from −37% in 2008 to 32.4% in 2013. On the other hand, 90-day U.S. Treasury bills, used here as a surrogate for cash, had the lowest average annual return of 2.3%; however its annual returns were always positive and they fluctuated

TABLE 7.4

Twenty-Year Returns of Investment Securities (1997–2016)

Investment Security	Average Annual Return (%)	Best Year	Worst Year
U.S. stocks	7.7	32.4% (2013)	−37% (2008)
International stocks	4.2	22.8% (2013)	−43.4% (2008)
U.S. bonds	5.3	11.6% (2000)	−2% (2013)
Cash	2.3	6.2% (2000)	0% (2015)

Note: U.S. stocks are represented by S&P 500 index; international stocks are represented by MSCI-EAFE index; U.S. bonds are represented by Barclay's U.S. Aggregate bond index; cash is represented by 90-day U.S. Treasury bills.

between 0% (2015) and 6.2% (2000). Hence, a prudent investor would construct an investment portfolio that not only maximizes the average annual return but also minimizes the year-to-year fluctuations in the annual returns, known as the *variance of the portfolio*. However, the portfolio return and the portfolio risk are two conflicting objectives, namely, a portfolio consisting mostly of stocks will have a higher average return but also will have a larger variance! Hence, determining an optimal investment portfolio is actually a bi-criteria optimization problem that we discussed in Chapter 4. Next, we discuss the bi-criteria optimization model for portfolio selection proposed by Markowitz (1952, 1959).

7.3 Markowitz's Mean Variance Model for Portfolio Selection

The main drawbacks of the simple LP model presented in Section 7.2 are as follows:

- Investment risk is ignored,
- Use of mean values masks the variability in the returns.
- Lack of diversification, that is, "all eggs in one basket."

7.3.1 Quantifying Investment Risk

We begin by discussing how to incorporate the "risk" associated with each security. Some securities, such as "speculative stocks," may have a larger price appreciation possibility and, hence, a greater average return, but they may also fluctuate more in value, increasing its risk. On the other hand, "safe" investments, such as savings accounts and bank CDs, do not fluctuate in their values but have smaller returns. We can measure the *investment risk* of a security by the amount of fluctuation in total return from its average

Financial Engineering

value during the past T years. Let σ_{jj}^2 denote the *investment risk* or *variance* of security j, which can computed as follows:

$$\sigma_{jj}^2 = \frac{1}{T-1}\sum_{t=1}^{T}[r_j(t)-\mu_j]^2 \qquad (7.15)$$

where $r_j(t)$ and μ_j are computed using Equations 7.1 and 7.2 from past data.

The *standard deviation of the return* of security j is the square root of its variance and is given as follows:

$$\sigma_j = \sqrt{\sigma_{jj}^2}$$

Table 7.5 gives the means and standard deviations of the returns for the four major categories of securities, computed over a 20-year period (1997–2016). Note that the U.S. stocks have the highest average annual return but also have a very large risk, represented by the standard deviation of the returns. Note also that international stocks pose more risk than the U.S. stocks. Cash has the lowest risk, but also has the lowest return!

In addition to the variance of individual security returns, a group of securities may be tied to a certain aspect of the economy, and a downturn in that aspect of economy would affect the prices of all the securities in that group. Automobile stocks, utility stocks, and oil company stocks are some examples of such security groups, which may rise or fall together. To avoid this kind of risk, we must diversify the investment portfolio so that the funds are invested in different groups of securities. This can be done by computing the relationships in the annual returns between every pair of securities. Such a relationship, known as the *covariance*, is denoted by σ_{ij}^2 and can be computed from past data as follows:

$$\sigma_{ij}^2 = \frac{1}{T-1}\sum_{t=1}^{T}[r_i(t)-\mu_i][r_j(t)-\mu_j] \qquad (7.16)$$

TABLE 7.5

Twenty-Year Security Returns and Risk (1997–2016)

Security Type	Average Annual Return (%)	Standard Deviation (%)
U.S. stocks	7.7	15.3
International stocks	4.2	16.9
U.S. bonds	5.3	3.4
Cash	2.3	0.7

Note that when $i = j$, the above equation reduces to the variance of security j, as given by Equation 7.15. Now, we can represent the variance or investment risk of a portfolio, denoted by V, as follows:

$$V = \sum_{i=1}^{N}\sum_{j=1}^{N} \sigma_{ij}^2 x_i x_j = x^T Q x \qquad (7.17)$$

where $Q_{(N \times N)} = [\sigma_{ij}^2]$ is the variance–covariance matrix of the returns of N securities.

Note that V is a quadratic function. The correlation coefficient of the returns between securities i and j, denoted by ρ_{ij}, is given by

$$\rho_{ij} = \frac{\sigma_{ij}^2}{\sigma_i \sigma_j} \qquad (7.18)$$

Since σ_{ij}^2 can be positive or negative, ρ_{ij}'s too can be positive or negative and take values between -1 and $+1$. A positive value of ρ_{ij} implies that the returns of securities i and j would rise or fall together. A negative value of ρ_{ij} implies that the return of security i would increase when the return of security j decreases and vice versa. The correlations between stock and bond returns are generally negative.

7.3.2 Excel Functions to Compute Investment Risk

Microsoft Office has special built-in functions to calculate the variance and covariance from sample data in its Excel program. They use Equations 7.15 and 7.16, respectively. The Excel functions can be accessed as illustrated below:

- Variance of security A
  = VAR (A1:A10)

  Here, $T = 10$ years, column A (cells A1 to A10) contain the annual returns from year 1 to year 10 for security A and Equation 7.15 is used to calculate the variance.

  - Covariance between securities A and B
    = COVAR (A1:A10, B1:B10)

  Here, columns A and B contain the annual returns from year 1 to year 10 for securities A and B, respectively, and Equation 7.16 is used to calculate the covariance.

Financial Engineering

7.3.3 Markowitz's Bi-Criteria Model

A rational investor would be interested in determining an investment portfolio that maximizes portfolio return but also minimizes the variance of the portfolio return. Markowitz (1956, 1959) formulated a bi-criteria optimization model to solve this problem as follows:

Maximize average annual return of portfolio:

$$Z_1 = \sum_{j=1}^{N} \mu_j X_j = \mu^T X \tag{7.19}$$

Minimize variance of portfolio return:

$$Z_2 = \sum_{i=1}^{N}\sum_{j=1}^{N} \sigma_{ij}^2 X_i X_j = X^T Q X \tag{7.20}$$

Subject to the constraints:

$$\sum_{j=1}^{N} X_j \leq C; \; X \geq 0 \tag{7.21}$$

$$\sum_{j \in J_1} X_j \leq b_1 \tag{7.22}$$

$$\sum_{j \in J_2} X_j \geq b_2, \tag{7.23}$$

where
X_j = dollars invested in security j
$X = (X_1, X_2, X_3, ..., X_N)^T$ = investor portfolio
μ_j = average annual return per dollar invested in security j

$Q_{(N \times N)} = [\sigma_{ij}^2]$

(*Note:* Diagonal elements of Q represent the variance of the security returns and off-diagonal elements represent the covariance of returns between pairs of securities.)
N = total number of securities available for investment
C = total capital available for investment

Equations 7.22 and 7.23 represent additional investment restrictions on a group of securities such as those given in the LP model of Section 7.2 (see Example 7.1). Since the portfolio return (Z_1) and its variance (Z_2) are conflicting objectives, there will be no portfolio that simultaneously maximizes the return and minimizes its variance (Risk). Recall that, in Chapter 4 (Section 4.4), we gave the definitions of dominated solution, efficient solution, ideal solution, and the efficient set for a bi-criteria optimization problem (see Example 4.4). Let us examine their meanings in the context of the bi-criteria portfolio selection model defined by Equations 7.19 through 7.23.

- Feasible Portfolio:

    A portfolio, $X = (X_1, X_2, ..., X_N)^T$, that satisfies the investment constraints defined by Equations 7.21 through 7.23 is called a feasible portfolio.

- Dominated Portfolio:

    A feasible portfolio, whose values of the investment return and its variance can be improved by another feasible portfolio. In other words, there exists another feasible portfolio that has a higher return for the same variance, a lower variance for the same return, or both.

- Efficient Portfolio:

    An efficient portfolio is a feasible portfolio that is not dominated. For an efficient portfolio, there exists no other feasible portfolio that can give a higher return for the same variance or the same return with a lower variance. In other words, to increase the average return of an efficient portfolio, an investor must be willing to accept an increase in the variance of the portfolio return. Similarly, to decrease the variance of an efficient portfolio, an investor has to accept a lower average portfolio return. Thus, an investor has to do a *trade-off analysis* between the average return and its variance in order to improve an efficient portfolio.

- Efficient Portfolio Frontier:

    The bi-criteria portfolio selection model has several efficient portfolios. The set of all efficient portfolios is called the *efficient portfolio frontier*. A rational investor should choose an efficient portfolio from the efficient frontier that meets the investor's *risk profile*, namely, the amount of investment risk the investor is willing to accept. For example, consider a young investor, who is selecting a portfolio for retirement. Since the investment horizon is long (30–40 years), the young investor can be more *aggressive* by choosing an efficient portfolio that not only has a very high average return but also has a large variance. However, a retiree, who needs the retirement funds for immediate living expenses, has to be *conservative* by choosing an efficient portfolio that has a much lower variance and consequently a lower average return. Thus, investors can choose the best portfolio that matches their tolerance with respect to investment risk. This can be done by using a risk–return graph as discussed in the next section.

7.3.4 Risk–Return Graph for Portfolio Selection

Figure 7.1 illustrates the set of all feasible portfolios (shown by hatched lines) for the bi-criteria optimization model defined by Equations 7.19 through 7.23. The *x*-axis (risk) represents the standard deviation (STD) of the portfolio return and the *y*-axis (return) represents the average portfolio return. For ease of understanding, both the return and risk are expressed as a percentage of the total investment and are computed as follows:

- Return (%) $= \left(\dfrac{Z_1}{C} \right) 100\%$ (7.24)

- Risk (%) $= \left(\dfrac{\sqrt{Z_2}}{C} \right) 100\%$ (7.25)

where Z_1 and Z_2 are given by Equations 7.19 and 7.20, respectively.

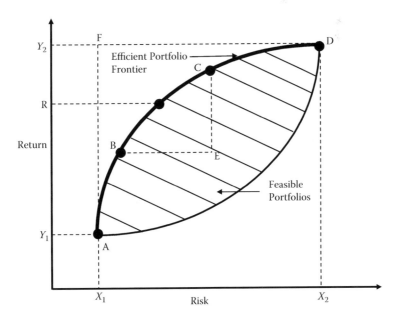

FIGURE 7.1
Risk–return graph for portfolio selection.

The efficient portfolios are defined by the points on the upper boundary A–B–C–D, shown in bold. It is called the efficient portfolio frontier. Any feasible portfolio, which is not in the efficient frontier, is a dominated portfolio, since it can always be improved with respect to return, risk, or both by another feasible portfolio. For example, a rational investor would always prefer efficient portfolios B and C over the portfolio E (see Figure 7.1) because portfolio B gives the same return for a lower risk, while portfolio C produces a higher return for the same risk!

Portfolio A in Figure 7.1 represents the most conservative portfolio, which produces the lowest return (Y_1) with minimum risk (X_1). On the other hand, portfolio D is the most aggressive portfolio, which produces the highest return possible, but with the maximum risk (X_2). Point F in Figure 7.1 corresponds to the point representing the minimum risk (X_1) and the maximum return (Y_2) for the portfolio selection problem. It is called the *ideal solution*, but it is *not achievable*, because point F is not feasible. Points A and D in Figure 7.1 can be used to set bounds on the feasible values of the return and risk as follows:

- Bounds on return: (Y_1, Y_2)
- Bounds on risk: (X_1, X_2)

To generate the efficient portfolio frontier (curve A–B–C–D), we can solve a sequence of single objective quadratic programming problems as follows:

$$\text{Minimize } Z_2 = \sum_{i=1}^{N}\sum_{j=1}^{N} \sigma_{ij}^2 X_i X_j = X^T Q\ X$$

Subject to

$$\sum_j \mu_j X_j = R \qquad (7.26)$$

+ other constraints given by Equations 7.21 through 7.23.

The optimal solution to the above problem gives the minimum risk portfolio for a given expected return R. This portfolio will be an efficient portfolio. By changing the value of R between Y_1 and Y_2 in Equation 7.26, we can generate the entire efficient portfolio frontier for the bi-criteria portfolio selection problem. It is from this set of efficient portfolios that a rational investor has to choose the "best" portfolio consistent with his or her investment philosophy and risk tolerance.

7.3.5 Use of Portfolio Mean and Standard Deviation

When an investor selects the "best" portfolio (p) from the risk–return graph (Figure 7.1), he has the expected annual return of the portfolio (μ_p) and its

Financial Engineering

risk, measured by its standard deviation (σ_p). If σ_p is large, then the actual returns will vary widely year to year. Since the actual return for any given year is the weighted sum of the returns of individual securities in the portfolio, it can be assumed to have a Normal distribution with mean μ_p and variance σ_p^2. Using the properties of the Normal distribution, we can then set probability estimates on the annual returns. For example, there is a 68% probability that the yearly returns will be within one standard deviation from the mean. Since an investor is mainly concerned with the returns that are lower than the mean, we can say that there is an 84% probability that the actual returns will be at least ($\mu_p - \sigma_p$). Thus, we can get probability estimates on the yearly returns as follows:

- 84% chance that the returns will be at least ($\mu_p - \sigma_p$)
- 97.5% chance that the returns will be at least ($\mu_p - 2\sigma_p$)
- 99% chance that the returns will be at least ($\mu_p - 3\sigma_p$)

For example, for a portfolio with an average return of 15% and a standard deviation of 5%, we can make the following predictions about the year-to-year performance of the portfolio:

- 68% chance that the returns will fall within 10% to 20%
- 84% chance that the return will be ≥10%
- 95% chance that the returns will be within 5% to 25%
- 99% chance that the returns will be within 0% to 30%
- Near-zero chance the annual return will ever be negative

7.3.6 Illustration of Markowitz' Model

We shall illustrate Markowitz's mean-variance model and the construction of the risk–return graph with a numerical example.

Example 7.2* (Markowitz's bi-criteria model)

An investor can invest in three securities. Based on the past 15 years of performance data, the means and standard deviations of the annual returns of the three securities have been computed as shown in Table 7.6. Note that security 1 has the highest average return of 14% per year, but its standard deviation of 20% makes it the most risky investment. The three securities are also positively correlated as shown in Table 7.6.

---

* Adapted from Winston and Albright (2001).

TABLE 7.6
Data on Security Returns and Correlations between Security Returns (Example 7.2)

Securities	Security Returns Mean	Standard Deviation
1	0.14	0.20
2	0.11	0.15
3	0.10	0.08
		Correlation Coefficient
Securities 1 and 2		0.6
Securities 1 and 3		0.4
Securities 2 and 3		0.7

The investor has $100,000 to invest and the investment portfolio should satisfy the following constraints:

i. No more than 50% should be invested on any one security.
ii. Invest all the $100,000 among the three securities.
iii. Invest at least $10,000 each in securities 1 and 3.

The objective of the investor is to achieve maximum portfolio return and minimum portfolio variance.

a. Formulate the above problem as a bi-criteria optimization problem.
b. Reformulate the problem if the objective were to select the minimum variance portfolio that yields an average portfolio return of at least 12% and determine the optimal portfolio.
c. Generate the risk–return graph and identify the efficient portfolio frontier.

SOLUTION TO PART (A)
Decision Variables

Let X_1, X_2, X_3 denote the dollars invested in securities 1, 2, and 3, respectively.

Constraints

The portfolio constraints can be written as follows:

i. No more than 50% investment in any one security:

$$0 \le X_1 \le 50,000 \qquad (7.27)$$

$$0 \le X_2 \le 50,000 \qquad (7.28)$$

$$0 \le X_3 \le 50,000 \qquad (7.29)$$

ii. Invest all the money available:

$$X_1 + X_2 + X_3 = 100,000 \tag{7.30}$$

iii. Minimum investment in securities 1 and 3:

$$X_1 \geq 10,000 \tag{7.31}$$

$$X_3 \geq 10,000 \tag{7.32}$$

Objective Functions

There are two conflicting objectives to the selection of the optimal portfolio. Objective 1 is to maximize the expected annual return of the portfolio, denoted by Z_1. Using the data given in Table 7.6, objective 1 becomes

$$\text{Maximize } Z_1 = 0.14X_1 + 0.11X_2 + 0.1X_3 \tag{7.33}$$

Objective 2 in the Markowitz's model is to minimize the variance of the portfolio return, given by Equation 7.20. It requires the calculation of the variance–covariance matrix $Q_{(3\times3)}$, whose diagonal elements are the variances of the security returns and the off-diagonal elements are the covariances between the security returns. The variances of the three security returns can be calculated from their standard deviations given in Table 7.6 as 0.04, 0.0225, and 0.0064, respectively. Using the formula given by Equation 7.18, the covariances between security returns can be calculated from the correlation coefficients and the standard deviations given in Table 7.6. The covariance between securities i and j (σ_{ij}^2) is given by

$$\sigma_{ij}^2 = \rho_{ij}(\sigma_i)(\sigma_j),$$

where ρ_{ij} is the correlation coefficient and σ_i and σ_j are their standard deviations.
Thus,

$$\sigma_{12}^2 = (0.6)(0.2)(0.15) = 0.018$$

$$\sigma_{13}^2 = (0.4)(0.2)(0.08) = 0.0064$$

$$\sigma_{23}^2 = (0.7)(0.15)(0.08) = 0.0084$$

The variance–covariance matrix is given by

$$Q_{(3\times3)} = \begin{bmatrix} 0.04 & 0.018 & 0.0064 \\ 0.018 & 0.0225 & 0.0084 \\ 0.0064 & 0.0084 & 0.0064 \end{bmatrix}.$$

Thus, objective 2, which is to minimize the variance of the portfolio return, is given by

$$\text{Minimize } Z_2 = (X_1, X_2, X_3) Q_{(3\times3)} \begin{pmatrix} X_1 \\ X_2 \\ X_3 \end{pmatrix}$$

$$= 0.04X_1^2 + 0.0225X_2^2 + 0.0064X_3^2 + 0.036X_1X_2 + 0.0128X_1X_3 + 0.0168X_2X_3 \tag{7.34}$$

The final bi-criteria model of Markowitz is to maximize the portfolio's average return Z_1 (Equation 7.33) and minimize the portfolio's variance Z_2 (Equation 7.34), subject to the constraints given by Equations 7.27 through 7.32.

Note that the objective function Z_1 and the constraints are linear functions, while the objective function Z_2 is a quadratic function.

SOLUTION TO PART (B)

Here, we are asked to find the minimum variance portfolio that will yield at least a 12% average annual return. Thus, objective Z_1 (Equation 7.3.3) becomes a constraint in this model as follows:

$$(0.14)X_1 + (0.11)X_2 + (0.1)X_3 \geq 12,000. \tag{7.35}$$

The model becomes a single-objective optimization problem, which is to minimize Z_2, subject to the constraints, given by Equations 7.27 through 7.32 and 7.35. Since the optimization model has a quadratic objective function with linear constraints, it is called a *quadratic programming* problem. Markowitz (1956) also proposed a specialized algorithm to solve the quadratic programming model. With the use of the optimization software Excel Solver, the optimal portfolio to the quadratic programming model is obtained as $X_p = (X_1, X_2, X_3) = (50,000, 0, 50,000)$, that is, split the money evenly between securities 1 and 3. Substituting the optimal values of (X_1, X_2, X_3) in Equations 7.33 and 7.34, we get the average return of the portfolio (Z_1) and its variance (Z_2) as follows:

$$Z_1 = (0.14)(50,000) + (0.1)(50,000) = 12,000$$

$$Z_2 = (0.04)(50,000)^2 + (0.0064)(50,000)^2 + (0.0128)(50,000)^2 = (0.0592)(50,000)^2.$$

Financial Engineering

Using Equations 7.24 and 7.25, we can convert the portfolio return and portfolio risk in percentages, where risk is measured by the standard deviation of the portfolio return as follows:

- Return (%) $= \left(\dfrac{12,000}{100,000} \right) 100\% = 12\%$

- Risk (%) $= \left(\dfrac{\sqrt{0.0592}\,(50,000)}{100,000} \right) 100\% = 12.17\%$

As discussed in Section 7.3.4, this portfolio, with a return of 12% and a risk of 12.17%, is an efficient portfolio. In other words, there does not exist a feasible portfolio that can yield a 12% return with a lower risk. Similarly, there is no feasible portfolio with a higher return for the same risk of 12.17%.

SOLUTION TO PART (C)

To generate the risk–return graph (see Figure 7.1), we need to determine the ideal solution to Example 7.2. The ideal solution corresponds to the portfolio with the maximum return (ignoring the variance) and the minimum variance (ignoring the return). It can be computed by solving two single-objective optimization problems as follows:

Problem P1: Maximize Z_1 (Equation 7.33) subject to the constraints (Equations 7.27 through 7.32).
Problem P2: Minimize Z_2 (Equation 7.34) subject to the constraints (Equation 7.27 through 7.32).

The portfolio returns and risk (standard deviations) in percentages, corresponding to the optimal solutions to problems P_1 and P_2, are given in Table 7.7.

Note that the portfolio with the maximum return of 12.4% also has the highest risk of 14.89%. Similarly, the portfolio with the lowest risk

TABLE 7.7

Ideal Solution (Example 7.2)

Problem	Return (%)	Risk [STD] (%)
P1 (max return)	12.4	14.89
P2 (min variance)	10.87	10.5

of 10.5% also has the lowest return of 10.87%. Thus, the bounds on the return and risk are as follows:

$$\text{Return} = (10.87\%, 12.4\%)$$

$$\text{Risk} = (10.5\%, 14.89\%)$$

Note that the ideal solution, which represents the maximum return and the minimum risk, is given by (12.4%, 10.5%). However, the ideal solution is not achievable because the two objectives are conflicting.

To generate the risk–return graph, we vary the portfolio return by choosing return values between 10.87% and 12.4% and resolving the single-objective quadratic programming problem given in part (b). Each quadratic program to be solved will have the same objective function Z_2 to minimize (Equation 7.34) and the constraints (Equations 7.27 through 7.32). However, the right-hand side of the constraint (Equation 7.35) will be varied from $10,870 to $12,400. The efficient portfolios, obtained by solving a sequence of quadratic programs for different values of return, are given in Table 7.8. The risk–return graph is illustrated in Figure 7.2. The efficient portfolio frontier is identified by the bold line.

TABLE 7.8

Efficient Portfolios (Example 7.2)

Return ($)	Return (%)	Risk (%)
12,400	12.40	14.89
12,000	12.00	12.17
11,750	11.75	11.54
11,500	11.50	11.05
11,250	11.25	10.70
11,000	11.00	10.53
10,870	10.87	10.50

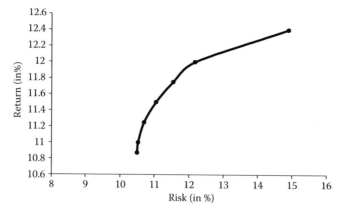

FIGURE 7.2
Risk–return graph (Example 7.2).

A "conservative" investor will choose an efficient portfolio from the lower part of the risk–return graph. An "aggressive" investor will choose an efficient portfolio from the upper part of Figure 7.2.

7.4 Sharpe's Bi-Criteria Model for Portfolio Selection

The major drawback of Markowitz's mean-variance model is that it requires a lot of computations to determine the variance–covariance matrix Q. When there are n securities, the determination of Q will require the calculation of $\frac{n(n+1)}{2}$ matrix elements. For example, for 1000 securities, it would require half a million calculations! In addition, to generate the efficient frontier shown in Figure 7.1, a sequence of quadratic programming problems have to be solved. Computing power in the 1950s was very limited and expensive. Only large universities and corporations had computers. Hence, the use of Markowitz's model in the financial industries was very limited at that time.

7.4.1 Market Risk of a Security

In 1963, William Sharpe simplified Markowitz's bi-criteria model by developing the concept of *market risk* of securities (Sharpe 1963). He showed that interrelationships of security returns (correlations) can be captured by using a linear relationship between the individual security return and the return of the U.S. Stock Market. Thus, an investment problem with n securities will require just n computations, as opposed to $\frac{n(n+1)}{2}$ in Markowitz's model.

Using the market risk of security, which he called the *Beta risk*, Sharpe developed a bi-criteria LP model for the portfolio selection problem. Thus, only a sequence of LP models has to be solved to generate the efficient portfolio frontier. He showed that 2000 securities can be analyzed at extremely low computing cost, namely, as little as 2% of the computational cost that would be required for the bi-criteria quadratic programming model of Markowitz!

Even though Sharpe's bi-criteria LP model was attractive to the investing community during the last three decades of the twentieth century, it is no longer true in the twenty-first century. As predicted by Moore's law, computing power has increased exponentially, along with an exponential decrease in cost. For example, consumers now have more computing power in their smartphones than those in the big mainframe computers used by universities and big corporations in the 1960s!

Hence, Markowitz's bi-criteria model has once again become attractive in the financial industries.

7.4.2 Meaning of the S&P 500 Index

Sharpe's market risk (Beta risk) requires the calculation of the average return of the U.S. Stock market. Standard & Poor's (S&P) 500 index is commonly used by the financial industries to gauge the return of the U.S. Stock Market. Before discussing the S&P 500 index, we introduce the concept of market capitalization of a public traded company.

The *market capitalization* of a company, also known as the *market cap* of a company, is given by

$$\text{Market cap} = (\text{Price per share})\ (\text{Number of outstanding shares}). \quad (7.36)$$

Depending on the market cap values, companies are classified as large, mid-size, or small companies as given below:

- Large cap: market cap > $6 billion
- Mid cap: market cap between $1 billion and $6 billion
- Small cap: market cap < $1 billion

For example, Apple Inc. is generally considered the most valuable company in the world. Its market cap reached $800 billion on May 15, 2017!

The S&P 500 index is the market cap weighted average of the stock prices of the 500 large U.S. companies. The index is calculated as follows:

$$\text{S \& P 500 Index} = \frac{\sum_{i=1}^{500}(MC_i)(P_i)}{\sum_{i=1}^{500} MC_i},$$

where
MC_i = market cap of company i
P_i = price per share of company i

The total market cap of the S&P 500 companies reached $21 trillion during May 2017 and represents more than 75% of the U.S. Stock Market. Hence, the change in the S&P 500 index over time is considered as a good barometer of how the U.S. Stock Market is performing and is used in the calculation of Sharpe's Beta risk (market risk) of a stock security.

In the determination of the Beta risk, the Beta value for the S&P 500 index is set to one. If the Beta value of security j (β_j) is 1.3, then security j is considered to be 30% more volatile than the U.S. Stock Market. In other words, when the S&P index returns are increasing ("up" market cycles, called the "Bull Markets"), the return of security j will be 30% more than the average market return. However, when the stock market is "down" with negative

Financial Engineering

returns (called the "Bear Markets"), the return of stock *j* will be 30% worse off than the market return. Thus, a Beta value of more than one indicates that the stock's return is more volatile and hence risky. On the other hand, a stock with a Beta of 0.7 indicates that it may not produce much higher returns during the "bull markets," but it will protect the investor from large losses during the "bear" markets!

7.4.3 Calculating Beta Risk

The Beta value of a security can be computed by comparing the monthly returns of that security with those of the S&P 500 index over several months as illustrated in Figure 7.3. The *x*-axis represents the monthly returns of the S&P 500 index. The *y*-axis represents the corresponding monthly returns of the security under consideration. Usually, the last 36 months of returns are used in estimating the Beta risk. Using a linear regression model, a linear equation is then fitted as given below (see Figure 7.3 also):

$$R = \alpha + \beta R_M \tag{7.37}$$

where
 R_M = return of the S&P 500 index
 R = security return under consideration

The slope of Equation 7.37 is the Beta risk and the intercept at the *y*-axis is called "Alpha" of the security.

While Beta gives the direct relationship between the security return and the market return, Alpha measures the performance of that security, independent of the market's performance. Thus, Beta measures volatility with respect to the market and Alpha measures the performance of a particular security that is independent of the market. Higher values of Alpha are

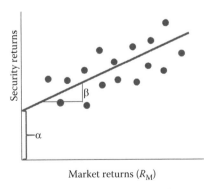

FIGURE 7.3
Plot of security returns against market returns.

better for investing. A positive value of Alpha provides higher returns than expected in the bull markets and a cushion protecting that security from bigger losses during the bear markets. For example, consider a stock with a Beta of 1.0. If the stock earns 12% and the S&P 500 index returns 8%, then its Alpha is 4%. We shall now present the bi-criteria LP model proposed by Sharpe (1963) for portfolio selection.

7.4.4 Sharpe's Bi-Criteria Model

Sharpe's bi-criteria model for portfolio selection is essentially similar to the bi-criteria model of Markowitz, discussed in Section 7.3.3 (Equations 7.19 through 7.23). The only difference is in the second objective function (Z_2), which is to minimize the portfolio risk. In the Markowitz model, portfolio risk is measured by the variance of the portfolio return (Equation 7.20). In Sharpe's model, portfolio risk is measured by the portfolio's Beta risk, which is computed as follows:

$$\text{Minimize } Z_2 = \frac{\left(\sum_{j=1}^{N} \beta_j X_j\right)}{C} \tag{7.38}$$

Note that objective Z_2 represents portfolio's "Weighted Beta Risk" and is a linear function. A portfolio with the weighted Beta of more than one will be considered a high-risk portfolio. Since the total investment C is a constant, it can be ignored in the optimization model. Thus, Sharpe's bi-criteria LP model can be stated as follows:

Maximize the average annual return of the portfolio

$$Z_1 = \sum_{j=1}^{N} \mu_j X_j, \tag{7.39}$$

Minimize the weighted Beta risk of the portfolio

$$Z_2 = \sum_{j=1}^{N} \beta_j X_j. \tag{7.40}$$

Subject to the constraints:

$$\sum_{j=1}^{N} X_j \leq C; \ X \geq 0 \tag{7.41}$$

Financial Engineering

$$\sum_{j \in J_1} X_j \leq b_1 \qquad (7.42)$$

$$\sum_{j \in J_2} X_j \geq b_2, \qquad (7.43)$$

where
X_j = dollars invested in security j
μ_j = average annual return per dollar invested in security j
β_j = Beta risk of security j
N = total number of securities considered for investment
C = total capital available for investment

Since the objective functions Z_1 and Z_2 are conflicting, there does not exist a portfolio that simultaneously optimizes both objectives. The definitions of dominated portfolios, efficient portfolios, and the efficient portfolio frontier are the same as those discussed in Section 7.3.3 for the Markowitz model, except for the fact that the portfolio risk is now measured by the portfolio's Beta risk. The entire set of efficient portfolios and the risk–return graph can be generated by solving the following single objective linear program:

$$\text{Minimize } Z_2 = \sum_{j=1}^{N} \beta_j X_j$$

Subject to

$$\sum_{j=1}^{N} \mu_j X_j = R \qquad (7.44)$$

+ constraints given by Equations 7.41 through 7.43, where R, the average return of the portfolio, is varied between its minimum and maximum values. The bounds on R can be obtained from the ideal solution to the bi-criteria model. The ideal solution represents the maximum portfolio return possible ignoring the Beta risk and the minimum Beta risk possible ignoring the portfolio return. Since the objectives Z_1 and Z_2 conflict with each other, the ideal solution is not achievable.

7.4.5 Illustration of Sharpe's Model

We shall illustrate Sharpe's bi-criteria LP model and the construction of the risk–return graph by extending the mutual fund selection problem we discussed in Example 7.1, with the addition of the Beta risk of the mutual funds.

Example 7.3 (Sharpe's bi-criteria model)

Mary has $100,000 to invest in mutual funds. She has identified 10 funds for possible investment. The performance data for the funds are given in Table 7.9. Refer to Example 7.1 for a description of growth, growth and income, and income funds. Even though the investor is interested in the "growth" of her portfolio, she has the following investment restrictions:

i. All the money should be invested.
ii. No more than $20,000 should be invested in any one fund.
iii. Between 20% and 40% should be invested in growth funds.
iv. At least $5000 should be available as ready cash in the money market fund.
v. At least 10% should be invested in income funds.
vi. At least 15% should be invested in growth and income funds.

    a. Formulate the problem of determining the optimal investment portfolio as a bi-criteria linear programming model, if the objectives were to maximize portfolio return and minimize portfolio's Beta risk.
    b. Determine the ideal solution and the upper and lower bounds on the portfolio return and risk.
    c. Determine the optimal portfolio that will give at least 12% return under minimum risk.
    d. Generate all the efficient portfolios for Mary.

TABLE 7.9

Mutual Fund Data for Example 7.3

Type of Fund	Average Annual Return (%)	Beta Risk
Growth funds		
Fund 1	30	1.4
Fund 2	20	1.2
Fund 3	15	1.0
Fund 4	25	1.3
Growth and income funds		
Fund 5	16	0.9
Fund 6	17	0.95
Fund 7	12	0.75
Income funds		
Fund 8	8	0.6
Fund 9	6	0.4
Money market fund		
Fund 10	4	0.0

Financial Engineering

SOLUTION TO PART (A)
Let X_j = dollars invested in fund j, for $j = 1, 2, \ldots, 10$.

Objective Functions
Maximize the average annual return of the portfolio

$$Z_1 = \frac{(30X_1 + 20X_2 + 15X_3 + 25X_4 + 16X_5 + 17X_6 + 12X_7 + 8X_8 + 6X_9 + 4X_{10})}{100,000}$$

(7.45)

Minimize weighted Beta risk of the portfolio

$$Z_2 = \frac{(1.4X_1 + 1.1X_2 + X_3 + 1.3X_4 + 0.9X_5 + 0.95X_6 + 0.75X_7 + 0.6X_8 + 0.4X_9 + 0.4X_{10})}{100,000}$$

(7.46)

Constraints

i. $\sum_{j=1}^{N} X_j = 100,000$ (Invest all the money)

ii. $0 \le X_j \le 20,000$ (Invest no more than $20,000 in any one fund)

iii. $20,000 \le \sum_{j=1}^{4} X_j \le 40,000$ (Invest 20% to 40% in growth funds)

iv. $X_{10} \ge 5,000$ (At least $5000 in cash)

v. $X_8 + X_9 \ge 10,000$ (At least $10,000 in income funds)

vi. $X_5 + X_6 + X_7 \ge 15,000$ (At least 15% in growth and income funds)

SOLUTION TO PART (B)
To determine the ideal solution, we have to solve two single-objective LP problems as follows:

LP-1: Maximize Z_1 (Equation 7.45), ignoring Z_2, subject to constraints (i) through (vi).
LP-2: Minimize Z_2 (Equation 7.46), ignoring Z_1, subject to constraints (i) through (vi).

The optimal solutions to LP-1 and LP-2 are given in Table 7.10. The optimal solution to LP-1 is an efficient portfolio with the maximum return of 19.2%, but also has the maximum Beta risk of 1.01. The optimal solution to LP-2 is the minimum risk efficient portfolio, with the lowest Beta risk of 0.55, but also has the lowest portfolio return of 9%. Thus, the ideal

TABLE 7.10

LP Optimal Solutions (Example 7.3)

Mutual Fund Type	Decision Variable	LP-1	LP-2	12% Return
Growth	X1	20K	0	20K
	X2	0	0	0
	X3	0	20K	0
	X4	20K	0	0
G&I	X5	20K	0	0
	X6	20K	0	0
	X7	5K	20K	20K
Income	X8	10K	20K	20K
	X9	0	20K	20K
Money market	X10	5K	20K	20K
Total portfolio		100K	100K	100K
Annual return (%)		19.20	9	12
Beta risk		1.01	0.55	0.63

solution (Z_1, Z_2) is (19.2%, 0.55) and the bounds on the portfolio return (Z_1) and portfolio risk (Z_2) are as follows:

$$9\% \leq Z_1 \leq 19.2\%$$

$$0.55 \leq Z_2 \leq 1.01$$

SOLUTION TO PART (C)

To determine the minimum risk portfolio that will provide 12% annual return, we have to solve the following LP problem:
 Minimize objective Z_2 (Equation 7.46).
 Subject to

- Constraints (i) through (vi)
- Z_1 (Equation 7.45) $\geq$ 12%

(7.47)

The optimal solution that gives 12% annual return at minimum risk is also given in Table 7.10. That portfolio's Beta risk is 0.63 and it is also an efficient portfolio.

SOLUTION TO PART (D)

To generate all the efficient portfolios, we have to solve a sequence of LP problems by varying the right-hand side of Equation 7.47 as follows:

Financial Engineering

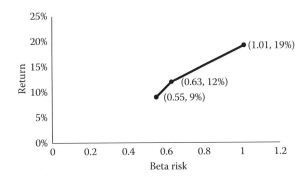

FIGURE 7.4
Risk–return graph (Example 7.3).

Minimize objective Z_2 (Equation 7.46).
Subject to

- Constraints (i) through (vi)
- Z_1 (Equation 7.45) $\geq R$, (7.48)

where R is percentage return required from the portfolio.

The LP optimal solutions for every value of R (Equation 7.48) between 9% and 19.2% will all be efficient portfolios. Figure 7.4 illustrates the efficient portfolio frontier for Example 7.3

7.5 Investing in Bonds

In this section, we focus on investing in bond securities. Bond is a debt security, a form of money borrowing. In other words, bond is an instrument of indebtedness of the bond issuer to the holders. In this, the issuer of the bond owes a debt to the bond holder and is obliged to pay the interest to the holder along with the principal amount on maturity. Bonds are generally used by corporations and governments whenever they need to raise capital. We will begin with a review of some of the key characteristics of bonds.

7.5.1 Understanding Bonds

Maturity Period: The date by which the issuer has to repay the principal amount (also known as face value) is called maturity period. For corporate bonds, maturity period ranges from 1 to 5 years. Maturity period for government bonds can range from short term (a few months) to long term (15 to

30 years). Bond issuers decide the maturity period, which then affects the bond's coupon (interest rate). If the maturity period is short, the company will have a high probability of repaying the bond amount, so the interest rate will be less.

Face Value: Face value is also known as "par value," which is the amount the issuer of the bond has to repay at the maturity date. For stocks, face value is the original cost of the stock. Therefore, face value is a poor indicator to calculate the actual worth of the stock. Bonds are quite different from stocks as the stockholders have equity stake in the company, whereas the bondholders have creditor stake in the company.

Coupon: The predetermined interest rate that the issuer agrees to pay to its bondholders is also known as coupon. It is calculated as the percentage of the face value. If the interest rate remains the same throughout the maturity period, then it is a fixed rate bond. If it varies over time, then it is a variable or adjustable or floating rate bond. There might also be the case when there is no interest paid on the bonds. These are called zero-coupon bonds. Here, the bonds are sold at a discounted price of their face values to be repaid at face value at the time of maturity, which is equivalent to the interest to be paid over the period of time. Zero-coupon bonds are issued by the U.S. government. Bond interests are generally paid quarterly, semi-annually, or annually.

Bond Price: The value of the bond changes with respect to changes in the prevailing interest rate in the economy. It is very important to note that the value of the bond varies inversely with the interest rate. This can be explained with an example below.

Example 7.4

Suppose an investor bought a bond of $100 with a fixed coupon rate of 10%, according to which it gets an annual interest of $10. After a while, due to changes in the economy, the prevailing interest rate in the market of such bonds decreases to 5%. However, the issuer of the old bond will continue to make coupon payments of $10 every year, which is clearly more valuable than any new bond issue, which will make coupon payments of $5 per year. Therefore, to compensate the effect of fluctuation of interest rate, the value of the old bond becomes $200 because its coupon payments of $10 will be in agreement with 5% of $200, the prevailing interest. Hence, a decrease in interest rate leads to increase in the value of the bonds and vice versa.

Yield: There is another parameter associated with the bonds known as *yield to maturity* (YTM). It is the interest rate at which the present value of the sum of all the coupon payments along with the final face value of the bond is exactly equal to the current price of the bond. It is used for long-term bonds and is calculated on an annual basis.

7.5.2 Price, Yield, and Coupons

Current price of a bond can be calculated using yield as follows:

$$\text{Bond Price} = \text{Coupon payment} * \frac{1 - \left(\frac{1}{(1+\text{interst rate})^n}\right)}{\text{interest rate}} \qquad (7.49)$$
$$+ \left[\text{Face value} * \frac{1}{(1+\text{interest rate})^n}\right]$$

In Equation 7.49, the parameter n denotes the remaining periods and interest rate is the yield to maturity. When yield is exactly equal to the coupon rate of a bond, such a bond is called a *par bond*.

Example 7.5

Consider a bond with a face value of $100, a coupon rate of 10%, and a maturity period of 30 years, that is, $10 will be paid once a year for the next 30 years. Suppose the prevailing interest rate or yield is 10%. Then using Equation 7.49, we obtain a bond price of $100. In this case, the coupon rate is the same as prevailing interest rate, so the price of the bond is the same as its par value.

Suppose, in Example 7.5, the coupon rate is 15% and the prevailing interest rate remains at 10%. In this case, the coupon payment will be $15 once a year for the next 30 years. Using Equation 7.49, we obtain a price of $147.13. This illustrates why bond prices change when prevailing interest rates change. After 1 year, parameter n becomes 29 and the bond price decreases to $146.84. Eventually, when n is 0, that is, there is no coupon payment remaining, only the face value matters, which is the final payment to be done, and the bond price comes down to $100. If we plot the bond price versus the time remaining until maturity, from the time of issue until it matures at the end of 30 years, we get a convex curve, called the *Price–Yield curve*, as shown in Figure 7.5. Price–yield curves are very important as this helps in getting a better understanding of the relationship between bond price, yield, and coupons. We illustrate this using Example 7.6.

Example 7.6

For a bond with a face value of $100 and a maturity period of 30 years, the change in the bond price for a varying yield and different coupon rates (0%, 5%, 10%, and 15%) is illustrated in Figure 7.6. We can see that for any specific coupon rate, the bond price decreases as the yield increases. Also, for a specific yield rate, the bond price increases as the coupon rate increases.

The maturity period of a bond also affects its price. To understand this relationship, suppose the face value of a bond is $100 and its coupon

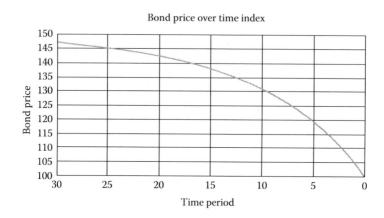

FIGURE 7.5
Change in bond price versus time remaining to maturity.

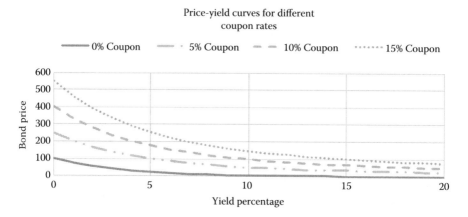

FIGURE 7.6
Change in bond price with varying yields for different coupon rates.

rate is 10%. Figure 7.7 illustrates the changes in the price of this bond for varying yield, for maturity periods of 10, 20, and 30 years. We can see that all the curves intersect at a point when the yield is 10%. This point is called the *par point*, when the yield is equal to the coupon rate. Also, the slope of the price–yield curve below the par point is steep, indicating that the bond prices are more sensitive to yield in this area. Furthermore, the price–yield curve gets steeper as maturity period increases, indicating that bonds with longer maturity periods are much more sensitive to yield below the par point.

Financial Engineering

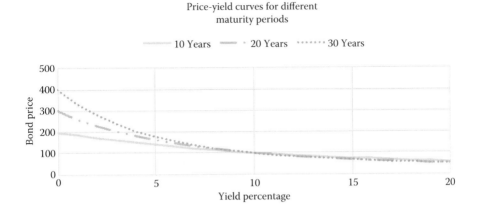

FIGURE 7.7
Change in bond price with varying yield for different maturity periods.

7.5.3 Types of Bonds

There are two major types of bonds, government bonds and corporate bonds, as explained below.

1. *Government Bonds*

    These bonds can be issued by the federal government (known as sovereign debt) and by the lower levels of government. U.S. government bonds are considered the safest bonds, which usually pay the lowest interest rates and these bonds are also considered as "risk-free." The U.S. government also issues zero-coupon bonds, known as *Treasury Strips*, in which there is no coupon; instead, they are sold at discounted rates, which can be redeemed at higher values at the time of maturity. Bonds issued by lower levels of governments, like state government or local government agencies, are called municipal bonds or *"munis"* for short and are considered riskier than the bonds issued by the federal government. The main advantage of a municipal bond is that the interest income may be free from federal income tax. Moreover, state and local governments may treat the interest incomes from their bonds non-taxable for the residents of the state, thus making some municipal bonds completely tax free, sometimes called as triple tax free. Therefore, the interest rates of municipal bonds will be comparatively less than those of the other bonds, because of their tax advantages.

2. *Corporate Bonds*

    Corporate bonds constitute the major share of the overall bond market. Companies can issue bonds of different coupons and different maturity periods. Based on the maturity period, bonds are of three types: short term (1–5 years), intermediate (6–14 years), and long term (15–30 years). These bonds have higher yields than the government bonds because they are riskier. Bonds associated with highly rated companies are known as "investment grade" bonds, while those below investment grade are called "high-yield" or "junk" bonds.

7.5.4 Bond Ratings

Bond rating is the grade given to bonds by rating organizations to indicate their credit quality. Some bonds may be subject to default due to financial problems of the issuer. Therefore, bond ratings are very important, which alert the investors about the quality and stability of the bonds. There are three major financial institutions, which publish the ratings of the bonds, Moody's, S&P, and Fitch. They are shown in Table 7.11.

Higher rated investment grade bonds are less risky. Hence, the bond issuer can pay a lower interest rate and still attract buyers. However, the lower rated speculative grade junk bonds, which are riskier, have to pay higher coupon rates in order to attract buyers. Because of the historically low interest rates in recent years, junk bonds have been purchased by investors to increase the return of their portfolios, even though they are very risky!

7.5.5 Bond Duration

Duration is the first derivative of how the bond price changes with respect to the changes in the interest rate. It is stated as number of years. There are many factors that affect the duration, such as coupon rate, yield, and

TABLE 7.11

Bond Ratings from Various Rating Agencies

Bond Ratings	Moody's	Standard & Poor's	Fitch	Risk	Grade
High grade	Aaa	AAA	AAA	Highest quality	Investment
	Aaa	AA	AA	High quality	Investment
Medium grade	A	A	A	Strong	Investment
	Baa	BBB	BBB	Medium grade	Investment
Speculative grade	Ba,B	BB, B	BB, B	Speculative	Junk
	Caa,Ca,C	CCC/CC/C	CCC/CC/C	Highly speculative	Junk
Default danger	C	D	D	In default	Junk

Financial Engineering

remaining time to maturity. There are various scenarios, which show how these factors are interrelated and affect the duration of the bond:

- If the coupon and yield are same, duration increases with time left to maturity.
- If maturity and yield are same, duration increases with a lower coupon.
- If the coupon and maturity are the same, duration increases with a lower yield.

Duration (D) can be calculated using the following formula:

$$D = \frac{\sum_{1}^{n} \frac{t \cdot CF_t}{(1+i)^t}}{P}, \qquad (7.50)$$

where i is the discount rate, CF is cash flow or the coupon amount, and P is the present value of the bond. Example 7.7 illustrates the calculations for computing the bond duration.

Example 7.7

Consider a $100 bond with a 5% coupon rate that is paid semiannually and maturity of three years. The bond will pay $2.50 every 6 months to the bond holder. The resulting cash flows are illustrated in Figure 7.8. Assuming a 7% discount rate, the bond duration can be determined as shown in Table 7.12, using Equation 7.50.

Duration is always less than the maturity period. If the coupon rate is zero, then the duration is equal to the maturity period.

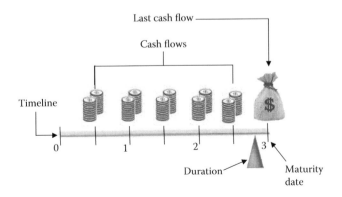

FIGURE 7.8
Weighted cash flows for Example 7.7.

TABLE 7.12

Calculating Bond Duration for Example 7.7

A Year	B Payments	C Discount Factor (at 7%)	D Present Value (B*C)	E (A*D)	Duration (E/D)
0.5	2.5	0.967	2.417	1.208	2.817
1	2.5	0.935	2.336	2.336	
1.5	2.5	0.903	2.259	3.388	
2	2.5	0.873	2.184	4.367	
2.5	2.5	0.844	2.111	5.277	
3	102.5	0.816	83.671	251.012	
Sum			94.977	267.589	

7.5.6 Bond Convexity

Bond convexity is the second derivative of the bond price with respect to the changes in the interest rate. Duration, being the first derivative, provides a linear measure, whereas convexity measures the curvature. Convexity is used as a tool for risk management. Duration is a measure of a bond's price sensitivity to small fluctuations in interest rates, whereas convexity is its measure of price sensitivity to large fluctuations in interest rates. Whenever the combined convexity and duration of a bond are high, so is the risk. Likewise, when the combined convexity and duration are low, there will be very little change in a bond's price even though there are fairly substantial interest rate movements.

In Figure 7.9, we can see that bond A has higher convexity than bond B. Therefore, the price of bond A will always be greater than the price of bond B, with everything else being equal, whether the interest rate increases or decreases. Generally, coupon rate and convexity are inversely related to each other. The higher the coupon rate, the lower the convexity of the bond and vice versa.

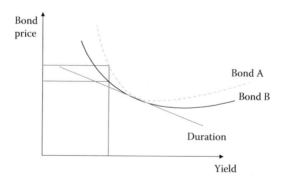

FIGURE 7.9
Convexity of bonds.

7.5.7 Immunization

Immunization is a strategy to ensure that a change in interest rate does not affect the net worth of the bond portfolio. This is used to construct a portfolio that has no interest rate risk. There are various methods that can be used for immunization, such as cash flow matching and duration matching, as explained below:

1. Cash flow matching

    This is probably the easiest way of immunization. For example, a company is obliged to pay $10,000 to a person in 10 years. Therefore, to immunize itself to this cash outflow, it can buy a zero-coupon bond, which has a value of $10,000 at its maturity period of 10 years. By doing this, any change in interest rate will not affect the ability to pay, as there is no interest associated with zero-coupon bond and it matches the total cash inflow and outflow.

2. Duration matching

    This is a more practical method of immunization. In this, the company matches the duration of its assets to the duration of its liabilities. This can be done in various ways. For the situation mentioned above, where a company is obliged to pay $10,000 in 10 years, there are various ways in which it can immunize itself using duration matching. First, it can buy zero-coupon bonds with a maturity period of 10 years and value of $10,000 at maturity period. Second, it can buy many bonds with a maturity period of 10 years and a total worth of $10,000. Third, it can buy many bonds, which has a total worth of $10,000 but an average duration of 10 years when viewed together.

7.5.8 Selecting a Bond Portfolio

The key strategy of any investor is always to maximize its profits and minimize the risks. Portfolio allocation is the financial strategy of investing money to balance risk and reward of the portfolio's assets according to the individual's goal, risk strategy, and investment horizon. As discussed in Section 7.1.4, asset allocation determines the amount of funds invested in equities, bonds, and cash. Selecting the individual securities for investment in bonds is illustrated in this section.

There are many options of investing in different types of bonds as given below:

- Treasury bonds
- Other U.S. government bonds
- Investment-grade corporate bonds

- High-yield corporate bonds (low quality bonds)
- International bonds
- Mortgage-backed bonds
- Municipal bonds

Now, there are four major strategies for managing the bond portfolio.

1. Passive Bond Strategy

    The main purpose of this strategy is to focus on the income-generating properties of the bonds. This strategy assumes that bonds are safe investments and predictable sources of income. Cash flow can be used for reinvestment in other bonds or other classes of assets. In this strategy, the future interest rates are not speculated. Moreover, the changes in current bond price due to change in interest rate are also not important. This strategy is effective in rough financial times.

2. Indexing Bond Strategy

    This is also known as quasi-passive strategy. Though it has many characteristics of passive bond strategy, it is more flexible. The bond portfolio under this strategy can imitate any published bond index. This strategy can be used easily by investing in bond index mutual funds described in Section 7.6.3.5.

3. Immunization Bond Strategy

    Immunization bond strategy is the most reliable strategy. It is the blend of both active and passive strategies. In the pure immunization strategy, bond portfolio does not depend on external factors such as changes in interest rates. Investments are done for a specific return and specific duration. The best suited bonds for this strategy are high-quality bonds with the least possibility of default. It is most commonly used by insurance companies, banks, and pension funds to match their cash flows with future liabilities.

4. Active Bond Strategy

    The objective of this strategy is to maximize the returns. But with maximum returns comes increased risks. In this strategy, forecasting is done about the future interest rates under different market financial scenarios. Based on that, decisions are taken, rather than opting for the passive strategy. This is not a feasible strategy for the individual investors.

Financial Engineering

Example 7.8

Consider the interest rate data (annualized in percentages) given in Table 7.13 for different types of bonds for the past 1 year (July 2016 to July 2017), along with the forecast of future interest rate. In Table 7.13, EE bonds and Inflation Protected bonds (I-bonds) are considered as U.S. savings bonds. The problem is to build a bond portfolio with maximum return and minimum risk, under the following constraints:

- Total investment is $1,000,000.
- Invest at least 10% in corporate bonds.
- Invest at least 20% in government securities.
- Invest at least 10% in municipal bonds.
- The purchasing limit of each U.S. saving bonds (EE bonds, I-bonds) is $10,000.
- The purchasing limit for 30-year fixed mortgage bonds is $400,000.

MODEL FORMULATION

We will use Markowitz's bi-criteria model, discussed in Section 7.3, to build the bond portfolio.

Let y_i be the actual money ($) invested in each kind of bond respectively for $i = 1$ to 6.

TABLE 7.13

Bond Data for Example 7.8

Timeline	Corporate	EE Bonds	I Bonds	Treasury	Municipal	Mortgage
Jul-16	4.09	0..1	0.1	2.23	2.57	3.44
Aug-16	4.10	0.1	0.1	2.26	2.77	3.44
Sep-16	4.22	0.1	0.1	2.35	2.86	3.46
Oct-16	4.31	0.1	0.1	2.5	3.13	3.47
Nov-16	4.63	0.1	0.1	2.86	3.36	3.77
Dec-16	4.75	0.1	0.1	3.11	3.81	4.2
Jan-17	4.59	0.1	0.1	3.02	3.68	4.15
Feb-17	4.63	0.1	0..1	3.03	3.74	4.17
Mar-17	4.68	0.1	0.1	3.08	3.78	4.2
Apr-17	4.58	0.1	0.1	2.94	3.54	4.05
May-17	4.54	0.1	0	2.96	3.47	4.01
Jun-17	4.32	0.1	0	2.8	3.06	3.9
Jul-17	4.42	0.1	0	2.88	3.03	3.97
Forecast	4.45	0.1	0	2.95	3	4

Objective functions

Maximize portfolio return:

$$\text{Max } Z_1 = (4.45y_1 + 0.1y_2 + 0y_3 + 2.95y_4 + 3y_5 + 4y_6)/100$$

Minimize Portfolio variance (Risk):

$$\text{Min } Z_2 = \sum_{i=1}^{6} \sigma_i^2 y_i^2 + \sum_{i=1}^{6}\sum_{j=1}^{6} q_{ij} y_i y_j \quad \text{for } i \neq j$$

where σ^2 is the variance of each bond return and q_{ij} is the covariance of returns between bonds i and j.

NOTE: Forecast returns are used for objective Z_1 and past data are used to estimate the variance–covariance matrix elements, using Equations 7.15 and 7.16, for objective Z_2.

Constraints

$y_1 + y_2 + y_3 + y_4 + y_5 + y_6 = 1,000,000$ (Total investment of 1,000,000)
$y_1 \geq 100,000$ (Corporate bonds at least 10%)
$y_2 + y_3 + y_4 \geq 200,000$ (Government securities at least 20%)
$y_5 \geq 100,000$ (Municipal bonds at least 10%)
$y_i \leq 10,000$ for $i = 2$ and 3 (Purchasing limit of each U.S. savings bonds is 10,000)
$y_6 \leq 400,000$ (Mortgage limit is 400,000)

When we optimize the model for maximum return irrespective of the risk, we get a return of 4.005%* with a risk of 0.066%.* Table 7.14 illustrates the optimal allocation. Since U.S. savings bonds have the lowest returns, the model does not allocate any amount in the optimal solution. Corporate bonds have the maximum return, so the model allocates the largest share of the portfolio to corporate bonds, while satisfying the other constraints.

On the other hand, when we optimize the model for minimum risk irrespective of the return, we get a risk of 0.063%, with a return of 3.947%. Table 7.15 illustrates the optimal allocation. Savings bonds have the minimum fluctuations in returns and are also the safest. Hence, under the minimum risk portfolio, the model allocates the maximum share of the portfolio to each savings bond.

---

* Values of Z_1 and Z_2 are converted to percentage return and percentage risk (STD) measures, using Equations 7.24 and 7.25.

TABLE 7.14

Maximum Return Portfolio (Example 7.8)

Bond Type	Amount Allocated (%)
Corporate	70
EE bonds	0
I bonds	0
Treasury	20
Municipal	10
Mortgage	0

TABLE 7.15

Minimum Risk Portfolio (Example 7.8)

Bond Type	Amount Allocated (%)
Corporate	70
EE bonds	1
I bonds	1
Treasury	18
Municipal	10
Mortgage	0

7.6 Principles of Investing

In this section, we will discuss some basic principles of investing in order to become a prudent investor. We begin with the difference between the average security return that we have been using in this chapter and the annualized return of a security that is more important to an investor.

7.6.1 Annualized Returns

In Section 7.2, we introduced the formula (Equation 7.2) to calculate the average annual return of a security, using the past performance data. Equation 7.2 represents the statistical average and is also used to calculate the variance of the portfolio return (Equation 7.17) in Markowitz's bi-criteria model. In other words, the variance of the portfolio return gives a measure of how the annual returns will fluctuate from its average value and hence is very useful in investing. However, the statistical average is not really the true return that will be experienced by an investor. Most investors leave their investments to grow over many years and the statistical average does not capture the compounding of the annual returns over the years.

To illustrate, consider an investor, who buys a stock at the beginning of the year for $100 and sells it after 2 years. Let us assume that the stock goes down by 20% in year 1 and goes up by 20% in year 2. Then, the statistical average return is

$$\frac{-20+20}{2} = 0\%.$$

This implies that the investor has not gained or lost any money over the 2 years. In reality, the investor actually has lost money over the 2 years, because the value of the stock at the end of year 1 is $80, and at the end of year 2, it is $96. In other words, the investor has *lost* $4 from his original investment of $100! The actual return over 2 years is really −4% and the true average annual return can be computed using the standard formula in engineering economy as follows:

$$F = P(1+i)^N \tag{7.51}$$

where
 P = initial investment
 F = future value after N periods
 i = return per period

Substituting $P = 100$, $F = 96$, $N = 2$ in Equation 7.49 and solving for i, we get $i = -2.02\%$ per year. This is called the *annualized return*, which is in fact the true annual return from owning that stock.

To generalize this concept, consider a security whose actual annual returns over N years is given as $R_1, R_2, R_3, ..., R_n$. If R_i is the return per dollar invested in year i, then the statistical average return per year is given by

$$= \frac{\sum_{i=1}^{N} R_i}{N} \tag{7.52}$$

However, the compounded annualized return is given by

$$= \left(\sqrt[N]{(1+R_1)(1+R_2)....(1+R_N)} \right) - 1. \tag{7.53}$$

Mutual funds, investing companies, and the financial media primarily report annualized returns over 3, 5, and 10 years as past performance data for the investors. They are in fact the actual returns experienced by the investors.

Example 7.9

Google (now Alphabet Inc.) stock went public in August 2004. Annual returns from owning the Google stock over 9 years (2005–2013) are given below:

Year	2005	2006	2007	2008	2009	2010	2011	2012	2013
Return (%)	115	11	50	−56	102	−4	9	10	58

Calculate the average annual return (statistical) and the annualized return over 9 years.

SOLUTION

Here, the values of $R_1, R_2, \ldots, R_9$ are as follows:

$R_1 = 1.15$, $R_2 = 0.11$, $R_3 = 0.5$, $R_4 = -0.56$, $R_5 = 1.02$, $R_6 = -0.04$, $R_7 = 0.09$, $R_8 = 0.1$, and $R_9 = 0.58$.

Using Equation 7.52, the average annual return is calculated as

$$\frac{1.15 + 0.11 + 0.5 - 0.56 + 1.02 - 0.04 + 0.09 + 0.1 + 0.58}{9} = 0.328 \text{ or } 32.8\% \text{ per year.}$$

Using Equation 7.53, the annualized (compounded) average return is given by

$$= \left(\sqrt[9]{\frac{(1+1.15)(1+0.11)(1+0.5)(1-0.56)(1+1.02)(1-0.04)}{(1+0.09)(1+0.1)(1+0.58)}} \right) - 1$$

$$= \left(\sqrt[9]{5.7863} \right) - 1 = 0.2154 \text{ or } 21.54\% \text{ per year.}$$

Even though the statistical average return was 32.8% per year, the actual return experienced by a Google investor was only 21.54% per year!

7.6.2 Dollar Cost Averaging Principle

An investor should buy a stock when the price is low and sell it when the price goes up, in order to make money. This is known as the "Buy-Low and Sell-High" strategy in investing. It is easier said than done. In practice, many investors just do the opposite, namely, "buy-high and sell-low," resulting in a net loss. For example, many investors bought Google stock in 2006, when they saw its return was 115% in 2005. During the financial crisis in 2008, Google stock lost 56% of its value and several investors panicked and sold their Google stock at a loss!

Dollar cost averaging is a systematic way to put the "Buy-Low and Sell-High" strategy in "auto-pilot." Under this strategy, a fixed amount of dollars is invested at regular intervals, monthly or quarterly. The fixed amount of investment buys more number of shares when the stock price is low and less number of shares when the price is high. This lowers the average cost of owning a stock and results in a positive return, when the stock is sold. Dollar cost averaging has proven to be a successful investment strategy for owning stocks. However, if a company continues to do poorly in the market place and its stock price is on a freefall, dollar cost averaging cannot help. It is best to sell the stock and cut your losses!

7.6.3 Investing in Mutual Funds

As discussed in Section 7.1.5, mutual funds are investment companies, whose shares can be bought and sold like regular securities by investors. The share price of a mutual fund, called its Net Asset Value, is determined by the value of the individual securities held by that mutual fund. Investing in mutual funds has become very popular among individual investors in the last three decades. They provide a simple low-cost approach to buy domestic and international stocks under professional management. Mutual funds can be bought with a minimum investment, as low as $1500. For their services, mutual funds charge a management fee, called the "Expense Ratio." The annual fees can vary between 0.02% and 3% of the invested amount. For example, if a mutual fund's expense ratio is 0.5%, then $50 of every $10,000 invested goes to cover the management fees. Naturally, investors should choose mutual funds with lower expense ratios in order to get higher returns. Many academic studies have shown that the Expense Ratio plays a key role in the performance of a mutual fund. In fact, expense ratios have been steadily decreasing over the years. The average fee of a stock mutual fund has fallen from 0.84% to 0.51% over the last two decades.

There are more than 5000 mutual funds available for investment, each specializing on a particular segment of stocks or bonds. We shall now give an overview of the different types of mutual funds available for investment in the United States.

7.6.3.1 Load versus No-Load Funds

Load funds charge a one-time fee called "load" to buy their shares. This fee is in addition to the annual fees charged for managing the funds. The load can be as high as 5% of the amount invested. For example, when you invest $10,000 in a load mutual fund, $500 is immediately deducted and only the remaining $9500 are actually invested. Load mutual funds are bought through security dealers and the load is essentially the broker's fee. In contrast, no-load mutual funds do not charge a fee to buy their shares. However, an investor has to buy the shares directly from the investment company.

When mutual funds became available in the market, they were primarily load funds. After the introduction of no-load mutual funds by the Vanguard mutual fund company, no-load funds have become very popular among individual investors. Several academic studies have shown that no-load mutual funds as a group have outperformed load mutual funds. Unless an investor needs professional help in buying mutual funds, there is really no advantage in owning load mutual funds.

7.6.3.2 Stock Mutual Funds

Stock mutual funds buy and sell stocks of companies in the United States and abroad. Individual investors can achieve maximum diversification of their stock portfolios by investing in these mutual funds. There are several types of stock mutual funds available for investment. Some invest solely in U.S. stocks. Among U.S. stocks, investors can choose to invest in all the companies or in specific sectors of the U.S. economy. For example, there are mutual funds investing only in healthcare, transportation, telecommunications, and financial companies.

Some mutual funds invest primarily in international stocks of developed countries or emerging markets. There are also country-specific international mutual funds. Investing in international funds is the easiest and best approach to take advantage of the globalization of the market place and its investment benefits.

7.6.3.3 Bond Mutual Funds

Bond mutual funds are also known as income funds and invest in bonds issued by corporations or governments. Based on the average maturity of the bonds in their portfolios, bond funds are also classified as long term (maturity ≥ 15 years), intermediate term (6–14 years), and short term (1–5 years). Government bond funds can be U.S. government, individual states, or municipal governments. Recently, international bond funds have also been introduced for potential investment.

7.6.3.4 Balanced Mutual Funds

Balanced funds are mutual funds with portfolios that are a mix of both stocks and bonds. The stock-to-bond ratio may vary from one fund to another. The managers of balanced funds may also have the flexibility to vary the stock-to-bond ratio based on the economic conditions.

7.6.3.5 Index Funds

Created in 1976 by John Bogle, founder of the Vanguard group of mutual funds, index funds have become extremely popular among individual and institutional investors. In general, a stock mutual fund is managed by a fund manager, with support from a team of stock pickers and market analysts. Naturally, this increases the fund's management cost and its expense ratio. The fund manager's performance is then compared against the stock market performance, generally represented by the S&P 500 index, discussed in Section 7.4.2. Over the years, it has been found that more than 80% of the stock fund managers do not perform any better than the S&P 500 index. Hence, Vanguard introduced the S&P 500 index fund, with the idea that instead of trying to beat the market (S&P 500 index), why not simply replicate the market by owning the 500 companies that are in the S&P index! The annual returns of the S&P 500 index funds will be very close to the market index. Since index funds require very little management expertise to buy and sell stocks, their expense ratios (management fees) are very low. The average expense ratio of index funds in 2016 was just 0.17%, according to Morningstar, the leading provider of mutual fund data. Actively managed funds have an average expense ratio of 0.75%. Hence, investing in index funds has now become extremely popular among individual investors.

Based on Vanguard's success with the S&P 500 index fund, there are now hundreds of index mutual funds available. They invest not only in the S&P 500 index but also in other market indices that represent both U.S. stocks and international stocks. Given below are examples of stock index funds offered by Vanguard:

- Vanguard Extended Market Index Fund (invests primarily in non S&P 500 stocks)
- Vanguard Total Stock Market Index Fund
- Vanguard Total International Stock Index Fund
- Vanguard European Stock Index Fund
- Vanguard Emerging Markets Stock Index Fund
- Vanguard Pacific Stock Index Fund

After the success with stock index funds, mutual fund companies began offering index funds for bonds as well. There are several types of bond index funds available in the market for individual investors. Given below are examples of bond index funds offered by Vanguard:

- Vanguard Total Bond Index Fund
- Vanguard Long-Term Bond Index Fund
- Vanguard Intermediate-Term Bond Index Fund
- Vanguard Short-Term Bond Index Fund

Financial Engineering

7.6.3.6 Life Cycle or Target Date Funds

As discussed in Section 7.1.4, asset allocation refers to the proportion of investment money allocated to equities, bonds, and cash. More than 90% of the portfolio's performance is directly correlated to asset allocation. A young investor, in the 30s, can be more aggressive and should have 70%–80% allocated to equities. On the other hand, an investor close to retirement should not have more than 40%–50% in equities. Thus, an investor has to adjust his asset allocation periodically as he gets older. Life cycle or target date funds essentially do this, on behalf of the investors, by adjusting their asset allocations as they get closer to their target retirement dates. For example, Vanguard offers target retirement funds for 2020–2060, in 5-year increments. A 33-year-old investor in 2017, who plans to retire at age 65, would choose a 2050 target retirement fund. With a minimum investment of $3000, a young investor can get a professionally managed diversified portfolio of stocks and bonds, with periodic adjustments to its asset allocation as he gets older.

7.6.4 Investment Strategies

We end this section with some common strategies a prudent investor should follow in order to maximize the portfolio's return and minimize the portfolio's risk:

1. Invest in diversified no-load mutual funds, including index funds and target date funds. Examples of fund families, with low expense ratios, include Vanguard, T. Rowe Price, Fidelity, and TIAA-CREF.
2. Avoid load mutual funds and actively managed funds with high management fees. Low expense ratio is one of the best predictors of a fund's relative performance.
3. Follow the Golden Rule of Investment of "Buy-Low and Sell-High." To achieve this, use the "dollar cost averaging" strategy discussed in Section 7.6.2.
4. Use the "Birthday Rule" for asset allocation, which states "own your age in bonds." For example, a 35-year-old should have 35% in bonds and 65% in equities.
5. Since asset allocation is the major predictor of the portfolio's performance, check the portfolio's asset allocation twice a year and make adjustments, if the allocation percentages change significantly.
6. Long-term investors should not be concerned with the day-to-day ups and downs of the stock market. Over the long run, stocks have always performed better than bonds.

7. The best-known and well-respected provider of historical performance data of individual stocks and mutual funds is the Morningstar Company. The data are available free to individual investors at www.morningstar.com. At the Morningstar website, investors can get detailed performance data on mutual funds and compare the performances of similar funds from different fund families.

Recent young graduates starting a career should plan to save at least 10%–15% of their after-tax salaries. Financial advisors recommend the following savings targets to reach over their lifetimes:

- Annual salary by age 35
- Three times the annual salary by age 45
- Five times the annual salary by age 55

Finally, there are several good books available to learn more about investing intelligently. Examples include Benz (2010), Bernstein (2001), Swedroe (1998), Malkiel (1999), and Lindauer et al. (2014). For a more quantitative treatment of investment models, the reader is referred to the books by Heching and King (2009), Luenberger (1998), and Best (2010).

Exercises

7.1 Explain the meaning of the following terms and discuss why they are important in investing:
   a. Asset allocation
   b. Dollar cost averaging
   c. Investment risk
   d. Mutual funds

7.2 Discuss the differences between the following terms:
   a. Equities and bonds
   b. Load and no-load mutual funds
   c. Markowitz's risk and Sharpe's risk
   d. Index funds and target date funds

7.3 Consider two similar stocks, S_1 and S_2. Both have averaged 10% annual return, and their risks in terms of standard deviations (STD) are also the same at 20%. The correlation coefficient between the two

Financial Engineering

stocks is 0.1. Prove that, by investing equal amount of money in S_1 and S_2, the portfolio risk can be reduced as compared to investing all the funds in either S_1 or S_2 alone.

7.4 Consider two mutual funds for investment. Fund 1 is a bond fund with 5% average return and 10% standard deviation (STD) risk. Fund 2 is a stock fund with 10% average return and 15% STD risk. The correlation between the two funds is −0.16.

a. Determine the portfolio's return and risk if all the funds are invested in the bond fund only.

b. Show that an investor can increase his portfolio's return and simultaneously reduce its risk by investing 90% in the bond fund and 10% in the stock fund.

7.5 Over the 20-year period (1997–2016), U.S. stocks have averaged 7.7% in annual return, with an STD risk of 15.3%. During the same period, U.S. bonds have averaged 5.3% in annual return, with an SD risk of 3.4%. An investor is considering a portfolio with proportion α in stocks and $(1 - \alpha)$ in bonds, where α varies between 0 and 1.

a. What will be the optimal portfolio that maximizes average return, ignoring risk? What will be the portfolio's average return and STD risk?

b. Suppose the investor splits the investment capital equally between stocks and bonds by choosing $\alpha = 0.5$. Determine the portfolio's average return and its STD risk.

c. Determine the optimal value of α that will produce a portfolio with minimum variance. Formulate this as an optimization problem and solve. What is the average return and STD risk of the minimum variance portfolio?

7.6 A young investor is considering three potential stock mutual funds for investment. Based on the last 20 years of performance data, the average returns and STD risks have been computed as shown in Table 7.16. The funds are also positively correlated as shown in Table 7.17.

TABLE 7.16

Data on Security Returns (Exercise 7.6)

Fund	Mean (%)	STD Risk (%)
1	5	10
2	10	20
3	15	30

TABLE 7.17
Correlation between Funds (Exercise 7.6)

Fund	Correlation
Funds 1 and 2	0.6
Funds 1 and 3	0.4
Funds 2 and 3	0.8

The investor has $50,000 to invest and has the following investment restrictions:

i. No more than 50% should be invested in any one fund.
ii. Invest at least $10,000 in Fund 1.
iii. Invest at least $15,000 in Fund 2.
iv. Invest no more than 70% in Funds 2 and 3.
v. Invest all the $50,000 among the three funds.

The objective of the investor is to achieve maximum portfolio return and minimum portfolio variance.

a. Formulate the above problem as Markowitz's bi-criteria optimization model.
b. Determine the minimum variance portfolio that will yield an average portfolio return of 12%.

7.7 Jill is interested in determining her asset allocation for her 401-K retirement plan. Her company has 10 mutual funds for investment. Their past performance data are given in Table 7.18.

Even though Jill is interested in the "growth" of her portfolio, she has the following investment restrictions:

i. No more than 20% should be invested in any one fund.
ii. 30%–45% should be invested in U.S. stock funds.
iii. 10%–25% should be invested in international funds
iv. At least 15% should be invested in bond funds.
v. Total investment in U.S. and international stock funds should not exceed 70%.
vi. At least 5% should be invested in money market fund.

a. Formulate Jill's asset allocation problem as Sharpe's bi-criteria linear program.
b. Determine the ideal solution and the upper and lower bounds on the portfolio's return and Beta risk
c. Determine the optimal portfolio that will give at least a 10% average return under minimum Beta risk

Financial Engineering

TABLE 7.18
Retirement Fund Data (Exercise 7.7)

Type of Fund	Average Annual Return (%)	Beta Risk Index
U.S. stock funds		
Fund 1	8	0.7
Fund 2	9	0.8
Fund 3	10	1.0
Fund 4	8	1.4
International funds		
Fund 5	11	1.2
Fund 6	15	1.7
Bond funds		
Fund 7	7	0.5
Fund 8	4	0.3
Fund 9	2	0.2
Money market fund		
Fund 10	1	0

7.8 XYZ stock is valued at $100 per share. It loses 20% in year 1, gains 20% in year 2, and gains 10% in year 3. What is the average annual return? What is the annualized (compounded) average return?

7.9 The annual returns of the Russell 2000 index (measures the performance of mid and small company stocks) for 2006–2010 are 18.37% in 2006, –1.57% in 2007, –33.79% in 2008, 27.17% in 2009, and 26.85% in 2010. Compute the average annual return and annualized average return.

7.10 (Case Study 1)

In this case study, you will be applying Markowitz's Bi-criteria Quadratic Programming model to generate the entire set of efficient portfolios. An investor is considering 10 potential securities for investment. They are listed in Tables 7.19 and 7.20. His objectives are to maximize the average portfolio return and, at the same time, minimize the portfolio risk as measured by the variance of the portfolio return. The securities include eight stocks in a variety of industries, a bond, and a 3-month Treasury bill for cash to achieve diversification. In addition, the investor wants to limit his investment to any one security to 30%. Table 7.21 gives the historical annual returns of the 10 securities for the 11-year period, 2003–2013.

Questions

a. Compute the average annual return (statistical average) and its standard deviation (STD) for each security expressed as a percentage.
b. Determine the variance–covariance matrix of the 10 securities.

TABLE 7.19

Description of Securities (Stocks)

Security	Stock	Symbol	Industry
1	Schlumberger	SLB	Energy
2	Qualcomm	QCOM	IT
3	Berkshire Hathaway	BRK-B	Financials
4	Honeywell Intl	HON	Industrials
5	Johnson & Johnson	JNJ	Healthcare
6	Wisconsin Energy Co	WEC	Utilities
7	Kimberly Clark	KMB	Consumer staples
8	Coach Inc.	COH	Consumer discretionary

TABLE 7.20

Description of Securities (Bond and Cash)

Security	Description	Symbol
9	Barclays bond	Bond
10	3-month Treasury bill	Cash

c. Formulate a bi-criteria model that will *maximize the average return and minimize the variance of the portfolio*. You must define all the decision variables and explicitly write out the constraints and the two objective functions.

d. Determine the ideal solution and the bounds on the portfolio return (%) and portfolio risk, represented by its standard deviation (STD%).

e. Determine the minimum variance portfolio that will achieve an average return of at least

(i) 10%   (ii) 12%   (iii) 14%   (iv) 16%   (v) 18%   (vi) 20%

For each part, you must give the optimal portfolio in terms of percent invested in each security, portfolio return (%), and portfolio risk (STD%).

f. Using part (e) results, draw the efficient frontier with *x*-axis representing portfolio risk (STD%) and *y*-axis representing portfolio return (%).

g. Based on the efficient portfolio frontier, what portfolio will you select and why?

TABLE 7.21
Annual Returns in Percentages (2003–2013)

				Securities					
Year	SLB (%)	QCOM (%)	BRK-B (%)	HON (%)	JNJ (%)	WEC (%)	KMB (%)	COH (%)	Bond (%)	Cash (%)
2003	30.01	48.20	16.18	39.29	−3.82	32.74	24.48	129.34	4.10	1.03
2004	22.35	57.24	4.30	5.92	22.76	0.78	11.37	49.40	4.34	1.23
2005	45.11	1.60	−0.02	5.20	−5.23	15.87	−9.36	18.23	2.43	3.01
2006	30.03	−12.28	24.89	21.45	9.85	21.51	13.91	28.85	4.33	4.68
2007	55.75	4.13	29.19	36.10	1.03	2.63	2.05	−28.82	6.97	4.64
2008	−56.97	−8.95	−32.14	−46.68	−10.30	−13.82	−23.94	−32.08	5.24	1.59
2009	53.77	29.11	2.24	19.40	7.66	18.70	20.80	75.88	5.93	0.14
2010	28.28	6.98	21.90	35.61	−3.97	18.12	−1.05	51.41	6.54	0.13
2011	−18.19	10.53	−4.76	2.24	6.03	18.79	16.69	10.36	7.84	0.03
2012	1.45	13.09	17.56	16.78	6.89	5.41	14.78	−9.06	4.22	0.05
2013	30.03	20.03	32.17	43.96	30.66	12.18	23.72	1.12	−2.02	0.07

7.11 (Case Study 2)

An investor has $250,000 to invest in mutual funds. She has selected 33 funds for possible investment. In addition, a Federal money market fund is available for holding cash reserve. Table 7.22 gives the following data for each fund based on their 6-year performance, during 2002–2007:

i. Name of fund
ii. Ticker symbol
iii. Type of fund: capitalization size (small, medium, large), strategy (value, growth, blend) or foreign
iv. Average annual return (%)
v. Market risk (Beta)

Assume that any money not invested in the mutual funds can be deposited in a money market fund that will yield 3% interest at zero risk. Even though the investor is primarily interested in "growth" of her portfolio, she has the following investment restrictions:

i. No more than $25,000 should be invested in any one fund (does not apply to the money market fund).
ii. Between 15% and 30% of her total investment should be in "Mid Cap" funds.
iii. No more than 70% of her investment should be in "Large Cap" and "Mid Cap" funds.
iv. At least 35% should be invested in "Large Cap" funds.
v. Investment in "International" funds should not exceed 30%.
vi. No more than 20% should be invested in "Small Cap" funds.
vii. Maintain a minimum cash reserve of $15,000 in the money market fund.

Questions

a. Develop the expressions for portfolio return and portfolio Beta risk.
b. Formulate a single-objective linear program (LP) for determining the optimal investment portfolio under the following criteria:
   – Maximize portfolio return only
   – Minimize portfolio risk only
c. Solve each of the LP problems formulated in part (b). Write down the optimal portfolios and their corresponding risk and return. Determine the ideal solution and the bounds on return and risk.
d. Formulate and solve the LP problems to determine the optimal portfolios that will provide the minimum risk, under a portfolio

Financial Engineering

TABLE 7.22

Performance Data on Mutual Funds (2002–2007)

Fund Name	Ticker	Type	Average Annual Return	Beta Risk
Allianz NFJ Small Cap Value	PCVAX	Small cap value	16.66	1.13
American Funds Amcap	AMPCX	Large cap growth	9.57	0.92
American Funds EuroPac	AEPGX	International	18.43	0.98
American Funds Fundamental	ANCFX	Large cap blend	14.45	1.01
BlackRock Equity	MADVX	Large cap value	14.13	0.82
Calamos Growth	CVGRX	Large cap growth	12.59	1.68
Columbia Acron USA	LAUAX	Small cap growth	14.72	1.25
DFA US Small Cap	DFSTX	Small cap blend	14.80	1.56
Fidelity Nordic	FNORX	International	23.41	1.42
First Eagle Fund of America	FEAMX	Mid cap blend	12.82	0.89
Frankling Bal Sh Investment	FCBSX	Mid cap value	14.55	1.09
Gabelli Equity Income	GCAEX	Large cap value	13.62	0.8
GMO Emerging Markets	GMOEX	International	33.48	1.46
GMO Foreign Small Co	GFSFX	International	26.74	1.06
Harbor International Inv	HIINX	International	21.07	1.12
Janus Sm Value	JSCVX	Small cap value	11.22	1.07
Legg Mason Emerging Market	LGEMX	International	31.88	1.65
Legg Mason Intl Eq FinInt	LGFEX	International	19.77	1.04
Legg Mason P Aggr Grow	SHRAX	Large cap growth	13.01	1.24
Longleaf Partners Sm-Cap	LLSCX	Small cap value	17.59	0.99
Lord Abbet Small Cap	LRSCX	Small cap blend	18.52	1.44
MainStay MAPA	MAPAX	Large cap blend	14.01	1.1
Marshall Mid Cap Val Adv	MVEAX	Mid cap value	13.94	1.05
Mass Mutual Prem Intl	MMIAX	International	16.36	1.14
Navellier Mid Cap Growth	NPMDX	Mid cap growth	10.34	1.49
Oakmark International	OAKIX	International	17.38	0.83
Old Mutual Mid Cap	OAMJX	Mid cap blend	13.00	1.3
T. Rowe Price New Horiz	PRNHX	Small cap growth	15.29	1.48
Van Kempen Growth & Income	ACGIX	Large cap value	12.12	0.86
Vanguard Euro Stock	VEURX	International	18.64	0.98
Victory Diversified Stk	SRVEX	Large cap blend	12.41	1.05
Waddell & Reed Adv Intl	UNCGX	International	13.23	1.08
WF Advanced Common Stk	SCSAX	Mid cap blend	15.37	1.26

      return of at least R%, where R varies between the bounds obtained in part (c) (vary R in increments of 2%). Plot a graph of return versus risk and identify the efficient portfolios.

   e. Based on the results in part (d), what will be your recommended portfolio?

References

Benz, C. 2010. *Thirty Minute Money Solutions*. Hoboken, NJ: Wiley.
Bernstein, W. 2001. *The Intelligent Asset Allocator*. New York, NY: McGraw-Hill.
Best, M. J. 2010. *Portfolio Optimization*. Boca Raton, FL: CRC Press.
Heching, A. and A. King. 2009. Financial engineering. In *Operations Research Applications*, ed. A. R. Ravindran. Chapter 7. Boca Raton, FL: CRC Press.
Lindauer, M., T. Larimore, and M. LeBoeuf. 2014. *The Bogleheads' Guide to Investing*. 2nd Edition. Hoboken, NJ: Wiley.
Luenberger, D. G. 1998. *Investment Science*. New York, NY: Oxford University Press.
Malkiel, B. G. 1999. *A Random Walk Down Wall Street*. New York, NY: Norton Company.
Markowitz, H. 1952. Portfolio selection. *The Journal of Finance*. 7(1): 77–91.
Markowitz, H. 1956. The optimization of a quadratic function subject to linear constraints. *Naval Research Logistics*. 3(1–2): 111–133.
Markowitz, H. M. 1959. *Portfolio Selection: Efficient Diversification of Investments*. New York, NY: Wiley.
Sharpe, W. F. 1963. A simplified model for portfolio analysis. *Management Science*. 9(2): 277–293.
Swedroe, L. E. 1998. *The Only Guide to a Winning Investment Strategy*. London, England: Truman Talley Books.
Winston, W. L. and S. C. Albright. 2001. *Practical Management Science*, 2nd Edition. Pacific Grove, CA: Duxbury.

8

Revenue Management

While designing the capacity of manufacturing systems, inventory is used as a buffer to account for the variability in customer demand and supply. However, perishability (no inventory) and simultaneity (direct customer interaction) make capacity management very difficult in service systems. Tools, such as revenue management and yield management, are specifically designed to address the challenge. In this chapter, we will discuss several strategies used by the service industries, under the umbrella of "revenue management/yield management." We will present mathematical models for "overbooking" and "differential pricing" strategies. Since revenue management originated in the airline industry, we will end this chapter with the discussion of different revenue management practices currently used by the airlines.

8.1 History of Revenue Management

Revenue management, also known as yield management, originated in the airline industry in the late 1970s. Deregulation of the U.S. airline industry in 1978 by President Carter was the catalyst for revenue management. Before the deregulation, airlines had restricted market access and price controls on airfares. After the deregulation, the airlines were able to set their own prices based on customer demands and their willingness to pay. Technically, airlines cannot discriminate by charging different prices to different customers for the same seat. However, an airline can take the same seat and package it multiple ways as different products and charge different prices for those products. For example, an "economy ticket" has a non-refundable lower fare, but requires Saturday night stay (to exclude business travelers), 21-day advance purchase, and change fees. A "full fare ticket" is higher priced but has no such restrictions and is fully refundable. By targeting the leisure travelers with economy fares, the airlines have segmented the market demand, even though they are selling the same airline seats. American Airlines pioneered revenue management with the introduction of 21-day advance purchase economy fares. Thus, the basic idea behind revenue management is to maximize the revenue by selling the fixed number of "perishable" airline seats at the right price, at the right time, to the right customer. Revenue management increased American Airline's revenues by $1.4 billion during the first 3 years of its introduction, as described in Smith et al. (1992).

After the success at American Airlines, other airlines followed with their own revenue management practices. In fact, revenue management strategies are not only applicable to the transportation sector but also to other service industries. In Section 8.3.3, we will discuss what ideal characteristics a service industry should possess, in order to successfully implement revenue management practices.

8.2 Difficulties in Managing Service Capacity

According to Haksever and Render (2013), one of the biggest challenges facing manufacturing and service industries is to plan capacity to match uncertain demand. The challenge is somewhat mitigated in manufacturing due to the non-perishability of their products and the use of inventory as a buffer. However, those tools are not available to services. Moreover, some of the unique characteristics of services pose additional problems in managing services as discussed below.

- Short-term imbalances are critical

    In manufacturing, inventory and backorders are used as buffers to smooth out imbalances between capacity and demand. Perishability and direct customer interaction require services to have sufficient capacity at all times to meet uncertain customer demand. For most services, backorders are not possible and short-term imbalances are costly, due to "lost sales" and the loss of "goodwill."

- Variability in service time

    Generally, cycle time to manufacture a product can be estimated. However, service times of customers vary widely, depending on their needs. Since short-term imbalances are critical, average service times do not help in capacity planning.

- Long run averages are not useful

    In manufacturing, monthly and quarterly demands can be used for capacity planning. When the actual demand is less, excess capacity is used to build inventory. When the demand is higher, inventory can be used to meet its demand. Since services cannot be inventoried, long run averages are worthless.

- Difficulties in Forecasting Demand

    In manufacturing, long-term forecasting can be used to do aggregate capacity planning. Unpredictable demands can be handled by

Revenue Management 409

overtime and additional production shifts. Because of direct customer interaction, service capacity has to match customer demand by time and day of the week. For example, for workforce planning, Taco Bell forecasts customer demands at 15-minute intervals during lunch time (11 a.m. to 2 p.m.) by day of the week (Hueter and Swart 1998).

- Fixed capacity of services

  Many service systems have limited flexibility to increase their physical facilities at short notice. Because of a sudden increase in ticket sales for a particular day and time, an airline can schedule a larger aircraft for that route. However, a hotel cannot increase the number of rooms it has in order to meet a big surge in demand for a football weekend! Revenue management strategies are particularly useful for services with fixed capacities. Metters et al. (2006) present a simple ice cream parlor example to illustrate the differences between manufacturing and services in capacity planning.

8.3 Revenue Management Strategies in Services

Under the rubrics of revenue management, we will discuss a number of strategies used by service industries in matching service capacity with customer demand. Fitzsimmons and Fitzsimmons (2006) and Davis and Heineke (2003) provide an extensive discussion of these strategies with illustrative examples. Strategies for matching service capacity and customer demand can be broadly classified into two groups:

- *Proactive strategies* that influence customer demand by changing their demand patterns. They help in reducing the wide demand fluctuations and smoothing the demand over time. This makes it easier to plan service capacity to fulfill the demands at the right time.
- *Reactive strategies* that provide flexibility to change service capacity so that the demands can be met when they occur.

8.3.1 Proactive Strategies to Manage Customer Demand

There are a number of strategies that can be used to influence customer behavior as listed below:

- Use of reservation/appointment systems
- Segmenting customers
- Differential pricing
- Sales promotions

8.3.1.1 Use of Reservations/Appointment Systems

Airlines, trains, and buses use the reservation system to distribute customer demand evenly over a time period. Hospitals and clinics also use the appointment systems to smooth out the patient demand.

Leisure industries have also started using the reservation system to reduce overcrowding at their popular attractions. Several years ago, Disney World was concerned about the long lines in the morning hours at their most popular attractions. This happens because the visitors want to make sure that they will see the popular sites by the end of the day. In order to spread the demand evenly, Disney introduced a reservation system called "Fast Pass" for their most popular attractions. As soon as the guests arrive at the Disney World, they can use the self-service kiosks to specify the attractions they want to see. They get reserved times to see those attractions. Then, they go to those attractions at their designated times and enter by a separate gate for "Fast Pass" holders. Other guests have to wait in long lines for "regular entry." Fast Pass helped Disney to spread the demand for most popular attractions evenly throughout the day. A side benefit of "Fast Pass" is that it has reduced overcrowding at the restaurants during lunch time because some of the "fast pass" guests are at the popular attractions.

A major problem with the reservation system is "No-shows," customers making reservations/appointments, but fail to show up. To reduce "No-shows," airlines sell nonrefundable economy fares with penalties for changing/cancelling reservations. Hotels require one night's deposit for holding reservations. Still, "No-show" is a problem for airlines for a variety of reasons, including family emergencies, flight delays due to weather and mechanical problems.

"Overbooking" is a strategy used by airlines and hotels to mitigate the revenue loss due to no-shows. An airline would sell more tickets than the number of seats they have with the expectation that there will be no-shows to offset the overbooking. However, this is a risky strategy if they overbook too much and there are no seats available for ticketed passengers. How much to overbook is a nontrivial optimization problem for which mathematical models exist to determine the optimal number of seats to overbook. We will discuss this in Section 8.4.

8.3.1.2 Customer Segmentation

Demand for services occurs from multiple sources with different customer needs. By segmenting the customers with some incentives, the fluctuations in demand can be reduced, resulting in a smoother demand flow. Disney's "Fast Pass" system is an example of customer segmentation. The benefits to "Fast Pass" holders are that they are guaranteed admission to the major attractions that day with no waiting. Hence, they can visit more number of attractions in a day. However, it reduces their flexibility. It is left up to the visitors to decide whether they want to use "Fast Pass."

Grocery stores also use customer segmentation to speed up the traffic at checkout lanes. Most stores now offer three options to their customers—regular checkout lanes, express lanes, and self-service lanes. Customers self-select the lanes they want to use even though the cost of the grocery items is the same in all the lanes. This results in less congestion at the checkout counters.

Use of E-Z pass at toll booths on state highways is another example of customer segmentation. Some motorists buy E-Z pass so that they can zip through toll booths without any wait. It reduces the cost of collecting tolls for the state, which can use fewer toll booths that are manned by toll collectors.

In the past, healthcare clinics and outpatient departments in the hospitals have used "pre-booked" appointments to see patients. "Walk-in" visits are permitted only in the emergency departments. Pre-booked appointments reduce demand uncertainty and maximize the utilization of healthcare providers. However, it increases the patient no-shows and the lead time to see a doctor. Under the "walk-in" system, a patient can see the doctor the same day but he or she may have to wait longer. From the hospital's perspective, the demand uncertainty is very high under the "walk-in" system and capacity planning is more difficult and expensive. Recently, a new system called "Open Access" has been introduced at some clinics and hospitals. Open Access provides same-day appointments to the patients. Srinivas (2017) has shown that the use of a hybrid appointment system, where some time slots are "Open Access" and others are "Pre-booked," helps both the patients and healthcare providers in terms of personnel utilization and patient satisfaction.

8.3.1.3 Differential Pricing

In all the customer segmentation examples cited in the previous section, the cost of service is the same irrespective of which group a customer joins. The only incentive to the customer is savings in time and convenience. Service providers can also use price incentives to influence the customer behavior. Telecommunication industries have been one of the earliest to use price differentials to segment customers. Telephone companies have used lower prices for calls made during off-peak hours (evenings and weekends). This has helped most residential customers to use off-peak hours at a lower price for calling their friends and relatives, while business customers have paid a high price for using their phones during the peak hours. The differential pricing helped telephone companies to smooth out the volume of calls during peak and off-peak time periods.

Parking garages have also used differential pricing to segment their customers as hourly, daily, and monthly. Another example of differential pricing is the Washington D.C. Metro Rail Subway System. Washington attracts a lot of tourists. In order to reduce the crowd at peak hours, which are used mostly by office personnel, the metro system offers lower ticket prices for travel during off-peak hours. This results in many tourists using the metro during off-peak hours to save money, which reduces the congestion at peak hours.

Lately, electric utilities have been testing differential pricing for electricity use based on the time when they use it. "Smart" meters have been introduced to measure not only the amount of electricity used but also when it is used. The electricity charges are higher during peak hours, which are primarily used by business customers. Lower prices motivate residential customers to use their major electrical appliances, such as electric ovens, dishwashers, and clothes dryers, during off-peak hours.

8.3.1.4 Sales Promotions

Both manufacturing and service industries have used sales promotions, including rebates, to increase customer demand. Offering sales promotions during low-demand periods has been found to be very effective for service industries. Promotions may also increase overall market share by stealing customers from the competitors. However, it has been found that when an airline introduces promotional fares in a particular route, other airlines follow suit immediately to match the lower price, nullifying the benefits of sales promotions.

8.3.2 Reactive Strategies for Capacity Planning

Reactive strategies, as the name implies, adjust service capacity to meet the current demand. They are more under the control of service providers. Some of these strategies are listed below:

- Workforce planning
- Use of part-time workers
- Cross-training of employees
- Promoting self-service
- Designing adjustable capacity

8.3.2.1 Workforce Planning

Workforce planning is used to determine the optimal number of workers to have at different times to meet the customer demand. This is a complex problem in service industries, where simultaneous presence of customer and worker is generally required and service is perishable. Insufficient number of workers would result in long lines, customer dissatisfaction, and loss of future business. More workers would lead to idle time and cost. An essential requirement for workforce planning is a good demand forecast by time of the day. Using the forecast, integer programming models have been successfully used to determine the optimal number of workers to employ at different periods. Example 3.5 in Chapter 3 illustrates a workforce planning model used in scheduling nurses in a hospital. United Airlines applied workforce planning models to improve the utilization of personnel at its reservation offices and airports by matching work schedules to customer needs more closely (Holloran and Byrn 1986). Taco Bell developed an integrated labor-management system to determine the number of workers needed in its restaurants at different times (Hueter and Swart 1998). Taco Bell achieved labor savings of more than $40 million over 3 years by using the integrated labor management system.

8.3.2.2 Use of Part-Time/Temporary Workers

It is generally more difficult in manufacturing companies to hire part-time or temporary workers on short notice, because of the required skills and training necessary to perform many of the manufacturing tasks. In comparison, many service tasks require less skills and training. Hence, it is easy to hire part-time or temporary workers and train them quickly for a particular service task. Brick-and-mortar retail stores, online businesses (e.g., Amazon), and logistics providers (e.g., FedEx, UPS) have used this strategy very effectively to handle the surge in demand during the Christmas season. According to Haksever and Render (2013), nearly 20% of the U.S. workforce in 2011 were employed as part-time or temporary workers.

8.3.2.3 Cross-Training of Employees

Many service operations do not require specialized skills. Hence, it is easy to cross-train employees so that they can perform multiple tasks. Supermarkets and grocery stores have used employee cross-training to create a flexible workforce, who can be assigned to different tasks depending on the current needs. For example, when the lines at the checkout lanes are longer, store room clerks are called in to run cash registers. Some may be asked to bag the groceries, so the customer service time can be reduced and more customers are served. Another advantage of cross-training is that it increases the skill set of the employees and gives them better job satisfaction.

8.3.2.4 Promoting Self-Service

Introduction of self-service at gas stations, retail stores, airports, and other locations has made customers to interact more actively for their services. In fact, customers are willingly working as "unpaid" employees of the company! When a customer orders a burger, fries, and a drink at McDonalds, he gets an empty cup for the drink. The customer has to fill it with the drink he wants, with or without ice, at the soda fountain. This significantly reduces the average service time and its variability at the counter, so that employees can serve more customers. The same thing is true, when airline passengers are encouraged to use online check-in and self-service kiosks at the airports. The airline agents have to assist with baggage check-in only. Banks have increased self-service with the use of ATMs and online banking. This has directly contributed to fewer tellers and reduced customer service times at the teller windows.

8.3.2.5 Designing Adjustable Capacity

Capacity planning in services is not limited just to the employed workers. Physical facilities also play a key role in many industries, where customer service is provided at a physical facility. Having adjustable capacity at the facility helps companies react quickly to make real-time changes to meet varying customer demand. The arrival area at an international airport is a good example of adjustable capacity. Most international airports use customer segmentation to divide the arriving passengers as "citizens" and "non-citizens," because their processing times for immigration are different; then, they use flexible partitions to designate their respective areas. Depending on which group is larger at a given time, the immigration personnel move the partitions accordingly to change the amount of waiting space for citizens and non-citizens.

8.3.3 Applicability of Revenue Management

As we discussed earlier, revenue management includes all the proactive strategies for demand management and reactive strategies for capacity planning. The main goal of revenue management is to maximize the revenue that is generated by a time perishable resource, such as airline seats and hotel rooms. This is done by selling the right capacity, to the right customer, at the right price. In this context, the term "right" is from the perspective of the service provider and not necessarily that of the customer!

Revenue management practices are not applicable to every industry. The following characteristics are essential for revenue management to be successful in an industry:

- Perishability
- Ability to segment customers
- Ability to sell in advance
- High fixed cost and low variable cost
- Relatively fixed capacity

Most service industries exhibit the above characteristics. Practically all services are perishable. In the earlier sections, we have discussed how service industries can segment the market through price differentials and other means. Most services have the ability to sell their services in advance through prior reservations and appointments. Of course, this leads to the problem of customer "no-shows" and we will discuss solutions to this in the next section.

Many services require big investments in physical facilities and equipment. Airlines have to purchase new planes to replace an aging fleet. Hotels have to invest in buildings and their maintenance. Car rental companies replace their fleet of cars every 2–3 years. Thus, in many services, the fixed cost is the same, irrespective of the customer demand. For example, whether or not the plane is full, the fixed cost of operating (planes and crews) is the same. The variable cost (fuel cost) changes slightly depending on the number of passengers on the plane. The same is true for hotel rooms also.

Most services also operate under relatively fixed capacity. Increasing the capacity at short notice may be prohibitively expensive. Examples are number of planes an airline has, limited number of guest rooms in a hotel, and the number of beds in a hospital. Universities are always faced with limited classroom space and dormitories. During periods of sudden enrollment increases, universities have used expensive private hotel rooms to house their freshmen students and temporary structures as classrooms.

8.4 Optimization Models for Overbooking

8.4.1 No-Show Problem

Airlines, trains, buses, hotels, restaurants, and car rental companies use the reservation system to lock in future sales. A major problem with the reservation system is "No-shows," customers holding reservations fail to show up. According to Cook (1998), airlines operate on a thin 3%–5% profit margin. Because the fixed cost of an airplane is much higher compared to its operating cost, empty seats cost the airlines dearly. According to Cross (1997), "no-shows" cost the world's airlines $3 billion annually. No-shows can range 10%–40% for restaurants and can be as high as 70% for car rental companies in Florida (Metters et al. 2006).

8.4.2 Practice of Overbooking

Overbooking is a key revenue management strategy used by airlines and other industries to mitigate the revenue loss due to no-shows. An airline would sell more tickets than the number of available seats, with the expectation that no-shows will offset the oversold tickets. Without overbooking, planes that are full now would be flying with an average of 15% empty seats (Metters et al. 2006).

To understand the consequences of overbooking, consider an airline with a 100-seat capacity that sells 120 tickets for a flight. If there are 22 no-shows for that flight, the airline would only have 2 empty seats as opposed to 22 empty seats without overbooking. On the other hand, if the no-shows are just 15, then the airline has to deal with 5 irate ticketed passengers demanding a seat on the plane. The Federal Aviation Administration (FAA) in the United States has stiff overbooking penalties for "bumping" ticketed passengers involuntarily. To avoid the stiff FAA penalties, the ticket agents at the gate ask for "volunteers," who would give up their seats for a price. Usually, a promise to put them on the next scheduled flight and give a free ticket for travel within the United States. When that does not get enough volunteers, airlines offer additional monetary inducements. Thus, overbooking creates the following two scenarios:

Scenario 1: No-shows exceed the number of overbooked seats, resulting in some empty seats and a revenue loss, called *spoilage*.

Scenario 2: No-shows are less than the number overbooked, resulting in "bumping" ticketed passengers and incurring *oversold penalty*.

The optimization problem is to determine the optimal number of seats to overbook such that the sum of the spoilage cost and oversold penalty is minimized.

8.4.3 Optimization Problem

In this section, we will present an optimization model for overbooking, using an airline application. An essential input to the overbooking model is the historical pattern of no-shows for a particular flight for a specific day and time.

8.4.3.1 Input Data

1. *No-shows*: Since the actual number of no-shows is an uncertain quantity, we treat it as a random variable. Let the random variable X denote the number of no-shows. Assume that X is a non-negative continuous random variable, whose density function is given by $f(x)$. Thus, the probability of having x or fewer no-shows is given by

$$\text{Prob}[X \leq x] = \int_0^x f(x)\,dx.$$

2. *Spoilage Cost*: When the no-shows exceed the number of overbooked tickets, there are empty seats on that flight, and the spoilage cost can be denoted by C_1 per unit.

Revenue Management

3. *Oversold Penalty*: When the no-shows are less than the number of overbooked tickets, passengers have to be denied boarding the plane (bumped). The penalty cost for the oversold situation can be denoted by C_2 per unit.

The decision variable in the overbooking model is the number of seats to overbook. Let this be denoted by the non-negative variable Q.

8.4.3.2 Objective Function

The optimization problem is to determine the value of Q such that the sum of the spoilage cost and the oversold penalty is minimized.

- Spoilage Cost

    When the no-show random variable X exceeds the value of Q, spoilage cost of C_1 per unit is incurred. In other words, the plane capacity has been "undersold." Thus, the expected cost of spoilage, denoted by EC_1, is given by

$$EC_1 = \int_Q^\infty C_1(x-Q)f(x)dx. \qquad (8.1)$$

In Equation 8.1, $(x - Q)$ represents the number of empty seats, when $x > Q$, and $C_1(x - Q)$ is the cost of empty seats or spoilage. The probability that the no-shows are larger than Q is given by $\int_Q^\infty f(x)dx$.

- Oversold Penalty

    When the no-show random variable X is less than Q, then passengers are bumped, incurring an oversold penalty cost of C_2 per unit. Thus, the expected cost of oversold penalty, denoted by EC_2, is given by

$$EC_2 = \int_0^Q C_2(Q-x)f(x)dx. \qquad (8.2)$$

The total expected cost, as a function of Q, denoted by $TEC(Q)$, is given by

$$\begin{aligned} TEC(Q) &= EC_1 + EC_2 \\ &= C_1\int_Q^\infty (x-Q)f(x)dx + C_2\int_0^Q (Q-x)f(x)dx. \end{aligned} \qquad (8.3)$$

8.4.3.3 Overbooking Policy for the Continuous Case

Figure 8.1 illustrates how EC_1, EC_2, and TEC change as Q increases from zero. $Q = 0$ represents no overbooking. Hence, there is no oversold penalty and $EC_2 = 0$. However, the spoilage cost will be at its maximum. As Q increases, there is more overbooking, which increases EC_2. But there will be less empty seats, and EC_1 decreases (see Figure 8.1). The total cost TEC(Q) is minimized at $Q = Q^*$, where the expected spoilage cost (EC_1) and the expected penalty cost (EC_2) are equal. To determine the optimal value Q^*, TEC (Equation 8.3) has to be minimized with respect to Q. To determine Q^*, we need to find where the first derivative of TEC(Q) with respect to Q is zero. In other words, solve

$$\frac{\delta TEC(Q)}{\delta Q} = 0$$

The differentiation of Equation 8.3 with respect to Q is done using the Leibniz's rule for differentiation under the integral sign. The final result is given below:

$$\frac{\delta\, TEC(Q)}{\delta Q} = (C_1 + C_2)\int_0^{Q^*} f(x)dx - C_1 = 0. \tag{8.4}$$

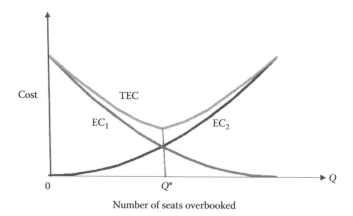

FIGURE 8.1
Impact of Q on EC_1, EC_2, and TEC.

Since $\dfrac{\partial^2 \text{TEC}(Q)}{\partial Q^2} \geq 0$, the value of Q that satisfies Equation 8.4 minimizes TEC. Thus, the optimal solution Q^* is given by

$$F(Q^*) = P(X \leq Q^*) = \int_0^{Q^*} f(x)\,dx = \frac{C_1}{C_1 + C_2} = \text{CF}, \qquad (8.5)$$

where CF is called the *critical fractile*, and it represents the ratio of spoilage cost to the sum of spoilage cost and oversold penalty. The optimal solution, represented by Equation 8.5, is the optimal *overbooking policy*, when the past data on no-shows follow a continuous distribution.

8.4.3.4 Overbooking Policy for the Discrete Case

The overbooking formula (Equation 8.5) has to be modified slightly if the probability distribution of the no-shows is a discrete distribution, given by p_x, where $p_x = \text{Prob}(X = x)$.

The total expected cost (TEC), in the discrete case, is given by

$$\text{TEC}(Q) = C_1 \sum_{x=Q}^{\infty} (x-Q)p_x + C_2 \sum_{x=0}^{Q} (Q-x)p_x \qquad (8.6)$$

The first difference of TEC(Q) with respect to Q, denoted by $\Delta\text{TEC}(Q)$, is given by

$$\Delta\text{TEC}(Q) = \text{TEC}(Q+1) - \text{TEC}(Q)$$
$$= (C_1 + C_2)\sum_{x=0}^{Q} p_x - C_1 \qquad (8.7)$$
$$= (C_1 + C_2)F(Q) - C_1.$$

Note that Equation 8.7 is negative at $Q = 0$, and as Q increases, $\Delta\text{TEC}(Q)$ increases. As long as $\Delta\text{TEC}(Q)$ is <0, TEC(Q) decreases with Q as shown in Figure 8.2. The minimum value of TEC(Q) occurs at Q^*, where $\Delta\text{TEC}(Q)$

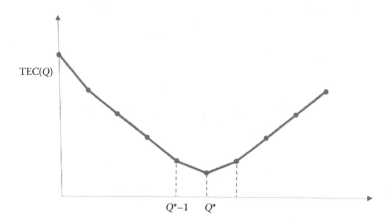

FIGURE 8.2
Total expected cost for a discrete distribution.

changes its sign from negative to positive (see Figure 8.2). Thus, the optimal Q^* satisfies the conditions given below:

$$F(Q^* - 1) < CF$$
$$F(Q^*) \geq CF, \qquad (8.8)$$

where $CF = \dfrac{C_1}{C_1 + C_2}$.

In other words, the overbooking policy, for the discrete case, is to find the *smallest value of Q*, such that

$$F(Q) = P(X \leq Q) \geq \dfrac{C_1}{C_1 + C_2} = CF. \qquad (8.9)$$

In many practical situations, past data on no-shows may not fit a standard probability distribution, such as Normal in order to use Equation 8.5. In such cases, it would be easier to generate an empirical frequency distribution of the no-shows and use Equation 8.9 to determine the optimal overbooking policy.

8.4.3.5 Illustrative Example

In this section, we will use several numerical examples to illustrate the use of overbooking models.

Revenue Management

Example 8.1

A regional airline is planning to develop an overbooking policy for one of its popular routes. Past data indicate that the number of no-show passengers follows a Uniform distribution over the interval (5, 25). The cost of an empty seat is estimated at $150. Each confirmed passenger, who is denied boarding due to overbooking, costs the airline $400. This includes real cost and an estimate of the goodwill loss. Determine the optimal number of seats to overbook.

SOLUTION

In this example, the distribution of the no-show random variable X is given as

$$X \sim U(5, 25).$$

The values of C_1 and C_2 are $150 and $400, respectively.

Recall from probability theory that if X is distributed Uniform over the interval (a, b), then

- $f(x) = \dfrac{1}{b-a}$ for $a \le x \le b$
- $F(x) = \dfrac{x-a}{b-a}$ for $a \le x \le b$
- $E(x) = \dfrac{a+b}{2}$
- $V(x) = \dfrac{(b-a)^2}{12}$.

Using Equation 8.5 for the continuous case, we have to solve for Q, such that

$$F(Q) = \frac{Q^* - 5}{25 - 5} = \frac{150}{150 + 350} = 0.3.$$

Solving for Q^*, we get

$$Q^* = (0.3)(20) + 5 = 11.$$

The optimal number of seats to overbook is 11, which is below the average number of no-shows, which is 15. One of the simplicities of the overbooking model is that it is not necessary to determine precisely the values of C_1 and C_2 in practice. Even an approximate estimate of the relative magnitudes of C_1 and C_2 is enough to apply the overbooking model. Example 8.2 illustrates this.

Example 8.2

A major hotel chain, located in a campus town, wants to develop an overbooking policy for football weekends. The hotel has 300 rooms. Past data indicate that the distribution of no-shows is Normal with a mean of 15 and a variance of 25. When the hotel has no rooms for confirmed guests, it accommodates them at another hotel in the next town at a much higher cost. It also pays for transportation to the stadium. The cost of an oversold room is approximately three times that of an empty room. Determine the optimal overbooking policy.

SOLUTION

In this example, the no-show random variable X has a Normal distribution, given by

$$X \sim N(\mu, \sigma^2) = N(15, 5^2).$$

The critical fractile is calculated as follows:

$$CF = \frac{C_1}{C_1 + C_2} = \frac{1}{1+3} = 0.25.$$

Using Equation 8.5, the optimal value of Q is obtained by solving

$$F(Q^*) = CF = 0.25$$
$$Q^* = F^{-1}(0.25).$$

Using Microsoft's Excel function for Inverse Normal, we get

$$Q^* = \text{NORMINV}(CF, \mu, \sigma)$$
$$= \text{NORMINV}(0.25, 15, 5)$$
$$= 11.62755.$$

Since Q^* has to be an integer, we round it "up" to the value of 12. In practice, the value of Q^* is usually rounded up or down to the nearest integer. Thus, the hotel should overbook by 12 rooms during the football weekends.

Example 8.3

The Tiger bus company operates two types of services between New York and Washington D.C. The non-stop express bus takes 3 hours and costs $75 per ticket. The multi-stop general bus takes 5 hours and costs $50 per ticket. The Tiger bus company is interested in developing an overbooking policy for its express bus service. The bus company has

Revenue Management

collected data on no-shows, which varies between 0 and 15. Table 8.1 gives the no-show values and their frequencies of occurrence. When the express bus is full, passengers with confirmed tickets are accommodated in their multi-stop general buses. For their inconvenience, the company gives them a travel voucher worth $35, which includes the difference in bus fares plus incidental expenses. Determine Tiger's overbooking policy.

SOLUTION

In this example, $C_1 = \$75$ and $C_2 = \$35$. Hence, the critical fractile,

$$CF = \frac{75}{75+35} = 0.68.$$

Since the no-show distribution, given by Table 8.1, is discrete, we use Equation 8.9 to determine the overbooking policy. The number of seats to overbook is the smallest value of Q such that

$$F(Q) = P(X \leq Q) \geq CF.$$

Using Table 8.1,

$$F(7) = 0.6 < 0.68$$
$$F(8) = 0.7 > 0.68$$

Hence, the optimal number of seats to overbook is 8.

TABLE 8.1

No-Show Data for Example 8.3

No-Shows (X)	Frequency (p_x)	Cumulative Probability F(x)
0	5%	0.05
1	5%	0.1
2	5%	0.15
3	5%	0.2
4	10%	0.3
5	10%	0.4
6	10%	0.5
7	10%	0.6
8	10%	0.7
9	10%	0.8
10	10%	0.9
11	2%	0.92
12	2%	0.94
13	2%	0.96
14	2%	0.98
15	2%	1.0

8.5 Revenue Management in Airlines

As discussed in Section 8.1, one of the first industries to broadly implement a systematic revenue management system was the airlines industry (Smith et al. 1992). This largely came about in a response to deregulation of that industry in the United States in the late 1970s. Before that time, airlines had to seek regulatory approval to provide service for a route. This allowed established carriers to effectively block competition since there were high barriers to entry for new carriers. After passage of the Airlines Deregulation Act in 1978, the industry significantly changed. One change was that established carriers moved from a point-to-point system to a "hub-and-spoke" network. In the hub-and-spoke model, a strategic airport was chosen as a hub to effectively handle traffic to origins and destinations. The difference between the two systems is shown in Figure 8.3. The primary benefit of the hub-and-spoke system is that it allows airlines to more effectively use their capacity. It also helped airlines to establish regional dominance in geographic regions around their hubs. Examples of hubs are Chicago for United Airlines and Atlanta for Delta Airlines.

A second change that resulted from deregulation was price competition. This included the introduction of "no-frills," low-cost carriers. One example of such a carrier was People Express (PEx) airlines, which was launched in 1981. In an effort to lure business travelers, PEx used a simple pricing structure, where all seats were sold at the same price. The prices on some flights were lower than the costs of the same route by some of the major carriers like American Airlines. American responded to PEx by introducing limited super-saver fares on flights with excess capacity at a price lower than PEx. American Airlines called the approach "yield management," and a year after implementation, revenues increased by 14.5% and its profits increased by 47.8% (Cross 1997).

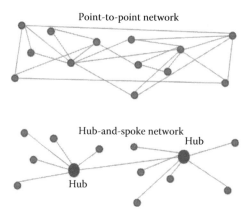

FIGURE 8.3
Point-to-point and hub-and-spoke networks.

8.5.1 Airline Fare Classes

Before going into details of airline revenue management, a little background is needed. When a customer purchases a ticket, they designate an origin (O) and destination (D). The customer may need to make stops on their route (O–D). For example, when flying from Atlanta, Georgia to State College, Pennsylvania on Delta Airlines, the route would typically be Atlanta (ATL) to Detroit (DTW) to State College (SCE). Each segment of the route is called a *leg*, and each leg will have a capacity based on the size of the plane. This is illustrated in Figure 8.4. Note that since DTW is a hub, passengers on the leg from ATL–DTW will have many different final destinations. The plane that the customer flies on may have various cabins, for example, first class and coach. In this section, we consider tickets sold in the coach cabin.

The idea of revenue management is to charge a different price (or *fare class*) for tickets in the same cabin depending on conditions. Example fare classes for coach tickets are shown in Table 8.2. Conditions are defined by the airlines and are called *fences*. Fences may include days out from the flight

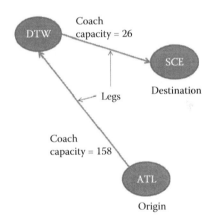

FIGURE 8.4
Multi-leg flight from Atlanta, Georgia to State College, Pennsylvania.

TABLE 8.2

Example Coach Fare Classes

Fare Class Code	Description
Y	Full coach fare
B, M	Slightly discounted coach fare
H, Q, K	Discounted coach fare
L, U, T	Deeply discounted coach fare

on date of purchase, Saturday night stay, ability to upgrade a ticket, and whether the ticket is flexible (i.e., customer can change the ticket at no cost). To receive a discounted fare class, several of these fences will need to be satisfied. The fewer the number of fences, the higher the price of the resulting fare class. For example, for one to receive a deeply discounted fare, it may require that a ticket is purchased at least 30 days in advance, it includes a Saturday night stay, it is not upgradable, and it cannot be refunded if the flight is missed.

Fences help to segment customers in order to price discriminate between them. We can illustrate by comparing a student going for spring break and an executive going to a business meeting. The average student will be much more price sensitive than a business executive. Further, the student will know when they want to take trip well in advance, will likely stay over a Saturday night on their trip, and will not likely cancel the trip. An executive will typically want a specific flight, may not know about the trip until just before it needs to be made, and wants flexibility in case the meeting gets changed at the last minute. The fences ensure that the business executive will not be able to pay the discount fare that the student does.

8.5.2 Setting Limits on Two Fare Classes

Because an airline does not want to sell too many tickets in lower fare classes, it needs to set limits for each fare class. In this section, we will discuss a simple case, where an airline has just two fare classes—full fare for business customers and discount fare for leisure travelers. It wants to determine how many full fare seats should be set aside for the business customers. Once the number of full fare seats is decided, the remaining seats are sold at a discounted rate, with restrictive conditions. The overbooking model we discussed in Section 8.4 can be applied to solve this problem.

8.5.2.1 Optimization Problem

Let the random variable X denote the full fare demand. Assume that the full fare price is F and the discount fare price is D. Let the decision variable Q denote the number of full fare seats reserved. Similar to the overbooking problem, we have the following two situations to consider:

i. *Case 1*: $X > Q$

Here, we have underestimated the full fare demand, and the actual demand for full fare exceeds the number of seats reserved for them. In other words, there are more business customers who want to travel, but the full fare tickets are sold out. In this model, we make this simple assumption that the discount seats are still available and these business customers will buy the discount ticket. Thus, the loss in revenue by not having enough full fare seats is $(F - D)$ per ticket. The expected cost function for this case is given by

$$EC_1 = (F - D)\int_Q^\infty (x - Q)f(x)dx$$

ii. *Case 2: X < Q*

Here, we have "overestimated" the full fare demand and set aside too many seats for full fare passengers. Hence, the excess full fare seats will go empty. Here, also, we make the simple assumption that these empty seats could have been sold to leisure travelers at the discount fare D. Thus, the expected cost function for this case is given by

$$EC_2 = D\int_0^Q (Q - x)f(x)dx.$$

Thus, the objective function to be minimized is the sum of EC_1 and EC_2 as given below:

$$TEC(Q) = EC_1 + EC_2$$
$$= (F - D)\int_Q^\infty (x - Q)f(x)dx + D\int_0^Q (Q - x)f(x)dx. \qquad (8.10)$$

Equation 8.10 is very similar to Equation 8.3 given in the overbooking model. The only difference here is that $C_1 = F - D$ and $C_2 = D$. Using the overbooking formula given in Equation 8.5, the optimal number of seats to be reserved for full fare passengers (Q^*) is given by

$$F(Q^*) = \int_0^{Q^*} f(x)dx = \frac{C_1}{C_1 + C_2} = \frac{F - D}{F - D + D} = \frac{F - D}{F}. \qquad (8.11)$$

If the full fare demand follows a discrete probability distribution, we can use the overbooking formula given by Equation 8.9, with the critical fractile CF calculated as $CF = \frac{F - D}{F}$. We shall illustrate the simple model for two fare classes with a numerical example.

Example 8.4

A regional airline, flying between Los Angeles and San Francisco, has two fare classes, a full fare of $350 and a discount fare of $150. The demand for full fare follows a Normal distribution with a mean of 150 and a standard deviation of 100. The airline wants to determine the number of full fare seats to set aside for the flight.

SOLUTION

We are given $F = 350$, $D = 150$, and $X \sim N(150, 100^2)$. The critical fractile

$$CF = \frac{F-D}{F} = \frac{350-150}{350} = \frac{200}{350} = 0.57.$$

Using Equation 8.11, the optimal value of Q is obtained by solving

$$F(Q^*) = CF = 0.57$$
$$Q^* = F^{-1}(0.57).$$

Using Microsoft's Excel function for Inverse Normal, we get

$$Q^* = \text{NORMINV}(0.57, 150, 100) = 167.6374.$$

By rounding the value "up" to the nearest integer, the airline should reserve 168 seats for full fare passengers.

Remarks: In practice, airlines do not set a fixed value for full fare seats. The number of full fare and discount fare tickets sold is updated daily based on the number of tickets sold for the flight at that point in time. If the demand is much higher than the historical values, they stop selling additional discount tickets. After a few days, if the demand pattern changes and the ticket sales fall below the historical norms, they start selling discount tickets again. That is why a flight with all discount fares sold out 10 days before the departure date may have discount fares available just 2 days before the departure! We discuss this more in detail in Sections 8.5.4 and 8.5.5.

8.5.3 Setting Limits on Multiple Fare Classes

When there are more than two fare classes, similar to Table 8.2, the overbooking model is no longer applicable. Another way to set limits is through a network optimization model. In this model, the total capacity on a leg can be segmented into several fare classes based on the price and expected demand. We illustrate this with a simple linear programming model. We make the following assumptions:

- The network has already been defined, including flights, departure times, and plane type.
- The fare classes and price for each class have been previously determined.

Revenue Management

- The expected demand for each fare class on each leg is known.
- The marginal cost of a flight is $0. The profit is therefore completely defined by the ticket price. This is without loss of generality.
- We are examining the network for departures over a given time interval (e.g., 9:00 a.m. to 10:00 a.m.).

We will use the following notation in the formulation:

c_i = total capacity on leg i {$i = 1, \ldots, M$}
d_j = expected demand for ticket j {$j = 1, \ldots, N$} at the posted price
p_j = price of a ticket j
a_{ij} = 1 if flight j crosses leg i and 0 otherwise
x_j = number of tickets of type j to sell

To simplify the indexing, the term ticket j includes the leg and fare class, which have been ordered. The formulation is as follows (x_j are the decision variables):

$$\text{Maximize } Z = \sum_{j=1}^{N} p_j x_j \tag{8.12}$$

Subject to

$$\sum_{j=1}^{N} a_{ij} x_j \leq c_i, \quad \forall i \text{ (Leg capacity)} \tag{8.13}$$

$$0 \leq x_j \leq d_j, \quad \forall j \text{ (Ticket demands)} \tag{8.14}$$

Example 8.5

Consider the network shown in Figure 8.5. The capacities on each leg are also given in this figure. Tickets (shown as O–D pairs) in the fare classes with prices and expected demand are given in Table 8.3.
   The linear programming formulation is

$$\text{Maximize } Z = 400x_1 + 300x_2 + 500x_3 + 400x_4 + 350x_5 + 600x_6 + 450x_7$$
$$+ 300x_8 + 200x_9 + 150x_{10} + 400x_{11} + 300x_{12} + 450x_{13}$$
$$+ 300x_{14} + 290x_{15} + 250x_{16} + 300x_{17} + 230x_{18}$$

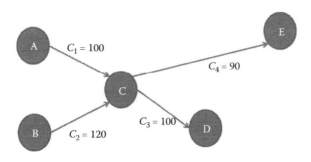

FIGURE 8.5
Airline network with leg capacities.

TABLE 8.3

Data for Example 8.5

Ticket Type	Origin–Destination	Class	Price	E [Demand]
1	A–C	Y	400	15
2	A–C	M	300	20
3	A–D	Y	500	20
4	A–D	M	400	30
5	A–D	Q	350	35
6	A–E	Y	600	15
7	A–E	M	450	20
8	B–C	Y	300	18
9	B–C	M	200	23
10	B–C	Q	150	30
11	B–D	Y	400	10
12	B–D	M	300	20
13	B–E	Y	450	18
14	B–E	M	300	26
15	C–D	Y	290	15
16	C–D	M	250	20
17	C–E	Y	300	20
18	C–E	M	230	25

Revenue Management

Subject to

$$x_1 + x_2 + x_3 + x_4 + x_5 + x_6 + x_7 \leq 100 \quad \text{(capacity of leg A–C)}$$

$$x_8 + x_9 + x_{10} + x_{11} + x_{12} + x_{13} + x_{14} \leq 120 \quad \text{(capacity of leg B–C)}$$

$$x_3 + x_4 + x_5 + x_{11} + x_{12} + x_{15} + x_{16} \leq 100 \quad \text{(capacity of leg C–D)}$$

$$x_6 + x_7 + x_{13} + x_{14} + x_{17} + x_{18} \leq 90 \quad \text{(capacity of leg C–E)}$$

$$x_1 \leq 15;\ x_2 \leq 20;\ x_3 \leq 20;\ x_4 \leq 30;\ x_5 \leq 35;\ x_6 \leq 15;$$

$$x_7 \leq 20;\ x_8 \leq 18;\ x_9 \leq 23;\ x_{10} \leq 30;\ x_{11} \leq 10;\ x_{612} \leq 20;$$

$$x_{13} \leq 18;\ x_{14} \leq 26;\ x_{15} \leq 15;\ x_{16} \leq 20;\ x_{17} \leq 20;\ x_{18} \leq 25$$

The solution is

$$Z^* = \$96,830;\ x_1^* = 15;\ x_2^* = 20;\ x_3^* = 20;\ x_4^* = 15;\ x_5^* = 0;\ x_6^* = 15;$$

$$x_7^* = 15;\ x_8^* = 18;\ x_9^* = 23;\ x_{10}^* = 30;\ x_{11}^* = 10;\ x_{12}^* = 20;$$

$$x_{13}^* = 18;\ x_{14}^* = 1;\ x_{15}^* = 15;\ x_{16}^* = 20;\ x_{17}^* = 20;\ x_{18}^* = 21$$

A key problem with using the linear programming formulation is that it does not account for any variation about the expected demand. For example, if the actual demand realized for ticket 1 in Example 8.5 was 17 rather than 15, we would not be able to get the higher price on the two additional units of demand since those seats had been allocated to a lower paying fare class. A dynamic approach that determines the fare classes to keep open and the ones to no longer offer as demand is realized over time would help address this issue. This is the idea to an approach called *bid price controls*.

8.5.4 Bid Price Controls

One of the important aspects of revenue management in the airlines is that different customers use the same capacity. In Example 8.5, for instance, customers flying from A to C and from A to D both use the capacity on leg A–C. We would like to be able to determine the value of the capacity to determine how to best use it. We can do this by looking at the dual formulation of the linear programming model. We will use the same notation as previously, with the exception of the dual variables.

y_i = dual variables for Constraint 8.13
v_j = dual variables for Constraint 8.14

The formulation of the dual linear program is

$$\text{Minimize } W = \sum_{i=1}^{M} c_i y_i + \sum_{j=1}^{N} d_j v_j \tag{8.15}$$

Subject to

$$\sum_{i=1}^{N} a_{ij} y_i + v_j \geq p_j, \forall j \tag{8.16}$$

$$y_i, v_j \geq 0, \forall i, j \tag{8.17}$$

The solution of dual variables y_i {i = 1, ..., M} gives the value of increasing the capacity on leg i by one unit. This is called the *bid price*. We can use the bid prices to determine which fare classes to keep open by using the following procedure:

- For each ticket j, determine the flight minimum value (V_j) by summing the bid prices of all of the legs that it crosses: $V_j = \sum_{i=1}^{M} a_{ij} y_i$.
- Close any fare class j where the price of the ticket is less than V_j.

We illustrate this bid pricing approach by using the data from Example 8.5.

Revenue Management

Example 8.6

Consider the data from Example 8.5. The dual formulation is

$$\text{Minimize } W = 100y_1 + 120y_2 + 100y_3 + 90y_4 + 15v_1 + 20v_2$$
$$+ 20v_3 + 30v_4 + 35v_5 + 15v_6 + 20v_7 + 18v_8$$
$$+ 23v_9 + 30v_{10} + 10v_{11} + 20v_{12} + 18v_{13} + 26v_{14}$$
$$+ 15v_{15} + 20v_{16} + 20v_{17} + 25v_{18}$$

Subject to

$$y_1 + v_1 \geq 400; \; y_1 + v_2 \geq 300; \; y_1 + y_3 + v_3 \geq 500;$$

$$y_1 + y_3 + v_4 \geq 400; \; y_1 + y_3 + v_5 \geq 350; \; y_1 + y_4 + v_6 \geq 600;$$

$$y_1 + y_4 + v_7 \geq 450; \; y_2 + v_8 \geq 300; \; y_2 + v_9 \geq 200;$$

$$y_2 + v_{10} \geq 150; \; y_2 + y_3 + v_{11} \geq 400; \; y_2 + y_3 + v_{12} \geq 300;$$

$$y_2 + y_4 + v_{13} \geq 450; \; y_2 + y_4 + v_{14} \geq 300; \; y_3 + v_{15} \geq 290;$$

$$y_3 + v_{16} \geq 250; \; y_4 + v_{17} \geq 300; \; y_4 + v_{18} \geq 230$$

The optimal solution to the dual problem is

$$W^* = \$96,830; \; y_1^* = 220; \; y_2^* = 70; \; y_3^* = 180; \; y_4^* = 230; \; v_1^* = 180; \; v_2^* = 80;$$

$$v_3^* = 100; \; v_4^* = 0; \; v_5^* = 0; \; v_6^* = 150; \; v_7^* = 0; \; v_8^* = 230;$$

$$v_9^* = 130; \; v_{10}^* = 80; \; v_{11}^* = 150; \; v_{12}^* = 50; \; v_{13}^* = 150; \; v_{14}^* = 0;$$

$$v_{15}^* = 110; \; v_{16}^* = 70; \; v_{17}^* = 70; \; v_{18}^* = 0$$

We would close class ticket type 5 since its price of $350 is less than the value of the capacity that it uses, which is $220 + $180 = $400. Note that this solution corresponds to $x_5^* = 0$ in the primal solution.

The real benefit of the bid price controls approach is that bid prices are updated as capacity is used. The process is therefore better able to adopt when demand varies from the expected value. For example, using the data in Example 8.5, suppose we sell 10 M-class tickets on B–E and everything else stays the same. This means that the capacities on legs B–C and C–E go to 110 and 80, respectively. Re-solving the dual gives the following bid prices: $y_1^* = 220; y_2^* = 120; y_3^* = 180; y_4^* = 230$. As a result, we would now close ticket type 14 in addition to ticket type 5.

In theory, we would update the bid prices after every ticket that is sold. In practice, however, airline networks are too large for this to be computationally feasible. Therefore, airlines typically update bid prices once or twice a day.

8.5.5 Booking Curves

Although the bid price controls approach presented in the last section can adopt somewhat if demands differ from expected values, its usefulness still depends to a certain extent on demand forecasts. Airlines typically have rather extensive data on past ticket sales. As such, they can estimate how cumulative demand grows for a flight as the departure date approaches. This is known as a *booking curve*. Two examples are shown in Figure 8.6. Part (a) represents a typical booking curve for business persons. Notice that most of the demand occurs close to the departure date. Part (b) represents passengers, who shop for discounts (e.g., non-business travelers such as students, tourists, or elderly passengers). They will book tickets much earlier in order to obtain a cheaper ticket price.

Statistical methods can be used to fit historical data to the booking curves. Let Y denote the total number of tickets that are sold and t the

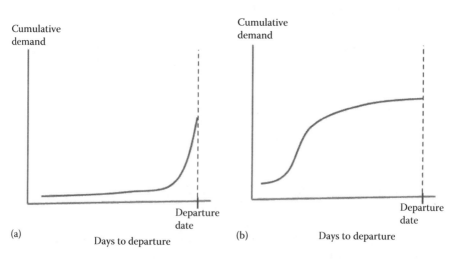

FIGURE 8.6
Examples of booking curve for non-business travelers (a) and business travelers (b).

Revenue Management

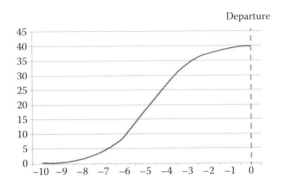

FIGURE 8.7
S-shaped booking curve ($a = 40$, $b = 0.2$, $c = 1$, $d = 5$).

days until departure. Curve (a) in Figure 8.6 can be modeled as an exponential curve:

$$Y = ae^{bt}, \tag{8.18}$$

where a is the estimated total demand and b ($b > 0$) defines the spread of the curve. Curve (b) in Figure 8.6 is known as the s-curve and can be modeled as*

$$Z = \frac{a}{1 + e^{b-c(t+d)}}. \tag{8.19}$$

The parameter c represents the spread of the data, b defines the shift of the curve, and d defines the time offset. The ratio of $b/c - t$ represents the point of the curve where half of the tickets are sold. Figure 8.7 shows a booking curve for $a = 40$, $b = 0.2$, $c = 1$, and $d = 5$. We can establish threshold curves based on the variability of ticket sales over time. If the actual demand falls outside of the threshold curves, then we can update our demand estimate, which effectively helps reallocate capacity. Figure 8.8 shows the case of tracking ticket sales over time compared to the booking curve. At 5 days out, the cumulative sales falls above the threshold. Therefore, the demand for business tickets should be increased. In the optimization model, this leads to an increase in the bid price on the capacities of legs that the flights on this ticket cross. Therefore, the capacity gets reserved for business travelers.

---

* Another common non-business booking curve is $Y = \exp\left(a - \frac{b}{t}\right)$, where Y is total tickets sold. This form is easily linearized.

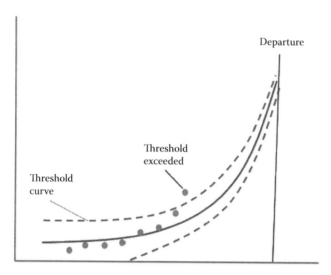

FIGURE 8.8
Cumulative demand exceeds the threshold.

8.5.6 Remarks on Airline Revenue Management

Although bid pricing controls is a dynamic strategy, its success does still depend on having reasonable forecasts. Part of the challenge of forecasting is that the demand is censored. That is, once a fare class is closed, it is difficult to determine what the remaining demand for that class is since it can't be directly observed. The consequences of this are discussed in Cooper et al. (2006) in what they call the *spiral down effect*. This occurs when incorrect assumptions about customer demand cause high-fare ticket sales to systematically decrease over time. In particular, they show that "if an airline decides how many seats to protect for sale at a high fare based on past high-fare sales, while neglecting to account for the fact that availability of low-fare tickets will reduce high-fare sales, then high-fare sales will decrease, resulting in lower future estimates of high-fare demand."

Another problem of using bid price controls is that it does not capture any of the stochastic nature of the problem. Historically, a dynamic programming value function has been used (McGill and van Ryzin 1999) to capture the stochastic demand component. The curse of dimensionality prevents direct solution of the dynamic program (DP) for reasonably sized problems. The bid price linear program presented here was developed as an approximation for it. It has been a very active area of research to extend this approach. For example, Adelman (2007) developed a functional transformation to the DP that allows for time-varying bid prices.

More recently, airlines have focused on other areas of revenue management beyond ticket sales. This includes new fees around luggage and the purchase of food. These have proven to be profitable endeavors for the

Revenue Management

airlines. An important issue in their implementation, however, is the consideration of consumer backlash to the practices. A popular article appeared in a travel magazine (Hess 1998) that parodies the situation of an airlines selling paint in a store and applying revenue management practices. The parody is perhaps even more *apropos* today.

Exercises

8.1 What are the pros and cons of overbooking as a revenue management strategy? Why is it necessary?

8.2 How does self-service help capacity planning in the service sector? Would you consider self-driving cars as an example of self-service? Why or why not?

8.3 Discuss the key benefits and limitations of revenue management for the airlines.

8.4 Compare and contrast the practices of Delta Airlines, which uses hub-and-spoke network, and Southwest Airlines, which uses point-to-point network. What do you believe are the advantages and disadvantages of each business model?

8.5 Consider the no-show data given in Table 8.1 for Example 8.3. Determine the overbooking policies for the following spoilage cost (C_1) and oversold penalty (C_2).
   a. $C_1 = \$100$, $C_2 = \$50$
   b. $C_1 = \$100$, $C_2 = \$100$
   c. $C_1 = \$100$, $C_2 = \$200$
   d. $C_1 = \$100$, $C_2 = \$300$
   e. $C_1 = \$100$, $C_2 = \$400$

8.6 Consider Exercise 8.5 again. What will be the total expected cost (TEC) under the following extreme cases, for the different values of (C_1, C_2).
   a. No overbooking
   b. Overbook to the maximum of 15

   Compare the total cost of the two extreme cases with the total cost of the optimal policies obtained in Exercise 8.5.

8.7 A family-operated inn in State College, Pennsylvania, is considering overbooking to minimize their revenue loss due to no-shows and last-minute cancellations. The cost of an empty room is estimated at $95 per night. When the inn is fully occupied, customers with guaranteed reservations are placed in nearby expensive hotel chains, averaging $165 per night. In addition to paying the difference in room rates, the inn also gives a $20 voucher for taxi fare. Determine the inn's overbooking policy, given that the distribution of no-shows is Uniform over the interval (5, 25).

8.8 Consider Exercise 8.7 again. What will be the overbooking policy for the inn if the no-shows follow a Normal distribution with a mean of 20 and a standard deviation of 5.

8.9 A small riverfront hotel is interested in developing an overbooking policy for its rooms. The no-show data are given in Table 8.4. The cost of an empty room is $100. The overbooking penalty cost is $75.

  a. Determine the total expected cost, if the hotel overbooks by 0, 1, 2, 3, 4, 5, 6, and 7 rooms. What should be the optimal number of rooms to overbook?

  b. Verify the optimal policy obtained in part (a) by using the overbooking formula, given by Equation 8.9.

8.10 A fly-by-night airline offers "red-eye" flights between San Francisco and Washington D.C. The full fare ticket price is $700 and the advance purchase discount fare is $400. The demand for full fare tickets is distributed Normal with a mean of 100 and a standard Hdeviation of 25.

How many full fare seats the airline should set aside in order to minimize the total expected cost?

8.11 Redo Exercise 8.10, assuming now that the demand for full fare passengers follows a Uniform distribution over the interval (50, 200).

8.12 For a given flight with a capacity of 85 seats, price and demand are as follows (assume a marginal cost is $0):

Price	$100	$130	$150	$200	$240	$300
Demand	80	70	65	50	37	29

  a. What is the maximum profit that can be achieved with only a single price?

  b. A fence can be established so that two prices can be charged. How does your answer to part (a) change?

  c. Repeat part (b) using three prices.

TABLE 8.4
No-Show Data for Exercise 8.9

No-shows	0	1	2	3	4	5	6	7
Frequency	8	7	6	5	4	3	2	1

Revenue Management

8.13 Consider the airline network with capacities as shown in Figure 8.9. The price and demand data are shown in Table 8.5.
   a. Formulate and solve the primal LP problem. How should the capacity be allocated?
   b. Formulate and solve the dual LP problem to determine the bid prices.
   c. What is the difference in the way that the two approaches handle demand?

8.14 For the data in Exercise 8.13, the ticket requests received are given in Table 8.6. Determine whether or not to accept each one, and update the bid prices after each transaction. Only accept a transaction if all tickets in the transaction will be sold.

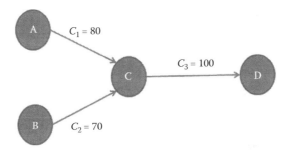

FIGURE 8.9
Capacity data for Exercise 8.13.

TABLE 8.5

Price and Demand Data for Exercise 8.13

Ticket Type	Origin–Destination	Class	Price	Demand
1	A–C	Y	400	15
2	A–C	M	280	30
3	A–D	Y	550	12
4	A–D	M	450	22
5	A–D	Q	300	38
6	B–C	Y	300	30
7	B–C	M	220	40
8	B–D	Y	550	15
9	B–D	M	450	22
10	B–D	Q	300	40
11	C–D	Y	320	15
12	C–D	M	250	22
13	C–D	Q	200	30

TABLE 8.6

Ticket Requests for Exercise 8.14

Transaction Number	Ticket Type	Number of Tickets
1	7	5
2	13	2
3	2	4
4	5	4
5	5	10
6	9	3
7	2	1

TABLE 8.7

Demand Data for Exercise 8.16

Day to Departure	−10	−9	−8	−7	−6	−5	−4	−3	−2	−1
Demand	2	4	3	8	12	9	3	2	1	1

8.15 Develop an approach to construct threshold curves for the business customers.

8.16 Given the demand data in Table 8.7, estimate the parameters for the non-business booking curve.

References

Adelman, D. 2007. Dynamic bid prices in revenue management. *Operations Research*. 55: 647–661.

Cook, T. 1998. SABRE Soars. *OR/MS TODAY* (June), 26–31.

Cooper, W. L., T. Homem-de-Mello, and A. J. Kleywegt. 2006. Models of the spiral-down effect in revenue management. *Operations Research*. 54: 968–987.

Cross, R. 1997. *Revenue Management: Hard-core Tactics for Market Domination*. New York: Broadway Books.

Davis, M. M. and J. Heineke. 2003. *Managing Services: Using Technology to Create Value*. New York: McGraw-Hill.

Fitzsimmons, J. A. and M. J. Fitzsimmons. 2006. *Service Management*. 5th Edition. New York: McGraw-Hill.

Haksever, C. and B. Render. 2013. *Service Management*. Upper Saddle River, NJ: Pearson.

Hess, A. 1998. If airlines sold paint. *Travel Weekly*, October 16: 67. Available at http://www.travelweekly.com/Agent-Life/If-airlines-sold-paint.

Holloran, T. J. and J. E. Byrn. 1986. United Airlines station manpower planning system. *Interfaces*. 16(1): 39–50.

Hueter, J. and W. Swart. 1998. An integrated labor-management system for Taco Bell. *Interfaces*. 28(1): 75–91.

McGill, J. I. and G. J. van Ryzin. 1999. Revenue management: Research overview and prospects. *Transportation Science*. 33: 233–256.

Metters, R. D., K. King-Metters, M. Pullman, and S. Walton. 2006. *Successful Service Operations Management*. 2nd Edition. Mason, OH: South Western.

Smith, B., J. Leimkuhler, and R. Darrow. 1992. Yield management at American Airlines. *Interfaces*. 22(1): 8–31.

Srinivas, S. 2017. Analytics-driven design of multi-phase multi-provider appointment system for patient scheduling. PhD dissertation. The Pennsylvania State University, University Park, PA.

9

Retail Engineering

Retailing firms primarily focus on selling goods and services to consumers through their stores and increasingly through the Internet. Retailers play a vital role in their supply chains by serving as the link between consumers and the rest of the supply chain as shown in Figure 9.1. The retail industry plays a significant part in modern economies. According to the National Retail Federation, the retail industry supports one in every four jobs in the United States and contributes $2.6 trillion to the gross domestic product. Worldwide retail industry revenues are estimated to be more than $22 trillion, and it supports billions of jobs. Ubiquitous Internet connectivity, especially wide use of smartphones, is rapidly reshaping consumer purchasing behavior. It is estimated that $390 billion worth of goods were purchased in the United States over the Internet in 2016, which is also referred to as electronic retailing or *e-tailing*. This trend in e-tailing is expected to steadily increase in the foreseeable future.

9.1 Introduction

The retail industry is dominated by large firms, especially in developed economies, which has made them household names. The size of a retail firm can be a significant advantage in negotiating favorable terms and prices with manufacturers of merchandise sold by the retailer. Size can also be beneficial for a retail firm in gaining brand recognition and trust among consumers. For example, Walmart, Target, Macys, and Amazon are household names in the United States, indicating their brand recognition and trust among consumers. From an engineering viewpoint, large size offers retailers economies of scale to better utilize distribution channels, thereby lowering operating cost. This also tends to make retailers regionally strong by leveraging

FIGURE 9.1
Retail firms link consumers and their supply chains.

the proximity of distribution centers to lower cost. Many retailers have successfully translated their processes and practices to different parts of the world to become global firms. Table 9.1 shows some major retail firms along with the number of countries they operate in, the format of their stores, and annual revenues.

Format of retail stores can vary by retailer and also by location, ranging from 200,000 square feet for a typical Walmart supercenter in suburban locations to 3000 square feet for convenience stores in dense city locations. Table 9.2 shows different retailing formats and their important characteristics. Store format is an important aspect of a retailer's strategy because it determines the cost of the floor space and the amount of merchandise that can be carried in a store. Another aspect of a retailer's strategy is the breadth of merchandise, called "variety," and the depth of merchandise, called "assortment." Loosely, SKUs that are substitutable constitute the assortment for that particular merchandise, whereas SKUs that are not substitutable represent a variety of merchandise at the retailer. Here, SKU (pronounced as "skew") stands for Stock Keeping Unit, each of which has a unique Universal Product Code or barcode.

TABLE 9.1

Major Retail Firms around the World (2015 Data)

Rank	Company	Country of Origin	Number of Countries	Typical Format	Revenue (in Millions)
1	Walmart	United States	30	Supercenter	$482,130
2	Costco	United States	10	Warehouse club	$116,199
3	Kroger	United States	1	Supermarket	$109,830
4	Schwarz-Gruppe (lidl)	Germany	26	Discount store	$94,448
5	Walgreens	United States	10	Pharmacy	$89,631
6	The Home Depot	United States	4	Home improvement	$88,519
7	Carrefour	France	35	Supercenter	$84,856
8	Aldi	Germany	17	Discount store	$82,164
9	Tesco	United Kingdom	10	Supercenter	$81,019
10	Amazon	United States	14	Online	$79,268

TABLE 9.2
Store Formats and Their Characteristics

Format	Variety	Assortment	Size (Thousand Square Feet)	Location	SKU (×1000)	Price	Service
Department stores	Broad	Deep to average	100–200	Regional malls	100	Average to high	Average to high
Discount stores	Broad	Average to shallow	60–80	Stand-alone, power strip centers	30	Low	Low
Specialty stores	Narrow	Deep to average	4–12	Regional malls	5	High	High
Category specialists	Narrow	Very deep	50–120	Stand-alone, power strip centers	20–40	Low	Low to high
Home improvement centers	Narrow	Very deep	80–120	Stand-alone, power strip centers	20–40	Low	Low to high
Drugstores	Narrow	Very deep	3–15	Stand-alone, power strip centers	10–20	Average to high	Average
Off-price stores	Average	Deep but varying	20–30	Outlet malls	50	Low	Low
Extreme-value retailers	Average	Average and varying	7–15	Urban strip	3–4	Low	Low

Source: Adapted from Levy, M. and B. A. Weitz. 2007.

9.2 Types of Retailers

Retailers can be divided into two broad categories: food and general merchandise. Over the last 25 years, food retailing has evolved considerably from being Mom & Pop stores to the supercenters that we see today. Some of the key characteristics of the food retailing industry are as follows:

- Competition from larger discount stores with scale, pricing, and lower cost advantages is putting pressure on the smaller retailers.
- Changes in consumer lifestyle toward eating out more than cooking at home have changed the buying behavior.
- Rapidly increasing e-tailing services is changing consumer expectations to include fresh produce, which has complex logistics requirements because of its perishable nature.
- Food retailing has to cope with the proliferation of new food products launched by manufacturers to cater to evolving consumer tastes and demands.

General merchandising retailers are experiencing major competitive pressures from e-tailers, primarily Amazon. For example, in 2017, many established retailers have announced store closings in the United States, which has led to increasing retail space vacancy along with mall closures. Next, we will briefly review various types of retailers along with their salient characteristics.

Mom & Pop Stores: These are small family-owned and -operated independent stores typically in a single location. Examples of such retailers include stand-alone restaurants, fresh produce, florist, and seasonal specialty stores.

Convenience Stores: These stores carry everyday items like packaged foods, snacks, few groceries, and beverages. They are typically characterized by long open hours, 24×7 being common, and are located along busy streets quite commonly as part of gas stations. A well-known brand in this format is 7-Eleven, which has 56,600 stores across 18 countries.

Supermarkets: These stores have wide aisles and carry a variety of household products, groceries, fresh produce, meats, and food items. They do not carry large appliances and other expensive consumer durables. They are predominantly located in residential areas and tend to be open 24×7. Goods sold in supermarkets tend to be lower priced than at the convenience stores nearby. Examples of supermarkets in the United States include Albertsons, Whole Foods, Kroger, and Aldi.

Supercenters: These are large supermarkets, which are part of a chain of stores and have a wide variety including household products, groceries, fresh produce, meats, food items, appliances, and other expensive consumer durables. For example, Walmart operates 2700 supercenters, accounting for 81% of supercenter sales in the United States.

Department Stores: Department stores are characterized by good ambiance and elaborate presentation of merchandise coupled with good customer service. They typically carry a deep assortment of clothing, soft goods such as bed linen, and home furnishing. Examples of prominent department stores in the United States include Macy's, Kohl's, J.C. Penney, and Sears.

Discount Stores: These retailers are characterized by low prices for clothing, household goods, and durables. For example, Walmart has "Every Day Low Pricing" (ELDP), which contributed to its dominating position. Examples include Walmart, Kmart, and Target. Over the last few years, extreme value retailers such as Dollar General and Dollar Tree/Family Dollar are expanding and gaining some market share from discount stores; they typically focus on low-income consumers.

Specialty Stores: These retailers focus on a specific category and offer very deep assortment and very knowledgeable staff to provide customer service. For example, Abercrombie & Fitch focuses on clothing for young adults and Foot locker focuses specifically on footwear. One strategy that has made specialty retailers such as Zara and H&M successful is called "fast fashion" in which 2 to 3 new products are introduced every week, compared to 12 to 14 per year by others. The success of this "fast fashion" strategy is built on their ability to go from concept to product on the store in 25 days, together with a supply chain that is equally responsive.

Drugstores: These are specialty stores with focus on pharmaceuticals, health, and hygiene products. Examples of such retailers include Walgreens, CVS, and Rite Aid. They face competition from pharmacies that are part of many supermarkets, superstores, and e-tailers.

Category Specialists: These are typically big-box stores for a specific category with a deep assortment, in which products have low price and high service. Their scale and deep assortment gives them pricing power. Examples of these retailers include BestBuy and Toys-R-Us.

9.3 Financial Strategy in Retailing

A retailer's strategy is closely tied to its finances, so in this section, we will review the basics of financial statements and focus on using the information contained in them to understand important concepts that drive decision making in retail businesses. In particular, we will review three financial statements: balance sheet, income statement, and cash flow statement (Graham and Meredith 1937). We will also review the "DuPont model" for evaluating company performance.

9.3.1 Financial Reporting

Publicly traded companies, namely, companies whose stock are traded publicly, are regulated by various agencies in the countries where their stocks are traded. In the United States, the Securities and Exchange Commission (SEC) is the federal agency with legal authority to oversee several aspects of publicly traded companies. One of the roles of the SEC is to set financial standards with the help of the Financial Accounting Standards Board, whose declarations are called Generally Accepted Accounting Principles. Having such standards for reporting and regulatory oversight provides the general public and investors' accurate and credible financial information about the companies. These reports are publicly accessible to anyone from SEC's EDGAR website (https://www.sec.gov/edgar.shtml). The key information that is contained in these reports is briefly described next.

9.3.2 Assets, Liabilities, and Stockholder's Equity

Assets are what a company owns, for example, inventory, cash, and accounts receivable. Liabilities are what a company owes, for example, accounts payable, wages payable, and tax payable. Claims of the owners against business are called stockholder's equity. Liabilities and stockholder's equity are together called Claims on Assets, which can be expressed as follows:

$$\text{Assets} = \text{Liabilities} + \text{Stockholder's Equity} \quad (9.1)$$

In Equation 9.1, it should be noted that the stockholder's equity will be positive, when assets exceed liabilities.

9.3.3 Balance Sheet

This is a financial statement that summarizes a company's assets, liabilities, and shareholders' equity at a specific point in time.

Table 9.3 lists different types of assets, liabilities, and stockholder's equity. It should be pointed out that stockholder's equity is a liability for the company because it represents a claim on the assets. As a convention, assets are

TABLE 9.3
Typical Items in a Balance Sheet

Assets	Liabilities and Stockholders' Equity
Current assets	Current liabilities
Cash	Short-term debt
Cash and cash equivalents	Capital leases
Receivables	Accounts payable
Inventories	Taxes payable
Prepaid expenses	Accrued liabilities
Other current assets	Other current liabilities
Total current assets	Total current liabilities
Non-current assets	Non-current liabilities
Property, plant and equipment	Long-term debt
Land	Capital leases
Fixtures and equipment	Deferred taxes liabilities
Other properties	Minority interest
Property and equipment, at cost	Other long-term liabilities
Accumulated depreciation	Total non-current liabilities
Property, plant and equipment, net	Total liabilities
Goodwill	Stockholder's equity
Other long-term assets	Common stock
Total non-current assets	Additional paid-in capital
Total assets	Retained earnings
	Accumulated other comprehensive income
	Total stockholders' equity
	Total liabilities and stockholders' equity

listed on the left side of the balance sheet and liabilities are listed on the right side. Another important practice is that the most liquid assets are listed at the top and the least liquid assets are listed at the bottom. An asset that is readily converted to cash without losing its value is considered to be liquid. Moreover, current assets are those that are expected to get converted to cash within a year, whereas fixed assets are those that will not be converted to cash within a year. Hence, we see cash, the most liquid of current assets, listed at the top with fixed assets, such as land and building, at the bottom of the asset column in the balance sheets. Generally, assets are listed at value of their purchase or their book value and not necessarily at market value.

Likewise, liabilities that require payments within a year are called Current Liabilities, and the ones that are due beyond 1 year are called Long-term

Liabilities. The earliest due liability is listed at the top of the liability column followed by ones that are due later.

9.3.4 Income Statement

This is a statement that reports a company's financial performance over a specific time period. Table 9.4 shows operating and non-operating expenses and revenue that are found in an income statement.

Among the three important financial statements, the income statement is the one that provides an overview of a company's net income and sales. As shown in Table 9.4, it begins with the sales at the top and ends at net income at the bottom. The operating part of the income statement consists of information about income and expenses that occur due to business operations. The non-operating part of the income statement consists of information about income and expenses that occur due to activities that are not directly linked to business operations. For example, if a company sells its investment securities, the profit gained is reported in the non-operating part of the income statement.

9.3.5 Cash Flow Statement

A cash flow statement is divided into three categories: operating, investing, and financial cash flows, as shown in Table 9.5. It reflects a company's liquidity by detailing how cash flows in and out of the company based on operations, investing, and financing.

TABLE 9.4

Typical Items in an Income Statement

Sales revenues
Cost of goods sold
Gross profit
Operating Income and Expenses
Selling expense
Administrative expense
Total operating expense
Operating income
Non-operating Income and Expenses
Interest revenues
Gain on sale investments
Interest expense
Total non-operating
Net Income

TABLE 9.5

Typical Items in a Cash Flow Statement

Cash Flows from Operating Activities
    Net income
    Depreciation and amortization
    Deferred income taxes
    (Gain) Loss from discontinued operations
    Accounts receivable
    Inventory
    Accounts payable
    Accrued liabilities
    Income taxes payable
    Other working capital
    Other non-cash items
    Net cash provided by operating activities

Cash Flows from Investing Activities
    Investments in property, plant, and equipment
    Property, plant, and equipment reductions
    Acquisitions, net
    Purchases of investments
    Other investing charges
    Net cash used for investing activities

Cash Flows from Financing Activities
    Short-term borrowing
    Long-term debt issued
    Long-term debt repayment
    Repurchases of treasury stock
    Cash dividends paid
    Other financing activities
Net cash provided by (used for) financing activities
    Effect of exchange rate changes
    Net change in cash
    Cash at beginning of period
    Cash at end of period

9.4 Financial Metrics in Retailing

Information from financial statements can be used to quickly compute various metrics to gain insights into the performance of a company and to compare companies. We will briefly review several such metrics, which are used commonly in the retail industry.

Debt/Equity Ratio: It is the ratio of a firm's total liabilities to its stock holder's equity. A D/E < 1 indicates relatively low debt and low risk; likewise, a D/E > 1 indicates relatively higher debt and higher risk.

Acid Ratio: It is the ratio of the sum of cash, accounts receivable, and short-term investments to the current liabilities; it indicates if the company has enough current assets to pay its immediate liabilities. As inventory is not included, it is also known as working capital ratio or quick ratio. If the ratio is less than 1, it means that the company does not have adequate assets to pay immediate liabilities.

Gross Profit Ratio: It is the ratio of gross profit to net sales. In retail industry, gross margin depends on markups and markdowns in pricing. Gross margins will suffer if initial markups are low or markdowns are high. A consistently increasing gross profit ratio implies that the company is likely to be able to meet its liabilities comfortably.

Inventory Turns: It is the ratio of annual sales to average inventory. Inventory turn indicates how quickly the inventory is sold. While high inventory turns are generally desirable, a retailer should avoid stockouts. Low inventory turns indicate potentially overstocked inventory, which may indicate the need to mark down and clear the stock. It is a good metric to compare performance of the retailers.

Asset Turnover: It is the ratio of net sales to total assets. Hence, the higher the ratio, the better the performance of the company.

Sales per square foot: It is the ratio of net sales to total square feet. This metric indicates how well the store area is used to generate sales. It is obviously most relevant to brick-and-mortar stores. With increasing e-tailing, this number has been decreasing across the industry.

9.5 Analysis of Balance Sheet

To analyze a company's performance, we can look at trends in its balance sheets over a period of time. Such trends can be compared across companies to get additional insights. To illustrate this, we will look at Walmart's balance sheets for 5 years from 2013 to 2017, which are summarized in Table 9.6. Trends in the financial metrics for Walmart are given in Figures 9.2 through 9.4. Similarly, Amazon's balance sheets for 5 years (2012–2016) are summarized in Table 9.7, and its trends in financial metrics are given in Figures 9.5 through 9.7.

TABLE 9.6
Balance Sheet of Walmart Stores, 2013–2017

WALMART STORES INC (WMT) BALANCE SHEET

Fiscal year ends in January. USD in millions except per share data.	2013	2014	2015	2016	2017
Assets					
Current assets					
Cash					
Cash and cash equivalents	7781	7281	9135	8705	6867
Receivables	6768	6677	6778	5624	5835
Inventories	43,803	44,858	45,141	44,469	43,046
Prepaid expenses	1588	1909	2224	1441	1941
Other current assets		460			
Total current assets	59,940	61,185	63,278	60,239	57,689
Non-current assets					
Property, plant, and equipment					
Land	25,612	26,184	26,261	25,624	24,801
Fixtures and equipment	43,699	45,756	47,851	49,950	51,843
Other properties	102,413	106,738	108,522	112,480	114,485
Property and equipment, at cost	171,724	178,678	182,634	188,054	191,129
Accumulated depreciation	−55,043	−60,771	−65,979	−71,538	−76,951
Property, plant and equipment, net	116,681	117,907	116,655	116,516	114,178
Goodwill	20,497	19,510	18,102	16,695	17,037
Other long-term assets	5987	6149	5671	6131	9921
Total non-current assets	143,165	143,566	140,428	139,342	141,136
Total assets	203,105	204,751	203,706	199,581	198,825
Liabilities and stockholders' equity					
Liabilities					
Current liabilities					
Short-term debt	12,392	11,773	6402	5453	3355
Capital leases	327	309	287	551	565
Accounts payable	38,080	37,415	38,410	38,487	41,433
Taxes payable	5062	3520	3613	3065	3737
Accrued liabilities	15,957	16,239	16,560	17,063	17,838
Other current liabilities		89			
Total current liabilities	71,818	69,345	65,272	64,619	66,928
Non-current liabilities					
Long-term debt	38,394	41,771	41,086	38,214	36,015
Capital leases	3023	2788	2606	5816	6003
Deferred taxes liabilities	7613	8017	8805	7321	9344
Minority interest	5395	5084	4543	3065	2737
Other long-term liabilities	519	1491			
Total non-current liabilities	54,944	59,151	57,040	54,416	54,099

(*Continued*)

TABLE 9.6 (CONTINUED)

Balance Sheet of Walmart Stores, 2013–2017

WALMART STORES INC (WMT) BALANCE SHEET

Total liabilities	126,762	128,496	122,312	119,035	121,027
Stockholders' equity					
Common stock	332	323	323	317	305
Additional paid-in capital	3620	2362	2462	1805	2371
Retained earnings	72,978	76,566	85,777	90,021	89,354
Accumulated other comprehensive income	−587	−2996	−7168	−11,597	−14,232
Total stockholders' equity	76,343	76,255	81,394	80,546	77,798
Total liabilities and stockholders' equity	203,105	204,751	203,706	199,581	198,825

Source: http://financials.morningstar.com/balance-sheet/bs.html?t=WMT.

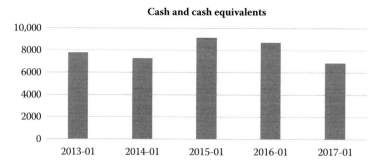

FIGURE 9.2
Trend in cash and cash equivalents (Walmart).

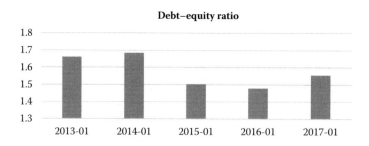

FIGURE 9.3
Trend in debt–equity ratio (Walmart).

Retail Engineering 455

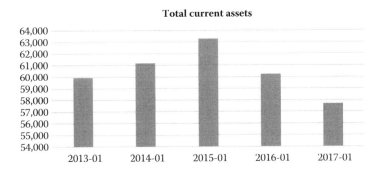

FIGURE 9.4
Trend in total current assets (Walmart).

A clear trend is the steady increase in Amazon's cash and cash equivalents. Notice that Amazon's total non-current assets are substantially smaller compared to Walmart's. However, this may be surprisingly high given that Amazon has no stores. Therefore, one could conjecture that much of the hard assets owned by Amazon are in their distribution operations and IT infrastructure. Another remarkable point is that Amazon's debt–equity ratio is twice that of Walmart. In the next section, we will review the "DuPont model," which is one of the ways we can analyze a company's performance.

9.6 DuPont Model

The original DuPont model combines key information from the balance sheet and income statement to provide additional insight into a company's performance. Specifically, this model quantifies operating efficiency using net profit margin, and asset efficiency by asset turnover. This original model has been extended to analyze the performance of a company from an investment perspective to calculate the return on equity (ROE), which can be mathematically represented as follows:

$$\text{ROE} = \text{Net Margin} \times \text{Asset Turnover Ratio} \times \text{Equity Multiplier} \quad (9.2)$$

TABLE 9.7

Balance Sheet of Amazon.com for Fiscal Years 2012–2016

AMAZON.COM INC (AMZN) BALANCE SHEET

Fiscal year ends in December. USD in millions except per share data.	2012	2013	2014	2015	2016
Assets					
Current assets					
Cash					
Cash and cash equivalents	8084	8658	14,557	15,890	19,334
Short-term investments	3364	3789	2859	3918	6647
Total cash	11,448	12,447	17,416	19,808	25,981
Receivables	3364	4767	5612	6423	8339
Inventories	6031	7411	8299	10,243	11,461
Deferred income taxes	453				
Total current assets	21,296	24,625	31,327	36,474	45,781
Non-current assets					
Property, plant, and equipment					
Land	2966	4584	7150	9770	13,998
Fixtures and equipment	6228	9505	14,213	18,417	25,989
Other properties	388	720	1367	1866	2454
Property and equipment, at cost	9582	14,809	22,730	30,053	42,441
Accumulated depreciation	−2522	−3860	−5763	−8215	−13,327
Property, plant and equipment, net	7060	10,949	16,967	21,838	29,114
Goodwill	2552	2655	3319	3759	3784
Intangible assets	725	645			
Deferred income taxes	123			1084	
Other long-term assets	799	1285	2892	2289	4723
Total non-current assets	11,259	15,534	23,178	28,970	37,621
Total assets	32,555	40,159	54,505	65,444	83,402
Liabilities and stockholders' equity					
Liabilities					
Current liabilities					
Accounts payable	13,318	15,133	16,459	20,397	25,309
Accrued liabilities	5684	6688	9807	10,384	13,739
Deferred revenues		1159	1823	3118	4768
Total current liabilities	19,002	22,980	28,089	33,899	43,816
Non-current liabilities					
Long-term debt	3084	3191	8265	8235	7694
Capital leases	746	1990	4224	5948	7519
Deferred taxes liabilities		571	1021		392
Other long-term liabilities	1531	1681	2165	3978	4696
Total non-current liabilities	5361	7433	15,675	18,161	20,301
Total liabilities	24,363	30,413	43,764	52,060	64,117

(Continued)

TABLE 9.7 (CONTINUED)
Balance Sheet of Amazon.com for Fiscal Years 2012–2016

AMAZON.COM INC (AMZN) BALANCE SHEET					
Stockholders' equity					
Common stock	5	5	5	5	5
Additional paid-in capital	8347	9573	11,135	13,394	17,186
Retained earnings	1916	2190	1949	2545	4916
Treasury stock	−1837	−1837	−1837	−1837	−1837
Accumulated other comprehensive income	−239	−185	−511	−723	−985
Total stockholders' equity	8192	9746	10,741	13,384	19,285
Total liabilities and stockholders' equity	32,555	40,159	54,505	65,444	83,402

Source: http://financials.morningstar.com/balance-sheet/bs.html?t=AMZN.

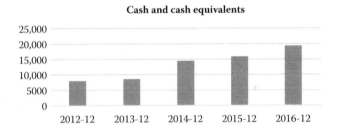

FIGURE 9.5
Trend in cash and cash equivalents (Amazon).

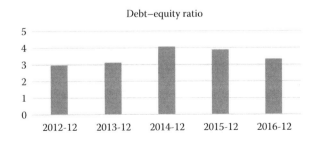

FIGURE 9.6
Trend in debt–equity ratio (Amazon).

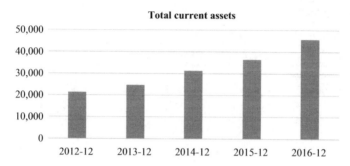

FIGURE 9.7
Trend in total current assets (Amazon).

$$\text{ROE} = \frac{\text{Net Income}}{\text{Net Sales}} \frac{\text{Net Sales}}{\text{Total Assets}} \frac{\text{Total Assets}}{\text{Shareholder's Equity}} \quad (9.3)$$

Net margin is the ratio of net income of a firm to its net sales. Net income is obtained by subtracting operating expenses, taxes, and interests from the company's total revenue. *Asset turnover ratio* is the ratio of net sales (or revenue) to total assets. This ratio conveys how well a company uses its assets to generate sales. *Equity multiplier* is the ratio of total assets to shareholder's equity. Companies can finance their asset purchases using equity or debt. Therefore, a higher equity multiplier ratio indicates that a company has high debts compared to its assets. *Return on assets* is the ratio of a company's net income to total assets, showing how efficiently it utilizes its assets for generating profits (Equation 9.4).

$$\text{Return on Assets (ROA)} = \text{Net Income}/\text{Total Assets} \quad (9.4)$$

Notice the close association between ROE and ROA. In particular, ROE = ROA × Total Assets/Equity.

The above metrics can be used to analyze a company's strategic profit model. Using the template given in Figure 9.8, one can readily visualize how the various inputs and policies lead to the company's ROA (Berman and Evans 2001; Levy and Weitz 2007).

The strategic profit model for Walmart for 2014 and 2017 is presented in Figures 9.9 and 9.10, respectively.

We can observe from Figures 9.9 and 9.10 that there is a significant decrease in ROA for Walmart from 2014 to 2017. We can attribute this to a decrease in net profit margin because there is no significant difference in asset turnover.

Retail Engineering 459

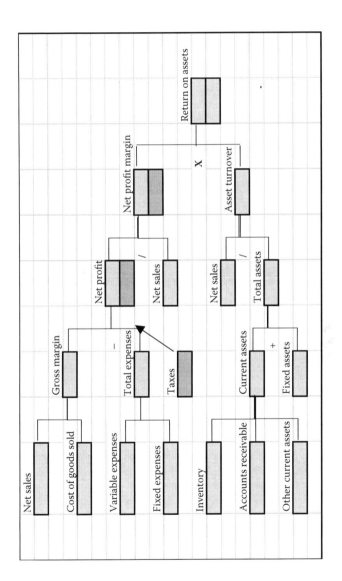

FIGURE 9.8
Strategic profit model layout.

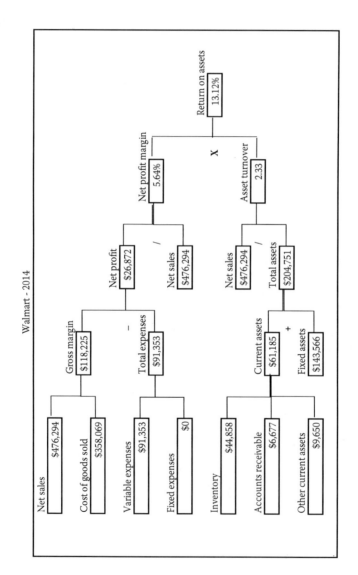

FIGURE 9.9
Strategic profit model of Walmart for 2014.

Retail Engineering 461

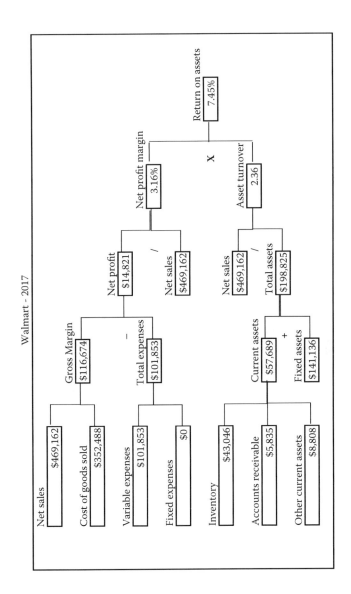

FIGURE 9.10
Strategic profit model of Walmart for 2017.

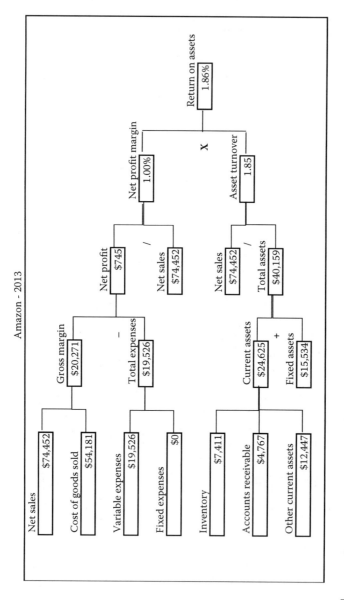

FIGURE 9.11
Strategic profit model of Amazon.com for 2013.

Retail Engineering

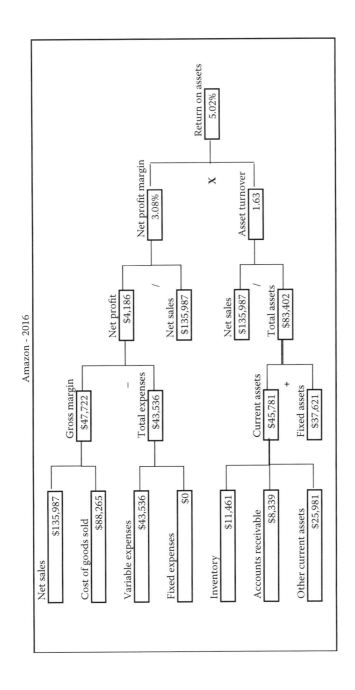

FIGURE 9.12
Strategic profit model of Amazon.com for 2016.

We can observe that the variable expenses have increased, which could be possibly attributed to the following factors:

- Increase in competition from other superstores
- Increase in advertising, sales material cost
- Change in customer behavior and habits
- Customers preferring E-commerce to store shopping

The strategic profit models for Amazon are calculated and shown in Figures 9.11 and 9.12 for 2013 and 2016, respectively.

We can observe a significant increase in ROA for Amazon from 2013 to 2016, primarily because of the increase in net profit margin. During this period, Amazon's asset turnover remained approximately the same, while the total assets more than doubled, indicating that Amazon was able to successfully scale their business while growing in size. The key factors that could have influenced Amazon's performance during this period can be summarized as follows:

- Increase in use of e-commerce by consumers.
- Amazon's fast delivery and customer service.
- Busy lifestyle of people makes them prefer shopping online.
- More efficient reverse supply chain.
- Broad variety and assortment of products.

As the above analysis of Walmart and Amazon illustrates, the DuPont model can be used as a strategic management tool by company owners, as well as managers, to understand the levers that are driving the return on equity.

9.7 Breakeven Analysis

The current trend in retail is to introduce private-label or store brand products to increase the profit margins. However, for every product the retailer launches, there would be a minimum number of units that would have to be sold before earning any profit, which is called the breakeven point. Traditionally, most manufacturers undertake breakeven analysis before launching a new product. However, with retailers introducing their own products, breakeven analysis is also important for retailers.

Some of the key variables and their definition used in breakeven analysis are discussed next.

Retail Engineering

Fixed Cost: It is the sum of all costs incurred before producing a single unit of the product. Therefore, this cost does not change with the volume of production. Typically, fixed cost includes design cost, costs for making any special tools, any marketing material, and non-recurring engineering cost.

Variable Cost: It is the cost of any materials and direct labor that go into making a unit of product. Therefore, this cost will increase as the volume of production increases.

Expected Sales: It is the number of units that are expected to be sold. Clearly, as the sales increase, the total variable cost will increase.

Unit Price: This is the expected selling price of the product. For breakeven analysis, it is assumed that the expected sales are independent of the price. This can be estimated based on the price of other comparable products that are already in the market and the features of the product being launched.

We can express the relationship among the above variables as follows:

$$\text{Total cost} = \text{Fixed cost} + (\text{Variable cost} \times \text{Expected sales}) \quad (9.5)$$

$$\text{Total revenue} = (\text{Expected sales} \times \text{Unit price}) \quad (9.6)$$

$$\text{Profit (loss)} = (\text{Total revenue} - \text{Total cost}) \quad (9.7)$$

We can use Equations 9.5 through 9.7 to solve for the number of units that need to be sold to match the total cost to total revenue, which gives us the breakeven point as shown in Equation 9.8.

$$\text{Breakeven point} = (\text{Fixed cost}/\text{Unit price} - \text{Variable cost}) \quad (9.8)$$

In any new product launch, there will be a lot of uncertainty as to what the fixed and variable costs may turn out eventually (Ravichandran 1993; Starr and Tapiero 1975). Therefore, in practice, it would be prudent to add a margin of safety to the breakeven point estimated in the above analysis.

Example 9.1

Cello is interested in developing a private label pen that will sell for $0.65. The fixed cost of designing and developing the pen is $500. This includes the cost of personnel salaries, benefits, and space for the members of the

466 Service Systems Engineering and Management

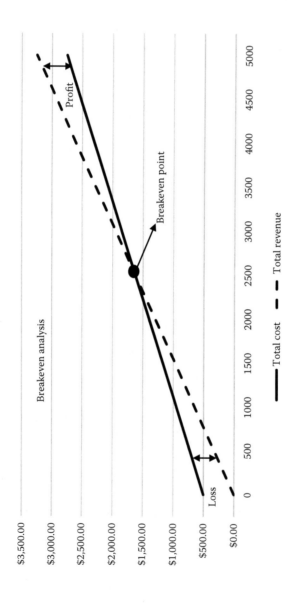

FIGURE 9.13
Breakeven analysis (Example 9.1).

design team. The variable cost of manufacturing the pen is $0.45. How many pens does Cello have to sell to breakeven?

Fixed cost = $500
Unit price = $0.65
Variable cost = $0.45

Breakeven = Fixed cost/(Unit price − Variable cost)
= 500/(0.65 − 0.45)
= 2500 units.

Figure 9.13 illustrates the manner in which total revenue and total cost vary as the sales increases. It can be seen that below the breakeven point of 2500 units, the total cost exceeds the total revenue, resulting in a loss. Beyond the breakeven point, the total revenue exceeds total cost, leading to a profit.

9.8 Checkout Process Staffing

Brick-and-mortar retail stores have several checkout counters, and one of the operational decisions store managers need to make is staffing these counters at any given time. If there are too many counters open, there may not be enough customers to keep the staff busy, which would be wasted staff salaries. On the other hand, if there are too few checkout counters that are staffed, then the wait times for customers might be unacceptably high (Lu et al. 2013; Mani et al. 2015). A major source of variable cost in retail is that of staffing; therefore, this decision also affects profitability of the retailer as discussed in the strategic profit model.

Checkout counters are commonly equipped with point-of-sale (POS) terminals, which are used by the staff to scan barcodes of the items being sold. Once a barcode is scanned, the POS obtains its price from the database in the retailer's information system. When all the items that a customer is purchasing on a visit are scanned, the POS can calculate the total payment due including discounts and applicable taxes. It is a common practice among retailers to offer discounts to customers based on loyalty cards. The customer's identification is captured by the POS, when the loyalty card is scanned. The purchase is complete, when the customer pays for the purchase using any of the accepted modes of payment, such as credit card, debit card, cash, or check. The different modes of payment are discussed in Chapter 11.

We can model the checkout process using the queuing theory that is discussed in Chapter 3 (Section 3.3). Let us consider one checkout counter in which a staff is using barcode scanners. The time it takes to serve a customer depends on two main factors: the rate at which the staff can scan the item and the number of items that the customer is purchasing. The number of items purchased by a customer in a particular visit is called the "basket size." The scan rate (number of items scanned per unit of time) can vary with the skill and experience of the staff. Scan rate can also vary depending on the item size and weight as well as the reliability of the barcode system. Both scan rate and the basket size can be modeled as random variables. The POS keeps a record of the payment transaction time, which we can model as the departure time of the customer at the checkout counter. However, there is usually no record of when a customer arrives at the checkout counter.

The time between departures (TBD) of two successive customers can be modeled as follows:

t_d = mean time between departures (TBD)

SR = mean scan rate of operators

t_e = mean service time

Let c_i be the ith customer's purchase completion time in minutes, then TBD can be calculated as follows:

$$t_{di} = c_i - c_{i-1} \quad i = 2,\ldots,n, \tag{9.9}$$

where t_{di} is the time between departure between the $(i-1)$th customer and the ith customer, and n is the number of customer visits. It should be noted that TBD of the first customer visit is zero, that is, $t_{d0} = 0$.

Similarly, the scan rate of an operator for the ith customer visit can be calculated by

$$SR_i = \frac{B_i}{t_{di}}, \tag{9.10}$$

where B_i is the number of items in the basket of the ith customer visit (basket size).

Consequently, the mean TBD and scan rate can be calculated using Equations 9.11 and 9.12.

$$t_d = \sum_{i=2}^{n} t_{di}/n, \tag{9.11}$$

$$SR = \sum_{i=1}^{n} SR_i/n. \tag{9.12}$$

There are two important practical considerations that arise in estimating the scan rate. The first consideration is that the time stamp recorded by

Retail Engineering

POS systems will have finite resolution (e.g., 1-minute resolution). In situations where two consecutive departure times were recorded to be equal, then the denominator of Equation 9.10 would be zero and the estimated scan rate would be infinite, which is absurd. The second consideration is that if the staff had idle time between two customers, then the scan rate would be underestimated. This can happen when the arrival time of the subsequent customer is greater than the departure time of the previous customer. The art of modeling the system would therefore entail judiciously filtering POS data to estimate the service rate. The checkout process can then be modeled as M/M/c or G/G/c queues as described in Section 3.3. One could approximate the arrival rate into this queuing system to be equal to the departure rate based on the filtered data to calculate various performance measures. Here, the decision is choosing a value of c, which is a trade-off in staff salary cost versus customer waiting time. Such queuing models can also be useful to evaluate cost-effectiveness of new technologies such as RFID to replace barcode scanning. An example, illustrating the staffing problem, is given in Chapter 11 (Example 11.1).

Exercises

9.1 Gather recent information about any two retailers that you consider to be competitors. Make sure at least one of these retailers is local! You should collect at least the following data:
   a. Annual revenue.
   b. Number of stores.
   c. Number of employees.
   d. Number of countries that the company operates around the world.
   e. Formats of retailing.
   f. Private label, if any.
   g. Gather and review last four quarters and annual performance based on published reports.
   h. Calculate the profit management path and asset management path for both companies and then develop the strategic profit model (SPM).
   i. Identify the similarities and differences between the two companies in terms of their SPM and productivity measures at the corporate level.
   j. Summarize this in a table for easy comparison. Briefly explain why you consider them as competitors.

9.2 Visit the stores or websites of the retailers you studied in Exercise 9.1 and observe the following for any specific category of products:
   a. In each of the stores, compare the four elements of the retail mix—the type of merchandise sold, the variety and assortment of merchandise sold, the level of service provided to customers, and the pricing. Indicate whether the store is an everyday low pricing store or offers frequent deals.
   b. Note the country of origin for the SKUs in the categories.
   c. Are there any significant differences in the layout of these stores?
   d. What is the multichannel/omnichannel capability of these retailers?
   e. Compare the variety and assortment offered in store versus non-store channel for the category you are considering.

9.3 Suppose a new product launch is estimated to have a fixed cost of $83,000, a planned selling price of $98/unit, and a variable cost of $36.5/unit. Calculate the number of units that need to be sold to break even. While the new product was being developed, it is learnt that a competitor will be launching a similar product at a price of $68/unit. If this price has to be matched, what is the new breakeven point? Illustrate the total cost and revenue for both prices in a graph.

9.4 Based on past customer traffic in a store, the store manager has to decide how many staff she will schedule to operate checkout counters. Using the past POS data, the parameters of the system have been estimated as shown in Table 9.8. Model this as an M/M/c system to determine the number of staff. Model this as a G/G/c system and see if the staffing decision changes. Discuss your findings. Assume that mean time between departures (TBD) is approximately equal to mean time between arrivals (TBA).

TABLE 9.8

Data for Exercise 9.4

Mean scan rate	3.77
Mean TBD (TBA) in minutes	2.01
Mean service time (minutes)	1.28
Mean arrival rate	0.497512
Stdev of TBD (TBA)	0.96
Stdev of service time	1.22
CV of interarrival time	0.477612
CV of service time	0.953125
Max number of checkout counters	5
Cashier wage ($/hour)	10.00
Cost for waiting line ($/customer/min)	0.30

References

Berman, B. and J. R. Evans. 2001. *Retail Management: A Strategic Approach.* Upper Saddle River: Prentice Hall International.

Graham, B. and S. B. Meredith. 1937. *The Interpretation of Financial Statements.* Vol. 4. New York: Harper.

Levy, M., and B. A. Weitz. 2007. *Retailing Management.* Boston, MA: McGraw-Hill/Irwin.

Lu, Y., A. Musalem, M. Olivares, and A. Schilkrut. 2013. Measuring the effect of queues on customer purchases. *Management Science.* 59(8): 1743–1763.

Mani, V., S. Kesavan, and J. M. Swaminathan, 2015. Estimating the impact of understaffing on sales and profitability in retail stores. *Production and Operations Management.* 24(2): 201–218.

Ravichandran, R. 1993. A decision support system for stochastic cost-volume-profit analysis. *Decision Support Systems.* 10(4): 379–399.

Starr, M. K. and C. S. Tapiero. 1975. Linear breakeven analysis under risk. *Journal of the Operational Research Society.* 26(4): 847–856.

10
Healthcare Delivery Systems

Healthcare makes up roughly 18% of U.S. gross domestic product (GDP), the largest component of any sector. Recent changes in the United States have led providers to an increased interest in not just how to increase revenue but also how to reduce costs and improve outcomes. As such, there are significant opportunities for the application of service engineering principles. In this chapter, we provide an overview of the healthcare delivery system in the United States and an introduction to how to start to address some of these challenges through effective capacity management, scheduling of resources, and analytics.

10.1 Introduction

Healthcare is the maintenance or restoration of the human body by the treatment and prevention of disease, injury, illness, and other physical and mental impairment. Healthcare is delivered by trained and licensed professionals in medicine, nursing, dentistry, pharmacy, and other allied health providers. The quality and accessibility of healthcare vary across countries and are heavily influenced by the health policies in place. It is also dependent on what are termed social determinants of health (Heiman and Artiga 2015), which are conditions of where people are born, live, and work. These include age, education, income, and access to healthcare. Note that these determinants can be the primary factors for health and well-being. Schroeder (2007) estimated that health and well-being as defined by risk of premature death is only 10% dependent on healthcare delivery, the remaining factors being 30% genetic, 40% individual behavior, and 20% social and environmental factors. The important point being that although we focus on the delivery component of healthcare in this chapter, even a perfectly designed delivery system will not ensure health and wellness of the population.

Healthcare engineering views healthcare as a set of complex systems and applies and innovates engineering design and analysis principles to these systems in order to improve the patient experience, reduce costs, and improve population health outcomes (Griffin et al. 2016). The service science component is an interdisciplinary approach to the study, design, and implementation of specific arrangements of people and technologies that take actions that provide value for others. Service may be thought of in this context as a provider and a client (e.g., doctor and patient) working together to create value. An important distinction is that whereas engineering systems typically examine the relationship between people and technology, service systems examine the relationship between people and other people.

Several important changes in healthcare have occurred in recent years. These include

- A focus on personalized and coordinated care
- A desire to implement evidence-based practices
- The implementation of outcomes-based financial incentives
- The use of multiple modes for delivery and patient interaction
- A broadening of efforts to improve and maintain health and wellness

Despite many positive changes, several challenges remain. The cost of healthcare is a significant burden for much of the U.S. population. Further, many individuals, including a large proportion of those covered by Medicaid or living in rural areas, either have no access to primary care or have to travel significantly farther than their privately insured counterparts. Despite countless publications and national health goals aimed at their reduction, disparities in health outcomes by race/ethnicity, gender, and geographic location persist. Furthermore, healthcare as an industry lags behind other industries (manufacturing, logistics, software technologies, finance) in its ability to deal with complexity.

In this chapter, we provide an overview of healthcare systems from an engineering perspective. We first discuss the important structures of healthcare, including how it is financed, information technology, and basic operations. We then discuss techniques for more effective use of healthcare resources, including scheduling and capacity management. We follow this with a brief discussion of health analytics. It is important to note that healthcare systems can vary significantly by country. We take a U.S. focus in this chapter. Readers interested in global issues in healthcare should consult text such as Holtz (2016).

10.1.1 U.S. Healthcare Costs

The costs of healthcare in the United States are, to say the least, significant. In 2015, total U.S. spending on healthcare was $3.2 trillion, or $9900 per person (Centers for Medicare and Medicaid Services [CMS] 2016). This makes up roughly 18% of the GDP. In the United States, spending per person rose from $3788 in 1995 to $9403 in 2015 (The World Bank 2016). Comparison to a small sample of countries is shown in Table 10.1.

Although not the highest, the United States spent significantly more per person in 2015 in healthcare than any other country in the world, with the exception of Norway ($9522) and Switzerland ($9674).

TABLE 10.1

Healthcare Spending per Person (U.S. Dollars) in 1995 and 2015

Country	Healthcare Spending in 1995 (per Person; U.S. Dollars)	Healthcare Spending in 2015 (per Person; U.S. Dollars)
Argentina	$613	$605
Canada	$1831	$5292
China	$21	$420
Denmark	$2835	$6463
Ethiopia	$4	$27
France	$2745	$4959
Germany	$3129	$5411
Japan	$2845	$3703
Norway	$2698	$9522
Global average (all countries)	$461	$1059

Source: The World Bank. 2016. *Health Expenditures per Capita.* Available at http://data.worldbank.org/indicator/SH.XPD.PCAP

The breakdown of U.S. healthcare spending is as follows (CMS 2016):

- Dental services: 4%
- Durable medical equipment (e.g., contact lenses, eyeglasses, hearing aids): 2%
- Home healthcare: 3%
- Hospital care: 36%
- Other health, residential, and personal care services: 5%
- Other non-durable medical products (e.g., over-the-counter medicines, medical instruments): 2%
- Other professional services (e.g., physical therapy, optometry, chiropractic): 3%
- Nursing care facilities and continuing care retirement communities: 5%
- Physician and clinical services: 20%
- Prescription drugs: 10%

There is great debate as to why the costs of healthcare in the United States are so high, and we will certainly not solve this issue here. However, some of the standard reasons given for these high costs include the following:

- The overuse of expensive equipment such as magnetic resonance imaging (MRI) machines; the United States performs 107 MRIs per 1000 persons annually compared to the OECD (Organisation

for Economic Co-operation and Development) median of 51 (Commonwealth Fund 2017).
- The use of expensive procedures when there is no evidence they are more effective than less expensive alternatives (e.g., the use of proton beam therapy for prostate cancer).
- The complicated financing system that separates the patient from the payer from the provider, high physician costs driven in part by the strong lobbying efforts of the American Medical Association (AMA) and the restricting of physician capacity; the United States has only 2.6 physicians per 1000 persons compared to the OECD median of 3.2 or Germany, which has 4.1 (Commonwealth Fund 2017).
- Allowing prescription drug companies to charge significantly higher prices in the United States than they sell the same drugs for in other countries.
- The overuse of specialty care (which is three to six times higher than in peer countries).
- The relatively small investment in supporting social services (e.g., housing, food programs, disability benefits); the United States spends 9% of its GDP compared to 21% in France, 20% in Germany, and 15% in the United Kingdom (Bradley and Taylor 2015).

In addition, some argue that a key reason for high costs is that it is a system that waits "until something is broken." That is, people do not receive appropriate preventive care because it isn't covered, they do not have access to preventive care from providers, and the system isn't incentivized to consider the wellness of the population.

10.1.2 Healthcare Outcomes

Although the United States pays a high cost per person for healthcare, it famously has some poor outcomes from this spending. For example, in 2013, the United States had the lowest life expectancy at birth in the OECD (78.8 years); the OECD median was 81.2 years by comparison (OECD 2017). Further, the United States had the highest infant mortality rate among OECD countries (6.1 deaths per 1000; the OECD median was 3.5 deaths). Other comparative outcomes are shown in Table 10.2 (Commonwealth Fund 2017).

It is important to note that the statistics in this table were "cherry picked" and that they are not necessarily the result of a poor healthcare system, but rather a broader political environment and cultural norms. In some areas, the United States performs well. One often cited example is that mortality rates from cancer in the United States are lower than the OECD median and have declined more quickly from 1995 to 2007 (Stevens et al. 2015). That said, it is hard to argue against the point that the U.S. healthcare system is both expensive and could improve in its outcomes.

Healthcare Delivery Systems

TABLE 10.2

Population Health Outcomes and Risk Factors in OECD Countries in 2013

	Life Expectancy at Birth	Infant Mortality per 1000 Live Births	% of Population over 65 with 2 or More Chronic Conditions	Obesity Rate	% of Population Who Are Daily Smokers	% of Population over 65
Australia	82.2	3.6	54	28.3	12.8	14.4
Canada	81.5	4.8	56	25.8	14.9	15.2
Denmark	80.4	3.5	–	14.2	17.0	17.8
France	82.3	3.6	43	14.5	24.1	17.7
Germany	80.9	3.3	49	23.6	20.9	21.1
Japan	83.4	2.1	–	3.7	19.3	25.1
Netherlands	81.4	3.8	46	11.8	18.5	16.8
New Zealand	81.4	5.2	37	30.6	15.5	14.2
Norway	81.8	2.4	43	10.0	15.0	15.6
Sweden	82.0	2.7	42	11.7	10.7	19.0
Switzerland	82.9	3.9	44	10.3	20.4	17.3
United Kingdom	81.1	3.8	33	24.9	20.0	17.1
United States	78.8	6.1	68	35.3	13.7	14.1
OECD median	81.2	3.5	–	28.3	18.9	17.0

Source: Commonwealth Fund. 2017. U.S. Healthcare from a Global Perspective. Available at http://www.commonwealthfund.org/publications/issue-briefs/2015/oct/us-health-care-from-a-global-perspective

At the provider level, the Institute for Healthcare Improvement (IHI) has promoted the "triple aim" as a framework to approach simultaneous healthcare systems improvement. The three aims are as follows:

- Improve the patient experience of care (quality and satisfaction)
- Improve the health of a population
- Reduce the costs of healthcare

Many hospitals also include the fourth aim of improving the work life of healthcare employees. Part of the challenge of the triple aim is measurement. Many of the measures rely on self-reporting, which has many known biases. The important point, however, is that most providers are now trying to continuously improve through the use of these three or four measures. They have also helped to change the focus from being transactions based to measures that consider their entire patient population. This has also led to an expansion of the area of *population health*, which is concerned with a group of health outcomes for a group of patients. This includes the consideration of genetics, physical environment, and individual behavior.

10.2 Healthcare Financing

Healthcare financing is concerned with who pays for which services and how. This includes insurance and provider compensation. In addition, providers need to develop a strategy for how to bill for the services provided.

10.2.1 Health Insurance

A key role of health insurance is to reduce risk to the individual; that is, it acts as a hedge against an unanticipated condition or event such as a heart attack, cancer, or an auto accident. Health insurance in the United States is provided through a variety of means including employer-based, patient-purchased, government exchange, or a government program. In the United States, healthcare spending by major sources of funds for healthcare spending has the following breakdown (CMS 2016):

- Medicaid: 17%
- Medicare: 20%
- Private insurance: 33%
- Out of pocket: 11%

When considering taxes, health insurance premiums, and out-of-pocket spending, the health spending by type of sponsor is 29% from the federal government, 28% from households, 20% from private businesses (employers), and 17% from state and local governments (CMS 2016).

Medicaid is jointly funded by the state and federal governments and provides insurance to qualified low-income individuals. The income thresholds for qualification are set by the state and depend on family size. A related program to Medicaid is the Children's Health Insurance Program (CHIP), which targets children, as the name implies, even if the family income is above the state Medicaid income threshold. In 2014, the United States spent $463 billion with the following breakdown: 21% to aged, 40% to disabled, 19% to adult, and 19% to children (Kaiser Family Foundation 2017).

Medicare is funded by the federal government and provides insurance to individuals over 65 years of age or those with a severe disability. Qualification for Medicare does not depend on income. Medicare spending in 2014 was $505 billion. Together, Medicaid and Medicare spending made up 23% of the federal budget (Kaiser Family Foundation 2017). Many individuals will purchase a supplement to Medicare (Medicare Supplement Insurance, Medigap) through private providers such as AARP. The supplementary insurance pays for components that Medicare doesn't cover such as deductibles.

Private insurance can come in many forms. Two broad categories are health plans and health insurance. Health plans are subscription-based services offered through a health maintenance organization (HMO) or a defined network of providers, preferred provider organizations, and point of service. In this case, the services offered are at the discretion of the plan (e.g., through a review nurse). In addition, if care is received by a provider that is not in the network, the patient will incur significantly higher costs. Health insurance is most commonly provided by the employer (though the percentage has been decreasing in recent years). The three largest health insurance payers in the United States in 2016 were United Health Group (net revenue = $185 billion), Anthem ($89 billion), and Aetna ($63 billion). Although this can vary significantly, the key components of employer-provided health insurance include the following:

- Employer contribution—monthly amount paid by the employer for the insurance
- Employee contribution (or premium)—monthly amount paid by the employee for the insurance
- Deductible—is the amount the individual pays for care before the insurance plan starts to pay
- Copayments (or copays)—a fixed amount paid for each service utilization (e.g., office visit, filled drug prescription)

Many employers have recently encouraged their employees to participate in what are known as high deductible plans. In this case, the monthly premiums are lowered but the deductible is significantly raised. This typically shifts costs away from the employer (and payer). In order to help employees to meet the high deductible, health savings plans can be used. The health saving plan program was initiated in 2003 and allows individuals to put salary into an account pretax that can be used for certain medical expenses.

10.2.2 Provider Compensation

As with the case of insurance, there are a variety of ways that providers are compensated in the United States. Two of the most common approaches are as follows:

- Fee for Service—providers are paid on a transaction basis for each service they provide. This would include a primary care physician office visit, a root canal, or a knee replacement. Of course, a single utilization can be made up of a variety of services. For example, a single visit to the dentist may be made of an exam, teeth cleaning, and x-rays, each of which would be billed individually.
- Capitation—a payment arrangement where providers are paid a flat fee per person enrolled in their system per period of time (e.g., per month). Even if an individual received no service, the providers are still reimbursed. In theory, capitation encourages the provision of preventive services and screenings (e.g., immunizations, prenatal care, and dental sealants) since keeping their members healthy means that they will utilize more costly services less frequently, and the provider will receive monthly reimbursement even if no services are used.

In recent years, incentives have played a large role in provider compensation. These can be offered from insurance companies or the government and can come in the form of a "carrot" or "stick." Two common examples are as follows:

- Pay for Performance (P4P)—insurance companies will offer a financial incentive to the provider if they are able to achieve a set of specified quality metrics for their patient population. The overall goal of P4P is to improve health outcomes at a lower cost.
- Shared Savings Plan (SSP)—this is an approach offered by CMS made up of two parts: a set of target health outcomes and a baseline for the cost of providing care are established. Providers that are able to meet the target outcomes at a cost lower than baseline are reimbursed a portion of the savings. SSPs encourage providers, therefore, to focus on reducing the costs in their system.

Healthcare Delivery Systems 481

Note that providers are actively involved in establishing contracts with the payers. In particular, providers negotiate reimbursement levels with different health plans. When patients receive care, providers use billing software to track episodes of patient care from appointments and registration to payment of a balance. This is known as the revenue cycle. Providers often exert significant effort to reduce the amount of time between when a service is provided to when payment is received, which is termed *revenue cycle management*. This is a very active area of effort, and the interested reader can learn more about the topic from Castro (2015) or Davis (2017).

10.2.3 Provider Cost Allocation

Determining how to allocate costs to be used for billing in healthcare can be a challenging task. In general, we can think of hospital functions as being revenue generating or non-revenue (cost) generating. Revenue-generating functions would include radiology, laboratory, and surgery while cost generating would include housekeeping and laundry. The purpose of cost allocation is to determine how to apportion the direct and indirect costs.

One approach for doing this is called the reciprocal allocation method. In this case, pre-allocation costs for each function are required. Functions are then assigned to be revenue or non-revenue generating, and the allocation of non–revenue-generating functions to the revenue-generating ones is performed. This can easily be done by solving a set of simultaneous equations.

Consider a very simple hospital system that consists of four functions (F1 to F4). In this case, F1 (housekeeping) and F2 (facilities) are non-revenue generating and F3 (surgery) and F4 (radiology) are revenue generating. Monthly pre-allocation costs and allocation percentages are provided in the table below:

	Services Allocated			
Source	Housekeeping (F1)	Facilities (F2)	Surgery (F3)	Radiology (F4)
Housekeeping (F1)	0%	20%	60%	20%
Facilities (F2)	10%	0%	50%	40%
Pre-allocation costs	$50,000	$90,000	$300,000	$200,000

The resulting set of simultaneous equations are as follows:

$$F1 = 50,000 + 0.1 F2$$

$$F2 = 90,000 + 0.2 F1$$

$$F3 = 300,000 + 0.6 F1 + 0.5 F2$$

$$F4 = 200,000 + 0.2 F1 + 0.4 F2$$

Solving the system yields the following: F1 = $60,204.08, F2 = $102,040.82, F3 = $387,142.86, and F4 = $252,857.15. Note that the sum of F3 and F4 equals the sum of the pre-allocation costs. Further, for surgery, the direct cost is $300,000 and indirect cost is $387,142.86 − $300,000 = $87,142.86, and similarly for radiology, the direct cost is $200,000 and indirect cost is $52,857.15.

Once the direct and indirect costs for the revenue-generating departments have been determined, then the service costs can be estimated (i.e., full costing) based on patient volumes and allocated costs. One approach to do this is to use the accounting practice of activity-based costing (ABC). Details of ABC accounting in healthcare may be found in Gapenski and Reiter (2015). One important aspect of ABC is the use of a financial driver (e.g., time required on a shared resource) in order to convert direct costs into *relative value units* (RVU). Different providers may employ unique methods to determine RVUs for services, but ultimately, they help providers determine which services to focus on and how to appropriately bill for those services.

10.3 Healthcare Components

In this section, we discuss the various components of the healthcare delivery system. We include a brief discussion on both physical units and types of flow. Table 10.3 (Griffin et al. 2016) provides a summary of the major types of components, including provider examples and the conditions they treat.

The key stakeholders in the healthcare system, including patients, caregivers, healthcare providers, insurers, and institutions, are shown in Table 10.4 (Agency for Healthcare Research and Quality [AHRQ] 2017).

We begin with a brief discussion of the types of ownership in healthcare. In general, we can define hospital ownership as falling into one of four types:

- Public Ownership—this includes state and local hospitals (i.e., safety net hospitals or long-term psychiatric) and federal hospitals (e.g., Veterans Health Administration and Department of Defense hospitals).
- Private Ownership—this includes not-for-profit hospitals that don't pay taxes, are often associated with a charity, and profits reinvested; secular hospitals, which are owned and operated by civic organizations; and religious hospitals.
- For Profit—these hospitals are owned by private corporations such as Hospital Corporation of America.
- Physician-Owned Hospitals—these hospitals tend to be ones with a single specialty (e.g., oncology or cardiology). There are recent restrictions on these types of hospitals.

TABLE 10.3

Delivery of Healthcare Services

Type	Delivery Focus	Providers	Conditions/Needs
Ambulatory or outpatient care	• Consultation, treatment, or intervention on an outpatient basis (medical office, outpatient surgery center, or ambulance) • Typically does not require an overnight stay	• Internal medicine physician • Endoscopy nurse • Medical technician • Paramedic	• Urinary tract infection • Colonoscopy • Carpal tunnel syndrome • Stabilize patient for transport
Secondary or acute care	• Medical specialties typically needed for advanced or acute conditions including hospital emergency room visits • Typically, not the first contact with patients, usually referred by primary care physicians	• Emergency medicine physician • Cardiologist • Urologist • Dermatologist • Psychiatrist • Clinical psychologists • Gynecologists and obstetricians • Rehabilitative therapists (physical, occupational, and speech)	• Emergency medical care • Acute coronary syndrome • Cardiomyopathy • Bladder stones • Prostate cancer • Women's health
Tertiary care	• Specialized highly technical healthcare usually for inpatients • Usually, patients are referred to this level of care from primary or secondary care personnel	• Surgeon (cardiac, orthopedic, brain, plastic, transplant, etc.) • Anesthesiologist • Neonatal nurse practitioner • Ventricular assist device coordinator	• Cancer management • Cardiac surgery • Orthopedic surgery • Neurosurgery • Plastic surgery • Transplant surgery • Premature birth • Palliative care • Severe burn treatment
Quaternary care	• Advanced levels of medicine that are highly specialized and not widely accessed • Experimental medicine • Typically, available only in a limited number of academic health centers	• Neurologist • Ophthalmologist • Hematologist • Immunologist • Oncologist • Virologist	• Multi-drug resistant tuberculosis • Liver cirrhosis • Psoriasis • Lupus • Myocarditis • Gastric cancer • Multiple myeloma • Ulcerative colitis

(Continued)

TABLE 10.3 (CONTINUED)

Delivery of Healthcare Services

Type	Delivery Focus	Providers	Conditions/Needs
Home and community care	• Professional care in residential and community settings • End-of-life care (hospice and palliative)	• Medical director (physician) • Registered nurse • Licensed practical nurse • Certified nursing assistant • Social worker • Dietician or nutritionist • Physical, occupational, and speech therapists	• Post-acute care • Disease management teaching • Long-term care • Skilled nursing facility assisted living • Behavioral and/or substance use disorders • Rehabilitation using prosthesis, orthotics, or wheelchairs

Source: Griffin, P. M., H. B. Nembhard, C. J. DeFlitch, N. D. Bastian, H. Kang, and D. Munoz. 2016. *Healthcare Systems Engineering.* Hoboken, NJ: John Wiley & Sons.

Hospitals are often part of a larger network. For example, IU Health in Indiana has 28 hospital locations that provide a variety of services and size. These range from small critical access hospitals (a designation given to rural hospitals with 25 or fewer acute care inpatient beds located at least 35 miles away from another hospital), such as IU Health Bedford, to children's hospitals, such as Riley Hospital in Indianapolis, to IU Health Indianapolis, which is the largest teaching hospital in Indiana.

Continuing with the IU Health discussion, it is also an example of what is called an accountable care organization (ACO). An ACO is a group of health providers that have a payment and delivery model that ties provider reimbursements to quality metrics for the patient population that they serve. The goal is to reduce total costs for care for the assigned group of patients and to provide coordinated care. An ACO is illustrated in Figure 10.1.

10.3.1 Hospital Units

A hospital provides many types of services. Figure 10.2 shows the relationship of these units (based on Hopp and Lovejoy 2013). Note that patients can present themselves to a hospital in one of three ways:

- Ambulatory (or walk-ins)—patients come of their own accord to the hospital. Ambulatory patients may arrive for a service that was scheduled at a particular time or they may come for an emergency (i.e., unscheduled) service.
- Ambulance—patient is brought to the hospital by ambulance or, in rare cases, by helicopter.
- Transfer—patient is sent from one provider to another. This would typically be done by ambulance.

TABLE 10.4

Stakeholder Groups

Stakeholders	Stakeholders' Perspective
Consumers, patients, caregivers, and patient advocacy organizations	It is vital that research answer the questions of greatest importance to those experiencing the situation that the research addresses. Which aspects of an illness are of most concern? Which features of a treatment make the most difference? Which kinds of presentation of research results are easiest to understand and act upon?
Clinicians and their professional associations	Clinicians are at the heart of medical decision making. Where is lack of good data about diagnostic or treatment choices causing the most harm to patients? What information is needed to make better recommendations to patients? What evidence is required to support guidelines or practice pathways that would improve the quality of care?
Healthcare institutions, such as hospital systems and medical clinics, and their associations	Many healthcare decisions are structured by the choices of institutional healthcare providers, and institutional healthcare providers often have a broad view of what is causing problems. What information would support better decisions at an institutional level to improve health outcomes?
Purchasers and payers, such as employers and public and private insurers	Coverage by public or private purchasers of healthcare plays a large role in shaping individual decisions about diagnostic and treatment choices. Where does unclear or conflicting evidence cause difficulty in making the decision of what to pay for? Where is new technology or new uses of technology raising questions about what constitutes a standard of care? What research is or could be funded?
Healthcare industry and industry associations	The manufacturers of treatments and devices often have unique information about their products.
Healthcare policymakers at the federal, state, and local levels	Policymakers at all levels want to make healthcare decisions based on the best available evidence about what works well and what does not. Comparative effectiveness research/patient-centered outcomes research can help decision makers plan public health programs, design health insurance coverage, and initiate wellness or advocacy programs that provide people with the best possible information about different healthcare treatment options.
Healthcare researchers and research institutions	Researchers gather and analyze the evidence from multiple sources on currently available treatment options.

Source: http://www.ahrq.gov/research/findings/evidence-based-reports/stakeholderguide/chapter3.html

Patients that arrive to the hospital utilize some or all of the following main categories of hospital units. Not all hospitals have all of these services, and specialized facilities may have units not described here.

- Emergency Department (ED)—provides for the acute care (e.g., heart condition, a cut, a car accident, an asthma event). Entry to the ED can come from any of the three means mentioned previously. Typically,

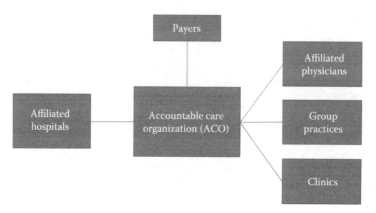

FIGURE 10.1
Accountable care organization.

there is a triage that first evaluates and categorizes patients based on vitals such as blood pressure, temperature, and interview answers. A common method to prioritize patients in the ED is the Emergency Severity Index (https://www.ahrq.gov/sites/default/files/wysiwyg/professionals/systems/hospital/esi/esihandbk.pdf), which has five levels: 1—resuscitation (patient is dying), 2—emergent (patient isn't dying, but shouldn't wait for care), 3—urgent (patient isn't dying or emergent, but has multiple vital signs in the "danger zone" and may require multiple resources such as imaging, laboratory, consult), 4—less urgent (patient isn't dying or emergent and only requires a single resource), and 5—(nonurgent). Triage for the ED is typically performed by a triage nurse.

- Admissions—this is the process by which a patient is admitted into the hospital. It can occur for patients from the ED (called emergent admissions) or when admission is from a physician order to an elective procedure such as a surgery (elective admissions), in which the patient would not come through the ED. During the admissions process, information about the patient including insurance and emergency contact information is collected.
- Observation Unit—these are adjacent to the ED and allow for extended observation of the patient. This is typically done if there is not enough information for the physician to determine if the patient should be admitted to the hospital. A patient normally stays in an observation unit for less than 24 hours. These are also known as clinical decision units.
- Intensive Care Unit (ICU)—serves patients with severe and life-threatening conditions. Patients in the ICU are closely monitored and often require special equipment such as ventilators to ensure

Healthcare Delivery Systems

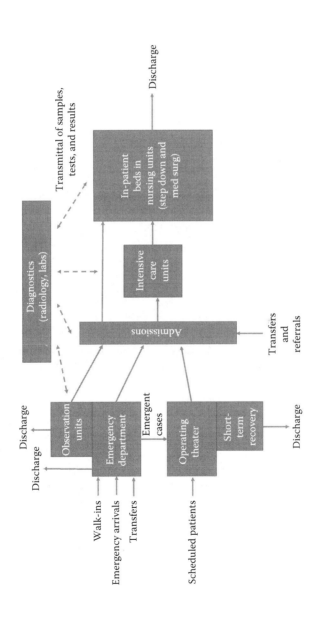

FIGURE 10.2
Hospital units and their relationships. (Based on Hopp, W. J. and W. S. Lovejoy. 2013. *Hospital Operations: Principals of High Efficiency Health Care*. Upper Saddle River, NJ: Pearson Education.)

normal function. There are several types of ICUs including neonatal (NICU), pediatric (PICU), post-anesthesia care (PACU), and coronary care (CCU). Patients are typically transferred into the ICU from the ED or from invasive surgery with complications. Common conditions in the ICU include shock (including septic shock), acute or chronic respiratory failure, renal failure, and neurological conditions such as stroke.

- Step Down Unit (SDU)—a patient in the ICU that improves is sent to an SDU. These provide intermediate levels of care and do not use invasive monitoring such as required in an ICU. However, patients in SDUs require a higher level of care than patients in the general ward.
- Med Surg (or nursing) Unit—also called general ward. These are areas that offer a broad range of care including behavioral health, women's services, and perioperative services.
- Diagnostic Services—these are made up primarily of radiology and laboratories. Radiology consists of various imaging equipment including x-rays, computed tomography (CT) scans, MRI, ultrasound, and positron emission tomography scans. These resources can, in some cases, be quite expensive. Laboratories are where tests are done on clinical specimens such as blood or urine.
- Operating Theater—this area is made of sterile-environment operating rooms (ORs) to perform surgeries. Surgeries can be scheduled or unscheduled. For a variety of reasons, scheduling surgeries is quite challenging, and a topic we pick up later in this chapter.
- Pharmacy—provides a wide range of medicines for patients. This can be in the form of pills, injections, or intravenous bags delivered through an infusion pump.
- Medical Devices—typical devices in a hospital include infusion pumps that deliver fluids and medications intravenously, ventilators that move air into and out of the lungs automatically, and bedside monitors that continuously monitor vital signs such as blood pressure and respiratory rate.

10.3.2 Patient Services

Patients that are formally admitted into a hospital with a doctor's order are called *inpatients*. The day before a patient is discharged is the last inpatient day. Patients that receive services, but are not formally admitted into the hospital, are called *outpatients*. Note that outpatients can receive services in a hospital and may even stay overnight in the hospital (e.g., in an observation unit). In addition, inpatient care can occur outside of a hospital.

The same services can be delivered to inpatients and outpatients. An example is a CT scan. An important issue is that the same procedure can be billed at two different rates if it is done as an inpatient or an outpatient as well as if it is in a hospital or a medical facility. For this reason, many hospitals provide outpatient services within the hospital rather than a clinic to be able to bill at this higher rate.

Patient services can also be defined by their clinical workflows in healthcare settings. We can classify these into seven types:

1. Primary Ambulatory Care—this includes routine and preventative care on an outpatient basis and ideally provides continuity across health services.
2. Specialty Ambulatory Care—this is provided on an outpatient basis and is most frequently triggered by referral. Specialty practices include cardiology, obstetrics, and dermatology.
3. Emergency Care—care is provided to a patient for a brief but severe illness or episode. This can include heart attacks, stroke, or car accidents. In recent years, emergency care has been used as a source of primary care for persons without insurance or access to primary care physicians.
4. Outpatient Surgical Center—as the name implies, this supports ambulatory care for surgeries. Typically, they do not involve an overnight stay. Examples include arthroscopy, cataract surgery, and caesarean sections.
5. OR Care—this is for major surgical procedures and typically occurs in a hospital setting. Most procedures will require an anesthesia provider in addition to the surgeon. OR care can be done on an emergent or scheduled basis.
6. Inpatient Care—this is hospital-based acute care. This would include the ICU, SDU, and med-surge unit. Admissions to inpatient care can occur from the ED or scheduled. Physician care is typically provided through rounding, which is the process of visiting each patient (typically in the morning) to review results and plan care.
7. Long-Term Care—this is provided to a patient with chronic conditions over a longer period of time. These include skilled nursing facilities (nursing homes), which provide 24-hour care for elderly or for younger patients with a permanent disability; long-term acute care hospitals, which serve patients with multiple, complex medical problems (stay for weeks to months; prolonged ventilator use, ongoing dialysis, etc.); and inpatient rehab facilities for patients who require comprehensive rehabilitation (e.g., strokes, serious fractures, joint replacement surgeries).

10.3.3 Patient Flow

Improving patient flow through a facility can help reduce wait times and overcrowding, promote better patient handoffs, improve financial conditions for the provider, and avoid delays. All of these can lead to improved patient outcomes and reduced capacity (e.g., workforce) requirements.

Several common metrics that providers use to measure patient flow are as follows:

- Average length of stay (LOS)—the average number of days an admitted patient stays in the hospital; this may also be broken down into wait and service times.
- Bed utilization—the average percentage of beds that are occupied.
- Left without being seen—the number (or percentage) of patients who arrive to the ED and leave before they can be seen by a certified physician. This is most commonly due to long wait times.
- Number of admissions.

Inefficient patient flow generally stems from variability within and between healthcare settings. Identifying and smoothing that flow is a key service engineering challenge. Value stream mapping (discussed in the next Section) can be a useful approach to start to improve patient flow.

Two important problems that can arise when patient flow becomes particularly bad are *boarding* and *diversion*. Patient boarding occurs when the ED experiences critical overcrowding because they are being held there due to a lack of inpatient beds in the ICU and make the ED essentially an extension of the ICU. When boarding occurs, it significantly reduces the efficiency of the ED by tying up precious resources. In addition, patient outcomes can be significantly reduced. The LOS for the patient will generally increase as a result, and several studies have found an increase in mortality as a function of the length of time they are boarded. For example, one study found that mortality increased from 2.5% for patients boarded less than 2 hours to more than 4.5% for those boarded 12 hours or more (Singer et al. 2011). In 2015, the median time from when a patient arrived to the ED to when they were placed in an inpatient bed (assuming they were admitted) was roughly 280 minutes; almost 100 minutes of this was simply waiting for a bed after they were admitted (Health Affairs 2016).

In cases of severe ED overcrowding over time, the hospital ED is put on diversion (or ambulance diversion). In these cases, patients that would have been brought into the ED by ambulance are sent to a different hospital ED, giving breathing room to the ED. The negative component of diversion is that it generally causes delays for arrival times for diverted patients. This delay in treatment (particularly for time-sensitive conditions such as strokes or acute myocardial infarction) can negatively affect outcomes. In addition, diversion tends to affect hospitals serving minority communities to a greater extent. In California, for example, hospitals servicing the highest percentages of minority patients were on diversion for 306 hours a year compared to 75 hours for hospitals serving the lowest percentage of minority patients (Health Affairs 2016).

The care of a patient may require changing health practitioners or settings (including the home) over time. These are known as *care transitions* (or handoffs) and are an important consideration for patient flow. Examples of care transitions include transfers with a hospital (e.g., ED to ICU), primary care to specialty care, inpatient settings to long-term facilities, and inpatient care to home. In order for care transitions to be successful, they require effective communications and adequate risk assessment. Consider the following examples (Chugh et al. 2009):

- "An 80-year-old retired school teacher visited the emergency department four times in a month for exacerbations to a mild heart failure condition, twice requiring hospitalization. When provided with discharge instructions, she is able to repeat them back accurately. However, she doesn't follow through with the instructions after returning home because she has not yet been diagnosed with dementia."
- "A 68-year-old man is readmitted for heart failure only one week after being discharged following treatment for the same condition. He brought all of his pill bottles in a bag; all of the bottles were full, not one was opened. When questioned why he had not taken his medication, he began to cry, explaining he had never learned to read and couldn't read the instructions on the bottles."

Both of these examples illustrate poor care transitions, which led to poor outcomes. In the United States, half of patients experience a medical error after discharge, 20% experience an adverse event within 3 weeks of discharge, and the vast majority of these errors and adverse events could have been avoided (AHRQ 2007). In order to improve care transitions, several protocols and re-engineering principals have been developed. One example is Project RED (Re-engineering Discharge) (Greenwald et al. 2007). The steps are as follows:

- Educate the patient throughout the hospitalization.
- Organize and schedule post-discharge appointments.
- Give the patient a written discharge plan that includes the reasons for hospitalization, instructions for following the medication regimen, recommendations on what to do if the condition worsens, follow-up appointments, and pending items (e.g., test results that have not returned at the time of discharge).

- Prepare a detailed discharge summary and deliver it promptly to the PCP.
- Reinforce the discharge plan and provide troubleshooting by telephone 2 to 3 days after discharge.

10.3.4 Value Stream Mapping

Process (or value stream) mapping is a useful framework for characterizing patient flow. It is essentially a graphical description of the patient flow process, which shows inputs, outputs, and transitions. It helps visualize the process to support the identification of steps that can be combined or eliminated to improve efficiency. It is also quite helpful to identify bottlenecks and redundancies, which lead to time savings. The basic steps of value stream mapping include the following:

- Define boundaries—define the scope of the process, including beginning and ending points.
- Process documentation—list and sequence all steps, identifying activities (such as taking a patient's blood pressure), decisions (such as determining whether a patient should be admitted or not), and flows (or transitions). Standard practice is for activities to be represented as rectangles and decisions by diamonds, and flows and arcs connect two steps.
- Clearly identify inputs and outputs for each step, as well as whether the input is controllable (can be changed), standard operating procedure, or uncontrollable (noise).
- Use the resulting diagram to consider the following questions:
  - What redundancies can be removed?
  - Does the order of steps make logical sense?
  - Are there missing steps?
  - What are the bottlenecks and where is time lost in the process?
  - What steps add no value?

It is important when doing value stream mapping that the result captures the actual practice and not the desired practice. An example of a value stream map for the ED process at the Hershey Medical Center in Pennsylvania is shown in Figure 10.3 (Kang et al. 2014).

Additional information about value stream mapping may be found in Jackson (2014), Jimmerson and Jimmerson (2009), and Pracek (2013). There are a large number of value stream mapping software programs to choose from. In practice, however, there is often significant value in going through the process with a large sheet of butcher paper and sticky notes. This is a very commonly used approach for lean/quality improvement teams in healthcare providers.

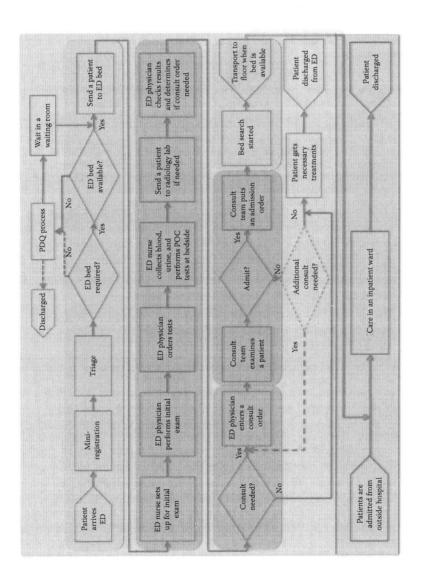

FIGURE 10.3
Example process map of the Hershey Medical Center in Pennsylvania. (Based on Kang, H., H. B. Nembhard, C. Rafferty, and C. J. DeFlitch. 2014. Patient flow in the emergency department: A classification and analysis of admission process policies. *Annals of Emergency Medicine*. 64: 335–342.)

10.4 Health Information Technology

Healthcare data are essential for the provision and implementation of healthcare. Unfortunately, healthcare is literally decades behind other industries in the area of information technology such as manufacturing.

The Department of Health and Human Services (HHS) uses the following taxonomy for health data (HHS 2007):

- Demographics and socioeconomic data—age, sex, race, ethnicity, education, and related demographic/socioeconomic variables
- Health status data—individual health status, including morbidity, disability, diagnoses, problems, complaints, and signs and symptoms, as well as behavioral and health risk factor data
- Health resources data—capacity and characteristics of the provider, plan, or health system
- Healthcare utilization data—nature and characteristics of the medical care visits, encounter, discharge, stay, or other use of healthcare services. Includes time, data, duration, tests, procedures, treatment, prescriptions, and other elements of the health encounter
- Healthcare financing and expenditure data—costs, prices, charges, payments, insurance status, and source of payment
- Healthcare outcomes—outcomes of prior or current prevention, treatment, counseling, or other interventions on future health status over time in a cyclical, longitudinal process
- Other factors—genes and proteins, environmental exposures

10.4.1 Medical Coding

The International Classification of Disease (ICD) was developed by the World Health Organization (WHO) in 1948 as a global standard for reporting statistics in morbidity and mortality. In 1967, the WHO stipulated that member states adopt the most current ICD version. The importance of the standard is that it allows for the consistent sharing of reporting across providers, no matter the location. ICD-10, the current standard used by most U.S. health providers, was endorsed in 1990, though it wasn't until 2014 that it was widely used in the United States, lagging most other WHO member companies.

The ICD-10 (or ICD-10-CM) diagnosis and procedure codes may be found at the WHO website http://apps.who.int/classifications/icd10/browse/2016/en.

There are more than 70,000 different ICD-10-CM codes. Examples include the following:

- A41.02—sepsis due to methicillin-resistant *Staphylococcus aureus*
- S82.0—fracture of patella
- B18.2—chronic viral hepatitis C

ICD-10-CM is alphanumeric code made up of seven characters, beginning with a letter. The basic structure of the ICD-10-CM code is (XXX.XXX X):

- First three characters: category (general type of injury or disease)
- Second three characters: subcategory (etiology, anatomic site, severity, other vital details)
- Last character: extension (initial encounter, subsequent encounter, sequela)

The following example for an ear wound illustrates the format:
S01—{Open wound of head}

- S01.3 Open wound of ear
    - S01.30 Unspecified open wound of ear
        - S01.301 Unspecified open wound of right ear
            - S01.301 A Unspecified wound of right ear, initial encounter

In the United States, the AMA developed the Current Procedural Terminology (CPT) coding as a standard for coding medical procedures. CPT codes identify the services that were provided to the patient, including diagnostic, radiology, surgical, and laboratory procedures. The CPT costs bridges the patient, physician, and payer. It is used by insurance companies to decide the fee that the physician receives for the service. CPT codes are more complex than ICD codes. Details for CPT codes may be found at the AMA website (https://www.ama-assn.org/practice-management/cpt-current-procedural-terminology).

10.4.2 Electronic Health Record

The electronic health record (EHR), also used synonymously with electronic medical record (EMR), is a digital record of a patient's medical history. The EHR is held by the provider and includes several types of information including the patient demographics (gender, age, etc.), measured vital signs, laboratory test results, radiology reports, immunizations, past history, and medical notes from physicians. The EHR can streamline the workflow of a physician, and being able to access a patient's medical records can help improve a patient's care by

- Sharing records between clinicians to help coordinate care
- Allowing quick access to records to support timely decision making
- Supporting safer prescribing practice, for example, by identifying potentially harmful drug–drug interactions
- Ensuring accuracy of records, as compared to written notes
- Supporting clinician–patient communications
- Supporting integration of data from multiple sources to improve clinical decision making by having a "complete" record

EHRs typically use encryption to help ensure protection of patient information, as defined by the Health Insurance Portability and Accountability Act (HIPAA). Several software vendors provide and support EHR systems. Some of the larger players are Cerner, MEDITECH, and Epic Systems, which, in 2016, comprised roughly 60% of the market share (Leventhal 2016).

Patients can also access at least a portion of the EHR through a patient portal. This also provides a way for the patient to communicate with the provider, track their information over time, and schedule future appointments.

The Health Information Technology for Economic and Clinical Health Act outlined plans for adoption of the EHR through what is called "meaningful use" (MU). In order to encourage broad EHR adoption, MU requires that providers demonstrate that they are using certified EHR technology in a way that ultimately improves patient outcomes. The CMS established a program with incentives and penalties that supported MU through three phases: (i) use EHR for data capture and sharing (e.g., track key clinical conditions, use a standardized format for electronic health information), (ii) advance clinical practice (e.g., provide patient-controlled data, transmission of electronic records over multiple settings), and (iii) improve outcomes (e.g., improved quality, safety, and efficiency).

10.4.3 Health Information Exchange

Electronic Health Information Exchange (HIE) was established as a critical component of MU. HIE is an intermediate repository on what health information can be accessed and shared in a secure fashion by multiple types of healthcare providers. There are three types of HIE:

- Directed exchange—information is shared between healthcare professionals who know and trust each other over the Internet in an encrypted and reliable way.
- Query-based exchange—allows providers to access and/or request information for a patient from other providers (often for unplanned or emergency care).
- Consumer-mediated exchange—allows patients to aggregate and manage their own health information online.

One of the key challenges of HIE is what is known as interoperability. Interoperability consists of two parts (according to IEEE Standard Computer Dictionary): (i) the ability of two of more systems to exchange information and (ii) the ability of these systems to use the information that has been exchanged. Key to interoperability is agreement on a standard. One popular standard developed for healthcare is Fast Health Interoperability Resources (FHIR), an Internet-based approach developed by Health Level Seven International (HL7), a not-for-profit organization. FHIR is working to allow developers to build standardized browser applications that are independent of the operating system (https://www.hl7.org/fhir/overview.html).

10.4.4 Claims and Billing

Once a patient is treated by the provider, a bill is typically sent to the payer for services provided. This is usually an insurance company or Medicaid/Medicare. This "claim" is then considered by the payer. There are typically three outcomes:

- The payer accepts the claim and pays the bill (though some of the costs may be passed to the patient per the insurance policy).
- The payer believes there is a billing error, in which case the claim is returned to the provider to be reconsidered/corrected.
- The payer rejects the claim, arguing that it is for services that are not part of the coverage. In this case, the patient is responsible for paying the bill.

The claims process itself is often done electronically. This points to the importance of standardized coding. For small practices, electronic billing has historically been challenging. However, recent Medicaid and Medicare billing requirements have "incentivized" most practices to now file electronically.

The time between when a patient makes an appointment or arrives for a set of services and when reimbursement is received by the provider is called the revenue cycle. Many hospitals put significant effort in managing the events that occur during the cycle in order to try to minimize the cycle length. This is called revenue cycle management. Some of the key steps include the following:

- Reduce the amount of paperwork when possible through technology or automation
- Better coordinate activities across the cycle through shared information and clearly defined workflows
- Properly capture patient/insurance information
- Determine reasons for claim returns or denials, and work to reduce their likelihood

Further information about revenue cycle management and reimbursement in general may be found in Castro (2015).

10.5 Resource Management

A key role of service engineering in healthcare delivery is guiding the effective use of resources. This includes the scheduling of workforce and rooms or equipment. It also includes capacity design, such as determining the impact of adding a bed to an ICU. In this section, we discuss four examples of resource management: (i) nurse scheduling, (ii) OR scheduling, (iii) capacity management of imaging equipment, and (iv) bed management. These are just a small, but important, subset of resource management examples.

10.5.1 Nurse Scheduling

Nurse scheduling (also called rostering) is concerned with finding the "best" way to assign a set of nurses to shifts subject to a set of constraints. These constraints may be "hard" in the sense that they can't be violated or the assignment is infeasible or "soft" in the sense that they can be violated, but at a cost. Examples of hard constraints include that each shift must have a minimum number of nurses or that a nurse can only be assigned for a certain number of hours in a week. An example of soft constraints is that we want to satisfy nurse preferences as much as possible.

Healthcare Delivery Systems

Many hospitals run three 8-hour shifts: 7 a.m. to 3 p.m., 3 p.m. to 11 p.m., and 11 p.m. to 7 a.m. It is also common for the 3 p.m. to 11 p.m. period to be broken into two 4-hour shifts: 3 p.m. to 7 p.m. and 7 p.m. to 11 p.m. This helps accommodate 12-hour nurse shifts: 7 a.m. to 7 p.m. and 7 p.m. to 7 a.m. For now, we will only consider 8-hour shifts. If the number of nurses required for each shift is known, then the staffing goal is to assign the smallest number of nurses to cover all the demand, subject to satisfying several constraints.

Consider the case where a nurse serves a rotation of 5 days on and 2 days off and serves the same shift for each day that they are on, and that the shift is an 8-hour shift. This problem can then be partitioned by shift, and the optimization model is a relatively simple one. Let d_i be the number of nurses required during the shift on day i (1 = Sunday, 2 = Monday, …, 7 = Saturday). Let x_i be the number of nurses that start their rotation on day i. We get the following optimization model for each shift:

$$\min \sum_{i=1}^{7} x_i \qquad (10.1)$$

Subject to

$$x_1 + x_4 + x_5 + x_6 + x_7 \geq d_1$$

$$x_1 + x_2 + x_5 + x_6 + x_7 \geq d_2$$

$$x_1 + x_2 + x_3 + x_6 + x_7 \geq d_3$$

$$x_1 + x_2 + x_3 + x_4 + x_7 \geq d_4$$

$$x_1 + x_2 + x_3 + x_4 + x_5 \geq d_5$$

$$x_2 + x_3 + x_4 + x_5 + x_6 \geq d_6$$

$$x_3 + x_4 + x_5 + x_6 + x_7 \geq d_7$$

$$x_i \in \{0, 1, 2, \ldots\} \quad \forall i$$

Suppose we now assign nurses to take 8-hour or 12-hour shifts. In particular, we allow a nurse to take five 8-hour shifts or three 12-hour shifts. Nurses are given the same shift time each day. In this case, we can no longer partition the problem by shift. Further, note that the nurses work a different number of hours. If we assume that all nurses would have the same pay (if broken down to an hourly rate), then it is 40/36 = 1.11 times more expensive to hire a nurse for three 12-hour shifts compared to five 8-hour shifts. Let d_{ij} be the demand for nurses on day i over shift j (j = 1, 2, 3), x_{ij} be the number of 8-hour nurses starting their rotation on day i during shift j (j = 1, 3, 4), and y_{ik} be the number of 12-hour nurses starting on day i during shift k (k = 1, 2). Our optimization model now becomes

$$\min \sum_{i=1}^{7}\sum_{j=1}^{3} x_{ij} + 1.11 \sum_{i=1}^{7}\sum_{k=1}^{2} y_{ik} \quad (10.2)$$

Subject to

$$x_{11} + x_{41} + x_{51} + x_{61} + x_{71} + y_{11} + y_{61} + y_{71} \geq d_{11}$$

$$x_{11} + x_{21} + x_{51} + x_{61} + x_{71} + y_{21} + y_{61} + y_{11} \geq d_{21}$$

$$x_{11} + x_{21} + x_{31} + x_{61} + x_{71} + y_{31} + y_{11} + y_{21} \geq d_{31}$$

$$x_{11} + x_{21} + x_{31} + x_{41} + x_{71} + y_{41} + y_{21} + y_{31} \geq d_{41}$$

$$x_{11} + x_{21} + x_{31} + x_{41} + x_{51} + y_{51} + y_{31} + y_{41} \geq d_{51}$$

$$x_{21} + x_{31} + x_{41} + x_{51} + x_{61} + y_{61} + y_{41} + y_{51} \geq d_{61}$$

$$x_{31} + x_{41} + x_{51} + x_{61} + x_{71} + y_{71} + y_{51} + y_{61} \geq d_{71}$$

$$x_{12} + x_{42} + x_{52} + x_{62} + x_{72} + y_{11} + y_{61} + y_{71} \geq d_{12}$$

$$x_{12} + x_{22} + x_{52} + x_{62} + x_{72} + y_{21} + y_{61} + y_{11} \geq d_{22}$$

$$x_{12} + x_{22} + x_{32} + x_{62} + x_{72} + y_{31} + y_{11} + y_{21} \geq d_{32}$$

$$x_{12} + x_{22} + x_{32} + x_{42} + x_{72} + y_{41} + y_{21} + y_{31} \geq d_{42}$$

$$x_{12} + x_{22} + x_{32} + x_{42} + x_{52} + y_{51} + y_{31} + y_{41} \geq d_{52}$$

$$x_{22} + x_{32} + x_{42} + x_{52} + x_{62} + y_{61} + y_{41} + y_{51} \geq d_{62}$$

$$x_{32} + x_{42} + x_{52} + x_{62} + x_{72} + y_{71} + y_{51} + y_{61} \geq d_{72}$$

$$x_{12} + x_{42} + x_{52} + x_{62} + x_{72} + y_{12} + y_{62} + y_{72} \geq d_{12}$$

$$x_{12} + x_{22} + x_{52} + x_{62} + x_{72} + y_{22} + y_{62} + y_{12} \geq d_{22}$$

$$x_{12} + x_{22} + x_{32} + x_{62} + x_{72} + y_{32} + y_{12} + y_{22} \geq d_{32}$$

$$x_{12} + x_{22} + x_{32} + x_{42} + x_{72} + y_{42} + y_{22} + y_{32} \geq d_{42}$$

$$x_{12} + x_{22} + x_{32} + x_{42} + x_{52} + y_{52} + y_{32} + y_{42} \geq d_{52}$$

$$x_{22} + x_{32} + x_{42} + x_{52} + x_{62} + y_{62} + y_{42} + y_{52} \geq d_{62}$$

$$x_{32} + x_{42} + x_{52} + x_{62} + x_{72} + y_{72} + y_{52} + y_{62} \geq d_{72}$$

$$x_{13} + x_{43} + x_{53} + x_{63} + x_{73} + y_{12} + y_{62} + y_{72} \geq d_{13}$$

$$x_{13} + x_{23} + x_{53} + x_{63} + x_{73} + y_{22} + y_{62} + y_{12} \geq d_{23}$$

$$x_{13} + x_{23} + x_{33} + x_{63} + x_{73} + y_{32} + y_{12} + y_{22} \geq d_{33}$$

$$x_{13} + x_{23} + x_{33} + x_{43} + x_{73} + y_{42} + y_{22} + y_{32} \geq d_{43}$$

$$x_{13} + x_{23} + x_{33} + x_{43} + x_{53} + y_{52} + y_{32} + y_{42} \geq d_{53}$$

$$x_{23} + x_{33} + x_{43} + x_{53} + x_{63} + y_{62} + y_{42} + y_{52} \geq d_{63}$$

$$x_{33} + x_{43} + x_{53} + x_{63} + x_{73} + y_{72} + y_{52} + y_{62} \geq d_{73}$$

$$x_{ij} \in \{0,1,2,\ldots\} \quad \forall i,j$$

$$y_{ik} \in \{0,1,2,\ldots\} \quad \forall i,k$$

Of course, many variations on this theme are easy to model in this way, including the use of overtime (at a higher rate), the use of temporary nurses with less restrictive labor requirements, though at a higher rate, and nurses that have different skill sets. Examples are provided in the exercises.

We can illustrate the use of "soft" constraints with a slightly more complicated example using goal programming. The following formulation is from Azaiez and Al Sharif (2005). In this case, we will balance between the nurse preferences and hospital objectives. In addition, there is a day shift and night shift and the schedule is for 4 weeks. Consider the following notation:

- n is number of days in a schedule
- m is number of nurses available
- i is index for days
- k is index for nurses
- D_i is staff requirement for day shift of day i
- N_i is staff requirement for night shift of day i
- P_j is penalty for goal j
- $XD_{ik} = 1$ if nurse k is assigned a day shift for day i, 0 otherwise
- $XN_{ik} = 1$ if nurse k is assigned a night shift for day i, 0 otherwise
- $XR_{ik} = 1$ if nurse k is assigned a day off for day i, 0 otherwise

In this example, five goals are set:

- Goal 1: All nurses are scheduled to have 15 days if possible in the 4-week schedule.
- Goal 2: Give preference to day shifts over night shifts.
- Goal 3: Avoid assigning a nurse to work a day shift and the night shift of the following day.
- Goal 4: Avoid on–off patterns of scheduling for nurses.
- Goal 5: Minimize isolated days off.

Healthcare Delivery Systems

The mathematical formulation works by using a "soft" constraint for each of the goals and then penalizing the violation of the soft constraint by a penalty. The formulation is as follows:

$$\min z = P_1 \sum_{k=1}^{m} d1_k + P_2 \sum_{k=1}^{m} d2_k + P_3 \sum_{i=1}^{n-1}\sum_{k=1}^{m} d3_{ik} + P_4 \sum_{i=1}^{n-2}\sum_{k=1}^{m} d4_{ik} + P_5 \sum_{i=1}^{n-1}\sum_{k=1}^{m} d5_{ik}$$

(10.3)

Subject to

$$\sum_{k=1}^{m} XD_{ik} \geq D_i \qquad \forall i$$

$$\sum_{k=1}^{m} XN_{ik} \geq N_i \qquad \forall i$$

$$XD_{ik} + XN_{ik} + XR_{ik} = 1 \qquad \forall i,k$$

$$XD_{i+1,k} + XN_{ik} \leq 1 \qquad \forall i \leq n-1, k$$

$$XR_{ik} + XR_{i+1,k} + XR_{i+2,k} + XR_{i+3,k} + XR_{i+4,k} \geq 1 \qquad \forall i \leq n-4, k$$

$$XR_{6k} + XR_{7,k} + XR_{13,k} + XR_{14,k} + XR_{20,k} + XR_{21,k} + XR_{27,k} + XR_{28,k} \geq 4 \qquad \forall k$$

$$\sum_{i=1}^{n} (XD_{ik} + XN_{ik}) \geq 14 \qquad \forall k$$

$$\sum_{i=1}^{n} (XD_{ik} + XN_{ik}) \leq 16 \qquad \forall k$$

$$\sum_{k=1}^{m} XN_{ik} \geq 4 \qquad \forall i$$

$$\sum_{i=1}^{n} \left(XD_{ik} + XN_{ik} - \left(s1_k^+ - s1_k^-\right)\right) = 15 \qquad \forall k$$

$$\sum_{i=1}^{n} XD_{ik} - \sum_{i=1}^{n} XN_{ik} - \left(s2_k^+ - s2_k^-\right) = 1 \qquad \forall k$$

$$XD_{ik} + XN_{i+1,k} - \left(s3_{ik}^+ - s3_{ik}^-\right) = 1 \qquad \forall i \leq n-1, k$$

$$XR_{ik} + XD_{i+1,k} + XN_{i+1,k} + XR_{i+2,k} - \left(s4_{ik}^+ - s4_{ik}^-\right) = 2 \qquad \forall i \leq n-2, k$$

$$XD_{ik} + XN_{ik} + XR_{i+1,k} + XD_{i+2,k} + XN_{i+2,k} - \left(s5_{ik}^+ - s5_{ik}^-\right) = 1 \qquad \forall i \leq n-1, k$$

In this formulation, the first nine constraints are "hard" constraints. The first two constraints satisfy daily staff requirements, the next two constraints avoid any two consecutive shifts, the fifth constraint ensures that there are no more than four consecutive days on, the sixth constraint ensures at least 4 days off in a 4-week schedule during weekends, the seventh and eights constraints enforce minimum (14) and maximum (16) working days per 4-week schedule, and the ninth constraint ensures the minimum night shifts in the scheduling period is 25% of the total. The last five constraints are for each of the five goals, consecutively. Note that the s variables in the goal (or "soft") constraints are the slacks (positive and negative deviations from the goal). These are penalized in the objective function. The emphasis on each of the goals is based on the penalties in the objective. The greater the penalty, the more emphasis that the soft constraint should be met. If the penalty were to be set to infinity, for example, then the soft constraint would become a hard constraint. If the penalty were set to zero, then this would be equivalent to having no constraint.

10.5.2 OR Scheduling

For many hospitals, the OR suite is a primary source of revenue generation. As such, it is important to effectively utilize this resource. OR scheduling, however, is a complicated process. Part of the complication is that the hospital will own the rooms, but often it is surgical groups (e.g., cardiology, neurology, and orthopedics) that are using the resource. One of the most common practices that hospitals use to schedule the suite of OR rooms is block scheduling. Under this framework, a time segment of the OR, called a "block," is reserved for the surgical group or individual surgeon. The surgical group then has the right to use that block as they see fit. The following are some of the cited benefits of block scheduling:

- It can help reduce changeover times between surgeries since it blocks like surgeries together, which typically require the same resource types. This can also help the surgeon establish a routine.
- It gives owner rights to the surgical group.
- It acts as an incentive to surgical groups to use the hospital facility.
- It simplifies planning since the blocks are known in advance.

Of course, blocks are not always well managed, which can still lead to poor OR utilization. One of the difficulties of scheduling within a block is the uncertainty in the length of a procedure, since complications can arise during a surgery. Further, at least a portion of the capacity must be set aside for emergency surgeries that cannot be planned in advance. Block scheduling is only effective for planning of scheduled (or elective) surgeries.

When block scheduling is used by a facility, there are two steps that need to be considered:

- *Step 1—Construct Blocks*: Allocate the capacity to be allocated to each surgical group based on historical needs, overtime considerations, and marginal revenue of the surgical group.
- *Step 2—Assign Surgeries to Blocks*: Assign the specific surgeries to the blocks with the goal of maximizing utilization.

Both of these steps can be solved through optimization.

Let us first consider the step 1 problem, that is, determine the block times form historical data. The following formulation for step 1 is from Hosseini and Taafe (2015). In this case, historical data are collected on K surgical groups using ORs over the past N weeks, where a week runs from Monday ($i = 1$) to Friday ($i = 5$). We assume the following notation:

- t^i_{jk} is the hours of surgery that group k performed on day i of week j.
- T_i is the maximum available number of OR hours on day i.
- u^i_{jk} is the undertime cost that is incurred if group k was assigned more than t^i_{jk} hours.

- o^i_{jk} is the overtime cost that is incurred if group k was assigned less than t^i_{jk} hours.
- h is the ratio of overtime to undertime costs.

The formulation for step 1 makes the block assignments by minimizing the sum of the overtime and undertime hours (with relative importance between them), based on historical data using the following formation:

$$\min z = \sum_{k=1}^{K} \sum_{j=1}^{N} \left(u^i_{jk} - h o^i_{jk} \right) \quad (10.4)$$

Subject to

$$\sum_{k=1}^{K} x^i_k \leq T_i$$

$$o^i_{jk} \geq t^i_{jk} - x^i_k \quad \forall j,k$$

$$u^i_{jk} \geq x^i_k - t^i_{jk} \quad \forall j,k$$

$$x^i_k, o^i_{jk}, u^i_{jk} \geq 0 \quad \forall j,k$$

This formulation is solved for each day of the week. The first constraint ensures that the total available time isn't exceeded. The second and third constraints ensure that overtime and undertime do not occur simultaneously. The last constraint is for non-negativity requirements.

Once the blocks are assigned to the surgical groups, then the surgeries can be assigned to the blocks (step 2). One approach is to determine the assignment of surgeries to blocks in a way that minimizes the unused OR time (Hans et al. 2008). In this case, we use the following notation:

- T_{max} is the maximum total time allocated to the surgical group (which is determined from step 1).
- c_l is the list of available durations of one block.
- S_i is the surgical cases to be performed in one block.
- d_i is the duration of surgery S_i.
- s_i is the number of times surgery S_i occurs.

- V_{ik} is the number of times that surgery i is assigned to block k.
- y_{kl} is a binary variable that equals 1 if a time duration from list L is chosen for block k.

The resulting formulation is

$$\min \sum_{k=1}^{K}\left(\sum_{l=1}^{L} c_l y_{kl} + \sum_{i=1}^{S} V_{ik} d_i\right) \quad (10.5)$$

Subject to

$$\sum_{k=1}^{K} V_{ik} = s_i \quad \forall i$$

$$\sum_{l=1}^{L} c_l y_{kl} - \sum_{i=1}^{S} V_{ik} d_i \geq 0 \quad \forall k$$

$$\sum_{k=1}^{K}\sum_{l=1}^{L} c_l y_{kl} \leq T_{max}$$

$$\sum_{l=1}^{L} y_{kl} \leq 1 \quad \forall k$$

$$y_{kl} \in \{0,1\}, V_{ik} \in \{\text{int}\}$$

In this formulation, the first constraint ensures that each surgery is planned. The second constraint ensures that the unused time in a block is non-negative. The third constraint ensures that the maximum total time isn't exceeded. The fourth constraint ensures that only one time is chosen from the list for the block duration.

In practice, open time blocks can be scheduled between assigned blocks. This can serve as a buffer to uncertainty such as surgery times or cancellations. Of course, adding buffers will typically negatively affect OR utilization.

10.5.3 Capacity Management for Imaging Equipment

Imaging equipment such as CT machines or MRI equipment are expensive resources. As such, providers wish to maintain high utilization for imaging equipment. Developing an effective schedule is often complicated by the fact that there are three types of patients that may use the equipment: (i) outpatients that are scheduled in advance, (ii) emergency (i.e., unplanned) patients, (iii) and inpatients, which may or may not be planned. In addition to having to effectively develop a schedule for planned use (which can come from the outpatient or inpatient side), there is the question of how to work in the unplanned patients, which typically have high priority.

Green et al. (2006) used a dynamic programming approach to show that a simple policy is "near" optimal. Because the mathematical development is beyond the level of this text, we simple present the results with some intuition. The policy essentially boils down to a simple prioritization if there are more than one patient waiting to utilize the equipment. The priority rule is

- If there is an emergency patient, they should be seen first (and if there are multiple emergency patients, choose the one with highest acuity).
- If there are both outpatients and inpatients, then serve inpatients first.

The essential reasoning of this policy is that if an inpatient is passed over for an outpatient, it can often lead to an increase of an extra day in the hospital. This LOS increase would in general be costlier than the revenue that would be generated from the outpatient. The policy is quite simple and therefore has been implemented in several hospitals.

If there are multiple patients of the same type waiting for a resource, it is typically better to first serve those with shorter processing times and less variability. This is because variability gets passed downstream. This tends to be a good (and easily implementable) practice in scheduling if the time and variation can be reasonably predicted.

10.5.4 Bed Management

A difficult issue in resource management is managing how patients are assigned to beds. As mentioned previously, a patient can transition from multiple departments. For example, a patient presenting with a myocardial infarction to the ED may use an ED bed, be transferred to the cardiac ICU, transferred again to the general ward, and then finally discharged. If the cardiac ICU is full when the patient arrives to the ED, then the patient may be boarded there, which can lead to negative outcomes such as increased likelihood of mortality.

A common approach to modeling bed management is through the use of queuing theory or simulation. In this section, we apply the queuing models presented in Chapter 3. In the simplest case (under the model assumptions), we can apply the M/M/c model.

Healthcare Delivery Systems

Consider a 10-bed unit with an average LOS in the unit of 2.5 days (i.e., $\mu = 1/2.5 = 0.4$). The table below shows several performance measures as a function of the arrival rate (λ) including the average customers in the queue (which would correspond to the number of patients boarded).

Arrival Rate (λ)	Utilization (ρ)	Average Time in Systems (LOS in days)	Average Number of Patients Boarded (L_q)
0.33	0.08	2.50	0.00
1.00	0.25	2.50	0.00
2.00	0.50	2.52	0.02
3.00	0.75	2.81	0.92
3.33	0.83	3.22	2.41
3.67	0.92	4.69	8.03
3.75	0.94	5.64	11.77
3.80	0.95	6.63	15.69
3.85	0.96	8.28	22.27

As expected, as the arrival rate increases, the three performance measures increase in a highly nonlinear fashion. In this case, when utilization is greater than 0.8, significant boarding can occur. When the average arrival rate is 3.67 patients per day (i.e., $\rho = 0.92$), we get an average of 4.69 patients boarded. In this case, their average time boarded (based on Equation 3.56 in Chapter 3) is $W_q = 2.19$ days, which would have a severely negative impact on morbidity and mortality.

When considering this as a G/G/c system (Section 3.3.5), and holding the average service rate constant, the performance measures are worse both in the average arrival rate and in the variability of the arrival and/or service rate. If we consider the same unit as above, but now model it as a G/G/c system, let us examine the impact of performance based on the variability of the service rate. Suppose the squared coefficient of variation of the arrival rate is 1.5 and that of the service rate is 2.0 and the average arrival rate is 3.50. Applying Equation 3.71 from Chapter 3, we get the following:

$$W_q = \left(\frac{1.5+2.0}{2}\right)\left(\frac{\left(\frac{3.50}{10(0.4)}\right)^{\sqrt{2(10+1)}-1}}{10\left(1-\frac{3.50}{10(0.4)}\right)}\right)\frac{1}{0.4} = 2.14 \text{ days}.$$

If we can reduce the service time squared coefficient of variation to 1.0, then the average boarding time reduces to 1.53 days, a 29% reduction. If, in addition, the average time in the system were reduced from 2.5 to 2.3 days, then the average boarding time decreases to 0.66 days. A relatively small change in the service time has a significant impact on boarding.

For this reason, carefully controlling bed management can greatly improve performance. This is particularly important since it is not easy for a hospital to add bed capacity, which is quite different compared to many manufacturing scenarios. There are several methods for achieving this. One common approach is to use a centralized bed authority (i.e., "bed czar"). In this case, an individual has the responsibility of processing all of the admissions and transfers into the units. The individual would normally hold a daily bed meeting to better understand the need, including demand, who is to be transferred, and who is to be discharged. This can help better plan in order to reduce any delays in admitting, transferring, or discharging, which affects both the average time in the ward and the squared coefficient of variation of service time. In general, it is most effective if the bed czar has control over all units in a hospital and not having a person for each unit.

Technology has also helped to support bed management. Many hospitals are starting to use RFID to track patients, workforce, and resources (e.g., TeleTracking Technologies). This also improves planning by knowing where everything and everyone are located.

The discharge process can keep patients in beds longer if not done effectively. Further, actively stepping down patients from ICUs to med-surge or general wards can help free up scarce capacity. Finally, the arrival rate can depend on readmissions. Therefore, allocating resource to help reduce the likelihood of a readmission can have a direct effect on the effective use of available capacity. A good overview of bed management, including decision support modeling to improve the practice, is found in Best (2015).

10.6 Health Analytics

An increasingly important area of healthcare is the use of health-related data to help support decision making. This is called *health analytics*. These data can come in various forms, including healthcare claims, health records from EHR data, medical device data, and real-time location systems. The types of decisions include how to most effectively use resources, personalize medicine, and improve the patient experience. In general, we can categorize health analytics into three areas:

- Descriptive (visual) analytics—the design and use of methods to represent data in a way that supports understanding. This helps reduce search, supports pattern recognition, and supports perceptual inference. Many commercial products such as Tableau (www.tableau.com) support visual analytics.

- Predictive analytics—the use of techniques that uses historical (e.g., observational) data to identify likely future outcomes. Outcomes of interest may include the demand for healthcare resources (e.g., ED arrivals), likelihood of a readmission, or prediction that a patient may become septic. Machine learning and data mining strategies, in particular, have become quite popular.
- Prescriptive analytics—uses methods to determine the best course of action. This includes showing the implications of various business options.

In this section, we will focus on a basic introduction to predictive analytics. In particular, we will introduce the topic of support vector machines (SVMs) for predictive analytics. The reader interested in more detailed coverage of health analytics could refer to several textbooks on the topic such as Ozcan (2017) or Yang and Lee (2016).

It is important to understand that although machine learning and other statistical techniques are actively being developed for healthcare applications, they have serious limitations. First and foremost, they are based on observational data. As we discuss in the next subsection, this limits the generalizability of their findings. In most cases, we can only make statements of correlation (i.e., variable X is related to variable Y), but not causation (e.g., the value of variable Y is a result of the value of variable X).* Therefore, before presenting the classification, we discuss two other topics: randomized controlled trials (RCTs) and measuring accuracy.

10.6.1 Randomized Controlled Trials

Suppose a hospital introduced a new protocol in the ICU to help reduce central line–associated bloodstream infections (CLABSI) in ICU patients. The hospital uses the measure of the total number of days that patients had CLABSI over a month i (CD_i). For example, if one patient had CLABSI for 3 days in June, a second for 2 days, and a third for 5 days, then $CD_{JUN} = 3 + 2 + 5 = 10$ CLABSI days. The protocol was introduced in June in the ICU. To determine whether the protocol worked, they measured CLABSI days in May (the month before the protocol was introduced) and September (two months after). They found that $CD_{MAY} = 13$ and $CD_{SEP} = 9$.

Although the metric decreased from 13 to 9, the hospital is not technically able to say anything about whether the protocol was effective. The reason is that there are several potential sources of bias. For example, the patient mix in September may have needed many fewer central lines as compared to June. In addition, there may have simply been fewer patients in the ICU

---

* It is the case that there is active development of techniques such as causal reasoning (Pearl et al. 2017), which relies on observational data. However, it is not widely practiced in healthcare at this time.

in September compared to June. Further, there may have been other changes in the ICU that could have led to the decrease. In order for the hospital to generate proper evidence that the intervention had a positive impact, a more careful experiment needs to be performed.

The gold standard for generating evidence in healthcare is the RCT. The idea of an RCT is to remove the impact of potential bias from confounders for a clinical intervention by randomly assigning individuals to a treatment or control group. The treatment group receives the intervention and the control group receives standard practice, a placebo, or no intervention at all. More formally, an RCT uses three steps:

- *Control Group* (to remove placebo effect or other effect not due to intervention, e.g., time)—the control group is a comparison group to the treatment group.
- *Randomization* (to avoid selection bias and balance the groups)—patients are randomly assigned to the control or treatment group.
- *Blinding* (to remove or equalize biases due to patient's desire to please and investigator's enthusiasm)—neither the patient nor the investigator knows which group patients are assigned until the end of the experiment.

In order to illustrate how an RCT is performed, we will use Salk's polio vaccine example, which at the time was the largest public health experiment ever performed. As a background, polio is an infectious disease that can invade the brain and spinal cord, leading to paralysis. There is no cure for polio. The polio epidemic hit the United States in the first half of the nineteenth century and was responsible for 6% of deaths in U.S. children aged 5 to 9, peaking in 1953. In 1954, American physician Jonas Salk developed a vaccine that he believed could be effective as initial research in animals looked promising. The basic steps of the experiment were as follows:

- Administer the experiment to first, second, and third graders (consent required from parents).
- Subjects were **randomly assigned** to treatment (200,000 children) and control groups (200,000 children).
- The control group was given a **placebo**, which was prepared to look exactly like the vaccine (note that the placebo-control group guards against the "placebo effect").
- Subjects and evaluators were "**blind**": they did not know to which group they were assigned (each vial was identified by a code number so no one involved in the vaccination or the diagnostic evaluation could know who got the vaccine).

The results of the experiment were as follows:

Group	Sample Size	Rate (per 100,000)
Treatment group	200,000	28
Control group	200,000	71

The potential bias was removed by randomization, and the difference between the groups was significant (although the statistical comparison is not shown here). As a result, the vaccine was used. By 1994, polio was virtually eliminated in the Americas. It is worth mentioning that there is an ethical issue about the appropriateness of Salk's experiment. Namely, is it ethical to not give children the vaccine (i.e., in the control)? In other words, does the benefit to society "outweigh" the risk to those children who did not get the vaccine? Since Salk's time, the National Research Act of 1974 and Belmont Reports have led to strict guidelines that must be followed to perform experiments in healthcare.

Although RCTs are an important approach to generate evidence, they have two important limitations. First, they are quite expensive and time consuming to conduct. Second, although randomization is used, it does not help determine if there are differences based on the heterogeneity of the population considered. In the Salk vaccine example, we would not be able to tell if the vaccine effect was different based on characteristics of the subject such as sex, race/ethnicity, or age. There are many historical examples where RCTs found good evidence for an intervention overall, only later to find that the intervention was actually harmful to a subset of the population.

10.6.2 Measuring Accuracy

Consider the case where we are trying to test for a disease or predict whether a person will get the disease. There is a population with the condition and a population without. Suppose we discriminate between the two based on a cutpoint on a particular test measurement. In this case, there are four outcomes as shown in following matrix:

	Test Result Positive	Test Result Negative
Has condition	True positive (TP)	False negative (FN)
Doesn't have condition	False positive (FP)	True negative (TN)

We can illustrate this graphically in Figure 10.4. Our goal is to determine the cutpoint that we should use that would help us to best distinguish between the presence and absence of the condition (e.g., disease).

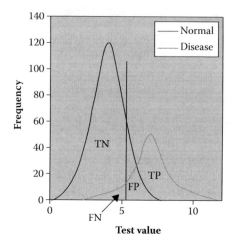

FIGURE 10.4
Outcomes from using a cutpoint of a test to determine the presence or absence of a condition.

In healthcare, there are four common measures used to measure accuracy:

- *Sensitivity (Se; true positive rate)*—if a person has a disease, how often will the test be positive (=TP/(TP + FN))?
- *Specificity (Sp; true negative rate)*—if a person doesn't have a disease, how often will the test be negative (=TN/(TN + FP))?
- *Positive predictive value*—if the test result is positive, what is the probability that the patient actually has the disease (=TP/(TP + FP))?
- *Negative predictive value*—if the test result is negative, what is the probability that the patient doesn't have the disease (=TN/(TN + FN))?

Consider the following example. We will use the qSOFA score to measure whether a patient in the ICU has sepsis. qSOFA is based on blood pressure, respiratory rate, and altered mentation (e.g., disorientation) and takes on a discrete value from 0 to 3, with 3 being the most serious. Suppose from 125 patients, we had the following results:

qSOFA Score	Sepsis or Septic Shock	Not Septic
3	18	1
2	7	17
1	4	36
0	3	39

If we use a qSOFA score of 3 for a cutpoint, we get the following:

- TP = 18; TN = 17 + 36 + 39 = 92; FP = 1; FN = 7 + 4 + 3 = 14
- Sensitivity = 18/(18 + 14) = 0.56
- Specificity = 92/(92 + 1) = 0.99

If, on the other hand, we use a qSOFA score of greater than or equal to 2, then we get

- TP = 18 + 7 = 25; TN = 36 + 39 = 75; FP = 1 + 17 = 18; FN = 4 + 3 = 7
- Sensitivity = 25/(25 + 7) = 0.78
- Specificity = 75/(75 + 18) = 0.81

Note that when we change from 3 to 2 for the cutpoint, it improves our sensitivity but decreases our specificity. We can't change the cutpoint to improve both the sensitivity and specificity together.

One common measure for a diagnostic test is called the area under the curve (AUC). We can determine this by plotting sensitivity (y-axis) versus 1 − specificity (x-axis) as shown in Figure 10.5. The curve is called the receiver operating characteristic (ROC) curve, and the area under the ROC is the AUC. The value of AUC will range from 0.5 to 1.0 (where the larger the number, the greater the diagnostic test). An AUC of 0.5 means that there was no difference between the diagnostic test and randomly guessing. The AUC is a commonly used measure for overall accuracy of a diagnostic test. A general rule of thumb is

- ACU in range of 0.90–1.00 = excellent (A)
- 0.80–0.90 = good (B)

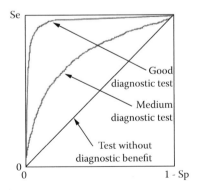

FIGURE 10.5
ROC curve.

- 0.70–0.80 = fair (C)
- 0.60–0.70 = poor (D)
- 0.50–0.60 = fail (F)

Going back to our qSOFA example, we would normally determine the AUC by integrating over the range (e.g., through software). However, we can approximate this value by summing the specificities and dividing by the number of cutpoints considered. We determined a Se = 0.56 when qSOFA = 3 and a Se = 0.78 when qSOFA ≥ 2. It is easy to determine that Se = 0.91 when qSOFA ≥ 1 and Se = 1.00 when qSOFA ≥ 0. Therefore, the AUC is given by (0.56 + 0.78 + 0.91 + 1.00)/4 = 0.81, which falls in the good range.

10.6.3 Prediction by Support Vector Machines

In this section, we introduce a common machine learning technique called support vector machines (SVMs). Note that this is only one approach to machine learning; there are numerous others including regularization algorithms, decision tree algorithms, Bayesian algorithms, clustering algorithms, artificial neural networks, and deep learning. Excellent textbooks on this topic include Goodfellow et al. (2016) and James et al. (2017). An excellent introduction for using the programming language Python for machine learning can be found in Raschka (2016).

SVMs are a form of "supervised learning" in that we use data for which we know the outcomes in order to "train." In this subsection, we will assume that we are trying to predict a binary classification (e.g., presence or absence of a disease) based on a set of features. This method works very well in high dimensions. However, a key drawback of the method is that it is not probabilistic in nature (i.e., it does not explicitly consider uncertainty).

Before presenting the SVM approach, some background in separating hyperplanes is useful. Consider a set of p features (e.g., blood pressure, temperature, age) that we will use for classification. Therefore, we will consider a p-dimensional feature space $\mathbb{R}^p$. A linear separating hyperplane is an affine $p - 1$ dimensional space embedded in the feature space. We can define an affine hyperplane by

$$b + w_1 x_1 + \ldots + w_p x_p = 0.$$

In dot product (or sometimes called inner product) form, this is represented by

$$\boldsymbol{w} \cdot \boldsymbol{x} + b = 0.$$

Note that bold type represents a vector. Consider the two-dimensional (2-D) example shown in Figure 10.6.

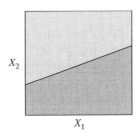

FIGURE 10.6
Two-dimensional feature space example.

In this case, we can classify an element by determining which side of the line it falls on. Elements in the upper half satisfy $w \cdot x + b > 0$, and elements in the lower half satisfy $w \cdot x + b < 0$. Note, therefore, that we can classify an element simply by the sign of the expression. This simple approach is the basis for SVM.

We will use $y^{(i)}$ to represent the "prediction" (classification) for observation (e.g., patient) i. For binary classification, then, we have one of two cases (i.e., $y^{(i)} = 1$ or $y^{(i)} = -1$). Examples of this type of classification are as follows: patient has type II diabetes or not, Medicaid claim was fraudulent or not, and LOS in a hospital will be greater than or less than 3 days. Our goal is to develop a classifier based on provided training observations that will correctly classify subsequent test observations using only their feature values.

Suppose we wish to classify septic (squares) versus non-septic patients (circles) as shown in Figure 10.7 using training data. Note that, in this 2-D case, a line (decision boundary) can be drawn that perfectly partitions the elements into two half-planes. The data in this case are called linearly separable. There are several choices that can be made, and one question is which decision boundary should be chosen?

The idea of an SVM is to pick the decision boundary that maximizes the boundary. This is illustrated in Figure 10.8. Here we are separating the circles from the squares. The examples of decision boundaries are shown. On the

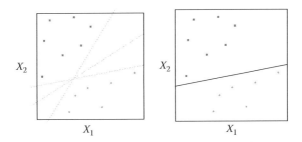

FIGURE 10.7
Septic patients (squares) and non-septic patients (circles). On left-hand side, several separating hyperplanes will linearly separate. The classifier chosen is shown on the right.

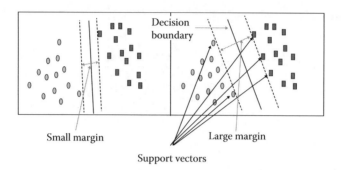

FIGURE 10.8
Support vectors and decision boundary.

left, notice that, for the decision boundary, there is a gap (or margin) for which parallel boundaries can be drawn and still separate. Similarly, on the right, there is a margin, but notice that it is wider. The width is "constrained" by the support vectors, as shown on the right. The SVM chooses the decision boundary that maximizes this margin.

Consider the positive and negative hyperplanes parallel to the decision boundary:

$$b + w \cdot x_{pos} = 1$$

$$b + w \cdot x_{neg} = -1$$

Subtracting the two linear equations from each other gives

$$w(x_{pos} - x_{neg}) = 2.$$

Normalizing the length of the w vector

$$\|w\| = \sqrt{\sum_{j=1}^{p} w_j^2},$$

which gives

$$\frac{w(x_{pos} - x_{neg})}{\llbracket w \rrbracket} = \frac{2}{\llbracket w \rrbracket}.$$

Healthcare Delivery Systems

Since our goal is to maximize the margin, we get the following optimization problem:

$$\min \frac{1}{2}[w]^2 \qquad (10.6)$$

subject to

$$b + w \cdot x \geq 1 \quad \text{if } y^{(i)} = 1$$

$$b + w \cdot x < -1 \quad \text{if } y^{(i)} = -1,$$

which is a quadratic program. We can transform the constraints to a single constraint:

$$y^{(i)}(b + w \cdot x^{(i)}) \geq 1 \quad \forall i.$$

This allows us to write the optimization model as a Lagrangian:

$$\mathcal{L}(\alpha, w, b) = \frac{1}{2}[w]^2 - \sum_i \alpha_i \left[y^{(i)}(b + w \cdot x^{(i)}) - 1 \right],$$

which is easy to solve.

In most cases, the data are such that it cannot be linearly separated. In Figure 10.9, for example, there is no straight line that can be drawn that

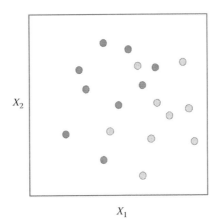

FIGURE 10.9
Example that is not linearly separable.

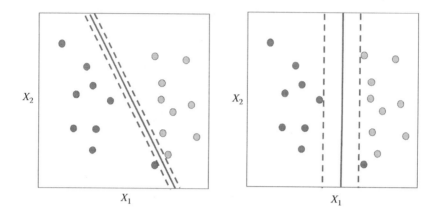

FIGURE 10.10
Decision boundary linearly separates on the left. Decision boundary on the right does not completely separate, but has a large margin.

separates the dark circles from the light circles. Further, it may be the case that we prefer a decision boundary that doesn't linearly separate the data, if we can achieve a large margin. This is illustrated in Figure 10.10. The reason for this is that the decision boundary is based on training data. When applied to actual data for classification, having a wide gap increases the likelihood of a proper classification.

We can modify our optimization model by using soft constraints (as discussed in Section 10.4.1). In the following model, we introduce slack variables ($\varepsilon^{(i)}$), and then penalize the slacks through C. This gives

$$\min \frac{1}{2}\|w\|^2 + C\left(\sum_i \varepsilon^{(i)}\right). \tag{10.7}$$

Subject to

$$b + w \cdot x^{(i)} \geq 1 \quad \text{if } y^{(i)} = 1 - \varepsilon^{(i)}$$

$$b + w \cdot x^{(i)} < -1 \quad \text{if } y^{(i)} = -1 + \varepsilon^{(i)}.$$

In practice, setting up the SVM requires "tuning" of C. If we chose large values of C, the optimization will choose a smaller-margin hyperplane if that hyperplane does a better job of getting all the training points classified correctly. The smaller the value of C, the more likely the result is a larger-margin separating hyperplane, even if that hyperplane misclassifies more points. A standard approach for tuning C is to use cross-validation. This involves the following steps:

- Divide the training set into k subsets of equal size.
- Sequentially, one subset is tested using the classifier trained on the remaining $k - 1$ subsets.
- Thus, each instance of the whole training set is predicted once so the cross-validation accuracy is the percentage of data that are correctly classified.
- This method is repeated for different values of C on same subsets (normally a grid search approach is used).

Finally, it may be possible to separate in a nonlinear way. Consider the 1-D example (Figure 10.11) with feature x. On the left, the light and dark circles are not separable. However, if we square each feature value x^2, then we get the figure on the right, which is linearly separable.

For a 2-D feature space, could map to a 3-D feature space:

$$(x_1, x_2) \xrightarrow{\Phi} (z_1, z_2, z_3)$$

For example:

$$\Phi(x_1, x_2) = (z_1, z_2, z_3) = \left(x_1, x_2, x_1^2 + x_2^2\right).$$

 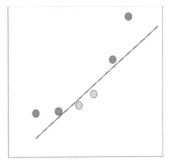

FIGURE 10.11
Light and dark dots are not linearly separable. Transforming by squaring the value of the feature yields the figure on the right, which is linearly separable.

While it is beyond the scope of this text, the dual of the Lagrangian relaxation has a dot (inner) product $(x_i \cdot x_j)$, which can be replaced by a kernel $K(x_i \cdot x_j)$. The most common kernel used is the radial basis function:

$$K(x_i, x_j) = e^{\left(-\gamma \|x_i - x_j\|^2\right)},$$

where γ is a parameter (that will need to be tuned like C).

Although we presented the mathematical foundations, it is easy to implement this in Python using the "scikit" learn package. To illustrate, consider the famous data set Iris, which can be downloaded at https://archive.ics.uci.edu/ml/datasets/Iris/. This data set has four features: sepal length, sepal width, petal length, and petal width. The goal is to classify whether or not a given object is an iris. The first step is to tune the values of C and γ using a radial basis function kernel. In Python, this is done using the following code:

```
import numpy as np
from sklearn.datasets import load_iris
from sklearn.cross_validation import StratifiedShuffleSplit
from sklearn.grid_search import GridSearchCV
iris = load_iris()
X = iris.data
y = iris.target
C_range = np.logspace(-2, 10, 13)
gamma_range = np.logspace(-9, 3, 13)
param_grid = dict(gamma=gamma_range, C=C_range)
cv = StratifiedShuffleSplit(y, n_iter=5, test_size=0.2,
       random_state=42)
grid = GridSearchCV(SVC(), param_grid=param_grid, cv=cv)
grid.fit(X, y)
print("The best parameters are%s with a score of%0.2f"
       % (grid.best_params_, grid.best_score_))
```

The result of this program will give $C = 1.0$ and $\gamma = 0.1$. Using these values, we can then use the SVM with the following code to predict:

```
clf = svm.SVC(kernel="rbf," C=1.0, gamma=0.10)
clf.fit(X, y)
classifiers.append((C, gamma, clf))
print(clf.predict([a1, a2, a3, a4]))
```

where a1 to a4 are the specific features. The output will be a prediction of whether the object with these features falls in the iris class or not.

10.6.4 Prediction Example: 30-Day Readmissions

As mentioned earlier, readmissions are problematic for hospitals. First, there may be associated penalties from CMS, and second, it can strain hospital capacity, which leads to poor patient outcomes. If hospitals can predict who the most likely patients that would be readmitted, then they can allocate additional resources for those identified patients to reduce the likelihood. Examples include the development of a follow-up plan before discharge, regular monitoring through telehealth, or detailed education programs.

Futoma et al. (2015) present a simple approach to predicting early hospital readmission (i.e., less than 30 days), including the use of SVMs. This was applied to the data from the New Zealand Ministry of Health, which consisted of roughly 3.3 million hospital admissions between 2006 and 2012. A binary classification was used, namely, whether or not a patient was readmitted to the hospital within 30 days of discharge. The data used for the classification included patient background (race, sex, age, LOS), facility type (public or private), number of hospital visits for the patient over the past year, whether a patient was a transfer or not, and disease group (which was broken into 280 different types).

By applying the standard SVM approach, the authors were able to achieve an AUC of 0.671. Note that several other researchers have been able to achieve higher AUCs by focusing on specific conditions rather than all readmissions.

10.7 Discussion

Healthcare makes up the largest component of the U.S. economy. In addition, it lags other industries such as manufacturing, distribution, or retail in terms of the level of sophistication of decision support and information technology. It is therefore ripe with opportunities for the application of service engineering principals.

There is no way to cover the topic in any level of detail in a single chapter. However, many of the key challenges are at least discussed here. Further details may be found in many of the provided references including Hopp and Lovejoy (2013), Griffin et al. (2016), and Yih (2016).

Exercises

10.1 Do some background searching and determine the main reasons for visiting a physician in the United States.

10.2 Look up ways that prostate cancer can be treated. What do you think are the key factors that determine the choice of treatment? What are ways that the patient could be empowered to help make this decision?

10.3 What are benefits and drawbacks of ACOs?

10.4 A major problem with EDs is the long wait times to be seen. This can lead to negative health consequences as well as patient dissatisfaction. What are the possible approaches to address this difficulty? Why do think it is a good idea? What would be the key challenge for implementation?

10.5 Construct a value stream map of getting ready to go to school or work.

10.6 Why do you think OR scheduling is challenging? What do you believe are the key factors for low utilization of OR rooms?

10.7 Suppose that you are worried that you might have a rare disease. You decide to get tested; the testing methods for this disease are correct 99% of the time (in other words, if you have the disease, it shows that you do with 0.99 probability, and if you don't have the disease, it shows that you do not with 0.99 probability). Suppose this disease is rare, occurring randomly in the general population in 1 of every 10,000 people. If your test results come back positive, what are your chances that you actually have the disease?

10.8 Consider the ICU discussed in Section 10.4.4. Determine the impact of the following (assuming M/M/c) assuming an arrival rate of 3.9: (i) adding another bed, (ii) reducing the squared coefficient of variation of arrivals by 25%.

10.9 For the ICU discussed in Section 10.4.4., suppose that the maximum number of boarded patients before the hospital is put on diversion is 6. What is the probability that the hospital goes on diversion if the arrival rate is 3.8?

10.10 Consider the publicly available heart disease data set at (http://archive.ics.uci.edu/ml/datasets/heart+Disease). Use the 13 attributes given to predict angiographic disease (binary classification) using an SVM approach. Train on half of the data, and validate on the second half.

10.11 For the data below, suppose we are using that a person tests positive for a disease if their value of $A \geq 4$. Determine the sensitivity and specificity of this test (note that sensitivity = TP/(TP + FN) and specificity = TN/(TN + FP)).

Value of A	No Disease	Disease
1	49	3
2	37	18
3	25	30
4	10	58
5	3	64
6	1	68

10.12 Consider a simple hospital system that consists of two non–revenue-generating functions (A and B) and two revenue-generating functions (C and D). Over the past month, the services provided are given in the table below. Give the set of simultaneous equations for the reciprocal allocation method (but do not solve).

	Services Allocated			
Source	A	B	C	D
A	0%	20%	40%	40%
B	10%	0%	60%	30%
Preallocation costs	$75,000	$50,000	$250,000	$140,000

10.13 An ICU has 20 beds. The average length of time a patient stays in the ICU once placed there is 5 days. The standard deviation of this time is 7 days. Patients arrive to the ICU at a rate of 3.7 per day. The standard deviation of interarrival time is 0.4 days. Answer the following:

a. What is the average number of people that would be boarded? Explain what boarding means.

b. Currently, the hospital uses a standard discharge time for step down. If instead they step down the patient at the time they are ready, they found that this reduced average time in ICU by 7 hours. How much would average boarding time decrease if they used this strategy? Also, give one other strategy that would reduce average time in the ICU. Finally, give a reason why staying longer than needed in the ICU is bad.

10.14 We developed a severity score (1 to 5) based on a set of measures (e.g., temperature, blood pressure) to determine if a patient has sepsis. Answer the following:

a. Suppose we use a score of 3 or higher as our prediction that the patient has sepsis. The table below shows the actual results (values are number of patients that fall under each category). Determine the sensitivity and specificity of this test:

Score	Has Sepsis	Doesn't Have Sepsis
1	4	39
2	7	33
3	18	12
4	29	8
5	43	4

b. Use the scores of 3 and 5 (don't worry about the others) and plot the ROC curve. Estimate the AUC and draw conclusions.

10.15 An IE has determined that a hospital ward has a need for the following number of nurses on the 7 a.m. to 3 p.m. shift for Sunday to Saturday, respectively:

12, 8, 7, 9, 10, 13, and 14.

If only full-time nurses are used, and they have a shift of 5 days on and 2 days off, write the optimization model to determine the staffing. Please write the full model (i.e., use the data provided).

10.16 In the formulation in Exercise 10.16, the implicit assumption made was that the problem was separable by shift. Why might this not be the case in practice? What other factors might affect the formulation?

References

Agency for Healthcare Research and Quality (AHRQ). 2007. Care Transitions. Available at https://psnet.ahrq.gov/perspectives/perspective/52

Agency for Healthcare Research and Quality (AHRQ). 2017. *The Effective Health Care Program Stakeholder Guide, Chapter 3.* Available at https://www.ahrq.gov/research/findings/evidence-based-reports/stakeholderguide/chapter3.html

Azaiez, M. M. and S. S. Al Sharif. 2005. A 0-1 goal programming model for nurse scheduling. *Computers & Operations Research.* 32: 491–507.

Best, T. J. 2015. Hospital Bed Capacity Management. Doctoral dissertation, The University of Chicago.

Bradley, E. H. and L. A. Taylor. 2015. The American Healthcare Paradox: Why Spending More is Getting Us Less. *Public Affairs.* March 3.

Castro, A. B. 2015. *Principles of Healthcare Reimbursement,* 5th Edition. Chicago, IL: American Health Information Management Association.

Centers for Medicare and Medicaid Services (CMS). 2016. National Health Expenditures 2015 Highlights. Available at https://www.cms.gov/Research-Statistics-Data-and-Systems/Statistics-Trends-and-Reports/NationalHealthExpendData/downloads/highlights.pdf

Chugh, A., M. V. Williams, J. Grigsby, and E. A. Coleman. 2009. Better transitions: Improving comprehension of discharge instructions. *Frontiers of Health Services Management.* 25: 11–32.

Commonwealth Fund. 2017. U.S. Healthcare from a Global Perspective. Available at http://www.commonwealthfund.org/publications/issue-briefs/2015/oct/us-health-care-from-a-global-perspective

Davis, N. A. 2017. *Revenue Cycle Management Best Practices,* 2nd Edition. Chicago, IL: American Health Information Management Association.

Futoma, J., J. Morris, and J. Lucas, J. 2015. A comparison of models for predicting early hospital readmissions. *Journal of Biomedical Informatics.* 56: 229–238.

Gapenski, L. C. and C. L. Reiter. 2015. *Healthcare Finance: An Introduction to Accounting and Financial Management,* 6th Edition. Washington DC: Health Administration Press.

Goodfellow, I., Y. Bengio, and A. Courville. 2016. *Deep Learning*. Boston, MA: The MIT Press.
Green, L. V., S. Savin, and B. Wang. 2006. Managing patient service in a diagnostic medical facility. *Operations Research*. 54: 11–25.
Greenwald, J. L., C. R. Denham, and B. W. Jack. 2007. The hospital discharge: A review of a high risk care transition with highlights of a reengineered discharge process. *Journal of Patient Safety*. 3: 97–106.
Griffin, P. M., H. B. Nembhard, C. J. DeFlitch, N. D. Bastian, H. Kang, and D. Munoz. 2016. *Healthcare Systems Engineering*. Hoboken, NJ: John Wiley & Sons.
Hans, E., G. Wullink, M. van Houdenhoven, and G. Kazeimier. 2008. Robust surgery loading. *European Journal of Operational Research*. 185: 1038–1050.
Health Affairs. 2016. Health policy brief: Ambulance diversion. *Health Affairs*. June 2.
Health and Human Services (HHS). 2007. HHS Data Council. Available at https://aspe.hhs.gov/health-and-human-services-hhs-data-council
Heiman, H. J. and S. Artiga. 2015. Beyond Healthcare: The Role of Social Determinants in Promoting Health and Health Equity. Issue Brief, Kaiser Family Foundation. Available at http://kff.org/disparities-policy/issue-brief/beyond-health-care-the-role-of-social-determinants-in-promoting-health-and-health-equity/
Holtz, C. 2016. *Global Healthcare: Issues and Policies*, 3rd Edition. Burlington, MA: Jones & Bartlett Learning.
Hopp, W. J. and W. S. Lovejoy. 2013. *Hospital Operations: Principals of High Efficiency Health Care*. Upper Saddle River, NJ: Pearson Education.
Hosseini, N. and K. M. Taaffe. 2015. Allocating operating room block time using historical caseload variability. *Health Care Management Science*. 18: 419–430.
Jackson, T. J. 2014. *Mapping Clinical Value Streams*. Boca Raton, FL: CRC Press.
James, G., D. Witten, T. Hastie, and R. Tibshirani. 2017. *An Introduction to Statistical Learning: with Applications in R*, 7th Edition. New York, NY: Springer Science.
Jimmerson, C. and A. Jimmerson. 2009. *Value Stream Mapping for Healthcare Made Easy*. Boca Raton, FL: CRC Press.
Kang, H., H. B. Nembhard, C. Rafferty, and C. J. DeFlitch. 2014. Patient flow in the emergency department: A classification and analysis of admission process policies. *Annals of Emergency Medicine*. 64: 335–342.
Kaiser Family Foundation. 2017. *State Health Facts*. Available at http://www.kff.org/statedata/
Leventhal, R. 2016. Three EHR vendors comprise majority of market share for hospitals participating in MU. *Health Care Informatics*. March 16. Available at https://www.healthcare-informatics.com/news-item/three-ehr-vendors-comprise-majority-market-share-hospitals-participating-mu
OECD. 2017. *OECD Health Statistics 2017*. Available at http://www.oecd.org/els/health-systems/health-data.htm
Ozcan, Y. A. 2017. *Analytics and Decision Support in Healthcare Operations Management*, 3rd Edition. San Francisco, CA: Jossey-Bass.
Pearl, J., M. Glymour, and N. P. Jewell. 2017. *Causal Inference in Statistics: A Primer*. Hoboken, NJ: John Wiley & Sons.
Pracek, R. 2013. *Value Stream Mapping for Healthcare*. Chelsea, MI: MCS Media.
Raschka, S. 2016. *Python Machine Learning*. Birmingham, UK: Packt Publishing.
Schroeder, S. A. 2007. We can do better—Improving the health of the American people. *New England Journal of Medicine*. 357: 1221–1228.

Singer, A. J., H.C. Thode Jr., P. Viccellio, and J. M. Pines. 2011. The association between length of emergency department boarding and mortality. *Academic Emergency Medicine*. 18: 1324–1329.

Stevens, W., T. J. Philipson, Z. M. Khan, et al. 2015. Cancer mortality reductions were greatest among countries where cancer care spending rose the most, 1995–2007. *Health Affairs*. 34: 562–570.

The World Bank. 2016. *Health Expenditures per Capita*. Available at http://data.worldbank.org/indicator/SH.XPD.PCAP

Yang, H. and E. Lee (eds.). 2016. *Healthcare Analytics: From Data to Knowledge to Healthcare Improvement*. Hoboken, NJ: Wiley Series in Operations Research and Management Science.

Yih, Y. (ed.). 2016. *Handbook of Healthcare Delivery Systems*. Boca Raton, FL: CRC Press.

11

Financial Services

Three thousand years ago, commerce involved bartering of goods, such as produce, grains, skins, and so on. As civilizations evolved, various objects were used as "money," including salt, shells, coins, silver, and gold (Weatherford 2009). The advent of credit cards, Internet, and electronic commerce has digitized money and given rise to new financial services and transformed many existing services such as banking, stock trading, and currency trading to name a few (Allen et al. 2002). In this chapter, we will survey the following financial services that form the backbone of modern economies around the world:

- **Payment systems** that are used in credit card, debit card, and check transactions
- **Banking** services that have been driven by advances in information technology and the need to reduce operating cost
- **Trading** in securities such as stocks and mutual funds
- **Derivatives** such as options are also surveyed from an introductory perspective

11.1 Payment Systems

There are more than 100 billion payment transactions made every year in the United States, using credit cards, debit cards, prepaid cards, and Automated Clearing House (ACH), which is widely used for direct bill payments, checks, and other cards. The share of these payments between 2000 and 2012 is shown in Table 11.1. In 2012, there were more than 300 million credit, debit, and prepaid cards in the United States, of which most are held by consumers, rather than businesses (Federal Reserve 2013). During 2012, there were an estimated 23.7 billion transactions using these cards, summing to a value of $2.2 trillion. Based on the trend in payments shown in Figure 11.1, it is clear that paper checks are being used much less compared to cards, with debit cards gaining rapidly in recent years.

Consumer's preference of payment mode for a transaction may depend on a variety of factors including sociocultural, value of transaction, the type

TABLE 11.1

Payment Modes in the United States, 2000–2012

Payment Modes	Percentage
Credit card	19
Debit card	28
Prepaid card	1
Other cards	6
ACH	15
Checks	32

Source: Federal Reserve. 2013. Federal Reserve Payments Study—Detailed Report. Retrieved on 7.4.2015. https://www.frbservices.org/files/communications/pdf/general/2013_fed_res_paymt_study_detailed_rpt.pdf

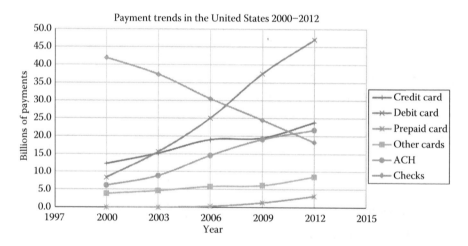

FIGURE 11.1
Trends in payments in the United States during 2000–2012. (Data from Federal Reserve. 2013. Federal Reserve Payments Study—Detailed Report. Retrieved on 7.4.2015. https://www.frbservices.org/files/communications/pdf/general/2013_fed_res_paymt_study_detailed_rpt.pdf)

of goods/services being purchased, and rebate incentives. Figure 11.2 illustrates the variation in payment modes used in several countries.

In order to understand the basic functioning of the prevailing payment systems, consider the four-corner payment model, which is used for credit cards, debit cards, ACH, and ATM transactions. A simplified version of this model is illustrated in Figure 11.3 (FFIEC 2010). In this model, the payer is depicted in the top left corner, the payee at the top right corner, financial institution for the payer ("issuer bank") is at the bottom left, and the financial institution ("acquirer bank") for the payee is at the bottom right corner.

Financial Services

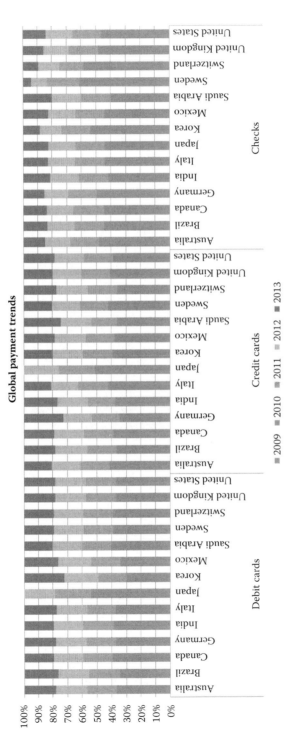

FIGURE 11.2
Trends in payment modes around the world. (Based on data from BIS. 2014. *Statistics on payment, clearing and settlement systems in the CPMI countries. Figures for 2013*, December 2014, Bank for International Settlements.)

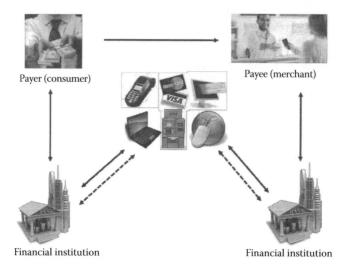

FIGURE 11.3
The four-corner payment model. Solid lines represent the flow of information and dashed lines represent the flow of funds.

When a payer initiates a payment transaction, a chain of information flow and funds flow is triggered across multiple financial institutions. Payment transactions are routed by payment networks or clearing house organization, which are shown in the center of the model (Figure 11.3).

A typical transaction in the four-corner payment model has the following sequence of information/funds flows:

1. Consumer initiates her purchase by presenting her card to the merchant.
2. When the merchant swipes the card, an electronic authorization request is sent to the issuer bank though the acquirer bank and payment network, along with details such as cardholder identity, card number, and the specific amount.
3. The issuer bank checks the details of the transactions in terms of active card accounts and available credit and sends back an electronic reply authorizing (or declining) the transaction.
4. The issuer bank posts the corresponding charge to the cardholder's account, and the acquirer posts the corresponding credit to the merchant's account. Payments and debits are consolidated across banks daily by the payment networks involved; this process is usually referred to as *clearing*. The actual transfer of funds is called *settlement* and can involve currency exchanges if the transaction involves foreign currency.

The specific payment network and clearing system used will depend on the particular type of transaction used. For instance, when a consumer uses a Visa-branded card to pay, then that transaction will use Visa's network, VisaNet, to route the transaction. Similarly MasterCard, American Express, and Discover have their own payment networks. American Express and Discover act as issuer and payment network, which makes the corresponding payment models a "three-corner" model rather than a four-corner model. These payment networks act as gateways for authorizing as well as setting the terms for fees and fraud liability. At any given time, payment networks may be simultaneously processing billions of transactions around the world, with each transaction typically taking a fraction of a second to complete. Similarly, there are separate networks for checks, ACH, and ATM transactions with similar flow of information and funds.

To consider the impact of payment systems, let us focus on grocery stores. It is estimated that the average cost of handling payments is about 1.7% of the revenue (Hahn and Anne 2006), which is comparable to the average pre-tax profit of 1% to 3% (Food Marketing Institute 2013). Therefore, there is clearly an incentive for retailers to process payments as efficiently as possible. Estimated cost of an e-payment is $1.34 per transaction compared to $2.97 for checks; about 40%–55% lesser than check (Humphrey et al. 2000). In Sweden, the cost of keeping cash is estimated at 0.26% of the GDP, compared to 0.19% and 0.09% for credit cards and debit cards, respectively, which is motivating a move toward cashless society in Sweden (Bloomberg 2013). Given the volume of payment transactions, the benefits of lowering the transaction costs are immense, and these benefits can be expected to accrue throughout the payment network. A recent report by the U.S. Federal Reserve identified key attributes for improving payment systems as speed, ubiquity, security, efficiency, and cross-border payments (Federal Reserve 2015).

Many innovations are afoot in payment systems, primarily driven by advances in new technology and e-commerce, as shown in Figure 11.4 (Bank for International Settlements [BIS] 2012). Some of the notable advances in payment technology is Near-Field Communication (NFC), mobile payments using cell phones/tablets, and electronic bill presentment and payments (EBPP). In some countries, electronic money is used widely by commuters for paying for public transport services. Here, electronic money means value stored electronically in a device such as a chip card or a computer. In many countries, innovation in payment systems is sought for financial inclusion of sections of the population without access to traditional banking and financial infrastructure. The market for new products for payment service needs to reach a critical mass for it to be viable. In other words, consumers would want it to be accepted by many merchants before adopting; likewise, merchants will want sufficient number of consumers to use it before adopting. A payment service product will be economically viable when the average cost per transaction decreases with the number of transactions processed, leading to economies of scale. Many times, the same infrastructure can be

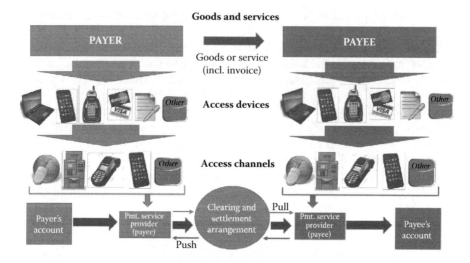

FIGURE 11.4
Payment process architecture. (Adapted from BIS. 2012. Innovations in retail payments. Report of the Working Group on Innovations in Retail Payments. May 2012, Bank for International Settlements.)

used for multiple products, thereby reducing the cost compared to having separate infrastructure, leading to economies of scope. Typically, ACH processes have good economies of scope because their infrastructure is used for different payments such as checks, credit transfer, direct debits, and so on. One of the advancements that have the potential for significant economies of scope is ISO 20022, a messaging standard for payment processing, card transactions, securities trade, trade finance, and currency trading.

11.2 Banking

Banks are a major component of the financial services industry and play a major role in the flow of money through modern economies as we have seen in the previous section. The key distinction between a bank and other financial service enterprises is that they can accept deposits. Therefore, there are stringent regulations for starting a bank because of the implied public trust. Typical deposit services that banks provide include savings account, checking account, certificate of deposits, and money market accounts. Typical credit services that banks provide include credit cards, car loans, home mortgage loans, home equity loans, home equity line of credits, and commercial loans to businesses. Banking infrastructure and access vary considerably around the world, with more access in higher-income regions as shown in Table 11.2.

TABLE 11.2

Selected Banking Infrastructure and Access Metrics

Region	Household Penetration: Deposit Accounts (%)	Deposit Accounts per 1000 Adults	Bank Branches per 100,000 Adults	ATMs per 100,000 Adults
High-income countries	91	2022	32	94
East Asia and Pacific	42	1756	15	11
Europe and Central Asia	50	1330	18	50
Latin America and Caribbean	40	1140	14	31
Middle East and North Africa	42	818	17	28
South Asia	22	317	7	4
Sub-Saharan Africa	12	163	3	5
All developing countries	NA	737	10	29

Source: Consultative Group to Assist the Poor and World Bank, *Financial Access 2010*.

Banks essentially earn profits by charging higher rates for loans compared to the interest paid on deposits. Over the last several decades, people have increasingly replaced bank deposits with mutual funds and other securities. Moreover, businesses have sought financing through commercial paper and corporate bonds. Advances in technologies and consolidation among banking enterprises have led to commoditization of traditional banking services, which has led to growing pressure to earn profits through new services.

Globalization of financial services has lowered barriers for financial flows. This, combined with a spate of mergers and acquisitions, has led to major changes in banking in the last 20 years (Claessens et al. 2002). As shown in Figure 11.4, there are a plethora of ways to access and conduct banking transactions. Hence, banks need to rely on cutting cost to produce "commodity services" and simultaneously provide personalization to attract more customers to be more profitable, fee-based services. It is estimated that, in 2006, cost of processing a paper check was 9.6 cents, which for electronic checks dropped to 2.5 cents by 2010, a 74% cost reduction (Humphrey and Hunt 2013). Retailers are now able to scan the checks at the point-of-sale terminals and electronically send the image to their bank over a communication network to speed up the process of collecting the payment. Likewise, banks provide their customers images of the checks they write over the Internet rather than mail canceled checks, which reduces mailing cost for banks.

11.2.1 Staffing and Customer Waiting Time

In order to provide good customer service, it is desirable to minimize waiting time at teller queues. Suppose the management sets the target of maximum queue length to be L_{max} customers, then we can use Little's law, discussed in Chapter 3, to determine the staffing at teller counters, if we know the customer arrival rate and teller service rate.

Example 11.1

Consider customer arrivals at a bank teller queue at 6.5/hour, but there is considerable fluctuation over the day shown in Figure 11.5. Suppose it takes an average of 4 minutes of teller's time to serve a customer. The management's service level objective is to limit the average queue length (L_q) to be less than 3. Using Equation 3.43 from Chapter 3, the average queue length is given by

$$L_q = \lambda W_q = \frac{\lambda^2}{\mu(\mu - \lambda)} = \frac{6.5^2}{15(15 - 6.5)} = 0.331 < 3.$$

This satisfies the service level objective on the average.

However, in banks, the demand for service may vary considerably over a period, which makes the system non-stationary (see Figure 11.5). Technically, we can deal with this by breaking the planning period into smaller intervals, and if service times are much smaller than the intervals, then good approximations can be readily obtained (Green et al. 2007).

Consider now the peak arrival rate of 14 customers/hour during the 9:00 a.m. to 10:00 a.m. interval (Figure 11.5). Then, the average queue length during this period becomes

$$L_q = \lambda W_q = \frac{\lambda^2}{\mu(\mu - \lambda)} = \frac{14^2}{15(15 - 14)} = 13 > 3.$$

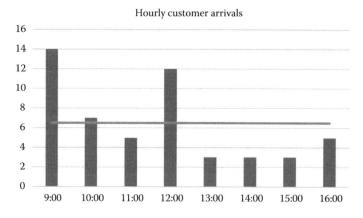

FIGURE 11.5
Customer arrival pattern at a bank (Example 11.1).

Financial Services 537

Since this violates the service level objective, one option is to increase the number of tellers. Suppose we use two tellers. Then, using the results of Section 3.3.5, we get

$$L_q = 0.26 < 3.$$

This implies that if we staff the bank with two tellers, then we can meet the service level objective at all times. Suppose it costs the bank $15/hour for a teller's salary. Then, having two tellers for 8 hours/day will cost $240/day. If one of the tellers were to be hired part time for 2 hours, from 9:00 to 10:00 and 12:00 to 13:00, when there is a spike in customer arrivals, then it would cost only $150/day to meet the service level objective.

11.2.2 Diversifying Banking Channels

Bank management needs to carefully consider investing in new facilities in order to ensure they are profitable and provide good customer service to ensure competitiveness. Desirable sites for new branches would have several characteristics including heavy vehicle or foot traffic, a surrounding with a large number of retail enterprises, and a growing population with average age over 45 years consisting of professionals and business owners (Rose and Hudgins 2014). Many banks have in-store branches that are located in the premises of other retail enterprises such as supermarkets, malls, and shopping centers with high traffic flows. Typically, these operations have one to three employees and offer a limited set of services focused on deposits and potentially upselling other services by taking advantage of face-to-face interactions. It is estimated that an in-person transaction at a branch may cost $1 but only $0.02 over the Internet. Similarly, an ATM transaction costs a fraction of in-person transaction. Banks need to provide their current and prospective customers adequately diverse channels and branch locations.

Example 11.2

Lion Country Bank is planning to expand into a new city by opening full-service branches, in-store branches, and ATMs. Management is willing to invest up to $7 million to establish a presence in this city if the overall return is better than 10% and need to know the investment risk. Based on market research and prior experience, they have estimated the required investments, expenses, and returns as shown in Table 11.3.

Geographical diversification offers potential for reducing risk. The variance–covariance matrix of returns on investment for the 11 potential sites is given in Table 11.4. Note that the first three indices are for full-service sites, followed by the next three for store sites, and the last five for ATM sites.

TABLE 11.3

Data for Example 11.2

Channel	Minimum	Maximum	Investment	Average Return
Full-service branches (index 1 to 3) FS1, FS2, FS3	1	3	$3,000,000	5%
In-store branches (index 4 to 6) SB1, SB2, SB3	0	3	$250,000	10%
ATM (index 7 to 11) ATM1, ATM2, ATM3, ATM4, ATM5	3	5	$50,000	12%

As discussed in Chapter 7, this problem can be modeled as a quadratic programming problem with the objective of minimizing risk characterized by the variance of returns on investment as follows:

$$\text{Minimize Risk} = x^T Q x = \sum_{I=1}^{N} \sum_{J=1}^{N} q_{IJ} x_I x_J \qquad (11.1)$$

$$\text{Subject to } \sum_{J=1}^{N} \mu_J x_J \geq \text{Ret} = 10\% \text{ (Average returns on investments)} \qquad (11.2)$$

$$\sum_{J=1}^{N} a_J x_J \leq B \text{ (Maximum investment)} \qquad (11.3)$$

In this case, x_J's are 0–1 binary variables, a_J is the investment in the *J*th site, and *B* is $7 million.

The above problem can also be modeled as a bi-criteria optimization problem:

Minimize Risk (Portfolio Variance)

$$= x^T Q x$$

Maximize average Annual Return

$$= \sum_{J=1}^{N} \mu_J x_J = \mu^T x$$

The solution to the problem is left as an exercise at the end of this chapter.

Financial Services

TABLE 11.4
Variance–Covariance Matrix of Returns on Investments (Example 11.2)

		FS1	FS2	FS3	SB1	SB2	SB3	ATM1	ATM2	ATM3	ATM4	ATM5
		1	2	3	4	5	6	7	8	9	10	11
FS1	1	0.0025	0.5	0.3	0.5	0.001	0.001	0.005	0.001	0.001	0.001	0.001
FS2	2	0.5	0.005	−0.01	−0.6	0.005	0.005	0.001	0.005	0.001	0.001	0.003
FS3	3	0.3	−0.01	0.007	0.5	0.005	−0.6	0.001	0.001	0.005	0.003	0.001
SB1	4	0.5	−0.6	0.5	0.08	0.005	0.005	0.001	0.005	0.001	0.001	0.003
SB2	5	0.001	0.005	0.005	0.005	0.1	0.005	0.001	0.001	0.001	0.001	0.001
SB3	6	0.001	0.005	−0.6	0.005	0.005	0.09	0.001	0.001	0.005	0.003	0.003
ATM1	7	0.005	0.001	0.001	0.001	0.001	0.001	0.02	0.001	0.001	0.001	0.001
ATM2	8	0.001	0.005	0.001	0.005	0.001	0.001	0.001	0.02	0.001	0.001	0.001
AMT3	9	0.001	0.001	0.005	0.001	0.001	0.005	0.001	0.001	0.02	0.003	0.001
ATM4	10	0.001	0.001	0.003	0.001	0.001	0.003	0.001	0.001	0.003	0.02	0.001
ATM5	11	0.001	0.003	0.001	0.003	0.001	0.003	0.001	0.001	0.001	0.001	0.02

11.3 Electronic Trading

Chapter 7 covered financial engineering in which we focused on selecting a portfolio of investments for a given objective. Investment securities such as stocks, bonds, and mutual funds held in such portfolios are bought and sold predominantly over a network of computer systems and is called electronic trading. Electronic trading involves several entities that constitute the securities industry as shown in Figure 11.6.

We briefly describe some of the major entities in the securities markets below (Simmons 2003).

- *Issuers*—These include companies that raise capital by issuing equity, which are sold to investors. Such companies are also called "publically traded companies" and they have to comply with government regulations and various reporting requirements. Other issuer entities are the ones that issue bonds, including sovereign governments and state and local governments including municipalities, universities, and international organizations.
- *Investors*—These can be individuals or institutions such as mutual funds, pension funds, hedge funds, and nonprofit organizations. Individual investors usually trade in the market through a web-hosted brokerage account with an intermediary, who provides access to the market. Such investors typically deal with their savings or individual retirement accounts. Institutional investors typically deal with large assets that range from billions or even trillions of

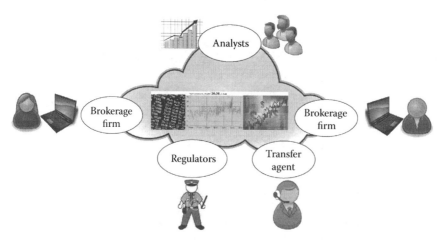

FIGURE 11.6
Overview of major entities in the securities industry.

dollars. Examples of such investors include mutual funds, brokerage firms, pension funds, and university endowments.
- *Brokers*—They are licensed to accept deposits from investors and to place investment orders on their behalf. There are two major types of brokers: (1) full-service brokers provide investment advice, while charging fees based on asset value or trades, and (2) discount brokers, who charge much smaller fees but offer very few services beyond being an intermediary that provides web access.
- *Exchanges*—Historically, these have been physical places, where securities have been traded. Now, exchanges are rapidly evolving into electronic platforms that can be accessed using computer networks. Exchanges are also regulated by government agencies to ensure fairness and transparency of prices of securities listed for trading. Individual exchanges impose a fairly comprehensive set of rules for companies to be listed for trading.
- *Registrars and Transfer Agents*—They keep electronic records of ownership of securities on behalf of issuers along with keeping track of transactions, mailing periodic statements to investors, and making sure investors get dividends when due. Additional services provided by these entities include communicating with investors any proxy and voting information on major corporate decisions as and when they arise.
- Regulators—They enforce government laws and set policies for the security industry that covers all major entities including exchanges, brokers, and investment advisors. Much of the regulations are intended to protect investors from fraudulent market manipulations by promoting full public disclosure of all relevant information.
- *Analysts*—These are individuals or companies that evaluate historic and current data to assess performance of individual companies and their economic sectors to discern business trends and recommend buying, selling, or holding specific securities. Such recommendations from various analysts are provided to brokers, investment advisors, and fund managers. An analyst may evaluate financial statements filed by a company to value its stock price or to forecast its stock price.

Since the 1990s, most securities tradings have moved from paper to automated electronic processes, leading to significant gains in process efficiencies, speed of trade execution, and other benefits. For example, such automated systems are estimated to have reduced trading transaction cost by about 30% in the United States (Allen et al. 2001). Today, it is common to pay a commission of less than $10 for a trade, compared to $45 or more in the 1980s.

Typically, an order has to be qualified in the following ways:

- *Buy* or *Sell*.
- *Market Order* specifies that the trade be executed at the best price prevailing in the market.
- *Limit Order* specifies the price, or better, at which to execute the trade; an order may not be executed if the specified price is never reached.
- *Stop Order* specifies the price at which the trade should be executed; in this case, the actual transaction price may be different from the specified price depending on the volatility and when the trade gets executed.
- *Day, Good-till-canceled, Fill or Kill* specify the time horizon of the order validity.

For example, let us consider purchasing 100 shares of Apple Inc. stock (Ticker Symbol AAPL). The simple moving average over the last 20 days is $115, which has fluctuated in a band of $120 to $112. It is quite common for stock prices to fluctuate by minute, or even faster. In this case, suppose we decided not to pay more than $117 for the stock and we are willing to wait the whole trading day to execute the trade. The first step is to go to a stock broker, which commonly involves logging into our account over the Internet with the brokerage firm. Such brokerage firms act as agents for individual investors to buy and sell shares and charge a fee for this service. Next, we specify the order for the desired trade as illustrated in Figure 11.7.

	SSEMBrokerage account (8148657601)	
Total account value	$12,000.00	As of 10/3/2016 1:20pm ET
Cash available to trade	$12,000.00	
Symbol	AAPL	
Action	Buy	
Order quantity	100	Shares
Order type	Limit order	
Limit price	$117.00	
Time in force	Day	
Order expiration	4:00pm EST	
Brokerage fee	$ 9.95	

FIGURE 11.7
Order specifications for an electronic trade.

Once the brokerage firm verifies the order, it is sent to brokers, who deal with the stock, perhaps to multiple stock exchanges to get the best price. For electronic trading to function reliably and smoothly, considerable business logic must be built into these systems. For instance, for the order specified in Figure 11.7, there is enough cash available to trade for purchasing the shares, which will be a maximum of $11,700, and the brokerage fees, $9.95.

To understand some of the intricacies of electronic trading, let us consider what happens in the above example once our brokerage firm sends our order to brokers, who deal with the stock. Such brokers deal with specific securities and act as market makers who constantly quote bid/ask rice and match buyers and sellers for the designated security. Market makers hold a large inventory of that security in order to maintain liquidity for any number of buy/sell orders. In other words, market makers take the opposite position of whatever is happening to their security at a given point of time in the market. If investors are buying a security, then its market makers have to sell the security; conversely, if investors are selling, then market makers have to buy the security, thereby maintaining liquidity for their specific security. Market makers earn a profit by maintaining a bid–ask spread, which represents the difference between their selling and buying prices.

Once a trade transaction is completed, security ownership is transferred along with corresponding funds and commissions, which is called the settlement. The period between transaction and settlement is usually called the settlement period, and it is tightly regulated to minimize risk of a party in the trade not fulfilling their obligation. For example, in the United States, most stock trades have a settlement period of 3 days, and this is abbreviated by the notation "T + 3," where "T" represents the transaction date. Current settlement processes require considerable manual effort, which cause delays leading to T + 3. There are ongoing initiatives to fully automate the trade and settlement process without requiring any manual intervention. This is referred to as "Straight-Through Processing," and is abbreviated as STP. It is anticipated that STP will reduce the settlement period to within the same day or even as short as a few seconds.

Many companies, such as hedge funds, are employing high-speed networks and powerful computers to analyze large volumes of market data and use algorithms to execute trade in fractions of seconds without any human intervention, called algorithmic trading. Specifically, this has given rise to automated trading of millions of securities that are bought and sold to exploit market trends over a very short time horizon. This is commonly referred to as high-speed trading and has generated debate in terms of its fairness and impact on security markets. However, the algorithms themselves are proprietary.

11.4 Valuing Companies and Their Stock Prices

Broadly, there are two approaches to determining when to buy stocks for a particular company:

1. Based on the fundamentals of the company's financial performance using techniques such as discounted cash flow
2. Based on technical analysis of the time series patterns of a stock's price

In this section, we will review discounted cash flows and how it is applied in fundamental analysis. In the next section, we will review some commonly used chart patterns.

11.4.1 Company Valuation Using Discounted Cash Flows

Discounted cash flows (DCF) is a technique to value a company based on its estimated future earnings (Bodie et al. 2012). If the valuation from DCF is greater than the current stock price, it is a good potential investment. Consider a stock that is modeled as below:

$$P_t = \frac{D_{t+1}}{(1+k)} + \frac{P_{t+1}}{(1+k)},$$

where P_t is the present price of the stock at period t, P_{t+1} is the future price of the stock at the end of period $t + 1$, D_{t+1} is the dividend paid at the end of period $t + 1$, and K is the discount rate.

For infinite periods, P_t can be stated as follows:

$$P_t = \frac{D_{t+1}}{(1+k)} + \frac{D_{t+2}}{(1+k)^2} + \frac{D_{t+3}}{(1+k)^3} + \ldots + \frac{D_{t+n}}{(1+k)^n} + \ldots \tag{11.4}$$

This means the value of the share of the stock is the present value of all the future dividends.

Equation 11.4 can be used in three ways:

- First, P_t can be found based on the estimates of future dividends (D_{t+i}) and discount rates (k). Thus, the estimate of the stock will give the direction of the value from its current market price.
- Second, the value of all the future dividends and the present market price of the stock can be used to calculate the discount rate k. Price may be adjusted to reflect the risk associated with the stock.

Financial Services

- Third, by dividing the above equation by earnings on both sides, this equation can be converted to a price–earnings ratio (P/E ratio, which is discussed later in this chapter).

In real life, dividends may grow or decline from year to year. Various forecasting techniques are used to estimate future dividends, which are then used to calculate the growth rate. Next, three models are discussed based on the growth rates of the dividends.

i. Constant Growth Model

When the growth rate is considered constant over indefinite future, the present value of the stock is given by

$$P_t = \frac{D}{(1+k)} + \frac{D(1+g)}{(1+k)^2} + \frac{D(1+g)^2}{(1+k)^3} + \ldots + \frac{D(1+g)^{n-1}}{(1+k)^n} + \ldots, \quad (11.5)$$

where g is the constant growth rate of the dividends.

Since Equation 11.5 is an indefinite geometric progression with first term as $\frac{D}{(1+k)}$ and common ratio $\frac{(1+g)}{(1+k)}$, which is less than 1,

$$P_t = \frac{D}{(k-g)}. \quad (11.6)$$

This is also called as the Gordon model, after Myron J. Gordon popularized this model.

The above equation can also be stated for rate of return, as follows:

$$k = \frac{D}{P} + g. \quad (11.7)$$

This constant growth model is derived from the assumption that the firm keeps its retention rate constant and earns fixed returns on new investments. Assume r be the fixed rate of return over new investment and b be the fraction of earnings retained for new investments, I_t be the investments in time period t, and E_t be the earning in time period t.

Then, total earnings in time period t, is given by

$$E_t = E_{t-1} + rI_{t-1}.$$

Since b is the fraction of earnings retained for new investments of the earnings, the above equation becomes

$$E_t = E_{t-1} + rbE_{t-1} = E_{t-1}(1+rb).$$

Growth in earnings can be calculated as the percentage change in earnings as follows:

$$g = \frac{E_t - E_{t-1}}{E_{t-1}} = \frac{E_{t-1}(1+rb) - E_{t-1}}{E_{t-1}} = rb. \tag{11.8}$$

Since dividends are paid as a fixed percentage of the earnings, that is, $D = (1-b)*E$, the growth rate of earnings is equal to the growth rate of dividends, which is also equal to the growth rate of the price of the stock. In other words,

$$g_E = g_D = rb. \tag{11.9}$$

Therefore, Equations 11.6 and 11.7 can be stated as follows:

$$P_t = \frac{D}{k - rb} \tag{11.10}$$

$$k = \frac{D}{P} + rb \tag{11.11}$$

The rate of return on new investments (r) can be written as the fraction of the rate of return of stock (k); therefore, $r = ck$, where c is the constant whose value is greater than 1.

Substituting $D = (1-b)*E$ and $r = ck$ in Equation 11.11, we get

$$k = \frac{(1-b)E}{(1-cb)P}. \tag{11.12}$$

ii. The Two-Period Growth Model

This model is an extension of the constant growth model. This model assumes a growth rate for N years after which it assumes constant growth rate for an indefinite period. Assume g_1 to be the growth rate for the first N years and P_N be the price of the stock at the end of the period N. After period N, assume g_2 to be the constant growth rate and it goes on for an indefinite period. Then,

$$P = \left[\frac{D}{1+k} + \frac{D(1+g_1)}{(1+k)^2} + \frac{D(1+g_1)^2}{(1+k)^3} + \ldots + \frac{D(1+g_1)^{N-1}}{(1+k)^N} \right] + \frac{P_N}{(1+k)^N}.$$

Using the sum of geometric progressions and substituting $P_N = \dfrac{D_{N+1}}{k-g_2}$ (from the constant growth model), where $D_{N+1} = D(1+g_1)^{N-1}(1+g_2)$, we get the following:

$$P = D\left[\dfrac{1-\left(\dfrac{1+g_1}{1+k}\right)^N}{(k-g_1)}\right] + \left[\dfrac{D(1+g_1)^{N-1}(1+g_2)}{(k-g_2)}\right]\left[\dfrac{1}{(1+k)^N}\right]. \quad (11.13)$$

This model is generally used in a different way. The P/E ratio (multiple) is used and the above equation is modified. Typically, after N years, since growth rate is constant, the price at which it sells (for P/E ratio) after N will be the same as the average P/E ratio (M_g) of the economy. Hence,

$$P_N = \dfrac{P_N}{E_N}(E_N) = M_g E_N.$$

Also, as per constant growth models, the growth rate for dividends is the same as the growth rate of earnings; hence, earnings in the Nth year is $E*(1+g_1)^{N-1}$. Thus, Equation 11.13 can be rewritten as

$$P = D\left[\dfrac{1-\left(\dfrac{1+g_1}{1+k}\right)^N}{(k-g_1)}\right] + \left[M_g E(1+g_1)^{N-1}\right]\left[\dfrac{1}{(1+k)^N}\right]$$

$$P = \dfrac{D}{(k-g_1)}\left[\dfrac{(1+k)^N - (1+g_1)^N}{(1+k)^N}\right] + \left[M_g E(1+g_1)^{N-1}\right]\left[\dfrac{1}{(1+k)^N}\right]. \quad (11.14)$$

When the dividends grow at some rate, say g_1 and g_2, as discussed above, prices grow at different rates. If g_1 is greater than g_2, then the price grows at a rate greater than g_2 but smaller than g_1.

iii. The Three-Period Model

Extending the two-period model, this model assumes that earnings continue to grow at a constant rate. After the end of the first period, say, after N years, in the second period, the earnings start to drop to some fixed value. After that, the model assumes the earnings to grow at a constant rate forever.

Example 11.3

Assume a company's stock grows at a constant rate of 15% for 4 years. Then, it declines linearly to 11% over a period of 3 years. After the seventh year, it is assumed that it grows at a constant rate of 11% throughout (see Figure 11.8). Assume the dividend to be $3 and the discount rate to be 17%.

For the first 4 years, since it is a constant rate, the present value will be

$$P = \frac{3}{(1.17)^1} + \frac{3(1.15)^1}{(1.17)^2} + \frac{3(1.15)^2}{(1.17)^3} + \frac{3(1.15)^3}{(1.17)^4}.$$

Now, it starts to decline to 11% till the end of year 7. Thus, the present value will be

$$P = \frac{3(1.15)^3(1.14)}{(1.17)^5} + \frac{3(1.15)^3(1.14)(1.13)}{(1.17)^6} + \frac{3(1.15)^3(1.14)(1.13)(1.12)}{(1.17)^7}.$$

After this, it grows at a constant rate forever, which is similar to the constant rate model. Hence,

$$P = \left[\frac{3(1.15)^3(1.14)(1.13)(1.12)(1.11)}{(0.17 - 0.11)} \right] \frac{1}{(1.17)^7}.$$

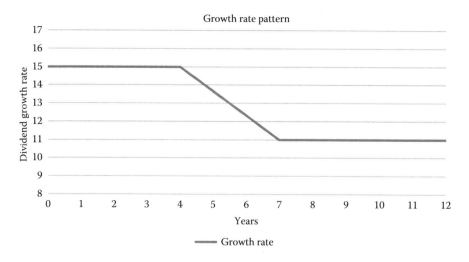

FIGURE 11.8
The three-period model.

Financial Services

Therefore, the value of the stock will be

$$P = \left[\frac{3}{(1.17)^1} + \frac{3(1.15)^1}{(1.17)^2} + \frac{3(1.15)^2}{(1.17)^3} + \frac{3(1.15)^3}{(1.17)^4}\right]$$

$$+ \left[\frac{3(1.15)^3(1.14)}{(1.17)^5} + \frac{3(1.15)^3(1.14)(1.13)}{(1.17)^6} + \frac{3(1.15)^3(1.14)(1.13)(1.12)}{(1.17)^7}\right]$$

$$+ \left[\frac{3(1.15)^3(1.14)(1.13)(1.12)(1.11)}{(0.17-0.11)}\right]\frac{1}{(1.17)^7} = 9.99 + 6.86 + 40.58 = 57.43$$

11.4.2 Price–Earnings Ratio

P/E ratio is the ratio of the company's stock price to its earnings per share (EPS).

$$\frac{P}{E} \text{ ratio} = \frac{\text{Market Price per Share}}{\text{Earnings per Share (EPS)}}$$

For example, if a company's stock price is $15 and its earnings per share is $1.25 over the past 1 year, then the P/E ratio = $15/$1.25 = 12. This means that the investor needs to invest $12 for every dollar of its earnings. Everything else being equal, one would prefer a stock with a lower P/E ratio. At certain points in time, a stock may have a higher P/E ratio if its investors are anticipating higher future growth.

11.4.3 PE Growth Ratio

PE growth (PEG) ratio is also used with P/E ratio to gain additional insight. Using P/E ratio for companies with a higher growth rate makes them look overvalued and thus makes it difficult to compare these companies.

$$\text{PEG ratio} = \frac{\frac{\text{Price}}{\text{Earnings}}\left(\frac{P}{E}\right)}{\text{Annual EPS Growth}}$$

The value of PEG ratio is interpreted as follows:

If the ratio is >1: company may be overvalued
If the ratio is =1: company is in line with the earnings growth
If the ratio <1: company may be undervalued

Example 11.4

Consider two companies A and B with P/E ratios of 40 and 15, respectively. Assume the growth rate of company A to be 15% and that of company B to be 10%. Calculating their PEG ratios, we get the following:

PEG ratio for company A = 40/15 = 2.67
PEG ratio for company B = 15/10 = 1.5

We can see that company A has a high P/E ratio of 40, which cannot be justified with its growth rate of just 15%, as compared to that of company B. Therefore, it can be inferred that company A's stock price is overvalued.

11.4.4 Earnings Estimates

Whenever anyone hears the phrase "beat the street," it means the earnings of a company exceeded the forecast made by analysts. Analysts use different forecasting models to estimate earnings. Consensus earnings estimate is the average or the median of the forecasted earnings by different analysts. Generally, consensus earnings are published for the next several quarters.

Analysts consider several factors in forecasting earnings (Elton et al. 2009). The paramount factor is the economy, such as inflation, GDP, rate of growth, and changes in employment. Analysts also consider how the earnings of a company are affected by the industry to which it belongs. For example, earnings for the industries such as automobiles, chemicals, and steel are considerably influenced by the economy; whereas earnings for the industries such as oil and rubber are influenced by the innovations in the industry itself. For industry-specific changes, analysts speak to customers, suppliers, and competitors to forecast the trends. Therefore, the forecast of the changes in the economy and specific industry changes may be useful in estimating the earnings of the companies.

11.4.5 Technical Analysis

Technical analysis looks at time-series patterns of a stock to determine whether to buy or sell (Investor's Business Daily 1996). There are certain patterns, which consistently reappear and produce the same results. For example, if the sentiments of the market shift from fear to optimism, a certain trend would appear before the investors will start buying the stock. These chart patterns are divided into two categories:

1. *Continuation Patterns* are those in which the trend will continue once the pattern appears. Some examples of continuation patterns are the following:
   - Flag, Pennant
   - Symmetrical Triangle

Financial Services

- Ascending Triangle
- Rectangle
- Cup with Handle
- Measured Move—Bullish/Bearish

2. *Reversal Patterns* are those in which the trend will reverse once the pattern is complete. Some examples of reversal patterns are given below:
   - Head and Shoulders
   - Falling Wedge
   - Double Top Reversal
   - Triple Top Reversal
   - Bump and Run Reversal
   - Rounding Bottom

These are examples of continuation and reversal patterns but many stocks can demonstrate both continuation and reversal based on the situation. Next, we will examine one pattern from each category.

11.4.6 Cup with Handle

Cup with handle is a continuation pattern, which consists of a drop in the price followed by the rise back to the previous value, followed by a smaller drop and then again a rise past the previous peak. This pattern is illustrated in the Figure 11.9. After an advance on the left, there is a "U"-shaped bottom. After the cup pattern is complete, there is a handle on its right side. The perfect pattern would have equal highs on both sides of the cup. Once the

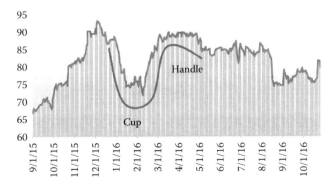

FIGURE 11.9
Cup and handle chart for TTL stock.

handle is complete, it might resume its higher trend. The depth of the cup should trace one-third or less than its previous advance. In the worst market, it may go up to one-half of its advance.

The cup generally lasts for about 1 to 6 months, whereas handle lasts for 1 to 4 months on weekly charts. One of the tests to verify the pattern is the midpoint test, in which the highest and lowest point of the handle are added and divided by two. A similar test is done for the cup. The midpoint for the handle should be greater than that of the cup. One of the important characteristics of the handle is that it must show a downward trend at least for a short period.

11.4.7 Head and Shoulders Pattern

This reversal pattern consists of a left shoulder, a head, and a right shoulder as shown in Figure 11.10. It is used to predict bullish to bearish trend reversal.

After a long bullish trend, the price rises to the peak and subsequently declines to the trough forming the left shoulder. The prices rise again to form another peak, higher than the previous peak and then falls again subsequently, thus forming the head. The advance from the low of the head forms the right shoulder, which is usually in line with the first peak (the left shoulder) and is less than the second peak, which is the head. There can also be a head and shoulders "mirror image" as shown in Figure 11.11.

Another important characteristic of this chart type is the neckline, which is the line that connects the lows of the peaks. Sometimes, more than two low points can also be used for the formation of the neckline. Depending on the orientation of the low points in the plane, the slope of the neckline can be upward, downward, or horizontal. The slope of the neckline is the measure of the bearishness or bullishness of the trends.

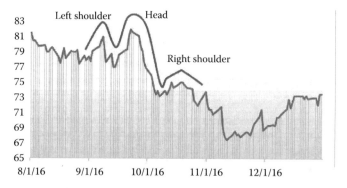

FIGURE 11.10
Head and shoulder chart for ETR.

Financial Services

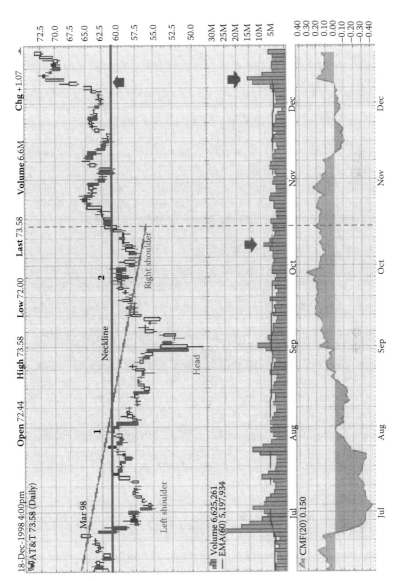

FIGURE 11.11
Head and shoulders mirror image. (Courtesy of Stockcharts.com)

Such patterns are used by stock traders to determine at which price point to buy or sell a particular stock. There are many such patterns that are automatically recognized by trading software by using very fast computers to automate stock trading.

11.5 Derivatives

A derivative is defined as a financial instrument, which represents a contract between two or more parties based on the value of an underlying asset. It is a security with a price, whose value depends on the values of other underlying variables such as asset, index, or interest rate. Derivatives are broadly classified into two groups of contracts:

- Over the counter (OTC): These are privately traded derivatives, which do not go through any exchange or intermediary.
- Exchange traded derivatives (EDT): These are traded through derivatives exchanges or other exchanges.

Some of the most common derivatives are *futures, forwards, swaps*, and *options*.

i. *Futures*: These are contracts for a standard volume of securities to be exchanged on a specific settlement date in the future. Future contracts are standardized for trading in markets like Chicago Mercantile Exchange. These future contracts could be for stocks, currencies, and commodities (metals, livestock, agricultural produce, etc.).
ii. *Forwards*: These are similar in concept to futures but are usually non-standard contracts between two parties, which are traded over the counter as opposed to standardized contracts traded in exchanges.
iii. *Swaps*: These are a series of forward contracts between two parties, who have complementary obligations. Swaps can be made for mitigating fluctuations in interest rates or currencies.
iv. *Options*: An option is a contract providing the right, but not an obligation, to buy or sell a financial security or assets at a fixed price during a certain period of time or date.

In the next section, we review *options* in greater detail because of its widespread use in investments.

Financial Services 555

11.6 Options

Some of the key terminologies associated with options are as follows (Hull 2006; Thomsett 2009):

- *Option*: It is the right to *buy or sell* underlying assets, at a fixed price, by a certain date in the future.
- *Call*: An option acquired by a buyer or granted by a seller, to *buy* underlying assets at a fixed price by a certain date.
- *Put*: An option acquired by a buyer or granted by a seller, to *sell* underlying assets at a fixed price by a certain date.
- *Expiration date*: The date at which an option becomes worthless, which is specified in the option contract. It is also called maturity date.
- *Exercise*: The act of buying a stock under the terms of the call option or selling a stock under the terms of the put option, at the specified price per share in the option contract.
- *Strike price*: The fixed price to be paid for underlying assets specified in the option contract, which will be paid or received by the owner of the option contract upon exercise, regardless of the current market value of the stock.
- *Underlying stock*: The stock that the option grants the right to buy or sell, which is specified in every option contract

An option is an intangible asset, unlike bonds or shares, which is guaranteed to be worthless or nonexistent in less than a year. Since prices of securities fluctuate in the market, an option gives a right to its owner to freeze the price to buy or sell the financial security in the future within some restrictions. First of all, this right to buy or sell in the future is not for an indefinite period but for a few months. A second important restriction is that each option contract is multiples of 100 of the underlying stock. For example, purchasing five call options contracts means that one is acquiring the right to buy 500 shares of that stock.

11.6.1 Call Option

When you buy a call option, it means "I am willing to pay the price being asked to acquire a contractual right. That right enables me to buy 100 shares of the stock at the specified fixed price per share, and I can buy those shares at that price at any time between now and the specified deadline" (Thomsett 2009). If the price of the stock rises after the purchase, the investor makes a profit because the option becomes more valuable. For example, the call option gives you a right to buy 100 shares at $50 (strike price) per share till the expiration date. Before the deadline, if the price of those shares increases to $65, the option gives you the right to buy at $50 per share, which is $15 profit per share. It should be emphasized that the investor has only the option to buy, and there is no obligation to exercise it by the expiration date.

Call options also cost money and the cost has to be taken into account before exercising an option. Consider an investor who buys a call option with a strike price of $50 to purchase 100 shares of company ABC and the expiration date is in 3 months. Also, the price of an option to purchase one share is $5, which means that the total initial investment will be $500 to buy this option. Suppose the current price of the stock is $48. Then, it is not profitable to exercise this option because the investor can buy the stock directly in the open market at a price cheaper than the strike price. In this case, the investor lost $500 by buying the option.

Suppose the current price reaches $50; the investor will still be making a loss of $5, which is the initial cost of the option. However, if the current price is more than the strike price, say $70, then the investor would be making a profit of $15 per share ($70–$50–$5), a total profit of $1500. Now, consider another scenario, where the current price of the share is $52, which means a loss of $3 per share considering the option cost of $5 per share, for a total loss of $300. If the investor does not exercise the option, then his total loss will be $500, which was the initial investment. Therefore, exercising the option will reduce the loss by $200. *Hence, call options should always be exercised at the expiration date, if the stock price is above the strike price!*

11.6.2 Put Option

When you buy a put option, it means "I am willing to pay the asked price to buy a contractual right. That right enables me to sell 100 shares of stock at the indicated price per share, at any time between now and the specified deadline" (Thomsett 2009). If the price of the stock decreases after the investor buys a put option, the investor makes a profit. For example, the put option gives you a right to sell 100 shares at $90 (strike price) per share till the expiration date. Before that date, assume that the price of the share decreases to $80 per share. The put option gives you the right to sell it for $90, which is $10 more than the original price.

This can be further explained by assuming another scenario. Suppose there is an investor who buys a put option to sell 100 shares of company ABC with a strike price of $80 and the expiration date is 3 months. Also, the price of an option to purchase one share is $7, which means an initial investment of $700 to buy this option. The current price of the stock is $70. Since the investor can buy a share for $70 and sell it for $80, making a total profit of $300 ($10 × 100 − $700). The investor will never exercise this option if the current price of the stock is greater than the strike price because it will lead to a loss that is greater than $700, which is the initial investment.

11.6.3 Option Valuation

There are a few major factors that affect the price of the options, namely, the stock price, the exercise price, the intrinsic value, the volatility, time to expiration, the interest rate, and the dividend rate of the stock. The intrinsic value of any given option is the payoff that could be obtained by the immediate exercise, which is given by the following formulae:

Intrinsic value (call option) = Underlying stock's current price − Call Strike price

Intrinsic value (put option) = Put Strike price − Underlying stock's current price

Security prices can change over time, and this price fluctuation is known as volatility of the security. The difference between actual call price and the intrinsic value is known as the **time value** of the option. The more time the option has to expire, the greater the probability for it to be profitable. There are several ways to value options. Next, two methods for valuing options are discussed: binomial option pricing and Black–Scholes.

11.6.3.1 Binomial Option Pricing

The binomial option pricing model was developed by Cox et al. (1979). To illustrate the model, consider Figure 11.12, in which the current price of the stock is S_0, which will either increase by factor "u" or decrease by factor "d," leading to uS_0 or dS_0. Similarly, after the first year, price may increase or decrease leading to three possibilities, which are u^2S_0, udS_0, and d^2S_0. An important thing to notice here is that there are two ways to reach udS_0. Similarly, for ud^2S_0, there are three possible ways to reach, which are udd, dud, and ddu. There are eight possible combinations over the three periods, namely, uuu, uud, udu, duu, udd, dud, ddu, and ddd. Each has the probability of 1/8, but

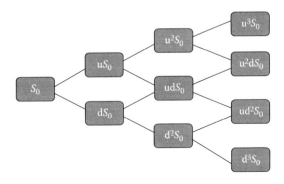

FIGURE 11.12
Binomial option pricing.

TABLE 11.5

Binomial Option Pricing Example

Combination	Event	Probability	Final Price
u^3 (uuu)	3 ups	1/8	$100 \times 1.05^3 = 115.76$
u^2d (uud, duu, udu)	2 ups and 1 down	3/8	$1.05^2 \times 0.96 \times 100 = 105.84$
ud^2 (udd, ddu, dud)	1 up and 2 downs	3/8	$1.05 \times 0.96^2 \times 100 = 96.77$
d^3 (ddd)	3 downs	1/8	$0.96^3 \times 100 = 88.47$

from the perspective of the final price, there are only four possibilities, which are u^3, u^2d, d^2u, and d^3.

For illustration, consider S_0, the current price of the stock, to be $100, and the increase to be 5% and the decrease to be 4%. The final prices and their probabilities are given in Table 11.5.

Now, assume that the stock sells at $100, with $u = 1.15$ and $d = 0.9$, which means that the stock could be priced at either $115 or $90. Consider a call option, which has the strike price of $110 and an expiration of 1 year. Therefore, if the price of the stock increases to $115 after 1 year, then the investor will profit by $5; there will be no profit if the price of the stock falls to $90. The possible trees for these two scenarios are as follows:

Stock price　　　　　　　　　　　　　　Call-option value

Now, compare the payoff of the call to that of the portfolio consisting of one share of the stock and the borrowing of $78.26 with an interest rate of 15%. Now, the value of the stock can be either $90 or $115, whereas the amount of the borrowings with interest rate at the end of the year would be $78.26 + 15% × 78.26 = $90. Thus, net gain would be $0 ($90 − $90) or $25 ($115 − $90). Therefore, the portfolio's value tree is as follows:

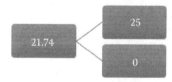

Hence, five call options are needed to make it equal to the payoff from the portfolio ($25). Having said that, five call options should be sold at the same price as the cost of establishment of the portfolio, which is

$$5C = 21.74.$$

Financial Services

This means that each call option should sell at $4.35. This ability to create a perfect hedge is the key feature to evaluate the call option. The hedge ratio for this example is 1/5, which is defined as the ratio of the range of the value of the option to the range of the values of the stock. For two state options, hedge ratio can be defined as

$$H = \frac{C_u - C_d}{uS_0 - dS_0}$$

where C_u is the worth of the call if stock price increases and C_d is the worth of the call if the stock price decreases. Thus, at every node, one could set up a portfolio that is perfectly hedged over next time interval. A continuous process of hedging at each node would help earn riskless returns and it is called *dynamic hedging*. This also helps to make the evaluation more precise.

11.6.3.2 Black–Scholes Option Valuation

This model was developed by Black and Scholes (1973) for the valuation of options earning. Merton (1973) also developed an option valuation model. Both Merton and Scholes received the Nobel Prize in economics in 1997 for their work. Underlying assumptions for both the models are the same, but the Black–Sholes has two additional assumptions: both the risk-free interest and time to expiration are constant over the life of the option. In addition, the binomial option model assumes that the movement of price follows a binomial distribution, whereas in Black–Scholes, for many trials, this binomial distribution approaches the lognormal distribution. Although the binomial pricing model is extremely flexible and is more precise, its computation is very slow as compared to Black–Scholes.

Black–Scholes Formula: This formula needs five parameters as input, namely, strike price, current price of the stock, expiration date of the option, risk-free rate, and volatility. The Black–Scholes formula for the call option is expressed as

$$C_0 = S_0 N(d_1) - Xe^{-rT} N(d_2) \tag{11.15}$$

where

$$d_1 = \frac{\ln\left(\frac{S_0}{X}\right) + \left(r + \frac{\sigma^2}{2}\right)T}{\sigma\sqrt{T}}$$

$$d_2 = d_1 - \sigma\sqrt{T}$$

and

C_0 = current call option price
S_0 = current stock price
X = exercise price
$N(d)$ = standard normal cumulative distribution function
e = exponential function
r = current continuously compounded risk-free interest rate
T = expiration date of the option in years
ln = natural logarithm function
σ = annual volatility of stock price (standard deviation of the annualized continuously compounded rate of return)

Example 11.5

We illustrate the Black–Scholes formula with a simple example.
Current stock price, $S_0 = \$100$
Exercise price, $X = \$90$
Risk-free rate, $r = 10\%$ per annum, that is, 0.1
Expiration date, $T = 0.5$ (half yearly)
Standard deviation, $\sigma = 0.5$
To calculate the present value of the call, first calculate d_1 and d_2

$$d_1 = \frac{\ln\left(\frac{S_0}{X}\right) + \left(r + \frac{\sigma^2}{2}\right)T}{\sigma\sqrt{T}}$$

$$= \frac{\ln\left(\frac{100}{90}\right) + \left(0.1 + \frac{0.5^2}{2}\right)0.5}{0.5\sqrt{0.5}}$$

$$= 0.61.$$

$$d_2 = d_1 - \sigma\sqrt{T}$$
$$= 0.61 - 0.5\sqrt{0.5}$$
$$= 0.26.$$

Now, $N(d_1) = N(0.61) = 0.73$ and $N(d_2) = N(0.26) = 0.60$.

Financial Services

This can be derived using the NORMSDIST function in Excel or using online calculators.

Substituting it in Equation 11.15, we get

$$C_0 = S_0 N(d_2) - Xe^{-rT} N(d_2)$$
$$= 100 * 0.73 - 90 * e^{-0.1*0.5} * 0.6$$
$$= 21.63.$$

Hence, the present value of the call should be $21.63.

Options are widely used to hedge against market swings in prices of various assets, such as stocks, currencies, commodities, and energy.

Exercises

11.1 Research any two major payment networks and compare them in terms of annual revenue, number of cardholders, the number of places where each one is accepted worldwide, and any consumer protection they offer.

11.2 Discuss various factors a consumer would consider in selecting a payment mode.

11.3 Discuss some of the incentives offered by card issuers to attract consumers.

11.4 Research Bitcoin and briefly describe payments using this currency.

11.5 Discuss NFC, an emerging payment mode.

11.6 Compare PayPal and Google Wallet from a consumer perspective and from a merchant perspective.

11.7 A gas station is forecasted to sell 4000 gallons per day. Approximately 70% of the customers use credit/debit card, and the rest use cash. It costs the gas station owner 5 cents per gallon to process the credit/debit card and gas price is at $2.50 per gallon. Cash processing cost is 0.04% of all the cash deposited. What should be the cash price for gas to make the net revenue comparable to credit price? What is the estimated monthly revenue for this station and its monthly payment processing cost?

11.8 Solve the quadratic programming model (Equations 11.1 through 11.3) formulated for Example 11.2. What is the optimal investment policy for the Lion County Bank?

References

Allen, F., J. McAndrews, and P. Strahan. 2002. E-finance: An introduction. *Journal of Financial Services Research.* 22(1–2): 5–27.

Allen, H., J. Hawkins, and S. Sato. 2001. Electronic trading and its implications for financial systems. In, ed. Bank for International Settlements, *Electronic Finance: A New Perspective and Challenges*, vol. 07, pp. 30–52. Bank for International Settlements, http://EconPapers.repec.org/RePEc:bis:bisbpc:07-04

Bank for International Settlements (BIS). 2012. Innovations in retail payments. Report of the Working Group on Innovations in Retail Payments. May 2012, Bank for International Settlements.

BIS. 2014. Statistics on payment, clearing and settlement systems in the CPMI countries. Figures for 2013, December 2014, Bank for International Settlements.

Black, F. and M. Scholes. 1973. The pricing of options and corporate liabilities. *Journal of Political Economy.* 81(3): 637–654.

Bloomberg. 2013. "Swedish Banks Make Money by Saying No to Cash," by N. Magnusson and K. Gustafsson, April 11, 2013. Retrieved 7.4.2015. http://www.bloomberg.com/news/articles/2013-04-10/era-of-paper-money-dies-out-in-sweden-as-virtual-cash-takes-over

Bodie, Z., A. Kane, and A. J. Marcus. 2012. *Essentials of Investments*, 9th Edition. McGraw-Hill.

Claessens, S., T. Glaessner, and D. Klingebiel. 2002. E-finance: Reshaping the financial landscape around the world. *Journal of Financial Services Research.* 22(1/2): 29–61.

Cox, J. C., S. A. Ross, and M. Rubinstein, M. 1979. Option pricing: A simplified approach. *Journal of Financial Economics.* 7(3): 229–263.

Elton, E. J., M. J. Gruber, S. J. Brown, and W. N. Goetzmann. 2009. *Modern Portfolio Theory and Investment Analysis.* John Wiley & Sons.

Federal Reserve. 2013. Federal Reserve Payments Study—Detailed Report. Retrieved on 7.4.2015. https://www.frbservices.org/files/communications/pdf/general/2013_fed_res_paymt_study_detailed_rpt.pdf

Federal Reserve. 2015. Strategies for Improving the U.S. Payment System. https://fedpaymentsimprovement.org/wp-content/uploads/strategies-improving-us-payment-system.pdf

FFIEC. 2010. FFIEC IT Examination Handbook Retail Payment Systems Booklet. Retrieved 8.5.2015. http://ithandbook.ffiec.gov/ITBooklets/FFIEC_ITBooklet_RetailPaymentSystems.pdf

Food Marketing Institute. 2013. Retrieved 7.5.2015. http://www.fmi.org/research-resources/supermarket-facts

Green, L. V., P. J. Kolesar, and W. Whitt. 2007. Time-varying demand and staffing requirements. *Production and Operations Management.* 16(1): 13–39.

Hahn, R. and L. F. Anne. 2006. The move toward a cashless society: Calculating the costs and benefits. *Review of Network Economics.* 5(2): 1–30.

Hull, J. C. 2006. *Options, Futures, and Other Derivatives.* Pearson Education India.

Humphrey, D. B., L. B. Pulley, and J. M. Vesala. 2000. The check's in the mail: Why the United States lags in the adoption of cost-saving electronic payments. *Journal of Financial Services Research.* 17(1): 17–39.

Humphrey, D. B. and R. Hunt. 2013. Cost savings from check 21 electronic payment legislation. *Journal of Money*. 45(7): 1415–1429.

Investor's Business Daily. 1996. *Investor's Business Daily Guide to the Markets*, ISBN: 978-0-471-15482-2. John Wiley & Sons.

Merton, R. C. 1973. Theory of rational option pricing. *The Bell Journal of Economics and Management Science*. 4(1): 141–183.

Rose, P. and S. Hudgins, S. 2014. *Bank Management & Financial Services*, 9th Edition. McGraw-Hill Education.

Simmons, M. 2003. *Securities Operations: A Guide to Trade and Position Management*. John Wiley & Sons.

Thomsett, M. C. 2009. *Getting Started in Options*, Vol. 82. John Wiley & Sons.

Weatherford, J. 2009. *The History of Money*. Crown Business.

Index

A

Accountable care organization (ACO), 484
ACH, *see* Automated Clearing House
AHP, *see* Analytic Hierarchy Process
Airlines, revenue management in, 424–437
    airline fare classes, 425–426
    bid price controls, 432–434
    booking curves, 434–435
    setting limits on multiple fare classes, 428–432
    setting limits on two fare classes, 426–428
Airlines sector, conflicting objectives in, 18
Allocative efficiency, 185
All-unit quantity discounts, 98–100
Analytic Hierarchy Process (AHP), 158
Ancillary and support services, 9–10
Annualized returns, 391–393
Area under the curve (AUC), 515
ARIMA (Auto Regressive Integrated Moving Average) method, 84
Asset allocation, 351–352
Asset turnover ratio, 458
Automated Clearing House (ACH), 529

B

Banking, 534–539; *see also* Financial services
    diversifying banking channels, 537–539
    globalization, 535
    staffing and customer waiting time, 535–537
Bar chart, 33
Batch picking, 323
BCM, *see* Buildable Combination Matrix
Best compromise solution, 163
Beta risk, 371, 373–374
Bi-criteria linear program (BCLP), 166
Bid price controls, 432–434
Binary variables, modeling with, 93–104
    capital budgeting problem, 93–96
    constraint with multiple right-hand-side constants, 97–98
    fixed charge problem, 96–97
    integer programming models, 93
    linear programming models, 93
    mixed integer programming models, 93
    net present value, 93
    nonlinear integer programs, 102–104
    pure integer programs, 93
    quantity discounts, 98–102
Bonds, investing in, 350, 379–391
    bond convexity, 386
    bond duration, 384–386
    bond ratings, 384
    immunization, 387
    price, yield, and coupons, 381–382
    selection of bond portfolio, 387–391
    strategies, 388
    types of bonds, 383–384
    understanding bonds, 379–380
Booking curves, 434
Bucket brigades, 303, 335–338
Buildable Combination Matrix (BCM), 115
Bundled products, 4

C

Cash, investing in, 350
Cash-to-cash cycle, 214
Cause-and-effect diagram, 36–37
Challenges posed by services, 3–4
    bundled products, 4
    impact of technology, 3–4
    managing capacity and demand, 4
    self-service, 4
    service quality, 4

Children's Health Insurance Program (CHIP), 479
Claims on Assets, 448
Clarke–Wright Savings Algorithm, 275
Classification of services, 5–12
  ancillary and support services, 9–10
  service industries, 5–9
  Service Process Matrix, 10–12
Clinical decision units, 486
Cluster picking, 323
Cobb–Douglas production function, 174
Cold chain, 307
Communication services, 5–6
Compromise programming, 161, 168–169
Compromise solution, 169
Constant level forecasting methods,
Constant returns to scale (CRS), 173
Consumer's risk, 24
Control charts, see Statistical process control
Control issues, 16
Convex hull, 177
Corporate bonds, 383–384
Cross docking, 338–343
CRS, see Constant returns to scale (CRS)
Current Liabilities, 449
Customer(s)
  abandonment, 135, 139
  demand, proactive strategies to manage, 409–412
  surveys, 49
  time between departures of, 468
Cycle stock, 277

D

Data envelopment analysis (DEA), 151, 171–180
  allocative and total efficiency in, 184–188
  decision-making unit, 172
  linear programming formulation, 180–183
  nonparametric models, 176–180
  practical considerations for, 183–184
  production function, 172–175
  productivity, 171
  software, 191–192
  stochastic, 188–191
  technology, 175–176
Days of inventory, 212
DC, see Distribution center (DC)
DCF, see Discounted cash flows (DCF)
DEA, see Data envelopment analysis (DEA)
Decision-making unit (DMU), 172
Decision Support System (DSS), 233
Delphi Method (forecasting), 48
Demand forecasting in services, 45–91
  ARIMA method, 84
  averaging method, 52
  bottom-up approach, 49
  computing optimal weights by Linear Programming model, 54–56
  constant level forecasting methods, 51
  Croston's method, 84–85
  customer surveys, 49
  Delphi Method, 48
  deseasonalized demands, 66
  executive committee consensus, 47–48
  exercises, 85–90
  exponential smoothing method, 56–57
  forecast horizon, 45
  forecasting errors, 73–76
  forecasting in practice, 81–83
  forecasting process, 46–47
  grassroots approach, 49
  last value method, 52
  Mean Absolute Deviation, 54
  monitoring forecast accuracy, 77–78
  multi-period forecasting, 70–73
  non-stationary data, 45
  periodicity, 69
  qualitative forecasting methods, 47–49
  quantitative forecasting methods, 49–57
  real-world applications, 81–82
  role of, 45–46
  seasonality in forecasting, incorporation of, 58–61
  seasonality index, 58, 66
  seasonality and trend in forecasting, incorporation of, 66–70

simple moving average method, 53
software, 79–81
stationary point, 62
survey results, 82–83
survey of sales force, 49
time series forecasting, 49–51
Tracking Signal, 77
trend in forecasting, incorporation of, 62–66
weighted moving average method, 53
Deseasonalized demands, 66
Design of service systems, 93–149
   binary variables, modeling with, 93–104
   Buildable Combination Matrix, 115
   call center as queuing system, modeling of, 134–142
   capital budgeting problem, 93–96
   closed network, 132
   constraint with multiple right-hand-side constants, 97–98
   exercises, 142–147
   fixed charge problem, 96–97
   G/G/c model, 129–131
   integer programming models, 93
   linear programming models, 93
   mixed integer programming models, 93
   M/M/1 model, 121–125
   M/M/c model, 125–127
   M/M/c/K model, 127–129
   net present value, 93
   nonlinear integer programs, 102–104
   offered load, 139
   Poisson processes, 120–121
   Pollaczek–Khintchine formulas, 129
   Prototype Optimization Model, 115
   pure integer programs, 93
   quality and efficiency driven operational regime, 139
   quantity discounts, 98–102
   queuing models in service systems, 119–142
   set covering and set partitioning models, 104–119
   state-dependent service, 120
   Station Manpower Planning System, 118
   steps of ISA, 147

Deviational variables, 165
Discounted cash flows (DCF), 544
Discrete order picking, 322–323
Distribution, *see* Warehousing and distribution
Distribution center (DC), 202, 303
DMU, *see* Decision-making unit (DMU)
Dollar cost averaging principle, 393–394
Dominated portfolio, 362
Dominated solution, 159
Double exponential smoothing, 64
DSS, *see* Decision Support System (DSS)
DuPont model, 455–464
   asset turnover ratio, 458
   equity multiplier, 458
   net margin, 458
   return on assets, 458
   return on equity, 455
Dynamic hedging, 559

E

Earnings per share (EPS), 549
EBPP, *see* Electronic bill presentment and payments (EBPP)
Economic order quantity (EOQ) model, 277
Economy, role of services in, 12–14
   economic sectors, 12
   top 10 world economies, 14
   U.S. employment by sector, 13
   U.S. GDP by sector, 13–14
Education sector, 6
Efficient frontier, 173
Efficient portfolio, 362
Efficient portfolio frontier, 362
Electronic bill presentment and payments (EBPP), 533
Electronic health record (EHR), 495–496
Electronic retailing (e-tailing), 443
Electronic trading, 540–543
Employment, by sector (U.S.), 13
Entertainment sector, 6
EOQ model, *see* Economic order quantity (EOQ) model
EPS, *see* Earnings per share (EPS)
Evaluation of service systems, 151–200
   allocative efficiency, 185
   Analytic Hierarchy Process, 158

best compromise solution, 163
bi-criteria linear program, 166
Borda method, 154–156
Cobb–Douglas production function, 174
compromise Programming, 161
compromise solution, 169
concept of "best solution," 152–153
convex hull, 177
data envelopment analysis, 171–180
DEA, allocative and total efficiency in, 184–188
DEA, practical considerations for, 183–184
DEA linear programming formulation, 180–183
DEA software, 191–192
decision-making unit, 172
decision variables, 159
deviational variables, 165
dominated alternative, 153
dominated solution, 159
efficient frontier, 173
efficient, non-dominated, or Pareto optimal solution, 159–161
exercises, 192–199
free disposability, 176
global criterion and compromise programming, method of, 168–169
goal programming, 163–168
ideal solution, 153
interactive methods, 169–170
isocost line, 185
MCMP methods, classification of, 162–163
MCMP problems, 158–162
multiple-criteria decision making, 151–152, 170–171
multiple-criteria ranking methods, 153–158
multiple-criteria selection problems, 152–153
non-dominated alternatives, 153
nonparametric models, 176–180
output elasticity, 174
pairwise comparison of criteria, 156–158
partitioning gradient-based algorithm, 167
pay-off matrix, 152
preference function, 163
problem solution, 167–168
production function, 172–175
productivity, 171
rating method, 154
returns to scale, 173
scaling criteria values, 158
stochastic DEA, 188–191
Tchebycheff Metric, 169
weighted methods, 153
weighted objective problem, 161
Exercises
  demand forecasting in services, 85–90
  design of service systems, 142–147
  evaluation of service systems, 192–199
  financial engineering, 398–405
  financial services, 561
  healthcare delivery systems, 523–526
  overview of service systems, 39–42
  retail engineering, 469–470
  revenue management, 437–440
  supply chain engineering, 287–298
  warehousing and distribution, 346–347
Expert Choice software (AHP), 158
Exponential smoothing method (forecasting), 56–57

F

Fast Health Interoperability Resources (FHIR), 497
FCC, *see* U.S. Federal Communications Commission (FCC)
Feasible portfolio, 362
Financial engineering, 349–406
  annualized returns, 391–393
  asset allocation, 351–352
  Beta risk, 371, 373–374
  bonds, investing in, 350, 379–391
  "Buy-Low and Sell-High" strategy, 393
  cash, investing in, 350

Index

dollar cost averaging principle, 393–394
dominated portfolio, 362
efficient portfolio, 362
efficient portfolio frontier, 362
Excel functions to compute investment risk, 360
exercises, 398–405
feasible portfolio, 362
ideal solution, 364
income-producing securities, 350
investing, basic concepts in, 349–352
investment strategies, 397–398
junk bonds, 384
market capitalization, 372
market risk of a security, 371
Markowitz's bi-criteria model, 361–363
modern portfolio theory, 349
mutual funds, 352, 394–397
Net Asset Value, 394
portfolio mean and standard deviation, 364–365
portfolio selection, Markowitz's mean variance model for, 358–371
portfolio selection, Sharpe's bi-criteria model for, 371–379
portfolio selection, simple model for, 352–358
price–yield curve, 381
principles of investing, 391–398
quadratic programming problem, 368
quantifying investment risk, 358–360
quasi-passive strategy, 388
risk–return graph for portfolio selection, 363–364
S&P 500 index, meaning of, 372–373
stocks, investing in, 350–351
trade-off analysis, 362
variance of the portfolio, 358
yield to maturity, 380
zero-coupon bonds, 380, 383
Financial industries, conflicting objectives in, 18
Financial services, 529–563
  Automated Clearing House, 529
  banking, 534–539
  binomial option pricing, 557–559
  Black–Scholes option valuation, 559–561
  call option, 555–556
  clearing, 532
  cup with handle, 551–552
  derivatives, 554
  discounted cash flows, company valuation using, 544–549
  diversifying banking channels, 537–539
  dynamic hedging, 559
  earnings estimates, 550
  electronic bill presentment and payments, 533
  electronic trading, 540–543
  exercises, 561
  globalization, 535
  head and shoulders pattern, 552–554
  hedge ratio, 559
  high-speed trading, 543
  Near-Field Communication, 533
  options, 555–561
  payment systems, 529–534
  PE growth ratio, 549–550
  price–earnings ratio, 549
  put option, 556
  settlement, 532
  staffing and customer waiting time, 535–537
  Straight-Through Processing, 543
  technical analysis, 550–551
  valuing companies and their stock prices, 544–554
Financial strategy (retailing), 447–451
  assets, liabilities, and stockholder's equity, 448
  balance sheet, 448–450
  cash flow statement, 450–451
  financial reporting, 448
  income statement, 450
Financing, healthcare, 478–482
Fishbone aisle layout, 321
Fishbone diagram, 36–37
Five dimensions of service quality, 20–21
Flowchart, 35
Flow-through racks, 310, 311

Flying-V cross lane, 321
Forecasting software, 79–81; see also Demand forecasting in services
　automatic software, 79
　manual software, 80
　semi-automatic software, 79
　user experience with, 81
Free disposability, 176

G

GDP, see Gross domestic product (GDP)
Generally Accepted Accounting Principles, 448
Golden zone, 329
Goods and services, 1–2
　differences between, 2–3
　intangibility, 2
　perishability, 2–3
　proximity, 3
　simultaneity, 3
Government bonds, 383
Government services, 7
Graduated quantity discount, 100–102
Gross domestic product (GDP), 533
　healthcare costs and, 473, 474
　manufacturing share in dollars, 15
　outsourcing and, 15
　by sector (U.S.), 13–14
　self-service and, 4
　supply chain logistics and, 273

H

Healthcare delivery systems, 473–528
　accountable care organization, 484
　bed management, 508–510
　capacity management for imaging equipment, 507–508
　care transitions, 491
　claims and billing, 497–498
　clinical decision units, 486
　components, 482–493
　delivery types, 483–484
　descriptive analytics, 510
　electronic health record, 495–496
　exercises, 523–526
　health analytics, 510–523
　healthcare engineering, 473
　healthcare financing, 478–482
　healthcare outcomes, 476–478
　Health Information Exchange, 496–497
　health information technology, 494–498
　health insurance, 478–480
　Health Insurance Portability and Accountability Act, 496
　health maintenance organization, 479
　high deductible plans, 480
　hospital units, 484–488
　inpatients, 488
　measuring accuracy, 513–516
　Medicaid, 479
　medical coding, 494–495
　Medicare, 479
　nurse scheduling, 498–504
　OR scheduling, 504–507
　outpatients, 488
　patient flow, 490–492
　patient services, 488–489
　prediction example, 522–523
　predictive analytics, 511
　prescriptive analytics, 511
　private insurance, 479
　provider compensation, 480–481
　provider cost allocation, 481–482
　randomized controlled trials, 511–513
　reciprocal allocation, 481
　resource management, 498–510
　revenue cycle management, 481, 498
　RFID, 510
　social determinants of health, 473
　stakeholder groups, 485
　support vector machines, prediction by, 516–522
　U.S. healthcare costs, 474–476
　value stream mapping, 492–493
Healthcare sector, conflicting objectives in, 18
Health Insurance Portability and Accountability Act (HIPAA), 496
Health maintenance organization (HMO), 479
Hedge ratio, 559
High-speed trading, 543

Index

Histogram, 33
Honeycombing, 310, 311
Hospitality and leisure industries, 7–8

I

Insourcing, 15
Insurance sector, 8
Integer Programming (IP), 93, 229
Inventory
　capital, 213
　modeling, 277–279
Investing, basic concepts in, 349–352
　asset allocation, 351–352
　bonds, investing in, 350
　cash, investing in, 350
　mutual funds, 352
　stocks, investing in, 350–351
Isocost line, 185

J

Junk bonds, 384

L

Labor efficiency, 333
Last value forecasting method, 52
Linear Programming (LP) model, 54–56, 93
Linear Weighted Point (LWP) method, 242
Line balancing, 335–338
Long-term Liabilities, 449–450

M

Manufacturing (U.S.), strength of, 14–16
Market capitalization, 372
MCDM, *see* Multiple-criteria decision making (MCDM)
MCMP problems, *see* Multiple-Criteria Mathematical Programming (MCMP) problems
Mean absolute deviation (MAD), 54, 74
Mean absolute percentage error (MAPE), 74
Mean squared error (MSE), 74
Medicaid, 479
Medical coding, 494–495

Medicare, 479
Mixed Integer Linear Programming (MILP), 230
Mixed integer programming (MIP) models, 93
Moving product, 308–309
MSE, *see* Mean squared error (MSE)
Multi-period forecasting, 70–73
　under constant level, 71
　problem, 70
　with seasonality, 71
　with seasonality and trend, 72–73
　with trend, 72
Multiple-criteria decision making (MCDM), 18, 151–152; *see also* Evaluation of service systems
Multiple-Criteria Mathematical Programming (MCMP) problems, 158–162
Multiple-Criteria Selection Problems (MCSP), 158
Mutual funds, investing in, 394–397
　balanced mutual funds, 395
　bond mutual funds, 395
　index funds, 395–396
　life cycle or target date funds, 397
　load versus no-load funds, 394–395
　Net Asset Value, 394
　stock mutual funds, 395

N

Naive Method (forecasting), 52
Near-Field Communication (NFC), 533
Net margin, 458
Net Neutrality, 9
Net present value (NPV), 93
Network design and distribution (SCE), 217–234
　AT&T, 233–234
　BMW, 233
　Hewlett–Packard, 232–233
　location–distribution with dedicated warehouses, 223–225
　location–distribution problem, 220–223
　multinational consumer products company, 229–231
　Procter and Gamble, 231–232
　real-world applications, 229–234

supply chain distribution planning, 217–220
supply chain network design, 225–229
Non-stationary data, 45
NPV, *see* Net present value (NPV)

O

Offered load (OL), 139
Options, 555–561
    binomial option pricing, 557–559
    Black–Scholes option valuation, 559–561
    call option, 555–556
    dynamic hedging, 559
    hedge ratio, 559
    option valuation, 557–561
    put option, 556
    time value of, 557
Order picking, 322–325
    batch picking, 323
    cluster picking, 323
    discrete order picking, 322–323
    S-shape heuristic, 324
    wave picking, 323
    zone picking, 323
Output elasticity, 174
Outsourcing decisions and supplier selection (SCE), 234–273
    comparison of ranking methods, 265–266
    criteria for selection, 238
    final selection, 241–242
    group decision making, 265
    importance of supplier selection, 234
    in-house or outsource, 236–237
    multi-criteria ranking methods for supplier selection, 245–258
    multiple sourcing methods for supplier selection and order allocation, 266–273
    prequalification of suppliers, 238–241
    scaling criteria values, 258–265
    single sourcing methods for supplier selection, 242–245
    sourcing strategy, 237–238
    supplier selection methods, 237–242
    supplier selection process, 235

Overbooking, optimization models for, 415–423
    continuous case, 418–419
    discrete case, 419–420
    illustrative example, 420–423
    input data, 416–417
    no-show problem, 415
    objective function, 417
    optimization problem, 416–423
    practice of overbooking, 415–416
Overview of service systems, 1–43
    bar chart, 33
    challenges posed by services, 3–4
    classification of services, 5–12
    differences between goods and services, 2–3
    exercises, 39–42
    fishbone diagram, 36–37
    flowchart, 35
    goods and services, 1–2
    histogram, 33
    insourcing, 15
    outsourcing, 15
    Pareto chart, 34
    quality of service, 19–21
    quality of service, methods for measuring, 21–37
    reshoring, 15
    role of services in the economy, 12–14
    run chart, 32
    scatterplot, 34–35
    Service Systems Engineering, 16–19
    SERVQUAL survey, 21–24
    statistical process control, 24–31
    strength of American manufacturing, 14–16

P

Pareto chart, 34
Partitioning gradient-based (PGB) algorithm, 167
Payment systems, 529–534; *see also* Financial services
Pay-off matrix, 152
PE growth (PEG) ratio, 549
Point-of-sale (POS) terminals, 467
Pollaczek–Khintchine (P–K) formulas, 129
POM, *see* Prototype Optimization Model

Index

Pooling and contracting, 280–287
　correlation and pooling, 282–283
　location pooling, 281
　postponement, 281
　product pooling, 281–282
　supply chain contracting, 283–287
Portfolio selection, Markowitz's mean variance model for, 358–371
　dominated portfolio, 362
　efficient portfolio, 362
　efficient portfolio frontier, 362
　Excel functions to compute investment risk, 360
　feasible portfolio, 362
　illustration, 365–371
　Markowitz's bi-criteria model, 361–363
　portfolio mean and standard deviation, 364–365
　quantifying investment risk, 358–360
　risk–return graph for portfolio selection, 363–364
　trade-off analysis, 362
Portfolio selection, Sharpe's bi-criteria model for, 371–379
　Beta risk, calculation of, 373–374
　illustration, 375–379
　market risk of a security, 371
　S&P 500 index, meaning of, 372–373
Portfolio selection, simple model for, 352–358
　drawbacks of LP model, 357–358
　expected return, 353
　linear programming model, 354–356
　variance of the portfolio, 358
Post-distribution cross docking, 330
Pre-distribution cross docking, 339
Price–earnings ratio, 549
Price–yield curve, 381
Producer's risk, 24
Production function, 172–175
Professional services, 8
Prototype Optimization Model (POM), 115
Pure integer programs, 93

Q

Quadratic programming problem, 368
Qualitative forecasting methods, 47–49
　customer surveys, 49
　Delphi Method, 48
　executive committee consensus, 47–48
　survey of sales force, 49
Quality and efficiency driven (QED) operational regime, 139
Quality of service (QoS), 19–21
　assurance, 21
　dimensions of service quality, 20–21
　empathy, 21
　measures, 19–20
　reliability, 21
　responsiveness, 21
　Service Quality Gap, 20
　tangibles, 21
Quality of service (QoS), methods for measuring, 21–37
　bar chart, 33
　fishbone diagram, 36–37
　flowchart, 35
　histogram, 33
　Pareto chart, 34
　run chart, 32
　scatterplot, 34–35
　statistical process control, 24–31
　X–Y chart, 34
Quantitative forecasting methods, 49–57
　averaging method, 52
　choice of smoothing constant, 57
　computing optimal weights by Linear Programming model, 54–56
　constant level forecasting methods, 51
　exponential smoothing method, 56–57
　last value method, 52
　Mean Absolute Deviation, 54
　simple moving average method, 53
　time series forecasting, 49–51
　weighted moving average method, 53
Quantity discounts, 98–102
　all-unit quantity discounts, 98–100
　graduated quantity discount, 100–102
Queuing models in service systems, 119–142
　call center as queuing system, modeling of, 134–142
　closed network, 132
　customer abandonment, 135, 139
　G/G/c model, 129–131
　M/M/1 model, 121–125
　M/M/c model, 125–127

M/M/c/K model, 127–129
Poisson processes, 120–121
Pollaczek–Khintchine formulas, 129
queuing networks, 132–134
service times, 137–139
staffing, 139–141
system utilization, 123
time-varying arrivals, 137
Quick ratio, 452

R

Randomized controlled trials (RCTs), 511–513
Rate of technical substitution (RTS), 173
Receiver operating characteristic (ROC) curve, 515
Reshoring, 15
Retail engineering, 443–471
  acid ratio, 452
  asset turnover, 452
  balance sheet, analysis of, 452–455
  breakeven analysis, 464–467
  category specialists, 447
  checkout process staffing, 467–469
  convenience stores, 446
  Current Liabilities, 449
  debt/equity ratio, 452
  department stores, 447
  discount stores, 447
  drugstores, 447
  DuPont model, 455–464
  electronic retailing, 443
  exercises, 469–470
  expected sales, 465
  financial metrics, 451–452
  financial strategy, 447–451
  fixed cost, 465
  gross profit ratio, 452
  inventory turns, 452
  Long-term Liabilities, 449–450
  mom & pop stores, 446
  quick ratio, 452
  sales per square foot, 452
  specialty stores, 447
  supercenters, 447
  supermarkets, 446
  types of retailers, 446–447
  unit price, 465

  variable cost, 465
  working capital ratio, 452
Retail and wholesale sectors, 8
Return on assets (ROA), 213, 458
Return on equity (ROE), 455
Returns to scale, 173
Revenue cycle management (healthcare), 481, 498
Revenue management, 407–441
  adjustable capacity, design of, 414
  airlines, revenue management in, 424–437
  applicability of, 414–415
  capacity planning, reactive strategies for, 412–414
  cross-training of employees, 413
  customer demand, proactive strategies to manage, 409–412
  customer segmentation, 410–411
  differential pricing, 411–412
  difficulties in managing service capacity, 408–409
  exercises, 437–440
  history of, 407–408
  no-show problem, 415
  overbooking, optimization models for, 415–423
  part-time/temporary workers, 413
  reservations/appointment systems, 410
  sales promotions, 412
  self-service, promotion of, 413–414
  strategies in services, 409–415
  workforce planning, 412–413
Right-hand-side (RHS) constants, 97–98
ROA, see Return on assets (ROA)
ROC curve, see Receiver operating characteristic (ROC) curve
ROE, see Return on equity (ROE)
Root cause analysis, 36–37
RTS, see Rate of technical substitution (RTS)
Run chart, 32

S

Scatterplot, 34–35
SCE, see Supply chain engineering (SCE)

Index

SCM, *see* Supply chain management (SCM)
Seasonality in forecasting, incorporation of, 58–61
Seasonality index, 58, 59, 66
Seasonality and trend in forecasting, incorporation of, 66–70
    static seasonality indices, method using, 66–67
    Winters method, 67–70
Securities and Exchange Commission (SEC), 448
Service industries, 5–9
    communication services, 5–6
    education sector, 6
    entertainment, 6
    financial services, 6
    government services, 7
    healthcare, 7
    hospitality and leisure industries, 7–8
    insurance sector, 8
    professional services, 8
    retail and wholesale sectors, 8
    transportation services, 9
    utilities sector, 9
Service Process Matrix, 10–12
    mass service, 11–12
    professional service, 12
    service factory, 11
    service shop, 11
    use of, 12
Service Quality Gap (SQG), 20, 23
Service Systems Engineering, 16–19
    control issues, 16
    design issues, 16
    engineering issues in service systems, 16
    engineering problems in services, 16–17
    Multiple-Criteria Decision Making in services, 17–19
SERVQUAL survey, 21–24
Set covering matrix, 105, 107
Set covering and set partitioning models, 104–119
    application to airline scheduling, 108–112
    application to warehouse location, 106–108
    Ford Motor Company, 115
    Mount Sinai Hospital, 118
    real-world applications, 114–119
    set covering matrix, 107
    set covering problem, 105–106
    set partitioning problem, 106
    United Airlines, 118–119
    United Parcel Service, 115–117
    VOLCANO model, 117
    Wishard Memorial Hospital, 117
    workforce planning, 112–114
Simple linear trend model, 62–64
Simple moving average forecasting method, 53
SKU, *see* Stock keeping unit (SKU)
Slotting, 325–331
    golden zone, 329
    packing SKUs, 330
    prioritizing SKUs, 328–330
    SKU profiling, 325–328
Smoothing Constant, 56
SMPS, *see* Station Manpower Planning System
Software
    DEA, 191–192
    Expert Choice (AHP), 158
    forecasting, 79–81
    MCDM, 171
Sorting product, 312
S&P 500 index, meaning of, 372–373
SQG, *see* Service Quality Gap
S-shape heuristic, 324
Standard deviation of forecast errors (STD), 74
State-dependent service, 120
Station Manpower Planning System (SMPS), 118
Statistical process control, 24–31
    attribute control charts, 28–29
    selection of sample size for control charts, 30–31
    variable control charts, 24–28
Stochastic DEA, 188–191
Stockholder's equity, 448
Stock keeping unit (SKU), 304
    golden zone, 329
    next fit strategy, 330
    packing, 330
    prioritizing, 328–330

profiling, 325–328
rationalization, 280
Universal Product Code, 444
Stocks, investing in, 350–351
Storing product, 309–311
Straight-Through Processing (STP), 543
Supplier selection, *see* Outsourcing decisions and supplier selection (SCE)
Supply chain engineering (SCE), 201–301
   assessing and managing supply chain performance, 208–211
   business financial metrics, 213–214
   cash-to-cash cycle, 214
   Clarke–Wright Savings Algorithm, 275
   conflicting criteria in supply chain optimization, 211
   correlation and pooling, 282–283
   cycle stock, 277
   days of inventory, 212
   decisions and design metrics, 201–217
   exercises, 287–298
   flows in supply chains, 203
   group decision making, 265
   importance of SCM, 215–217
   Integer Programming, 229
   inventory capital, 213
   Linear Weighted Point method, 242
   location–distribution with dedicated warehouses, 223–225
   location–distribution problem, 220–223
   location pooling, 281
   Mixed Integer Linear Programming, 230
   multi-criteria ranking methods for supplier selection, 245–258
   multinational consumer products company, 229–231
   multiple sourcing methods for supplier selection and order allocation, 266–273
   network design and distribution, 217–234
   operational decisions, 205
   outsourcing decisions and supplier selection, 234–273
   pooling and contracting, 280–287
   postponement, 281
   prequalification of suppliers, 238–241
   product pooling, 281–282
   product sourcing, 231
   ranking methods, comparison of, 265–266
   real-world applications, 229–234
   relationship between supply chain metrics and financial metrics, 211–214
   scaling criteria values, 258–265
   single sourcing methods for supplier selection, 242–245
   SKU rationalization, 280
   sourcing strategy, 237–238
   strategic decisions, 204
   supplier selection importance, 234
   supplier selection methods, 237–242
   supplier selection process, 235
   supply chain contracting, 283–287
   supply chain distribution planning, 217–220
   supply chain drivers, 206–208
   supply chain efficiency, 209–210
   supply chain enablers, 205–206
   supply chain logistics, 273–279
   supply chain responsiveness, 210
   supply chain risk, 210–211
   tactical decisions, 204–205
   third-party logistics provider, 207
   Total Cost of Ownership, 243
   traveling salesperson problem, 274
   vehicle routing, 274–277
   wholesale price contract, 286
   working capital, 214
Supply chain management (SCM), 18, 201, 203
Support vector machines (SVMs) (healthcare), 516–522

T

Tchebycheff Metric, 169
Technology, impact of, 3–4 (fa subs)
Third-party logistics provider, 207
Time between departures (TBD), 468
Time series forecasting, 49–51
Total Cost of Ownership (TCO), 243
Tracking product, 312–313
Tracking Signal, 77
Trade-off analysis, 362

Transportation services, 9
Traveling salesperson problem (TSP), 274
Treasury Strips, 383
Trend adjusted exponential smoothing, 64
Trend in forecasting, incorporation of, 62–66
    Holt's method, 64–66
    simple linear trend model, 62–64
Trip Evaluation and Improvement Program (TRIP), 112
TSP, see Traveling salesperson problem (TSP)
Two-stage cross dock, 340
Type I and Type II errors, 24

U

United Parcel Service (UPS), 112
Unit load, 309
Universal Product Code, 444
U.S. Federal Communications Commission (FCC), 9
Utilities sector, 9

V

Value stream mapping (healthcare), 492–493
Variable returns to scale (VRS), 173
VOLCANO (Volume, Location and Aircraft Network Optimizer) model, 117

W

Warehousing and distribution, 303–348
    aisle layout, 321
    batch picking, 323
    bucket brigades, 303
    cluster picking, 323
    cold chain, 307
    cross docking, 338–343
    discrete order picking, 322–323
    exercises, 346–347
    fishbone aisle layout, 321
    flow-through racks, 310, 311
    Flying-V cross lane, 321
    forward–reserve, 331–335
    honeycombing, 310, 311
    line balancing and bucket brigades, 335–338
    material flow and staffing, 316
    moving product, 308–309
    order picking, 322–325
    slotting, 325–331
    sorting product, 312
    S-shape heuristic, 324
    storage strategies, 316–317
    storing product, 309–311
    tracking product, 312–313
    unit load, 309
    warehouse design, 314–321
    warehouse equipment, 307–313
    warehouse functions, 303–307
    warehouse layout, 318–321
    warehouse location and inventory, 343–345
    wave picking, 323
    zone picking, 323
Weighted moving average forecasting method, 53
Weighted objective problem, 161
Wholesale price contract, 286
Working capital, 214
Working capital ratio, 452
Work-in-progress (WIP), 203
Worldwide Inventory Network Optimizer (WINO), 232

X

X–Y chart, 34

Y

Yield management, see Revenue management
Yield to maturity (YTM), 380

Z

Zero-coupon bonds, 380, 383
Zone picking, 323

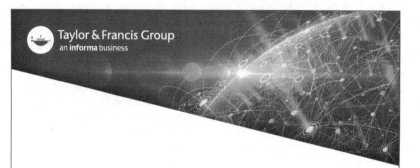